Cell Imaging Techniques

METHODS IN MOLECULAR BIOLOGY™

John M. Walker, Series Editor

337. Ion Channels: *Methods and Protocols,* edited by *J. D. Stockand and Mark S. Shapiro, 2006*

336. Clinical Applications of PCR: *Second Edition,* edited by *Y. M. Dennis Lo, Rossa W. K. Chiu, and K. C. Allen Chan, 2006*

335. Fluorescent Energy Transfer Nucleic Acid Probes: *Designs and Protocols,* edited by *Vladimir V. Didenko, 2006*

334. PRINS and *In Situ* PCR Protocols: *Second Edition,* edited by *Franck Pellestor, 2006*

333. Transplantation Immunology: *Methods and Protocols,* edited by *Philip Hornick and Marlene Rose, 2006*

332. Transmembrane Signaling Protocols: *Second Edition,* edited by *Hydar Ali and Haribabu Bodduluri, 2006*

331. Human Embryonic Stem Cell Protocols, edited by *Kursad Turksen, 2006*

330. Nonhuman Embryonic Stem Cell Protocols, Vol. II: *Differentiation Models,* edited by *Kursad Turksen, 2006*

329. Nonhuman Embryonic Stem Cell Protocols, Vol. I: *Isolation and Characterization,* edited by *Kursad Turksen, 2006*

328. New and Emerging Proteomic Techniques, edited by *Dobrin Nedelkov and Randall W. Nelson, 2006*

327. Epidermal Growth Factor: *Methods and Protocols,* edited by *Tarun B. Patel and Paul J. Bertics, 2006*

326. *In Situ* Hybridization Protocols, *Third Edition,* edited by *Ian A. Darby and Tim D. Hewitson, 2006*

325. Nuclear Reprogramming: *Methods and Protocols,* edited by *Steve Pells, 2006*

324. Hormone Assays in Biological Fluids, edited by *Michael J. Wheeler and J. S. Morley Hutchinson, 2006*

323. Arabidopsis Protocols, *Second Edition,* edited by *Julio Salinas and Jose J. Sanchez-Serrano, 2006*

322. *Xenopus* Protocols: *Cell Biology and Signal Transduction,* edited by *X. Johné Liu, 2006*

321. Microfluidic Techniques: *Reviews and Protocols,* edited by *Shelley D. Minteer, 2006*

320. Cytochrome P450 Protocols, *Second Edition,* edited by *Ian R. Phillips and Elizabeth A. Shephard, 2006*

319. Cell Imaging Techniques, *Methods and Protocols,* edited by *Douglas J. Taatjes and Brooke T. Mossman, 2006*

318. Plant Cell Culture Protocols, *Second Edition,* edited by *Victor M. Loyola-Vargas and Felipe Vázquez-Flota, 2005*

317. Differential Display Methods and Protocols, *Second Edition,* edited by *Peng Liang, Jonathan Meade, and Arthur B. Pardee, 2005*

316. Bioinformatics and Drug Discovery, edited by *Richard S. Larson, 2005*

315. Mast Cells: *Methods and Protocols,* edited by *Guha Krishnaswamy and David S. Chi, 2005*

314. DNA Repair Protocols: *Mammalian Systems, Second Edition,* edited by *Daryl S. Henderson, 2006*

313. Yeast Protocols: *Second Edition,* edited by *Wei Xiao, 2005*

312. Calcium Signaling Protocols: *Second Edition,* edited by *David G. Lambert, 2005*

311. Pharmacogenomics: *Methods and Protocols,* edited by *Federico Innocenti, 2005*

310. Chemical Genomics: *Reviews and Protocols,* edited by *Edward D. Zanders, 2005*

309. RNA Silencing: *Methods and Protocols,* edited by *Gordon Carmichael, 2005*

308. Therapeutic Proteins: *Methods and Protocols,* edited by *C. Mark Smales and David C. James, 2005*

307. Phosphodiesterase Methods and Protocols, edited by *Claire Lugnier, 2005*

306. Receptor Binding Techniques: *Second Edition,* edited by *Anthony P. Davenport, 2005*

305. Protein–Ligand Interactions: *Methods and Applications,* edited by *G. Ulrich Nienhaus, 2005*

304. Human Retrovirus Protocols: *Virology and Molecular Biology,* edited by *Tuofu Zhu, 2005*

303. NanoBiotechnology Protocols, edited by *Sandra J. Rosenthal and David W. Wright, 2005*

302. Handbook of ELISPOT: *Methods and Protocols,* edited by *Alexander E. Kalyuzhny, 2005*

301. Ubiquitin–Proteasome Protocols, edited by *Cam Patterson and Douglas M. Cyr, 2005*

300. Protein Nanotechnology: *Protocols, Instrumentation, and Applications,* edited by *Tuan Vo-Dinh, 2005*

299. Amyloid Proteins: *Methods and Protocols,* edited by *Einar M. Sigurdsson, 2005*

298. Peptide Synthesis and Application, edited by *John Howl, 2005*

297. Forensic DNA Typing Protocols, edited by *Angel Carracedo, 2005*

296. Cell Cycle Control: *Mechanisms and Protocols,* edited by *Tim Humphrey and Gavin Brooks, 2005*

295. Immunochemical Protocols, *Third Edition,* edited by *Robert Burns, 2005*

294. Cell Migration: *Developmental Methods and Protocols,* edited by *Jun-Lin Guan, 2005*

293. Laser Capture Microdissection: *Methods and Protocols,* edited by *Graeme I. Murray and Stephanie Curran, 2005*

292. DNA Viruses: *Methods and Protocols,* edited by *Paul M. Lieberman, 2005*

291. Molecular Toxicology Protocols, edited by *Phouthone Keohavong and Stephen G. Grant, 2005*

METHODS IN MOLECULAR BIOLOGY™

Cell Imaging Techniques

Methods and Protocols

Edited by

Douglas J. Taatjes

Brooke T. Mossman

Department of Pathology
University of Vermont
Burlington, VT

HUMANA PRESS ✵ TOTOWA, NEW JERSEY

999 Riverview Drive, Suite 208
Totowa, New Jersey 07512

www.humanapress.com

This publication is printed on acid-free paper. ∞
ANSI Z39.48-1984 (American Standards Institute)Permanence of Paper for Printed Library Materials.

Cover design by Patricia F. Cleary

Cover illustration: *Background:* whole-cell experiment combined with a pre-embedding method: CLS-RCM image of a 90-nm Epon section (Chap. 18, Fig. 15D, p. 397). *Inset, upper left:* electron micrograph depicting a porosome close to a microvillus at the apical plasma membrane of a pancreatic acinar cell (Chap. 15, Fig. 2B, p. 303). *Upper right:* NSOM image of an antibody-labeled mouse macrophage (Chap. 14, Fig. 3C, p. 282). *Lower left:* metaphase analysis with multiple single-gene probes (Chap. 12, Fig. 1, p. 238). *Lower right:* adeno-associated virus serotype 5 (Chap. 7, Fig. 9, p. 161).

For additional copies, pricing for bulk purchases, and/or information about other Humana titles, contact Humana at the above address or at any of the following numbers: Tel.: 973-256-1699; Fax: 973-256-8341; E-mail: orders@humanapr.com; or visit our Website: www.humanapress.com

Printed in the United States of America. 10 9 8 7 6 5 4 3 2 1

eISBN 1-59259-993-1

ISSN 1064-3745

Library of Congress Cataloging-in-Publication Data

Cell imaging techniques: methods and protocols / edited by Douglas J. Taatjes, Brooke T. Mossman.
p. cm. -- (Methods in molecular biology ; 319)
Includes bibliographical references and index.
ISBN 1-58829-157-X (alk. paper)
1. Cytology. 2. Microscopy. 3. Molecular biology. 4. Imaging systems in biology. I. Taatjes, Douglas J. II. Mossman, Brooke T., 1947- III. Methods in molecular biology (Clifton, N.J.) ; 319

QH585.C463 2005
611'.0181--dc22

2005046145

Preface

In 1665, a book was published that inaugurated the use of the microscope to investigate the natural world. The author was Robert Hooke, a talented artist, architect, and amateur scientist. Hooke wrote *Micrographia: Or Some Physiological Descriptions of Minute Bodies Made by Magnifying Glasses with Observations and Inquiries Thereupon,* at the behest of the newly chartered Royal Society in London, for whom he was working as curator of scientific experiments. In *Micrographia,* he presented the first detailed observations of everyday objects made with his self-constructed light microscope. Although this book contains a treasure-trove of drawings (in Hooke's own hand) of the appearance of various animate and inanimate specimens as viewed in magnified form, one of the drawings and its associated description stands out as particularly germane to our present topic of cell imaging techniques. In his description of a thin piece of clear cork cut with a penknife and observed with his microscope, Hooke described the honey-comb-like appearance of the cork with "pores" or "cells" representing the basic structural unit. This represents the first printed reference of the term "cell" to describe a unit structure of an organism.

In the 340 years since the publication of *Micrographia,* a multitude of new microscopy-based systems have evolved for the observation of cells. Indeed, many of these techniques have been developed in the past few decades. Recent books have sought to present single volumes detailing methods for specific types of microscopy, such as confocal scanning laser microscopy, atomic force microscopy, and electron microscopy. In the present book, we have sought to present an eclectic collection of what we consider some of the essential state-of-the-art methods for imaging cells and molecules. *Cell Imaging Techniques: Methods and Protocols* has been organized to begin with light microscopic methods to observe molecules such as mRNA, calcium, and collagen. Chapters covering confocal scanning laser microscopy, quantitative computer-assisted image analysis, laser scanning cytometry, laser capture microdissection, microarray image scanning, near-field scanning optical microscopy, atomic force microscopy, and reflection contrast microscopy follow. The book then finishes with chapters on preparative methods for transmission electron microscopy of particles and cells.

We have tried to arrange the chapters in a logical format, beginning with light microscopy techniques, proceeding through scanning probe-type techniques, and ending with electron microscopy. The chapter on reflection contrast microscopy serves as a link between light and electron microscopy. Although *Cell Imaging Techniques: Methods and Protocols* is primarily intended to convey detailed methods and protocols for cell imaging, we have also included some review-type chapters to set the stage for the protocol-driven chapters. Moreover, given the incredible breadth of microscopy-based imaging techniques available today, we tried to include many that might not have been covered in detail in previous books. By necessity, we have had to exclude many valuable and marvelous techniques (multiphoton confocal microscopy, for one), but are secure in the knowledge that they have been comprehensively treated in other volumes of the Methods in Molecular Biology series.

We believe that *Cell Imaging Techniques: Methods and Protocols* will be useful for those involved in seeking a variety of microscopy-based techniques for imaging cells and molecules. With the proliferation of core "cell imaging facilities" at universities, hospitals, and pharmaceutical and biotechnology companies throughout the world, this volume should provide a handy reference or starting point for researchers seeking the latest information and protocols for a wide variety of cell imaging techniques. We hope that readers will find value in the techniques presented herein and might even be tempted to try some techniques they had not considered previously.

Finally, we would like to thank those associated with the production of this book. First, the authors themselves for agreeing to take the time to prepare their chapters in a timely manner and in a form filled with technical details not usually present in a research publication. It was not an easy task, and we thank them for their efforts. Second, we would like to thank Professor John Walker, the series editor, for his helpful insights, interest in the book, and his timely response to our queries. Third, we would like to express our appreciation to Craig Adams and the staff at Humana Press for their patience and editorial efforts in the production of the book, and to Marilyn Wadsworth at the University of Vermont for her invaluable assistance in our editorial tasks. Finally, the color reproduction of images, so important in a volume like this, would not have been possible without the generous financial support of the Optical Analysis Corporation (Nashua, NH), JMAR Technologies, Inc. (South Burlington, VT), and the Department of Pathology, University of Vermont (Burlington, VT).

Douglas J. Taatjes
Brooke T. Mossman

Contents

Contributors

VYTAUTAS P. BINDOKAS • *Department of Neurobiology, Pharmacology, and Physiology, University of Chicago, Chicago, IL*
DIANA P. BRATU • *Department of Molecular Genetics, Public Health Research Institute, Newark, NJ; Program of Molecular Medicine, University of Massachusetts Medical School, Worcester, MA*
SANG-JOON CHO • *Department of Physiology, Wayne State University School of Medicine, Detroit, MI*
I. CORNELESE-TEN VELDE • *Department of Pathology, Leiden University Hospital, Leiden, The Netherlands*
KEVIN D. COSTA • *Department of Biomedical Engineering, Columbia University, New York, NY*
STACY COWHERD • *Laboratory of Pathology, Center for Cancer Research, National Cancer Institute, National Institutes of Health, Bethesda, MD*
GUY COX • *Electron Microscope Unit, University of Sydney, Sydney, Australia*
ZBIGNIEW DARZYNKIEWICZ • *The Brander Cancer Research Institute, New York Medical College, Valhalla, NY*
E. DE HEER • *Department of Pathology, Leiden University Hospital, Leiden, The Netherlands*
DOUGLAS J. DEMETRICK • *Departments of Pathology and Laboratory Medicine, Oncology and Biochemistry and Molecular Biology, University of Calgary, Alberta, Canada*
PETER EGGLI • *Anatomical Institute, University of Bern, Bern, Switzerland*
VIRGINIA ESPINA • *Center for Applied Proteomics and Molecular Medicine, George Mason University, Manassas, VA*
JEROME F. FIEKERS • *Department of Anatomy and Neurobiology, University of Vermont, Burlington, VT*
WERNER GRABER • *Anatomical Institute, University of Bern, Bern, Switzerland*
KARL-OTTO GREULICH • *Department of Single Cell and Single Molecule Techniques, Institute of Molecular Biotechnology, Jena, Germany*
MICHAEL HAUSMANN • *Kirchoff Institute of Physics, University of Heidelberg, Heidelberg, Germany*
ELENA HOLDEN • *CompuCyte Corporation, Cambridge, MA*
BHANU P. JENA • *Department of Physiology, Wayne State University School of Medicine, Detroit, MI*

ELEANOR KABLE • *Electron Microscope Unit, University of Sydney, Sydney, Australia*

DAVINDER KAUR • *Laser Cytometry Facility, Glenfield Hospital, University of Leicester, UK*

LANCE A. LIOTTA • *Center for Applied Proteomics and Molecular Medicine, George Mason University, Manassas, VA*

CHRISTOPHER B. MANNING • *Department of Pathology, University of Vermont, Burlington, VT*

JOHN MILIA • *Arcturus Bioscience Inc., Mountain View, CA*

THOMAS O. MONINGER • *Central Microscopy Research Facility and Department of Internal Medicine, University of Iowa, Iowa City, IA*

KENNETH C. MOORE • *Central Microscopy Research Facility, University of Iowa, Iowa City, IA*

BROOKE T. MOSSMAN • *Department of Pathology, University of Vermont, Burlington, VT*

SABITA K. MURTHY • *Head of Division, Medical Genetics, Al Wasl Hospital, Dubai, United Arab Emirates*

RANDY A. NESSLER • *Central Microscopy Research Facility and Department of Pediatrics, University of Iowa, Iowa City, IA*

BIRGIT PERNER • *Department of Single Cell and Single Molecule Techniques, Institute of Molecular Biotechnology, Jena, Germany*

LOUIS H. PHILIPSON • *Department of Medicine, University of Chicago, Chicago, IL*

PIOTR POZAROWSKI • *The Brander Cancer Research Institute, New York Medical College, Valhalla, NY, and Department of Clinical Immunology, School of Medicine, Lublin, Poland*

F. A. PRINS • *Department of Pathology, Leiden University Hospital, Leiden, The Netherlands*

MICHAEL RADERMACHER • *Department of Molecular Physiology and Biophysics, University of Vermont, Burlington, VT*

LATHA RAMDAS • *Cancer Genomics Core Laboratory, University of Texas M.D. Anderson Cancer Center, Houston, TX*

ALEXANDER RAPP • *Department of Single Cell and Single Molecule Techniques, Institute of Molecular Biotechnology, Jena, Germany*

DAVID A. REW • *Southampton University Hospitals, Southampton, UK*

MICHAEL WM. ROE • *Department of Medicine, University of Chicago, Chicago, IL*

TERESA RUIZ • *Department of Molecular Physiology and Biophysics, University of Vermont, Burlington, VT*

HARRY SCHERTHAN • *Max-Planck-Institute for Molecular Genetics, Berlin, Germany*

David J. Schneider • *Department of Medicine, University of Vermont, Burlington, VT*

Burton E. Sobel • *Department of Medicine, University of Vermont, Burlington, VT*

Maria Stern • *Department of Pathology, University of Vermont, Burlington, VT*

Daniel Studer • *Anatomical Institute, University of Bern, Bern, Switzerland*

Douglas J. Taatjes • *Department of Pathology, and Microscopy Imaging Center, University of Vermont, Burlington, VT*

Wendy Tra • *Department of Pathology, and Microscopy Imaging Center, University of Vermont, Burlington, VT*

Dimitri Vanhecke • *Anatomical Institute, University of Bern, Bern, Switzerland*

Hans van Hirtum • *Department of Pathology, and Microscopy Imaging Center, University of Vermont, Burlington, VT*

Marilyn P. Wadsworth • *Department of Pathology, and Microscopy Imaging Center, University of Vermont, Burlington, VT*

Leo Wollweber • *Department of Single Cell and Single Molecule Techniques, Institute of Molecular Biotechnology, Jena, Germany*

Gerrit Woltmann • *Glenfield Hospital, University of Leicester, UK*

Glendon Wu • *Laboratory of Pathology, Center for Cancer Research, National Cancer Institute, National Institutes of Health, Bethesda, MD*

Wei Zhang • *Cancer Genomics Core Laboratory, University of Texas M.D. Anderson Cancer Center, Houston, TX*

Robert M. Zucker • *Reproductive Toxicology Division, National Health and Environmental Effects Research Laboratory, US Environmental Protection Agency, Research Triangle Park, NC*

Color Plates

Color Plates follow p. 274.

1

Molecular Beacons

Fluorescent Probes for Detection of Endogenous mRNAs in Living Cells

Diana P. Bratu

Summary

A novel approach for detecting nucleic acid in solution has been adopted for real-time imaging of native mRNAs in living cells. This method utilizes hybridization probes, called "molecular beacons," that generate fluorescent signals only when they are hybridized to a complementary target sequence. Nuclease-resistant molecular beacons are designed to efficiently hybridize to accessible regions within RNAs and then be detected via fluorescence microscopy. The target regions chosen for probe binding are selected using two computer algorithms, *mfold* and OligoWalk, that predict the secondary structure of RNAs and help narrow down sequence stretches to which the probes should bind with high affinity in vivo. As an example, molecular beacons were designed against regions of *oskar* mRNA, microinjected into living *Drosophila melanogaster* oocytes and imaged via confocal microscopy.

Key Words: Molecular beacons; fluorescent probes; hybridization; secondary structure prediction; RNA localization; live-cell imaging.

1. Introduction

The direct visualization of specific mRNAs in living cells has been desirable for some time for accelerating the studies of intracellular RNA trafficking and localization, just as green fluorescent protein has stimulated the study of specific proteins in vivo. Tyagi and Kramer have developed a general method using hybridization probes called "molecular beacons," which generate fluorescence signals as they hybridize to complementary nucleic acid target sequences (*see* **Fig. 1**) *(1)*. Unbound molecular beacons are nonfluorescent and it is not necessary to remove excess probes to detect the hybrids. These probes bind to their targets spontaneously at physiological temperatures; thus, their introduction into cells is sufficient to illuminate target mRNAs *(2–4)*.

From: *Methods in Molecular Biology, vol. 319: Cell Imaging Techniques: Methods and Protocols*
Edited by: D. J. Taatjes and B. T. Mossman © Humana Press Inc., Totowa, NJ

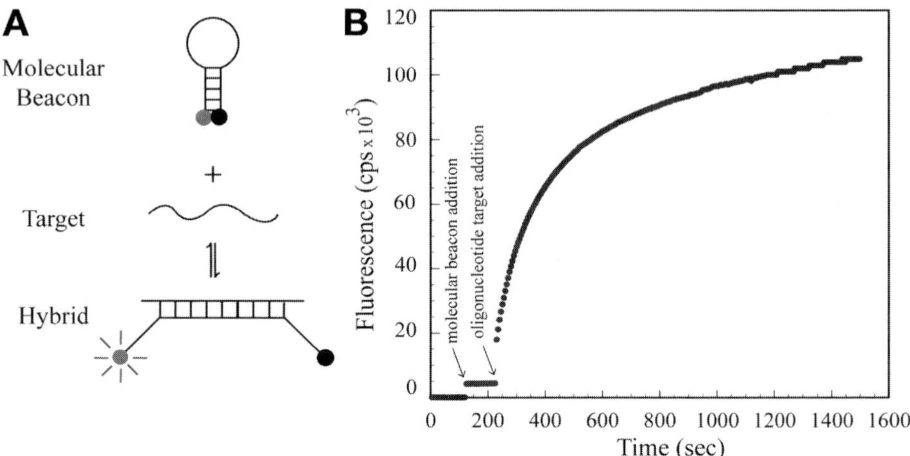

Fig. 1. Principle of operation. Molecular beacons are oligonucleotides that possess complementary sequences on either end of a probe sequence, enabling the molecule to assume a hairpin configuration, in which a fluorophore and a quencher are held in close proximity. When these probes hybridize to a complemetary target, the formation of a probe–target hybrid disrupts the hairpin stem, removing the fluorophore from the vicinity of the quencher, and restoring the probe's fluorescence.

Theoretically, any sequence within a target RNA can be chosen as a site for molecular beacon binding. The endless possibilities give one the confidence that such regions are easily identified. However, the extent of target accessibility is primarily a consequence of complex secondary and tertiary intramolecular structures, which are difficult to predict and can mask many of these regions. Furthermore, inside the cell, mRNAs exist in association with proteins that further occlude parts of the mRNA. Although regions involved in protein binding can only be identified by experimental analysis, reasonable attempts can be made to predict the regions that are not involved in tight secondary structures. So far, several in vitro assays and theoretical algorithms are available to help identify putative target sites within mRNA sequences, as well as probes with high affinity for binding ***(5–8)***.

The *mfold* RNA-folding algorithm is used to predict the most thermodynamically stable secondary structure along with an ensemble of suboptimal structures ***(9)***. Because none of these structures can be considered to represent the naturally occurring conformation, the parameters that describe the entire ensemble are analyzed. The number of candidate sites is winnowed down by employing a second algorithm. OligoWalk scans the folded RNA sequence for regions to which various-length oligonucleotides are capable of binding ***(10)***. With consideration of the base composition of each oligonucleotide and of the predicted secondary structure of the RNA, the output provides information about the stability of the expected hybrid and thus identifies potential target regions.

Once identified, molecular beacons specific for those regions are designed and synthesized. To detect an intracellular target, these probes need to be stable inside the cell and not perturb or destroy its target. Conventional molecular beacons synthesized from deoxyribonucleotides are not suitable for use inside living cells because the single-stranded probe sequence of conventional molecular beacons are subject to digestion by single-strand-specific cellular deoxyribonucleases. Therefore, for in vivo studies, molecular beacons are synthesized from unnatural nucleotides possessing an oxymethyl group in place of a hydrogen atom at the 2′-position of their pentose moiety. These modified molecular beacons are resistant to cellular nucleases, and upon binding to RNA, they do not form a substrate for ribonuclease H, thus maintaining the integrity of the target molecule ***(2)***.

This chapter describes the steps necessary to find accessible probe-binding target sites within an RNA sequence and to design efficient molecular beacons for RNA detection in vivo. The last section provides an example of RNA imaging in living *Drosophila* oocytes.

2. Materials

1. RNAstructure (free download on http://rna.chem.rochester.edu/index.html for PC only).
2. Spectrofluorometer (Photon Technology International, South Brunswick, NJ).
3. Spetrofluorometric thermal cycler (Applied Biosystems Prism 7700).
4. Nuclease-free water (Ambion Inc., Austin, TX).
5. TE buffer: 1 m*M* EDTA, 10 m*M* Tris-HCl, pH 8.0.
6. 10X Phosphate-buffered saline (PBS) with $CaCl_2$ and $MgCl_2$ (Sigma, St. Louis, MO). Dilute to 1X with nuclease-free water or diethyl pyrocarbonate (DEPC)-treated water. Store at room temperature.
7. Molecular beacon. Dissolve stock solutions in nuclease-free water and store at –20°C. Dilute working solutions in 1X PBS, keep protected from light, and store at –20°C for 1 mo.
8. Oligonucleotide target complimentary to the probe sequence of the molecular beacon. Dissolve in TE buffer and store at –20°C.
9. In vitro-transcribed RNA.
10. Hybridization buffer: 1 m*M* $MgCl_2$, 20 m*M* Tris-HCl, pH 8.0.

3. Methods

3.1. Selection of RNA Target Regions

3.1.1. Secondary Structure Prediction Using mfold

Input the mRNA sequence on the *mfold* RNA server (http://www.bioinfo.rpi.edu/~zukerm/) and fold it using the default settings ***(9)***. An immediate output is obtained unless the RNA is longer than 800 bp, in which case it is folded as a "batch." The completed job is posted within a couple of hours, depending on the server's availability. The algorithm predicts the most thermodynamically stable secondary structure along with an ensemble of suboptimal structures (the maximum limit can be chosen prior to folding the RNA; a default value of 50 is sufficient for this analysis). Because none of these structures can be considered to represent a naturally occurring conformation, the parameters describing the entire ensemble must be analyzed. The first parameter is ss-count, which defines the probability of a nucleotide to be single-stranded. The second parameter is P-num, a value that denotes the total number of different basepairs that can be formed by a particular base within the set of foldings. These values are assigned to each nucleotide in the mRNA sequence (*see* **Note 1**).

Once the set of foldings is generated, obtain the following files to be analyzed.

- To view the predicted secondary structure of the RNA, select the jpg option of the first listed structure (the most stable structure in the set).
- To view this folded structure representing the ss-count values, choose the "ss-count" annotation, where each nucleotide (annotated as a dot or character) will be indicated by a color representing the value found in the "ss-count table." Clicking on the structure image converts this to a color-coded fold (*see* **Note 2**).
- This window also offers access to the "ss-count table" file. To obtain the actual values, open the file and save it. Open this file in Microsoft Excel; this will provide the opportunity to graph the ss-count value for each nucleotide.
- The P-num values are found in a file accessible from the first output screen. Follow the same steps as for the "ss-count table" to view the secondary structure, save the file, and plot the graph of P-num values for the entire sequence.

3.1.1.1. Criteria for Selecting Accessible Target Regions

The ss-count/P-num graph indicates regions of structural plasticity.

- Plot both ss-count and P-mm for each number nucleotide. Adjust ss-count values by x before plotting. The ss-count values are adjusted in order to make them comparable with the range of values for P-num.

$$x = \text{P-mm}_{max} / (\text{ss-count}_{max} + 1)$$

$$\text{ss-count}_{adj} = (\text{ss-count} + 1) \times x$$

- Evaluate the graph and choose regions with the following criteria:
  ss-count = high (single-stranded)
  P-mm = low (well determined) OR high (poorly determined)

Values for ss-count of more than 50% (probability of being single-stranded) are considered high and are represented by warm colors in the color-annotated secondary structure. Nucleotides involved in double-stranded regions are represented by cool colors. If these regions are well determined, the P-num values for the nucleotides involved are very low. If a region is flexible, meaning the nucleotides are not found in the same basepair but they form several kinds of basepairs within the predicted structures, then the P-mm is high, representing a poorly determined structure. Such regions are easily determined when evaluating the P-num color-annotated secondary structure (*see* **Note 3**).

3.1.2. RNAstructure: OligoWalk

The RNAstructure program is a faithful recreation of *mfold* for Windows ***(10)***. The OligoWalk program overcomes the need to synthesize a large number of oligonucleotides by using thermodynamic data to find oligonucleotides that bind strongly to a target RNA while offering a fast search interface. Utilizing the most current set of thermodynamic parameters for nucleic acid secondary structure, this algorithm calculates the equilibrium affinity of each set-length complementary oligonucleotide and predicts its overall free energy of binding while taking into account duplex stability, local secondary structure within the target RNA, as well as intermolecular and intramolecular secondary structures formed by the oligonucleotide ***(10–15)***.

For each oligonucleotide chosen, the following thermodynamic parameters (ΔG) are calculated in kcal/mol:

$\Delta G_{\text{overall}}$: net ΔG of oligonucleotide–target binding (with consideration of all contributions, including breaking target structure and oligonucleotide-self structure)
ΔG_{duplex} ($\Delta G_{\text{binding}}$): ΔG of the oligonucleotide–target hybrid from unstructured states
$\Delta G_{\text{break target}}$: energy penalty resulting from the breaking of intramolecular target basepairs when the oligonucleotide is bound
$\Delta G_{\text{oligo-self}}$: ΔG of the intramolecular oligonucleotide structure
$\Delta G_{\text{oligo-oligo}}$: ΔG of the intermolecular oligonucleotide structure

Before running the OligoWalk module, several selections must be made. First, a set of optimal and suboptimal structures of the RNA is predicted, using the folding feature of RNAstructure, creating a '.ct' file (connect table) that contains the sequence and basepair information (the sequence must be in caps). Then, the "mode" that determines the types of ΔG calculation for a given sequence is chosen. Use the default mode, "break local structure." In this mode, the target sequence breaks wherever the oligonucleotide binds ($\Delta G_{\text{break target}}$), oligonucleotides lose pairs in self-structures ($\Delta G_{\text{oligo-oligo}}$ and $\Delta G_{\text{oligo-self}}$) and gain pairs in oligonucleotide–target binding (ΔG_{duplex}). The option to include all suboptimal structures of the target is chosen to determine the total free-energy loss of the target. Each structure's free-energy loss is weighed according to the free energy of the structure.

Once these options are set, OligoWalk calculations are run for an oligonucleotide concentration of 100 ng/μL, ranging in length from 18 to 25 nucleotides. These oligonucleotides are "walked" along the RNA sequence, one nucleotide at a time, until the criteria below are met (*see* **Note 4**). It is possible for a longer region of the target to be chosen as favorable, not just the region comprising the length of the oligonucleotide. $\Delta G_{oligo\text{-}self}$ and $\Delta G_{oligo\text{-}oligo}$ are disregarded, as the molecular beacon structure is analyzed separately using *mfold* (*see* subheading *3.2.1.*).

3.1.2.1. Criteria for Selecting Target Regions

*$\Delta G_{overall}$ and ΔG_{duplex} = most negative (–) kcal/mol values;
*$\Delta G_{duplex} - \Delta G_{overall} \leq -10$ kcal/mol;
*ΔG_{break} = lowest positive (+) kcal/mol values.

3.2. Molecular Beacons

3.2.1. Design

The probe length comprising the hairpin loop can range between 15 and 25 nucleotides. This is dependent on the target region length chosen to be accessible for probe binding. Longer sequence stretches permit probe design flexibility. The %GC composition of the probe sequence should range between 40% and 55%. Two arm sequences designed to be complementary to each other are then added at the respective ends of the probe sequence. Because the rate of hybridization of a molecular beacon is influenced by the stability of its stem, the composition of these sequences is also very important. In practice, the length of the arm sequences is four or five nucleotides and are composed mostly of G/C's (*see* **Note 5**). Using Zucker's *mfold* folding program, predict the structure of the molecular beacon. The sequence should not form unwanted secondary structures (*see* **Note 6**).

3.2.2. Signal-to-Background Ratios

Signal-to-background ratio measurements indicate the purity of a molecular beacon preparation. Because of the presence of oligonucleotides that possess only a fluorophore and not a quencher, or of inefficient quenching between the two moieties, the fluorescence signal resulting from the presence of the target is obscured by background fluorescence, leading to inaccurate hybridization measurements ***(16)***. The fluorescence of the molecular beacon in a 150-μL solution of 1 m*M* $MgCl_2$, 20 m*M* Tris-HCl, pH 8.0 is determined (F_{buffer}) using the optimal excitation and emission wavelength of the fluorophore. After 10 μL of a 0.1 μ*M* molecular beacon solution is added, a new fluorescence level is recorded (F_{closed}). The increase in fluorescence is monitored after a fivefold molar excess of a complementary oligonucleotide is added to the solution. F_{open} is the maximum level of fluorescence reached after the reaction is complete.

The signal-to-background ratio equals $(F_{open} - F_{buffer})/(F_{closed} - F_{buffer})$. Good molecular beacons generate a fluorescent signal at least 30 times more intense than in the absence of target.

3.2.3. Determination of Thermal Denaturation Profiles

The design of molecular beacons is highly permissive. They can be designed so that they are stable and function under a broad set of conditions. Combinations of loop and stem lengths can be chosen to detect various nucleic acid targets (*see* **Fig. 2**).

The fluorescence of molecular beacon solutions should be measured over a wide range of temperatures. Equimolar molecular beacon solutions are prepared in the hybridization buffer. Add a fivefold molar excess of an oligonucleotide that is perfectly complementary to the molecular beacon's probe sequence. In a spectrofluorometric thermal cycler, decrease the temperature from 95°C to 25°C in increments of 1°C. To ensure that equilibrium is reached at each temperature step, the steps should last at least 30 s.

The melting profiles of the probes alone and of their hybrids should indicate correct molecular beacon characteristics (*see* **Fig. 2**).

3.2.4. Testing Putative Probes: In Vitro Hybridization to RNA

Prior to using molecular beacons in living cells, in vitro-transcribed full-length RNA is used as a target molecule for in vitro hybridization reactions to determine whether the molecular beacons that were predicted to work well by the computer programs were in fact able to bind to their intended target sequences. The intensity of the fluorescence signal and the rate of hybridization are measured in a spectrofluorometer. These reactions are performed with "naked" RNA under plausible physiological conditions and are not carried out in the presence of a cellular extract that contains the various factors that might have an affinity for the mRNA. Each molecular beacon should be tested for its ability to hybridize to the mRNA in vitro. Even though each molecular beacon could bind spontaneously to an oligonucleotide that is complementary to its probe sequence, only a few might bind efficiently to full-length transcripts.

1. Prepare full-length RNA by in vitro transcription.
2. Measure the fluorescence intensity of a 125-µL solution containing 1 m*M* $MgCl_2$, 20 m*M* Tris-HCl, pH 8.0, and 80 n*M* of molecular beacon at 25°C until no change in fluorescence occurs.

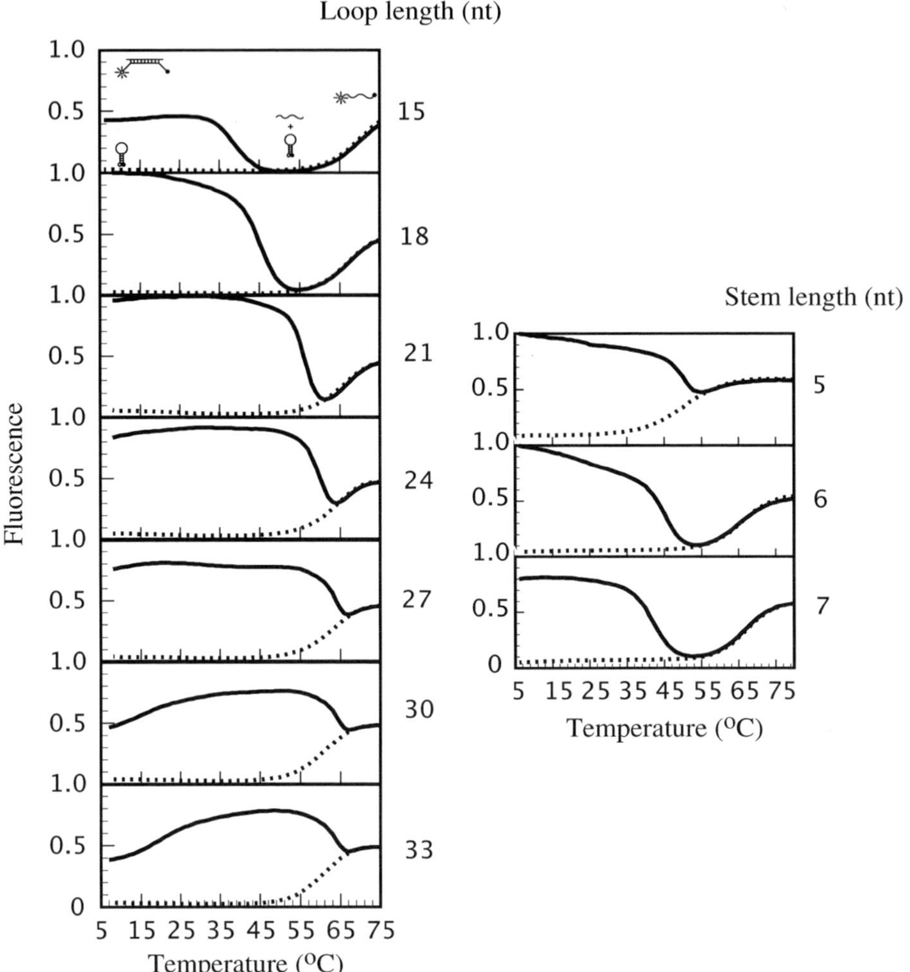

Fig. 2. Thermal denaturation profiles of DNA molecular beacons, alone (dotted lines) or bound to target (continuous lines), where fluorescence intensity was measured as a function of temperature. **(A)** The loop length of a molecular beacon, comprising the probe sequence, was increased in increments of three nucleotides at a time while maintaining a constant stem length of six nucleotides. The melting transition of the probe–target hybrids shifted toward higher temperatures, indicating an increase in the stability of the hybrids. **(B)** The stem length was increased 1 nucleotide at a time while maintaining a constant loop length of 15 nucleotides. The melting transition of the probe–target hybrids shifted toward lower temperatures, indicating a destabilization of the probe–target hybrids, while the melting transition of the molecular beacon stem shifted toward higher temperatures.

3. Add a 5-µL aliquot of 100 n*M* transcribed mRNA to this solution and measure the subsequent change in fluorescence as a function of time at 25°C (*see* **Note 7**). The time-course of hybridization is recorded at the optimal excitation and emission wavelengths of the fluorophore coupled to the molecular beacon (*see* **Note 8**).

3.3. *In Vivo Detection of* Oskar *mRNA*

The mRNA sequence encoded by the *Drosophila melanogaster* maternal gene *oskar* is shown as a model target for in vivo detection with molecular beacons. *Oskar* mRNA localizes during mid-oogenesis in a tight sliver at the posterior pole of the oocyte. Utilizing the RNA secondary structure prediction program *mfold* and applying the aforementioned criteria for selection of accessible target regions, sequences within the mRNA were identified. These regions are either single-stranded or paired with distant sequences in most of the thermodynamically favorable foldings and contained both flexible and stable structures.

Next, using OligoWalk's thermodynamic outputs, which reflected the degree of stability of all RNA–probe hybrids formed from scanning oligonucleotides, the choice of target sequences was narrowed down to 11 regions (*see* **Fig. 3**). Among the 11 regions selected for *oskar* mRNA, 4 regions (osk 62–87, osk 964, osk 2209, and osk 2579) had high values for both ss-count and P-num. The threshold value for the selection of these regions was 220, which represents a 50% probability that the respective region is single-stranded and is a measure of how "well determined" the structure formed by the sequence is. The remaining seven regions had low P-num values, indicating better determined and more stable regions. This theoretical analysis helped reduce the number of potentially accessible target regions and, with good confidence, provided sequences for *oskar* mRNA-specific molecular beacons.

A 100 ng/µL-solution of the *oskar* mRNA-specific molecular beacon dissolved in nuclease free water was microinjected into a stage 9 *Drosophila melanogaster* oocyte (*see* **Fig. 4**) *(17)*. The images acquired at various time intervals revealed a homogenous distribution of the molecular beacons with an increased fluorescent signal at the posterior end of the oocyte (*see* **Note 9**) *(2)*.

4. Notes

1. *mfold* values:
   - *ss*-count is the number of times that the *ith* base is single-stranded in *n* predicted foldings.
   - P-num $(i) = \sum_{k<i} \phi\,(\Delta G(k,i) \leq \Delta G + \Delta\Delta G) + \sum_{i<j} \phi\,(\Delta G(i,j) \leq \Delta G + \Delta\Delta G)$, where i, j, and k are bases and ϕ is defined as 1 when expression is true and 0 otherwise.

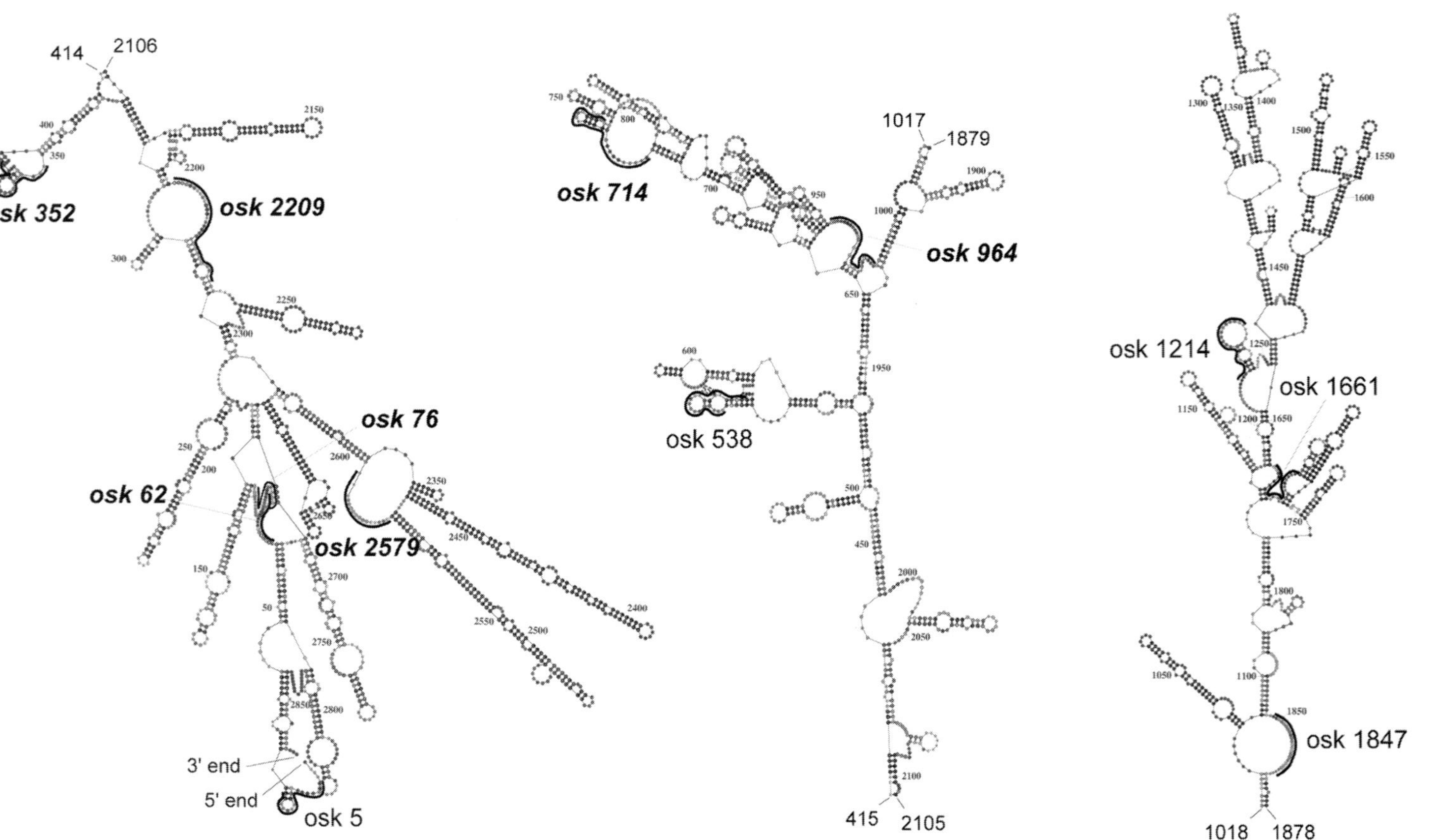

Fig. 3. Most stable secondary structure of *oskar* mRNA showing the location of molecular beacon target sequences. The computer-folded structure is depicted in three segments, with the connecting nucleotides indicated at the break points. The bold nomenclatures represent regions that were most accessible by the molecular beacons.

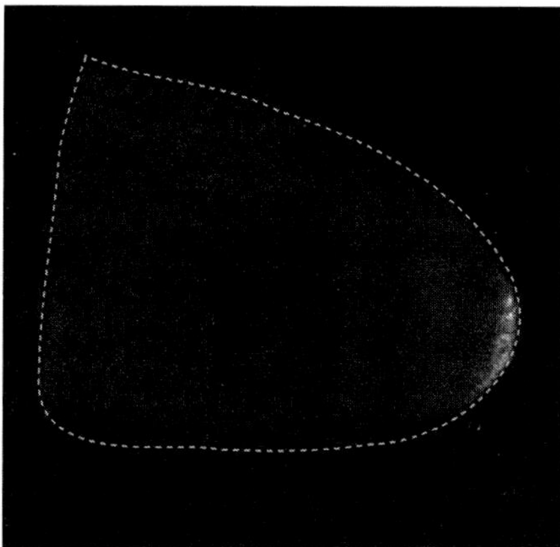

Fig. 4. Stage 9 *Drosophila* oocyte injected with a molecular beacon solution specific for *oskar* mRNA. After 10 min, an image was acquired using a confocal microscope. Anterior is to the left and posterior is to the right of the image.

- ΔG is the overall minimum free-energy of any folding that contains the i,j basepair:

$$\Delta G = \min_{1 \le i < j \le n} \Delta G(i,j).$$

- $\Delta\Delta G$ is a user-selected free-energy increment set at $1 \le \Delta\Delta G \le 12$ (kcal/mol)

$$\Delta\Delta G = \Delta G \times P/100.$$

- P is the parameter that controls the value of the free-energy increment $\Delta\Delta G$. The default value is 5.

2. To save this structure as a ".pdf" file, choose the output option as a "postscript" file. Click on the structure and the option to "save as" is available in a pop-up window. Save this as a ".cgi" file, which can then be converted to a ".pdf" file with Acrobat Distiller. View and manipulate in Adobe Illustrator.
3. Unlike predictions of local hairpins, long-range interactions or multibranch junctions remain poorly determined. Such predicted structures could provide insights into regions of potential structural plasticity within an RNA molecule and thus reveal potential accessible sites for antisense probes. Therefore, when considering probes for targeting an RNA sequence, it is very helpful to pay close attention to the information offered by a secondary structure fold.
4. In vivo, the mRNA structure is dependent on the intracellular environment. For example, mRNA accessibility becomes less predictable, as the structures of actively translated mRNAs are transiently altered by passing ribosomes. Also, the stability of the duplexes formed by probes and complementary nucleic acids is influenced by the presence of RNA-binding proteins. These might mask target

sites, preventing hybridization from occurring. Therefore, a number of different regions distributed throughout the mRNA sequences should be considered as target sites for molecular beacons.

5. Because the 2´-*O*-methyl backbone renders molecular beacons capable of forming stronger stems, the sequence composition of the complementary arms are either four nucleotides for shorter-length probe sequences or five nucleotides for longer hairpin loops. For example, molecular beacons with 15–18 nucleotides composing the hairpin loops should have stems such as CGGC (or a variation of this order). For longer loops, design stems such as GATCC or GGAGC. A 100% GC five- nucleotide stem is too strong and the molecular beacon would not open at room temperature under cellular conditions.
6. The structures generated by the folding program should have the intended hairpin structure. Any other structure that does not form the set-length stem hybrid will generate inefficient binding probes (i.e., the stem is longer, or the nucleotides within the hairpin loop form basepairs) or high-background signals (when the nucleotides at the 5´ and 3´ ends of the sequence do not form a pair, the fluorophore is removed from the vicinity of the quencher). In either instance, both the probe and stem sequences can be manipulated. If the chosen region on the RNA permits, shift the frame of the probe along the target sequence until a probe structure with minimal self-complementarity is obtained. For a longer hairpin loop, a small interior stem of two to three nucleotides long does not significantly affect the performance of a molecular beacon. If the alternative structures arise from the choice of the stem sequence, change its composition to ensure a hairpin conformation. The stem can also be formed by terminal nucleosides of the probe sequence.
7. For each molecular beacon tested, the kinetic characteristics of hybridization as well as the level of fluorescence generated might be different. The most accessible target sequences induce the fastest increase in fluorescence. However, the efficiency of hybridization and stability of binding can depend on more than the secondary structure of the RNA. The probe's size and sequence composition are important determinants of the stability of the resulting hybrid. For *oskar* mRNA, each probe has approx 40% GC composition and is 18 to 25 nucleotides in length. The rate of binding might also be influenced by the tertiary structure of the RNA. The sites accessible for targeting could be masked by the three-dimensional folding of the molecule or by other tertiary interactions, such as prestacking of the single-stranded regions. For these reactions, the molecular beacon molar concentration exceeds that of the target RNA to ensure that all RNA molecules are bound by a probe.
8. Molecular beacons can be labeled with a wide range of fluorophores, which can be efficiently quenched by the same quencher, such as dabcyl, BHQ1, or BHQ2 ***(16,18)***. A list of companies licensed to synthesize molecular beacons as well as a list of the available fluorescent labels can be found at http://www.molecular-beacons.org/.
9. When images are acquired by a confocal fluorescence microscope, the background signal from other focal planes generated by the nonspecific binding of cellular components to probes that changes their conformation and causes them to fluoresce is

eliminated. When conventional fluorescence microscopy is implemented, the images can be enhanced via deconvolution analysis, where the fluorescence contribution from other focal planes is removed, or by ratio imaging, where the intensity of the fluorescence generated by the specific molecular beacon is divided by the intensity of the spectrally distinguishable fluorescence of a second, nonspecific molecular beacon that is coinjected in an equimolar cocktail with the specific molecular beacon.

Acknowledgments

The author thanks Sanjay Tyagi and Fred Russell Kramer for their assistance and advice and Salvatore A.E. Marras for help with molecular beacon synthesis. This work was supported by National Institutes of Health grants ES-10536 and EB-00277.

References

1. Tyagi, S. and Kramer, F. R. (1996) Molecular beacons: probes that fluoresce upon hybridization. *Nature Biotechnol.* **14**, 303–308.
2. Bratu, D. P., Cha, B. J., Mhlanga, M. M., Kramer, F. R., and Tyagi, S. (2003) Visualizing the distribution and transport of mRNAs in living cells. *Proc. Natl. Acad. Sci. USA* **100**, 13,308–13,313.
3. Matsuo, T. (1998) In situ visualization of messenger RNA for basic fibroblast growth factor in living cells. *Biochem. Biophys. Acta* **1379**, 178–184.
4. Sokol, D. L., Zhang, X., Lu, P., and Gewirtz, A. M. (1998) Real time detection of DNA.RNA hybridization in living cells. *Proc. Natl. Acad. Sci. USA* **95**, 11,538–11,543.
5. Southern, E. M., Milner, N., and Mir, K. U. (1997) Discovering antisense reagents by hybridization of RNA to oligonucleotide arrays. *Ciba Found. Symp.* **209**, 38–44; discussion 44–46.
6. Ho, S. P., Bao, Y., Lesher, T., et al. (1998) Mapping of RNA accessible sites for antisense experiments with oligonucleotide libraries. *Nature Biotechnol.* **16**, 59–63.
7. Lima, W. F., Mohan, V., and Crooke, S. T. (1997) The influence of antisense oligonucleotide-induced RNA structure on *Escherichia coli* RNase H1 activity. *J. Biol. Chem.* **272**, 18,191–18,199.
8. Milner, N., Mir, K. U., and Southern, E. M. (1997) Selecting effective antisense reagents on combinatorial oligonucleotide arrays. *Nature Biotechnol.* **15**, 537–541.
9. Zuker, M. (2003) Mfold web server for nucleic acid folding and hybridization prediction. *Nucleic Acids Res.* **31**, 1–10.
10. Mathews, D. H., Burkard, M. E., Freier, S. M., Wyatt, J. R., and Turner, D. H. (1999) Predicting oligonucleotide affinity to nucleic acid targets. *RNA* **5**, 1458–1469.
11. Peyret, N., Seneviratne, P. A., Allawi, H. T., and SantaLucia, J., Jr. (1999) Nearest-neighbor thermodynamics and NMR of DNA sequences with internal A.A, C.C, G.G, and T.T mismatches. *Biochemistry* **38**, 3468–3477.

12. SantaLucia, J., Jr. (1998) A unified view of polymer, dumbbell, and oligonucleotide DNA nearest-neighbor thermodynamics. *Proc. Natl. Acad. Sci. USA* **95**, 1460–1465.
13. SantaLucia, J., Jr., Allawi, H. T., and Seneviratne, P. A. (1996) Improved nearest-neighbor parameters for predicting DNA duplex stability. *Biochemistry* **35**, 3555–3562.
14. Sugimoto, N., Nakano, S., Katoh, M., et al. (1995) Thermodynamic parameters to predict stability of RNA/DNA hybrid duplexes. *Biochemistry* **34**, 11,211–11,216.
15. Xia, T., SantaLucia, J., Jr., Burkard, M. E., et al. (1998) Thermodynamic parameters for an expanded nearest-neighbor model for formation of RNA duplexes with Watson-Crick base pairs. *Biochemistry* **37**, 14,719–14,735.
16. Marras, S. A., Kramer, F. R., and Tyagi, S. (2002) Efficiencies of fluorescence resonance energy transfer and contact-mediated quenching in oligonucleotide probes. *Nucleic Acids Res.* **30**, e122.
17. Bratu, D. P. (2003) *Imaging Native mRNAs in Living Drosophila Oocytes Using Molecular Beacons*. New York University, UMI Dissertation Service, New York, New York.
18. Tyagi, S., Bratu, D. P., and Kramer, F. R. (1998) Multicolor molecular beacons for allele discrimination. *Nature Biotechnol.* **16**, 49–53.

2

Second-Harmonic Imaging of Collagen

Guy Cox and Eleanor Kable

Summary

Molecules that have no center of symmetry are able to convert light to its second harmonic, at twice the frequency and half the wavelength. This only happens with any efficiency at very high light intensities such as are given by a pulsed laser, and because the efficiency of the process depends on the square of the intensity, it will be focal plane selective in exactly the same way as two-photon excitation of fluorescence. Because of its unusual molecular structure and its high degree of crystallinity, collagen is, by far, the strongest source of second harmonics in animal tissue. Because collagen is also the most important structural protein in the mammalian body, this provides a very useful imaging tool for studying its distribution. No energy is lost in second-harmonic imaging, so the image will not fade, and because it is at a shorter wavelength than can be excited by two-photon fluorescence, it can be separated easily from multiple fluorescent probes. It is already proving useful in imaging collagen with high sensitivity in various tissues, including cirrhotic liver, normal and carious teeth, and surgical repair of tendons.

Key Words: Collagen; second harmonic; structural proteins; biological imaging; 3D imaging; non-linear microscopy; matrix.

1. Introduction

1.1. Second-Harmonic Generation

Theodore Maiman won a much publicized race to make a working laser when he built a pulsed ruby laser in 1960. This was the world's first laser and it produced pulses of deep red (697 nm) light, just within the visible range. Almost immediately, it was found that shining pulses of ruby laser light through a quartz crystal produced near-ultraviolet light at 348 nm, the second harmonic of the original light *(1)*.

Second-harmonic generation (SHG) takes place when the electric field of the exciting light is sufficiently strong to deform a molecule. If the molecule is not symmetrical, the resulting anisotropy creates an oscillating field at twice the

From: *Methods in Molecular Biology, vol. 319: Cell Imaging Techniques: Methods and Protocols*
Edited by: D. J. Taatjes and B. T. Mossman © Humana Press Inc., Totowa, NJ

frequency, the second harmonic *(2)*. This means that the ability to generate second harmonics is peculiar to molecules that are not centrosymmetric. SHG will also take place at interfaces where there is a huge difference in refractive index, such as metal surfaces. For a more technical description, *see* **Note 1**.

The resulting beam of light at twice the frequency/half the wavelength usually travels in the same direction as the incident beam and in phase with it, although it commonly has a different plane of polarization. Different samples and illumination conditions will modify this behavior; the whole question of beam propagation is discussed in more detail in **Subheadings 2.3.** and **3.1**. Crystals of nonsymmetric molecules such as potassium deuterium hydrogen phosphate are very effective generators of second harmonics, and particularly when the angle of the crystal lattice is carefully matched to the incoming beam, they can give extremely high second-harmonic yields. The production of suitable crystals (potassium, cesium, or rubidium titanyl phosphate or arsenate are other common examples) is an important industry; and the crystals are in everyday use as frequency doublers in the laser industry—even humble items such as green laser pointers contain a frequency-doubling crystal.

Second-harmonic generation was first used in microscopy as long ago as 1974 *(3)* (*see* **Note 2**), but the more practically useful technique of scanned second-harmonic microscopy was first achieved 4 yr later, by Gannaway and Sheppard *(4)*, using a continuous-wave laser. All modern applications stem from this work. As they pointed out, in the scanning mode the image is focal plane selective because the signal depends on the square of the incident beam power. It is, thus, effectively equivalent in imaging properties to confocal microscopy and two-photon fluorescence; the fact that it is coherent rather than incoherent does cause small differences, but they are only of theoretical interest so far as biological applications are concerned. Scanning the beam across the sample means that high intensities only have to be present in a very small region at any one time, but, even so, the continuous-wave laser used by Gannaway and Sheppard necessitated power densities at the sample that could not be tolerated by biological specimens. It was clear even then that pulsed lasers would make the technique more practicable, and the introduction of second-harmonic microscopy as a practical technique had to await the availability of pulsed lasers with very short pulse lengths. These now offer very high instantaneous powers in conjunction with low total power averaged over time, at a repetition rate fast enough not to restrict scan speed.

Second-harmonic-generation microscopy has many features in common with two-photon fluorescence (TPF) microscopy, and the hardware required is very similar, but there are some key differences (summarized in **Table 1**). Fluorescence always involves some loss of energy in the sample, and the fact that electrons are raised to excited states means that bleaching is inevitable.

Table 1
Comparison of SHG and TPF

SHG	TPF
Exactly double the original frequency	Spectrum of frequencies less than two times the original
Largely frequency independent	Strongly frequency dependent
Virtually instantaneous, approx 1 fs	Lifetime in nanoseconds
Propagated forward	Propagated in all directions
Coherent (in phase with exciting light)	Incoherent
No energy loss or damage	Always energy loss and associated damage
Requires short laser pulses	Requires short laser pulses

SHG dissipates no energy in the sample and there is no excitation or bleaching. Of course, that does not imply that no damage can occur; there are several mechanisms through which intense laser pulses can damage the sample, but these are totally independent of the imaging process.

Serious biological use of SHG microscopy has been a recent phenomenon, with few papers dated prior to 2001. Gauderon et al. *(2)* were able to image the DNA of polytene chromosomes of *Drosophila*. Campagnola et al. *(5)* applied the known SHG properties of styryl potentiometric dyes to the microscope, continuing from earlier work in which these properties had been investigated in the cuvet. They also imaged collagen and other structural proteins in a variety of tissues at resolutions of up to 1 μm *(5,6)*. Membranes have been labeled with second-harmonic generating dyes, providing a sensitive probe of membrane separation *(7,8)*.

1.2. Collagen and SHG

Individual noncentrosymmetric molecules will generate a second-harmonic signal, but molecules arranged in a crystalline array will give a very much stronger response. Hence, any biological material that is both crystalline and noncentrosymmetric is likely to be suitable for SHG microscopy. Furthermore, SHG microscopy of such materials should be capable of providing information about orientation and crystallinity, as well as morphology.

Collagen is the most important extracellular structural protein of the vertebrate body, making up around 6% of the body mass, mainly in bone, cartilage, skin, interstitial tissues, and basal laminae. The collagen molecule is nonsymmetric and is arranged in a triple helix *(9)*. There are many different forms of collagen, coded separately in the genome. Minor amino acid changes affect the final conformation and, therefore, give the different collagen types different functions in the living organism. Four types (I, II, III, and V) form fibrils, type

IV forms sheets in basal laminae, and types VI and IX act as links, binding fibrillar collagen to other cell components. Type I, in particular, is highly crystalline and is very important as a structural component of the skeleton, cartilage, and soft tissue.

This unusual structure makes type I collagen an effective generator of second harmonics, a fact that has been appreciated for over 20 yr ***(10)***. The signal is, to a large extent, wavelength independent over a wide range of wavelengths in the infrared region ***(11,12)***. Furthermore, it is possible from this signal to determine the polarity of the collagen helix ***(13)***. Differences in signal between normal and experimentally damaged collagen have also been reported ***(14)***, raising the hope that pathological changes might be detected. Very recently, spectral changes in the second-harmonic signal from normal and cancerous tissue have been described, opening up a possibility that it might become a useful diagnostic tool in oncology ***(15)***. We have shown that SHG can be an exquisitely sensitive tool for detecting and imaging collagen at high resolution, with diffraction-limited resolution (sub-300 nm) easily achievable ***(12,16)***. There seems to be the potential to distinguish between different collagen types, with highly crystalline type I collagen giving a much stronger second-harmonic signal than type III, even though both stain identically with collagen-specific dyes ***(12)***.

2. A Microscope for SHG Imaging

The basic requirements for SHG microscopy are those for TPF microscopy: a scanning microscope (usually a confocal microscope, although confocal optics are irrelevant) coupled to a pulsed infrared laser. However, the different nature of signal generation in SHG microscopy means that some simple and straightforward modifications to a normal multiphoton microscope will be needed.

2.1. The Laser

A pulsed titanium–sapphire (Ti-S) laser is the normal choice, although neodymium yttrium aluminium garnet (Nd YAG) lasers have been used ***(11)***, and, in principle, other lasers such as chromium forsterite are possible and might come into wider use in the future. Because SHG depends on the square of the flux density, shorter pulses will mean a substantial reduction in the average power required. Therefore, in principle at least, a "femtosecond" Ti-S laser (typically delivering 100- to 200-fs pulses) will usually be preferred over a "picosecond" version, which delivers 1- to 2-ps pulses.

However, to gain the full benefit of short pulses, it is essential that they remain short. Light travels approx 30 µm during a 100-fs laser pulse, so the pulse will contain fewer than 40 waves at 800 nm. With such a small number of waves in a pulse, the wavelength cannot be specified with great precision and, consequently, there will be a range of wavelengths present in a Gaussian distribution with a full width at half-maximum (FWHM) range of about 2 nm. This makes no difference in terms of biological imaging, as the spread is narrower than the band passed by current detection systems, whether filter based or spectrometric. It is also trivial in terms of TPF excitation. However, it is very much wider than the completely monochromatic light produced by a continuous-wave laser. There is an important practical consequence of this wavelength spread: It will broaden the pulse if it travels through a dense medium such as glass. All glass has dispersion; that is, its refractive index is wavelength dependent, so that longer wavelengths travel more slowly than shorter ones. In a very short laser pulse, therefore, the different wavelengths will travel at different speeds and the pulse will broaden. Longer pulses have a much smaller wavelength spread, so they will not be affected to the same extent; in practical terms, pulse broadening is only a problem in the femtosecond region.

A consequence of this is that the laser needs to be directly coupled to the microscope; transmitting the light through an optical fibre will broaden the pulse to an unacceptable extent (*see* **Note 3**). Any additional optical elements in the beam path—even such useful items as beam expanders—will lengthen the pulse, as will objective lenses; so the less glass in the objective, the better. This implies that there is little point in using complex lenses such as apochromats, especially because color correction is irrelevant (only one wavelength has to be brought to a focus). Fluorite lenses typically offer comparable numerical apertures with many fewer elements, and because their basis is the low-dispersion mineral fluorite (calcium fluoride), pulse broadening is minimal.

Femtosecond Ti-S lasers fall into three categories:

1. Fully tuneable between 700 and 1000 nm. These lasers typically have a range of manual adjustments that need attention in normal operation, and more than one set of optical components might be needed to obtain the widest tuning range, although 700 to 950 nm or 750 to 1000 nm are typically now available with a single optics set.
2. Tunable over a limited range, but fully automated so that no external adjustments need to be made. The tunable range available with these is increasing rapidly, and although fully tuneable lasers still offer a wider range at the time of writing (June 2003), the automated models are now very close to them in performance and might well match them before long.
3. Fixed wavelength, commonly 800 nm.

Fixed-wavelength lasers are not very suitable—unless set to a rather longer wavelength than 800 nm—because detection of the second harmonic at 400 nm is often blocked or attenuated by other optical elements in the microscope. Automated lasers are probably now the most effective solutions, unless other usage requirements dictate a wider tuning range. Because the SHG signal from collagen shows little wavelength dependence within the tuning range of a Ti-S laser, wider range lasers do offer the facility of optimizing laser wavelength for multiphoton fluorescence excitation without compromising SHG detection. In particular, effective two-photon excitation of many members of the GFP (green fluorescent protein) family requires long wavelengths that lie beyond the range of automated lasers (*see* **Note 4**).

Tunable lasers are tuned by moving a birefringent element, but this, in turn, will require adjustments to a prism to compensate group velocity dispersion—ensuring that all wavelengths remain together within the laser cavity—and there will also be alignments to mirrors and the slit that acts to impose mode locking on the pulses. Hence, a fully tunable laser requires not only an investment in technology but also an investment in expertise to keep it correctly adjusted.

2.2. Ancillaries

The coupling system to the microscope, whether lab built or supplied by the microscope maker, will need to divert part of the beam to diagnostic equipment. A spectrum analyzer is the most useful ancillary, showing not only the actual wavelength tuned but also the spread of wavelengths present. This is a very effective visual indication of the pulse length, and software is available that will compute an estimated pulse length from the spread, although this does depend on some assumptions about the pulse shape. An autocorrelator will give a more precise measure of the pulse length, but is a fiddly and complex instrument to use, although, again, rapid strides are being made in improving usability for the biologist. Recently, models have been produced that provide a remote head to measure the actual pulse width emerging from the objective. A power meter is also valuable, even if the diagnostics built into the laser itself gives a power reading, because an independent power meter can pick up losses further along the beam path. Often, a decline in power will be a useful warning that either the laser is not perfectly aligned or that the internal optics are in need of cleaning.

As discussed in **Subheading 2.1.,** the beam delivery to the microscope must be direct, not through an optical fiber. Normally, therefore, the microscope and laser will both be rigidly mounted to an optical table. Microscope manufacturers might have some degree of standardization in their hardware, but these systems are still largely custom installations, so the user will have a fair degree of choice. Points to consider include the following:

1. The range and convenience of alignment of the beam, because the pointing stability of Ti-S lasers is often rather worse than is normal for continuous-wave lasers.
2. The control over beam power. Although neutral density filters will provide attenuation, an electronic control such as a Pockel Cell or electro-optic modulator (EOM) will allow beam blanking on flyback (thus reducing the total beam exposure of the sample) as well as patterned irradiation.

3. The efficiency of the total throughput. Often much light is lost before it even reaches the microscope. Many partial beam splitters (used to divert part of the beam to diagnostic equipment) have a surprisingly high absorbance in the 700- to 900-nm range. A piece of plain, uncoated glass is often at least as effective as a custom beam splitter. If much power is disappearing, it is worth carrying out an audit with a power meter to see where the losses are.

2.3. The Microscope

Whereas second harmonic light is generally propagated forward, geometric considerations (the Gouy phase shift in a focused spot) might mean that much of the flux emanating from the scanning spot is in the off-axis directions *(7,8)*. Therefore, for effective collection, the condenser lens must have a numerical aperture (NA) equal to or larger than that of the objective. An oil-immersion condenser is a necessity. Equally important is that Köhler illumination be set up with scrupulous precision. Without this, a large proportion of the second-harmonic light will never reach the detector. Having a totally immersed system also reduces the amount of stray room light that can enter, an important consideration with any form of nondescanned detector.

Although they provide geometrically perfect focusing, galvanometer stage movements do not function well in this environment. The combination of short working distances and viscous drag from the immersion media impede the free motion of the stage, so that it often fails to follow the instructions given by the controlling computer. Even though, in theory, moving the sample between a stationary objective and condenser achieves optical perfection, it is preferable in practice to focusing with the nosepiece; a high precision is achieved with electronic or piezo-driven nosepieces. Generally, the precision with which a series can be collected will be very high, but the reproducibility will be less so; thus, there might be visible errors in returning to a given focal plane.

Unless one wishes to work in total darkness, every attempt should be made to prevent room light from reaching any of the nondescanned detectors. **Figure 1** shows the Leica screening arrangements as fitted to our microscope. Around the condenser, we have a large (10 cm) matt-black disk, which serves a dual function in excluding room light as well as safeguarding against accidental laser exposure. From the condenser to the lamp housing is a close-fitting matt black telescopic tube, which keeps out all room light above the condenser.

The slider that blocks off the wide-field epifluorescence illuminator (high-pressure mercury or argon arc lamp) often passes a surprisingly large amount of light when closed. Although this is not noticeable in conventional epifluorescence microscopy, it can interfere substantially with two-photon or second-harmonic detection. Because it is often not convenient to turn off the mercury lamp, a blanking plate should be made up to fit in the lamp housing. We use a simple

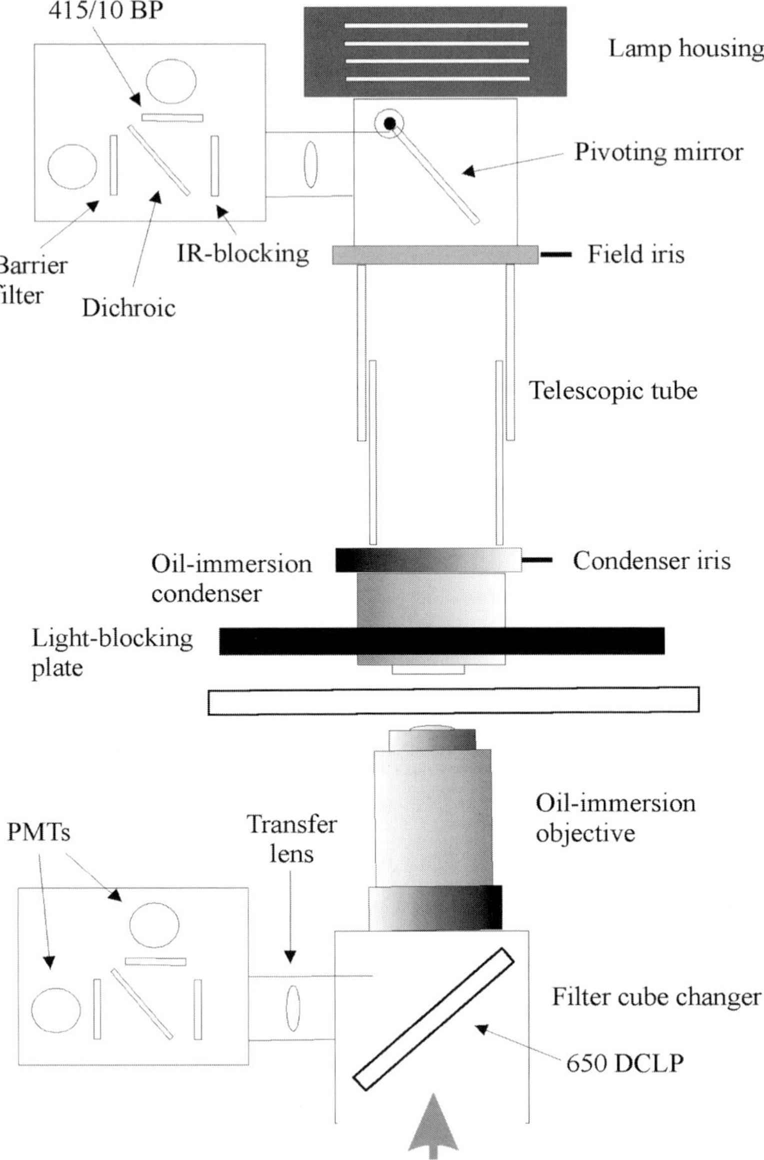

Fig. 1. Schematic arrangement of lens and detector arrangements on our Leica TCS2 MP/DMIRB inverted-microscope system. The two dual-channel detectors are identical and have removable filter cubes, so that different combinations are easy to assemble. The 650DCLP dichroic is mounted in a cube of the wide-field epifluorescence filter changer and directs all fluorescence returning through the objective to the detector. When this is rotated out of position, the normal confocal detection system can be used.

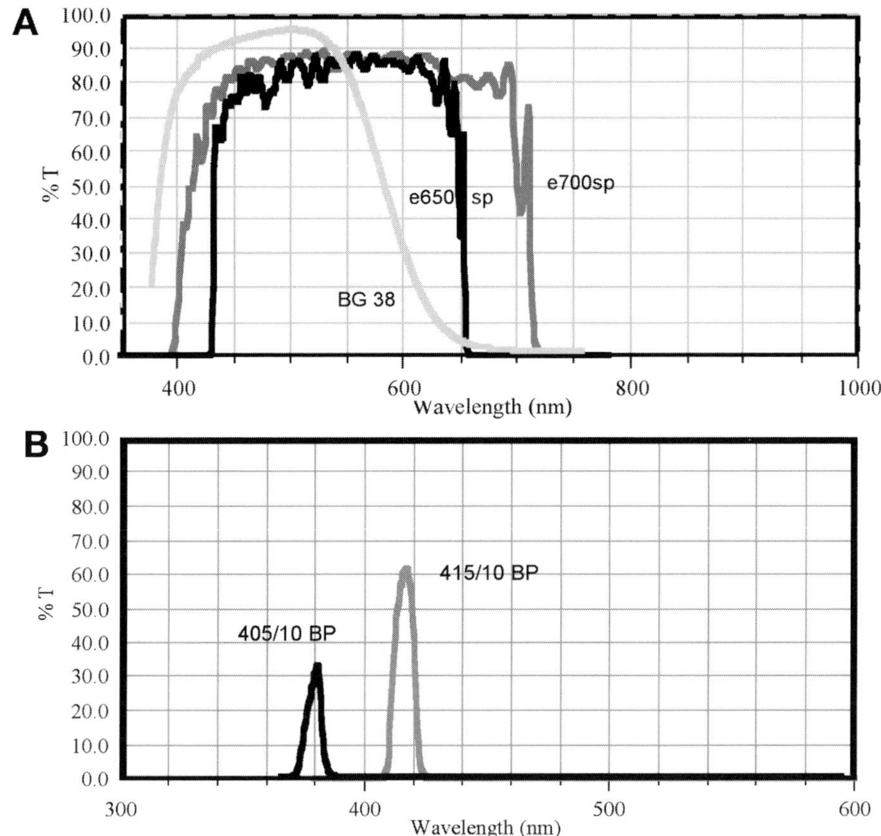

Fig. 2. **(A)** Curves of several infrared-blocking filters. The interference filters provide a sharp cutoff and good transmission. However, the e650 SP not only provides total blocking of the Ti-S laser light but also cuts off even the violet part of the visible spectrum. The e700 SP provides all of the visible spectrum but still cuts off at 400 nm and will also allow some laser light through at the short end of the tuning range. The colored glass filter BG38 is obviously very imperfect, allowing some leakage from 750 nm on and severely attenuating the red part of the visible spectrum; nevertheless, it transmits 80% at 400 nm and a usable fraction a little further into the ultraviolet. **(B)** Curves of two narrow-band filters suitable for detecting SHG signals. Because of the absorbance of the coating materials, only approx 35% transmission is attainable at 405 nm, so it will be preferable to tune the laser to 830 nm and use the 415-nm filter, which passes more than 60%. All spectra are reproduced by kind permission of Chroma Technology Corp.

aluminum disk, painted matt black, mounted in a filter holder, which will slip into one of the ventilation slots in front of the housing.

Filters for SHG detection need some careful consideration. As **Fig. 2A** shows, most infrared-blocking filters (essential to block transmitted laser light)

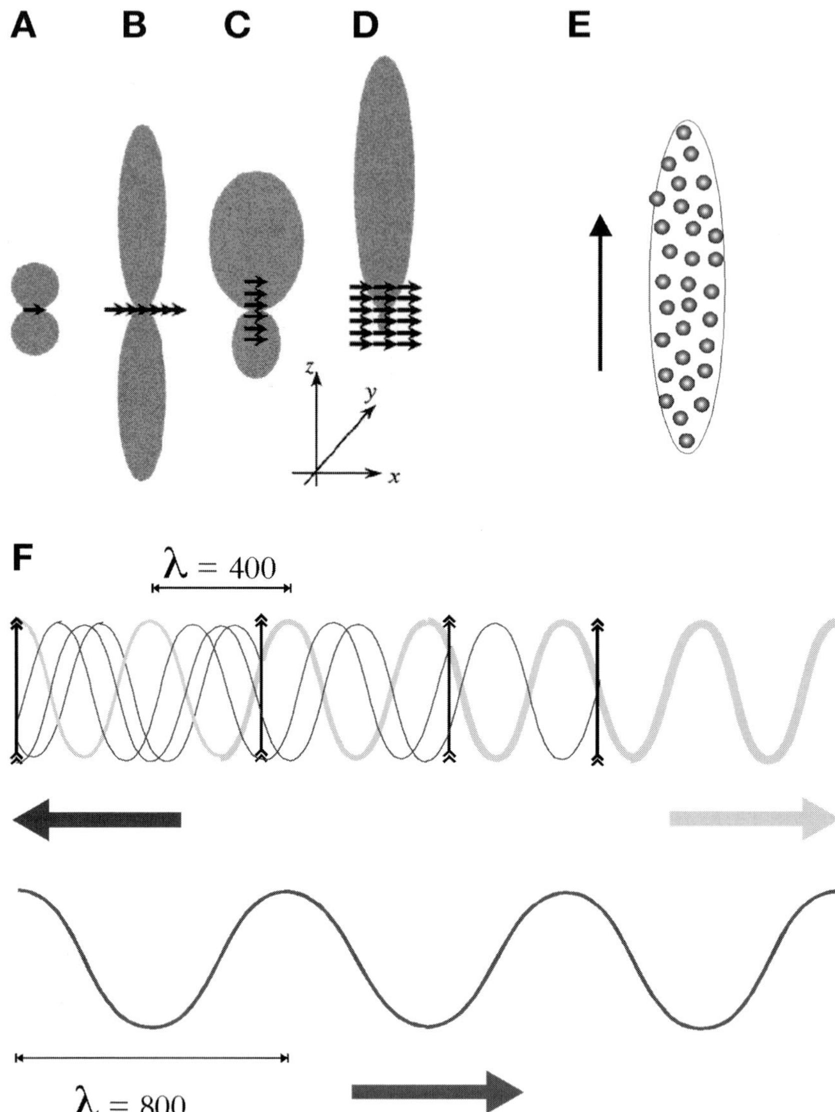

Fig. 3. Patterns of radiation from **(A)** single dipole, **(B)** array normal to the incident beam, **(C)** array in line with the incident beam, and **(D)** a bulk array. (D, from **ref. *18*** by permission of the authors and *The Biophysical Society*). **(E)** Shape of the excitation point-spread function (psf) superimposed on typical collagen spacing in connective tissue. (A–D, from **ref. *12*** with permission from Elsevier). **(F)** Randomly spaced fibers will propagate the second harmonic in a forward direction. Lower (red) wave is the excitation beam at 800 nm, propagating in the direction of the red arrow. The upper row shows second-harmonic radiation generated by the randomly spaced dipoles indicated by vertical arrows. Light propagated in the forward direction will be in step both with

also block light shorter than 400 nm. Therefore, even though most lasers can be tuned to wavelengths shorter than 800 nm, it will not be easy to detect the SHG signals in this part of the spectrum. Furthermore, the coatings used in most interference filters start to have strong absorption at wavelengths shorter than about 420 nm. At 415 nm, a 10-nm bandpass filter will transmit about 60% at its peak; at 405 nm, the equivalent filter will only transmit 30% (*see* **Fig. 2B**). Therefore, other things being equal, it is best to use an excitation wavelength of 830 nm or longer. We normally use 830-nm excitation and a 415-nm barrier filter.

3. Methods

3.1. Signal Propagation and Properties

Typically, depending on the extent to which the sample scatters light, between 80% and 90% of the SHG signal from collagen in a tissue sample will be propagated forward *(12)*. **Figure 3** shows the direction of propagation of the signal from dipoles that are small compared to the wavelength propagated. Parts A–D were drawn to represent coherent anti-Stokes Raman scattering (CARS) microscopy *(18)* but are equally applicable to SHG. A single dipole **(Fig. 3A)** radiates in all directions *except* normal to the incident beam. A single, 40- to 50-nm collagen fiber would be expected to behave in this way. A planar array normal to the incident beam will radiate strongly forward and backward, because waves in both of these directions will be in phase **(Fig. 3B)**. There will be little lateral propagation because in this direction, the wave from one dipole will not be in phase with the wave from its neighbor. An array orientated along the direction of the beam **(Fig. 3C)** will propagate strongly forward, because, regardless of the actual spacing (and even if it is completely irregular), all dipoles will have the same phase relationship to the exciting beam in the forward direction but will be randomly out of phase in the reverse direction **(Fig. 3F)**. This same effect will be even more marked in a bulk array of dipoles **(Fig. 3D)**, which will propagate the harmonic virtually exclusively forward. In SHG microscopy, the excited volume is elliptical, with the long axis along the direction of the beam, so that in any grouping of collagen fibers, there will be more excited dipoles in line with the beam than across it **(Fig. 3E)**. Thus, whereas an isolated individual fiber may radiate its signal both forward and backward, overall the signal is predominantly propagated forward. In experiments with cryo-sections of human endometrium, we have found that only

Fig. 3. *(Continued)* the exciting radiation and light from other dipoles, whereas in the reverse direction, radiation from each dipole will have a random phase relation with any other, so little propagation will take place.

10–12% of the detectable SHG signal is propagated back through the objective lens ***(12)***, the remainder passing through the condenser to the transmitted light detector.

Although SHG uses long wavelengths, the resolution achievable is quite respectable. Calculated resolution values at 830 nm are around 250 nm ***(19)***, and measurements close to that have been reported in practice ***(12,16,19)***. **Figure 5C** shows resolution of this order in a histological section of skin. The depth resolution will be identical to that of TPF, and under best conditions, around 800 nm should be achieved. This is more than adequate for effective three-dimensional reconstructions (*see* **Fig. 6**).

3.2. Signal Detection and Imaging

3.2.1. Suitable Samples

The SHG signal from type I collagen is very strong—typically stronger than most two-photon excited fluorescence in biological samples—so that low excitation levels can be used. When fluorescence is being detected at the same time, the laser intensity needed will generally be determined by the fluorescent signal. SHG from collagen seems to be unaffected by most preparation techniques. Good images can be obtained from histological paraffin sections, whether unstained (*see* **Fig. 4**) or stained with routine histological stains such as hematoxylin and eosin or Masson's trichrome (**Figs. 5,6; refs. *12,16***).

Sirius Red, which stains collagen and also enhances its birefringence ***(20,21)***, also has no effect on the SHG, although it is noticeable that the second-harmonic signal does not exactly colocalize with two-photon excited fluorescence from Sirius Red ***(12)*** or with the autofluorescence from aldehyde-fixed collagen (*see* **Fig. 4**). We have suggested that only the highly crystalline collagen gives the SHG signal and stains (and aldehydes) have more effect on surrounding less crystalline material; colocalization analyses support this ***(22)***.

One common preparation technique does seem to compromise the SHG signal from collagen. It is severely depleted in epoxy resin-embedded blocks of tissue fixed in glutaraldehyde and osmium tetroxide for electron microscopy, even when sectioned at 500 nm thickness for the optical microscope. Even though some signal is still present, on the basis of our preliminary trials we cannot recommend this material for SHG imaging; a disappointing conclusion for those needing to combine optical and electron microscope imaging of collagen.

The image formation in three dimensions is effectively equivalent to that in multiphoton fluorescence; therefore depth penetration in turbid and scattering samples will be substantially better than in single-photon confocal microscopy, the predominantly forward propagation of the signal does impose some restrictions on sample thickness for optimal imaging. Whereas in multiphoton mode one can simply probe down into a thick specimen until the point of no signal is

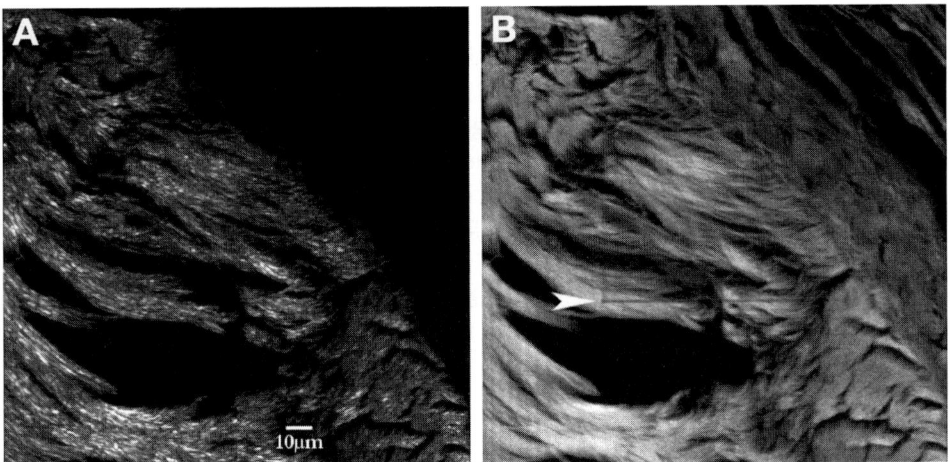

Fig. 4. Micrographs of kangaroo-tail tendon transplanted into rat muscle. Projection from 24 optical sections. Aldehyde-fixed, unstained paraffin section cleared in xylene and permanently mounted. For more details of the sample, *see* **refs. *12*** and ***18***. The implanted tendon is seen in the lower left half of the picture, and the host tissue (including some type III collagen from the host response) is seen in the upper right. **(A)** Forward-propagated signal, almost exclusively SHG at 410 nm, although detected over a wide spectral range (400–550 nm). **(B)** Back-propagated signal in the spectral range 500- to 550-nm showing aldehyde-induced autofluorescence. The photomultiplier tube voltage was set 100 V higher than in (A) in order to pick up the much weaker autofluorescence. Previous collection of an image at a higher zoom setting has caused slight bleaching in the fluorescent image (arrowed), but the second-harmonic image is unaffected. Note that no SHG signal is visible in the type III collagen (upper right), implying that if it either produces no, SHG signal or a much weaker, SHG signal.

reached, for optimal imaging in SHG microscopy the signal has to be able to pass through the entire sample. Thus, in either case, the penetration depth might be 200 μm; however, in SHG, the sample itself cannot be too much thicker than this. With thicker samples, one often can image down to a considerable depth at the edge of the specimen when no signal can be detected in the center—a consequence of the large angle through which the signal is generated and propagated. Alternatively, one can simply use the backscattered image; even though up to 90% of the signal might be lost, in a suitable sample it might still be possible to collect a good image ***(23)***.

3.2.2. Detection Strategies

Because the SH signal is by definition shorter than any two-photon excited fluorescence, a 10-nm FWHM filter will completely exclude any fluorescent signal. **Figure 6** shows SH images collected in this way. Although this does

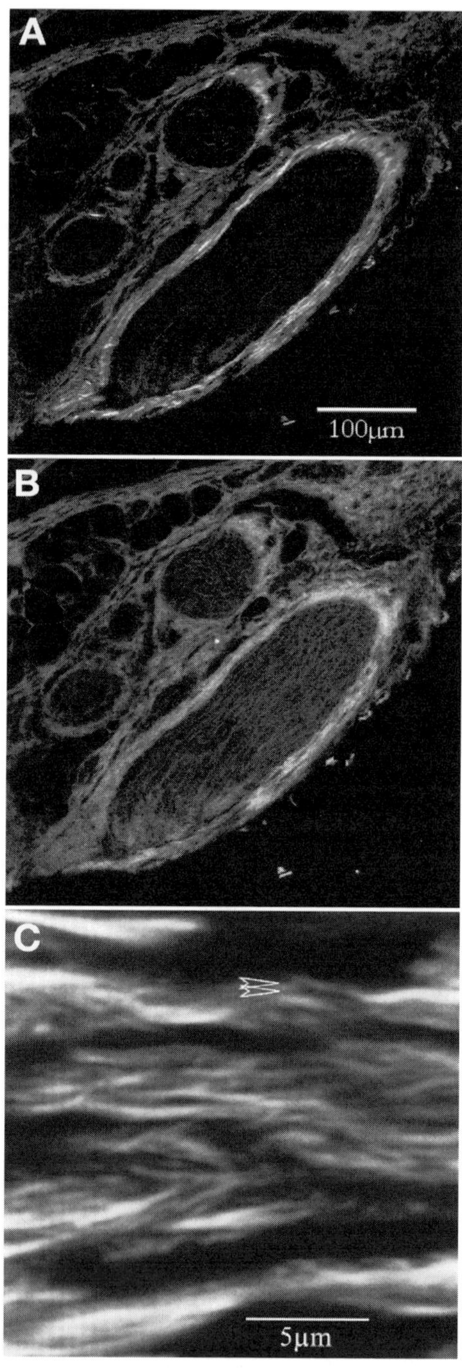

Fig. 5. Histological paraffin section of skin, stained with Masson's trichrome. **(A)** and **(B)** taken with 20× NA 0.5 lens, using 800-nm excitation and a broad-band collection

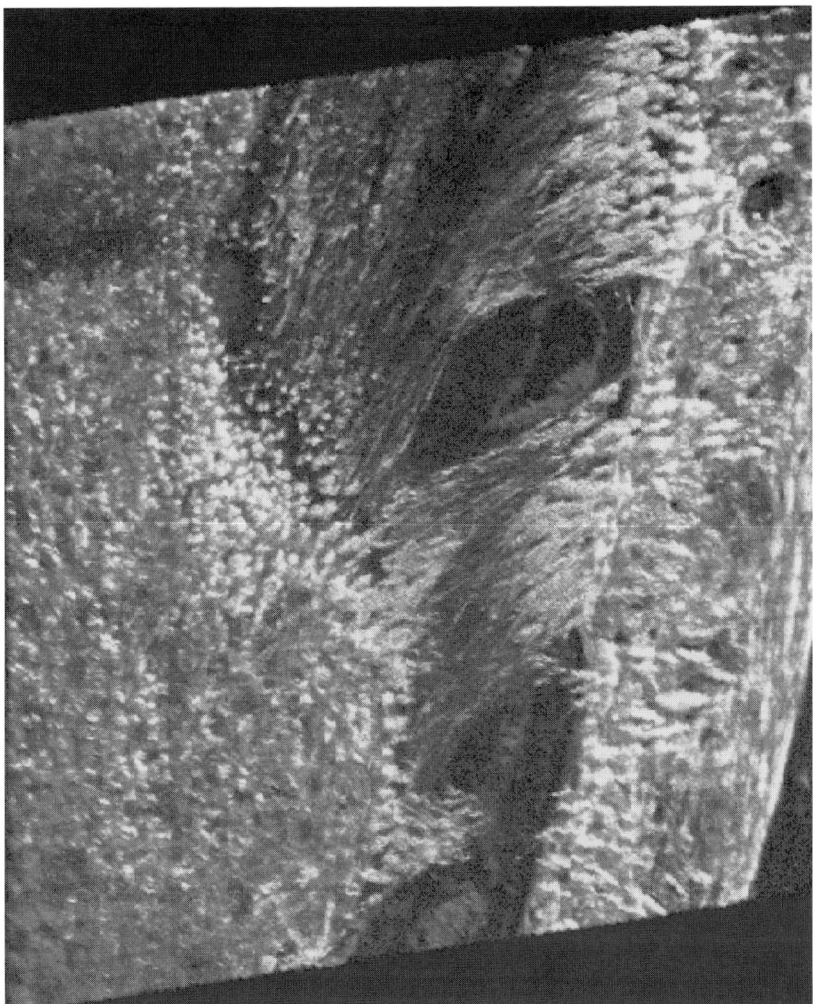

Fig. 6. Rendered image (simulated fluorescence projection at an oblique angle) from 20 optical sections through a histological section of the periodontal membrane of a developing tooth. SHG image, showing collagen only, captured with BP 415/10 filter, 25× NA 0.75 oil-immersion objective, 830-nm excitation.

Fig. 5. *(Continued)* filter. **(A)** Transmitted image; **(B)** back-propagated image of collagen around hair follicles in mouse skin. (A) shows both fluorescence and SHG, but the SHG is substantially brighter and is easily recognisable without any formal separation. **(C)** High-power image (×100 NA 1.4 objective) taken using 830-nm excitation and a BP 415/10 filter (Fig. 2) to exclude all fluorescence. Only collagen is seen and the resolution is excellent (arrowheads are spaced 330 nm apart).

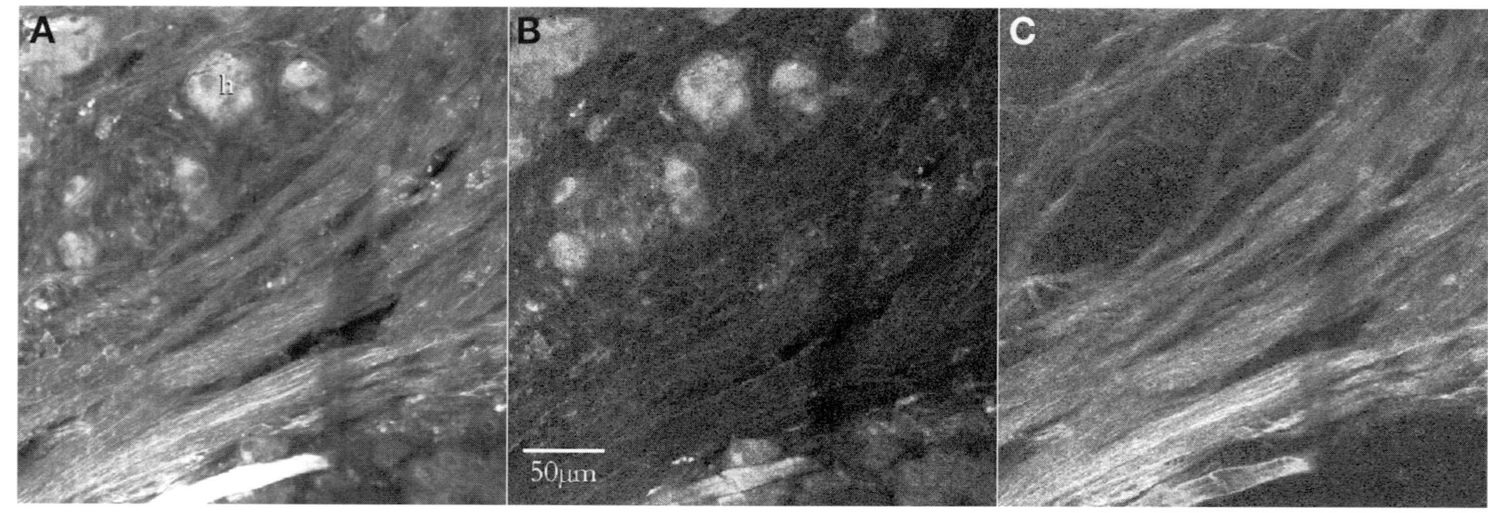

Fig. 7. Separation of signals by image subtraction, human cirrhotic liver biopsy. **(A)** Signal collected between 400 and 560 nm in the transmission detector, containing SHG and fluorescence. **(B)** Signal collected between 500 and 560 nm in the back-propagated nondescanned detector, showing only fluorescence. **(C)** Image B scaled to match intensities on the hepatocytes (h) and subtracted from image A, giving an image that shows only the SHG signal.

make detection absolutely unambiguous, it restricts excitation to wavelengths for which the appropriate filter is available.

Sometimes excitation of other fluorochromes makes it desirable to tune the laser to wavelengths for which the user does not have a suitable filter or for which a good filter is unavailable (such as close to 800 nm to excite DAPI or Hoechst). Also, a researcher wishing to find out whether SHG imaging can be useful in a particular project will probably not want to purchase special filters initially.

Fortunately, the fact that the wavelength must always be shorter than that of any two-photon excited fluorescence (as fluorescence always involves a Stokes shift) means that it is often quite easy to distinguish an SHG signal without custom filters. This is particularly handy for those cases in which one is not looking at the SHG signal in isolation. Often, an experiment will require one or more fluorochromes to be imaged in the same sample. Because the SHG signal from collagen can be excited anywhere within the tunable range of a Ti-S laser, it is often desirable to optimize the excitation for the fluorochrome rather than having the collagen signal at the wavelength of a particular bandpass filter. Particularly if there is not much autofluorescence (which tends to have a broad spectrum) present, one might be able to use the shorter wavelength of the collagen signal to distinguish it with nothing more than a conventional short-pass/dichroic/long-pass combination. For example, in a specimen stained with routine fluorochromes (FITC, TRITC, etc., without DAPI), there will be very little fluorescence (even autofluorescence) shorter than 500 nm, so a 500 DCLP dichroic in the detector will give a signal that is predominantly SHG in the short-wavelength channel. **Figure 4** shows an example of this; even with a 560 nm dichroic, the short part of the spectrum is totally dominated by the SHG signal while a strong fluorescent signal is detected in the range 560–650 nm. (In this sample, which is unstained, this is partly a function of the strength of the SHG signal from the highly crystalline kangaroo-tail collagen and partly a reflection if the aldehyde-induced autofluorescence tends to peak around 580 nm).

Where fluorescence is stronger in the short wavelengths, so that this strategy will not work, it is simple to use the directionality of the SHG signal to detect it, because fluorescence will propagate equally in all directions. If matching filters are used, the transmitted detector will show a combination of fluorescence and SHG signals **(Fig. 5A)**, whereas back-propagated detectors (*see* **Fig. 1**), whether confocal or nondescanned, will show mostly or entirely fluorescence **(Fig. 5B)**. Therefore, subtracting the "reflected" signal from the "transmitted" signal will leave only the SHG signal. **Figure 7** shows an example of this. The sample is a cirrhotic liver, which has a very high level of autofluorescence. Cirrhosis (fibrosis) of the liver is a proliferation of collagen fibers through the liver, and SHG microscopy is proving to be a powerful tool to study it *(12,23)*. The transmitted detector collects both SHG and fluorescent signals (*see* **Fig. 7A**), whereas the back-propagated signal has only the fluorescence (*see* **Fig. 7B**). The simple procedure to isolate the SH signal can be done in any image analysis or image processing program. The steps are as follows:

1. Measure intensity in both images of an area known to show only fluorescence (in this example, the hepatocytes labeled h).
2. Scale the intensity of the fluorescence-only image (**Fig. 7B**) so that it matches the corresponding area in the combined image **(Fig. 7A)**. In this case, the image in **Fig. 7B** needs to be scaled up by approx 10%.
3. Subtract the image, in **Fig.7B** from the image in **Fig. 7A**. The resulting image **(Fig.7C)** shows only the SHG signal.

The two-color combination of this SHG-only image with the fluorescence-only image can be seen as Fig. 7B of **ref.** ***12.***

In principle, one could also use fluorescent lifetime to separate the SH signal from fluorescence, although this does not seem to have been done in practice. Generation of the SH signal is virtually instantaneous, whereas, typically, fluorescence does not appear for several hundred picoseconds and then decays over 3–12 ns. With a suitably high time resolution, the signals should therefore be quite distinct.

3.3. Conclusion: Applications for SHG Imaging of Collagen

Although the ability of collagen to excite second harmonics has been known for 20 yr *(10)*, and putative uses for the technique have been proposed as long as 16 yr ago *(13)*, it has only been in the past 2 yr that advances in microscope and laser technology have brought SHG microscopy at high resolution into the hands of histologists and cell biologists. At this stage, it might be premature to predict what its final use will be.

Two broad areas seem likely: the use of SHG imaging as a tool to study the three-dimensional architecture of collagen in fine detail at the microscopic level in both healthy and diseased tissue, and its use to investigate changes in collagen in disease. On the former front, SHG has already been used to study the arrangement of collagen in human endometrium as part of a long-term study of the glandular and vascular structure in three dimensions *(12,16,24)*. It is also being developed as a method for detecting and quantifying collagen invasion in cirrhotic liver *(12,25)*.

We have shown *(12)* that SHG imaging appears to be able to distinguish between at least some collagen types (*see* **Fig. 4**). Current research in our laboratory aims to quantify and characterize this by using pure and well-defined collagen samples *(22)*. SHG has also been used as a tool to detect the polarity of collagen in connective tissue *(13)* and this might also become applicable at higher resolution with current developments in instrumentation. Experimentally induced damage has been shown to affect the SH signal from collagen ***(14)*** and, very recently, pathologic changes have also been shown to be detectable by the SHG properties ***(15)***. Teeth have a well-characterized collagen structure, and the SH signal is detectable in sectioned tooth material with and without decalcification (*see* **Fig. 6**). A study is currently in progress using SHG to study changes in collagen associated with dental caries.

Hereditary collagen diseases are another area, as yet unexplored, in which SHG imaging could have a significant impact. Now that multiphoton microscopes are in wide use, it seems inescapable that the associated technique of SHG microscopy will grow steadily more important in future years.

4. Notes

1. The process of second harmonic generation (SHG) has been summarized in a recent paper by Gauderon et al. ***(2)***. Briefly, as electromagnetic radiation propagates through matter, the electric field (*E*) exerts forces on the sample's internal charge distribution. The consequent redistribution of charge generates an additional field component. The resultant dipole moment per unit volume is referred to as the electric polarization (*P*) and can be expressed as a sum of linear and nonlinear terms. The nonlinear components only become significant at very high light intensities. The primary nonlinear effect is a polarization of second order in the electric field and is given by ***(27)***

$$P_i^{2\omega} = \chi_{ijk}^{2\omega} E_j^{\omega} E_k^{\omega} \tag{1}$$

   where subscripts denote Cartesian components and superscripts denote the relevant frequencies. Is a (3×3×3) third-rank tensor, termed the second-order nonlinear optical susceptibility, whose elements sum to zero for material with inversion symmetry.
2. In 1974, Hellwarth and Christiansen ***(3)*** looked at crystals illuminated by focused laser light in a conventional wide-field microscope. This is not an effective use of the nonlinear imaging properties of SHG; it is not depth-selective and the photon flux required to excite the entire visible field at the same time is extreme. Further, the difficulty of excluding exciting light from the image is considerable. It is, therefore, not a practically useful technique, although it does have the advantage, compared to scanning techniques, that the wavelength of the emitted light (the second harmonic) determines the resolution.
3. "Prechirping" (i.e., retarding the shorter wavelengths before the pulse enters the fiber) has been used to compensate for this but has the practical inconvenience that the "prechirp" must be adjusted for different wavelengths and pulse lengths. This, in turn, is likely to mean realigning the beam entering the microscope. Changing wavelengths—one of the key merits if the Ti-S laser—is thus made difficult and tedious; therefore, this solution has not gained general acceptance.
4. Two-photon excitation maxima for common members of the Clontech GFP family are as follows: eCFP = 860 nm, eGFP = 920 nm, eYFP = 960 nm, Ds-Red = approx 975 nm ***(17)***.

Acknowledgments

We are very grateful to Allan Jones, Frank Manconi, Anne Swan, and Mark Gorrell for samples, discussion, and collaboration. We also owe a deep debt to Colin Sheppard for introducing us to second-harmonic microscopy, and to Régis Gauderon and Paul Xu for working to develop and evaluate the technique. The microscope was purchased through a Research Infrastructure (Equipment and Facilities) grant from the Australian Research Council. We thank Sunney Xie and colleagues, the *Biophysical Journal*, and the Biophysical Society for allowing us to reproduce Fig. 3 A–D, and the *Journal of Structural Biology* for allowing us to reproduce Fig. 3E. We likewise thank Chroma Technology Corp. for permission to reproduce the filter curves shown in Fig. 2. We are very grateful to José Feijó for critically reading and substantially improving the final manuscript.

References

1. Franken, P. A., Hill, A. E., Peters, C. W., and Weinreich, G. (1961) Generation of optical harmonics. *Phys. Rev. Lett.* **7,** 118–119.
2. Gauderon, R., Lukins, P. B., and Sheppard, C. J. R. (2001) Simultaneous multichannel nonlinear imaging: combined two-photon excited fluorescence and second harmonic generation microscopy. *Micron* **32,** 685–689.
3. Hellwarth, R. and Christensen, P. (1974) Nonlinear microscopic examination of structure in polycrystalline ZnSe. *Opt. Commun.* **12,** 318–322.
4. Gannaway, J. N. and Sheppard, C. J. R. (1978) Second harmonic imaging in the scanning optical microscope. *Opt. Quant. Electron.* **10,** 435.
5. Campagnola, P., Clark, H. A., Mohler, W. A., Lewis, A., and Loew, L. M. (2001) Second-harmonic imaging of living cells. *J. Biomed. Opt.* **6,** 277–286.
6. Campagnola, P. J., Millard, A. C., Terasaki, M., Hoppe, P. E., Malone, C. J., and Mohler, W. A. (2002) Three-dimensional high-resolution second-harmonic generation imaging of endogenous structural proteins in biological tissues. *Biophys. J.* **81,** 493–508.
7. Mertz, J. and Moreaux, L. (2001) Multi-harmonic light microscopy: theory and applications to membrane imaging, in *Multiphoton Microscopy in the Biomedical Sciences* (A. Periasamy and P.T.C. So, eds.), *Proceedings of SPIE* **4262,** 9–17.
8. Moreaux, L., Sandre, O., Charpak, S., Blanchard-Desce, M., and Mertz, J. (2001) Coherent scattering in multi-harmonic microscopy. *Biophys. J.* **80,** 1568–1574.
9. Lodish, H., Berk, A., Lipursky, S. L., Matsudaira, P., Baltimore, D., and Darrell, J. (2000) *Molecular Cell Biology,* 4th ed., W. H. Freeman, New York.
10. Roth, S. and Freund, I. (1981) Optical second-harmonic scattering in rat-tail tendon. *Biopolymers* **20,** 1271–1290.
11. Georgiou, E., Theodossiou, T., Hovhannisya, V., Politopoulos, K., Rapti, G. S., and Yova, D. (2000) Second and third optical harmonic generation in type I collagen, by nanosecond laser irradiation, over a broad spectral region. *Opt. Commun.* **176,** 253–260.

12. Cox, G., Kable, E., Jones, A., Fraser, I., Manconi, F., and Gorrell, M. (2002) 3-dimensional imaging of collagen using second harmonic generation. *J. Struct. Biol.* **141,** 53–62.
13. Freund, I., Deutsch, M., and Sprecher, A. (1986) Connective tissue polarity. Optical second-harmonic microscopy, crossed-beam summation, and small-angle scattering in rat-tail tendon. *Biophys. J.* **50,** 693–712.
14. Kim, B. M., Eichler, J., Reiser, K. M., Rubenchik, A. M., and Da Silva, L. B. (2000) Collagen structure and nonlinear susceptibility: effects of heat, glycation, and enzymatic cleavage on second harmonic signal intensity. *Lasers Surg. Med.* **27,** 329–335.
15. Deng, X., Williams, E. D., Thompson, E. W., Gan, X., and Gu, M. (2002) Second harmonic generation from biological tissues: effect of excitation wavelength. *Scanning* **24,** 175–178.
16. Cox, G. C., Manconi, F., and Kable, E. (2002) Second harmonic imaging of collagen in mammalian tissue. *Proc. SPIE* **4620,** 148–156.
17. Blab, G. A., Lommerse, P. H. M., Cognet, L., Harms, G. S., and Schmidt, T. (2001) Two-photon excitation action cross-sections of the autofluorescent proteins. *Chem. Phys. Lett.* **350,** 71–77.
18. Ji-Xin Cheng, Kevin Jia, Y., Gengfeng Zheng, and Sunney Xie, X. (2002) Laser-scanning coherent anti-Stokes Raman scattering microscopy and applications to cell biology. *Biophys. J.* **83,** 502–509.
19. Cox, G., Kable, E., Sheppard, C. J. R., and Xu, P. (2002) Resolution of second harmonic generation microscopy. Durban, South Africa. *Proc. 15th Int. Cong. Electron Microscopy* **2,** 331–332.
20. Schräpler, V. R., Schmidt, T., Schultka, R., and Hepp, W-D. (1991) Farbstoffanalytische Untersuchungen zum polarisationsmikropischen Nachweis von Kollagen mit Solaminrot 4B (Teil II). *Acta Histochem.* **90,** 75–85.
21. Milthorpe, B. K. (1994) Xenografts for tendon and ligament repair. *Biomaterials* **15,** 745–752.
22. Cox, G. C., Xu, P., Sheppard, C. J. R., and Ramshaw, J. (2003) Characterization of the second harmonic signal from collagen. *Proceedings of SPIE* **4963,** 32–40.
23. Zipfel, W. R., Williams, R. M., Christie, R., Nitikin, A. Y., Hyman, B. T., and Webb, W. W. (2003) Live tissue intrinsic emission microscopy using multiphoton excited native fluorescence and second harmonic generation. *Proc. Natl. Acad. Sci. USA* **100,** 7075–7080.
24. Manconi, F., Cox, G., Kable, E., Markham R., and Fraser, I. S. (2001) Computer-generated three-dimensional reconstruction of uterine histological parallel serial sections displaying microvascular and glandular structures in human endometrium. *Micron* **32,** 449–453.
25. Gorrell, M. D., Wang, X. M., Levy, M. T., et al. (2003) Intrahepatic expression of collagen and fibroblast activation protein (FAP) in hepatitis c virus infection. *Adv. Exp. Med. Biol.* **524,** 235–243.
26. Yariv, A. (1967) *Quantum Electronics,* Wiley, New York.

3

Visualizing Calcium Signaling in Cells by Digitized Wide-Field and Confocal Fluorescent Microscopy

Michael Wm. Roe, Jerome F. Fiekers, Louis H. Philipson, and Vytautas P. Bindokas

Summary

Calcium (Ca^{2+}) is a fundamentally important component of cellular signal transduction. Dynamic changes in the concentration of Ca^{2+} ($[Ca^{2+}]$) in the cytoplasm and within organelles are tightly controlled and regulate a diverse array of biological activities, including fertilization, cell division, gene expression, cellular metabolism, protein biosynthesis, secretion, muscle contraction, intercellular communication, and cell death. Measurement of intracellular $[Ca^{2+}]$ is essential to understanding the role of Ca^{2+} and for defining the underlying regulatory mechanisms in any cellular process. A broad range of synthetic and biosynthetic fluorescent Ca^{2+} sensors are available that enable the visualization and quantification of subcellular spatio-temporal $[Ca^{2+}]$ gradients. This chapter describes the application of wide-field digitized video fluorescence microfluorometry and confocal microscopy to quantitatively image Ca^{2+} in cells with high temporal and spatial resolution.

Key Words: Intracellular calcium; signal transduction; cellular imaging; fluorescence; confocal microscopy; Fura-2; Fluo-3; cameleon; pericam; biosensor; transfection.

1. Introduction

Calcium is an essential second-messenger signal in all cell types. Many cellular processes, including mitosis, gene transcription, protein biosynthesis and processing, energy metabolism, membrane electrical activity, exocytosis, motility, receptor-mediated signal transduction, and intercellular communication, are regulated by highly coordinated, spatio-temporal gradients of the concentration of intracellular Ca^{2+} ($[Ca^{2+}]_i$), which can be transient, sustained, and oscillatory. Ca^{2+} homeostasis and signal transduction are precisely controlled by transport mechanisms and Ca^{2+}-binding proteins. Crucial to the regulation of Ca^{2+} signaling are mechanisms that control Ca^{2+} fluxes in the plasma membrane, endoplasmic reticulum, mitochondria, and Golgi complex.

From: *Methods in Molecular Biology, vol. 319: Cell Imaging Techniques: Methods and Protocols*
Edited by: D. J. Taatjes and B. T. Mossman © Humana Press Inc., Totowa, NJ

Understanding the role of Ca^{2+} in cell physiology and elucidation of the underlying regulatory mechanisms requires measurements of $[Ca^{2+}]_i$. In recent years, new technical developments in microscopy, digital cameras, and Ca^{2+} indicators have made the study of high-resolution, rapid changes in subcellular $[Ca^{2+}]$ possible. In this chapter, we describe the methods we employ to visualize Ca^{2+} signals in the cytoplasm and within organelles of electrically excitable endocrine cells in vitro.

Since the early 1980s, fluorescent synthetic Ca^{2+} dyes have been the most frequently used indicators to detect and visualize spatio-temporal $[Ca^{2+}]$ gradients in the cytoplasm and organelles of mammalian cells. Commercially available synthetic indicators such as Fura-2, Fluo-3, MagFura-2, and Rhod-2 can be easily loaded into cell suspensions and attached cells. The biophysical properties of synthetic fluorescent Ca^{2+} indicators are summarized in **Table 1**.

The general procedure for loading cells with a synthetic Ca^{2+} indicator involves incubation of cells in culture medium or physiological buffer that contains a membrane-permeable, acetoxymethyl ester (AM) precursor of the dye. Cellular nonspecific esterase activity facilitates removal of the ester and unmasks the negatively charged Ca^{2+}-binding site of the indicator. Consequently, the Ca^{2+}-sensitive anionic dye is trapped within the cytoplasm. This procedure, originally described by Tsien and his colleagues ***(1–5)***, allows labeling of large populations of cells with fluorescent Ca^{2+} indicators. However, because the lipophilic derivative of the indicator can partition across any membrane-surrounded organelle, it is extremely difficult to control the compartmentalization of the dye. This problem has long been recognized and methods for optimizing and evaluating dye loading conditions involving the evaluation of the effects of incubation time, temperature, and dye precursor concentration are described in detail elsewhere ***(6,7)***.

Notwithstanding these refinements, employing lipophilic precursors to load Ca^{2+} dyes into specific subcellular compartments remains problematic. Specific labeling of the cytoplasm with a fluorescent Ca^{2+} dye can be achieved by microinjection of individual cells with a membrane-impermeable form of the dye, reversibly permeabilizing the plasma membrane of a population of cells in the presence of the potassium salt form of the dye using scrape loading or osmotic techniques (*see* **ref.** *7* for a review). Microinjection is labor intensive, technically demanding, and does not allow the study of a large number of cells simultaneously. The latter approaches risk damaging cells and do not enable loading of subcellular compartments such as organelles with Ca^{2+} indicators.

Beginning in the 1990s, genetically targeted recombinant chemiluminescent and fluorescent biosynthetic Ca^{2+} sensors have been used with great success to study Ca^{2+} signaling ***(8–10)***. By utilizing specific targeting sequences that direct expression of the sensors to various locations within cells, Ca^{2+} biosensors have provided measurements of $[Ca^{2+}]$ in the cytoplasm, mitochondrial

Table 1
Properties of Synthetic Fluorescent Calcium Indicators

Target	Indicator	Type	K'_d (μM)	X_1 (nm)	X_2 (nm)	M_1 (nm)	M_2 (nm)	Dynamic range
Cytosol	Fura-2	DWX	0.15–0.22	340	365; 380	510		13–25
	Indo-1	DWM	0.23	340		400	475	20–80
	Fluo-3	SW	0.4	488; 505		530		40–100
	Fluo-4	SW	0.34	488; 495		> 510		> 100
	Fura Red	DWX; DWM	0.14	420	480; 488	> 640		5–12
	Calcium Green	SW	0.19	488; 505		530		~ 14
	Calcium Orange	SW	0.185	550		575		~ 3
	Calcium Crimson	SW	0.185	590		615		~ 2.5
Endoplasmic reticulum	MagFura-2 (Furaptra)	DWX	25	340	365; 380	> 505		6–30
	MagFluo-4	DWX	22	488; 490		> 510		
Mitochondria	Rhod-2	SW	1	550		575		14–100
Plasma membrane	Fura-2FF	SW	38	340	365; 380	> 505		
	Calcium Green C18	SW	0.28	488; 510		530		~ 8

Abbreviations: DWX, dual-wavelength excitation ratiometric dye; DWM, dual-wavelength emission ratiometric dye; SW, single-wavelength dye.

Note: The dynamic range is the fold change in fluorescence intensity or the ratio of fluorescence values measured under conditions of saturating and low calcium concentrations. Note that Fura red is normally used as DWX (ratio 420/480) but can be DWM when coloaded into cells with Fluo-3 or Fluo-4 (ratio 530/640).

Source: Adapted from **refs.** ***7***, ***25***, and ***26***.

matrix, endoplasmic reticulum lumen, Golgi complex, nucleus, secretory granule surface, caveolae, and cytoplasmic surface of the plasma membrane ***(9,11–23)***. Genes encoding biosensors based on recombinant aequorin, a Ca^{2+}-sensitive bioluminescent protein, as well as mutants of green fluorescent protein can be readily transfected into cells. **Table 2** summarizes the biophysical properties of genetically targeted fluorescent Ca^{2+} biosensors.

Constructs encoding recombinant aequorin chimeras are available commercially and can be targeted to specific locations within cells. Aequorin measures [Ca^{2+}] across a wide range (0.1–100 μ*M*). Unfortunately, the brightness of aequorin emission is very low compared with fluorescent indicators. Experiments are generally performed using cell suspensions or by integrating light signals from many cells grown as a monolayer. Imaging aequorin signals in single cells with current technology provides poor spatial and limited temporal resolution. Detection of aequorin luminescence requires photomultiplier tubes housed within light-tight containers or expensive high-sensitivity photon counting arrays. The light-generating reaction of Ca^{2+} with aequorin depends on coelenterazine, a cofactor that must be present during the experiments. In addition, aequorin molecules are irreversibly consumed by the reaction with Ca^{2+}. This raises two problems. First, prior to the Ca^{2+} measurements, cells must be incubated for prolonged periods of time in solutions that do not contain Ca^{2+}; this prevents consumption of aequorin by the high [Ca^{2+}] in extracellular solutions. Second, irreversible consumption of the indicator induces a measurement artifact that mimics lowering of [Ca^{2+}], thus limiting the amount of time for an experiment, especially when measuring Ca^{2+} signals from subcellular regions, which have high [Ca^{2+}] such as the lumen of the endoplasmic reticulum.

Genetically targeted biosynthetic fluorescent Ca^{2+} sensors overcome many of the limitations encountered with synthetic and chemiluminescent Ca^{2+} indicators. First described in 1997 by Tsien and his colleagues, cameleons provide investigators the tools necessary to record Ca^{2+} signals from specific subcellular compartments ***(9)***. Cameleons are fluorescent biosynthetic Ca^{2+} indicators constructed by inserting a Ca^{2+} sensor (*Xenopus laevis* calmodulin and M13, a calmodulin-binding protein) between two mutated forms of green fluorescent protein (GFP). Cameleon fluorescence is affected by differences in the concentration of Ca^{2+} that alter the amount of fluorescence resonance energy transfer (FRET) between mutant forms of GFP. This process is influenced by a Ca^{2+}-induced change in the conformation of calmodulin–M13, which, consequently, alters the relative angular displacement between the two mutant GFPs, bringing them closer together, for example, following an increase in [Ca^{2+}]. The increase in FRET is a direct function of [Ca^{2+}]. In contrast to cameleons, camgaroos and pericams do not utilize FRET; Ca^{2+} binding results in a direct electrostatic change in the environment of these circularly permuted fluorescent proteins, causing a Ca^{2+}-dependent shift in

Table 2
Biosynthetic Fluorescent Calcium Sensors

Target	Biosensor	Type	K'_d (μ*M*)	*n*	K'_d (μ*M*)	*n*	Dynamic range	Ref.
Cytoplasm	Cameleon-1	FRET	0.07	1.8	11	1	1.7–1.8	***9***
	Cameleon-1/E104Q	FRET			4.4–5.4	0.76	1.7–1.8	***9***
	Cameleon-1/E31Q	FRET	0.083	1.5	700	0.87	1.7–1.8	***9***
	Cameleon-2	FRET	0.07	1.8	11	1	1.7–1.8	***9***
	Yellow Cameleon-2 (YC2)	FRET	0.07	1.8	11	1	1.5	***9***
	Yellow Cameleon-2 (YC2)	FRET			1.24	0.79		***34***
	Yellow Cameleon-2.1 (YC2.1)	FRET	0.1	1.8	4.3	0.6	2	***15***
	Yellow Cameleon-2.1 (YC2.1)	FRET	0.2	0.62				***16***
	Yellow Cameleon-2.3 (YC2.3)	FRET					2	***13***
	Yellow Cameleon-2.12 (YC2.12)	FRET						***19***
	Yellow Cameleon-3.1 (YC3.1)	FRET			1.5	1.1	2	***15***
	Yellow Cameleon-3.2 (YC3.2)	FRET			1.5	1.1	1.85	***24***
	Yellow Cameleon-3.3 (YC3.3)	FRET			1.5	1.1	2	***13***
	Yellow Cameleon-6.1 (YC6.1)	FRET	0.11				2	***22***
	Yellow Red Cameleon-2 (YRC2)	FRET	0.2–0.4				0.1–0.4	***17***
	Cyan Red Cameleon-2 (CRC2)	FRET	0.2–0.4				0.1–0.4	***17***
	Sapphire Red Cameleon-2 (SapRC2)	FRET	0.2–0.4				0.1–0.4	***17***
	Camgaroo-1	CPSW			7	1.6	8	***24***
	Camgaroo-2	CPSW			5.5	1.24	7	***13***
	Ratiometric-Pericam (RPC)	CPDWX			1.7	1.1	10	***18***
	Flash-Pericam	CPSW	0.7	0.7			8	***18***
	Inverse-Pericam	CPSW	0.2	1			7	***18***

(Continued)

Table 2 ***(Continued)***

Target	Biosensor	Type	K'_d (μ*M*)	*n*	K'_d (μ*M*)	*n*	Dynamic range	Ref.
	G-CaMP	CPSW	0.235	3.3			4.5	***20***
	pGA	CRET			10			***8***
Endoplasmic Reticulum	Yellow Cameleon-3er (YC3er)	FRET			4.4–5.4	0.76	1.6	***9***
	Yellow Cameleon-4er (YC4er)	FRET	0.083	1.5	700	0.87	1.4	***9***
	Yellow Cameleon-4er (YC4er)	FRET	0.039	0.57	292	0.6		***34***
	Yellow Cameleon-4er (YC4er)	FRET			292	0.6	1.7	***12***
	Yellow Cameleon-6.2er (YC6.2er)	FRET						***22***
Mitochondria	Yellow Cameleon-2mt (YC2mt)	FRET			1.26	0.79	1.3	***34***
	Yellow Cameleon-2.1mt (YC2.1mt)	FRET						***34***
	Yellow Cameleon-3mt (YC3mt)	FRET						***34***
	Yellow Cameleon-3.1mt (YC3.1mt)	FRET			3.98	0.67	1.3	***34***
	Yellow Cameleon-4mt (YC4mt)	FRET						***34***
	Yellow Cameleon-4.1mt (YC4.1mt)	FRET	0.105	0.81	104	0.62	1.3	***34***
	Camgaroo-2mt	CPSW						***13***
	Ratiometric-Pericam-mt (RPC-mt)	CPDWX			1.7	1.1	10	***18***
Golgi	Galactosyltransferase-Yellow Cameleon-3.3 (GT-YC3.3)	FRET			1.5	1.1	2	***13***
Nucleus	Cameleon-2nu	FRET	0.07	1.8	11	1	1.7–1.8	***9***
	Yellow Cameleon-3.1nu (YC3.1nu)	FRET			1.5	1.1	2	***16***
	Yellow Cameleon-6.1nu (YC6.1nu)	FRET						***22***

	Ratiometric-Pericam-nu (RPC-nu)	CPDWX			1.7	1.1	10	***18***
Secretory Granules	Phogrin-Yellow Cameleon-2 (phogrin-YC2)	FRET						***11***
Caveolae	Caveolin-1-Yellow Cameleon-2.1 (CYC2.1)	FRET	0.26	1				***14***
Plasma Membrane	Neuromodulin-Yellow Cameleon-2.1 (NYC2.1)	FRET	0.26	1				***14***
	Ratiometric-Pericam-synaptosome associated protein of 25 kDa (RPC-SNAP25)	CPDWX			1.7	1.1		***21***

Abbreviations: CPSW, circularly permuted, single-wavelength excitation; CPDWX, circularly permuted, dual-wavelength excitation; CRET, chemilumenescence resonance energy transfer.

Note: Subcellular location, apparent dissociation constant (K'_d) for Ca^{2+} and Hill coefficient (n) is listed for the sensors. Cameleon-1, Cameleon-1/E31Q, Cameleon-2, Cameleon-2nu, YC2.1, YC4er, and YC4.1mt display biphasic Ca^{2+}-binding curves and high- and low-affinity Ca^{2+} binding. The reader should note that in some cases, investigators have determined K'_d and n values in vitro and in vivo and should consult the references for additional details. An estimation of the dynamic range of the sensors is provided. The dynamic range in this case is defined as the fold change in fluorescent ratio or intensity of the sensor exposed to nominally 0 M [Ca^{2+}] and saturating [Ca^{2+}] (10–20 mM). Biosensors constructed from circularly permuted mutations of EYFP, Camgaroo-1, Camgaroo-2, Pericams, and G-CaMP have a much higher dynamic range than the fluorescence resonance energy transfer (FRET) sensors.

fluorescence emission ***(13,18,20,22–24)***. Ca^{2+} biosensors exhibit different biophysical properties, such as pH sensitivity and Ca^{2+} affinities, that enable measurement of Ca^{2+} signals in different cellular compartments using single- and dual-wavelength emission and excitation microfluorometry.

The choice of the most appropriate indicator or indicators and measurement systems for the experiments is dictated by several considerations, including the source or location of the Ca^{2+} signal to be studied and whether direct visualization of spatial changes in subcellular [Ca^{2+}] is necessary ***(7,25–27)***. In addition, certain constraints specific to the cell type or tissue under study might be important. For example, relatively homogeneous loading of populations of dispersed adherent cells with synthetic Ca^{2+} indicators is readily accomplished using membrane-permeable fluorescent derivatives, but loading thick biological tissues such as explants of islets of Langerhans, which consist of clusters of 1000–3000 cells, is problematic. Dye compartmentalization can be a significant problem (ability to cross one membrane implies multiple membranes might be crossed if esterase activity is low; targeted probes could offer an advantage here). Moreover, in some cases, loading primary cultures of dispersed cells has proven difficult. In this chapter, we describe protocols we employ to measure [Ca^{2+}] spatio-temporal gradients in endocrine cell lines and islets of Langerhans.

2. Materials

What follows is a list of the basic items needed to perform quantitative imaging of Ca^{2+} signals in cells. We find that two laboratory areas are required for these experiments: an area to perform cell culture and preparation of solutions used in the experiments, and a separate area in the laboratory or small room devoted to housing the microscope and imaging system. For our experiments, the imaging studies are performed in a darkened room (~10 m^2) adjacent to the main laboratory. If space is limited, a small region of the laboratory can be darkened using black curtains suspended from the ceiling or a light-tight box can be constructed or purchased to fit over the microscope.

1. Laminar flow, tissue culture hood with ultraviolet (UV) light.
2. Temperature-regulated, humidified incubator.
3. 95% O_2: 5% CO_2 gas tanks (for bubbling bicarbonate-buffered perifusion solutions).
4. Complete cell culture medium.
5. Trypsin-EDTA cell dissociation solution.
6. Phosphate-buffered saline (PBS) without Mg^{2+} and Ca^{2+}.
7. Hemocytometer.
8. Tissue culture flasks or circular cell culture plates.
9. Six-well tissue culture plates.
10. Glass cover slips and coating reagent if required.

11. Pipet tips and micropipets.
12. Siliconized, 1.5-mL Eppendorf microcentrifuge tubes.
13. Serological pipets and handheld dispenser (1, 5, 10, 25 mL).
14. Fine-point jeweler's forceps.
15. Microperifusion system and chamber.
16. Temperature-regulated water bath and thermometer.
17. Peristaltic pump and tubing for perifusion system (various sizes; chemically resistant).
18. Vibration-free isolation table for microscope.
19. Pipette Aid motor or aquarium bubbler.
20. Two 500-mL vacuum Ehrlenmeyer flasks and rubber stoppers and desiccant.
21. Inverted or upright microscope equipped for epifluorescence (for standard wide-field fluorescence imaging), confocal laser scanning microscope, or spinning disk optical microscope.
22. Fluorescence objectives (×10, ×20, ×40, ×60, ×100).
23. Fluorescence light source (xenon or mercury) or laser.
24. Optical filters (neutral density, excitation, dichroic, and emission).
25. High-speed excitation and emission filter changing devices with computer-interfaced controller.
26. High-sensitivity, high-resolution cooled digital camera (charge-coupled device [CCD]).
27. Computer workstation with a 17- to 20-in. monitor, local area network access, high-capacity data storage peripheral devices (0.06–1 Tbyte capacity) for image and data archiving, and color printer.
28. Image acquisition and analysis software with appropriate video card and driver for CCD.
29. Chemical salts and buffers to prepare perifusion solutions, glass bottles, and conical centrifuge tubes.
30. Fluorescent Ca^{2+}-sensitive dyes (membrane permeable as well as potassium salt forms for performing in vitro calibration).
31. Mammalian expression plasmids encoding biosynthetic Ca^{2+} sensors.
32. Competent *Escherichia coli* strains for plasmid transformation, growth medium, sterile flasks, Luria–Bertani (LB) agar plates, transfer loop, bunsen burner, rotating temperature-regulated culture chamber.
33. Plasmid cDNA preparation kits.
34. Mammalian cell transfection reagents.
35. Ca^{2+} calibration reagents and buffers.

3. Methods

3.1. Tissue Culture

Our experiments are performed on endocrine cell lines and primary cells. For imaging studies using fluorescent synthetic or biosynthetic Ca^{2+} indicators, we seed cells onto uncoated glass cover slips in six-well tissue culture plates

24–72 h before the experiments; cells are maintained under normal culture conditions (*see* **Note 1**). Typically, we seed at a cell density to achieve 60–80% confluence by the day of the experiment in order to image [Ca^{2+}] transients in single cells and also to have sufficient cell-free regions to obtain background fluorescence measurements.

3.2. Synthetic Ca^{2+} Indicators

Lipophilic (membrane permeable) and hydrophilic (membrane impermeable) derivatives of synthetic fluorescent Ca^{2+} dyes can be obtained from many commercial sources. Membrane-impermeable Ca^{2+} indicators can be loaded into cells by microinjection, osmotic methods, or scrape-loading and remain in compartments into which they are introduced. They can also be used for in vitro calibrations. For our cell imaging experiments, we load cells with AM derivatives of the indicators (*see* **Note 2**).

3.3. Loading Cells With Synthetic Ca^{2+} Dyes

The basic protocol to load neuroendocrine cells with Fura-2, Fluo-3, Fluo-4, Calcium Crimson, Fura Red, Calcium Orange, Calcium Green, MagFura-2, and Rhod-2 is outlined here.

1. Using fine-point jeweler's forceps, we gently remove a cover slip from the six-well tissue culture plate and transfer it into a microperifusion chamber (Harvard Apparatus).
2. Cells are incubated in the chamber with Krebs-Ringer bicarbonate (KRB) or Krebs-Ringer HEPES (KRH) buffer solution containing 5 μ*M* AM form of the dye and 0.0125–0.025% Pluronic F127 (*see* **Note 3**). Cell loading is conducted for 15–20 min in a humidified CO_2/O_2 incubator at 37°C. Following this period of time, the microperifusion chamber is mounted onto the temperature-controlled specimen stage of an inverted fluorescence microscope equipped for either standard wide-field epifluorescence imaging or optical spinning disk laser confocal microscopy. The cells are rinsed for 15–30 min with KRB or KRH to remove all extracellular AM form of the dye and to provide time for complete de-esterification.

Evaluation of the quality of dye loading is essential. Whereas the use of AM membrane-permeable forms of fluorescent Ca^{2+} indicators enables large numbers of cells to be loaded simultaneously, assessment of the compartmentalization of the dye is essential for achieving reproducible results and for reliable interpretation of the imaging data. Dye loading is affected by AM dye concentration, loading temperature and time, and cell type. Several reviews are available that discuss in detail the evaluation of dye loading and optimization of loading conditions *(6,7)*. Optimal conditions for dye loading vary depending on cell type and must be determined empirically.

3.4. Expressing Biosynthetic Ca^{2+} Biosensors in Cells

Fluorescent Ca^{2+} biosensors are not yet available commercially. It has been our experience that, without exception, investigators who initially designed, constructed, and characterized a biosensor will provide samples of the plasmid constructs. In some cases, completing an interinstitutional materials transfer agreement will be required before obtaining the biosensors. The reader is urged to directly contact the investigator for procedures to obtain the constructs.

Following receipt of samples (usually 1–5 µL of a 1- to 2-µg/µL stock cDNA solution, we use commercial kits to produce larger volumes of the plasmid cDNA (Qiagen Maxi Prep) and to transiently transfect cells using liposomal methods. Cells are seeded onto 15-mm or 25-mm glass cover slips 1 d prior to transfection and incubated in complete growth medium. On d 2, cells are transfected with 1–2 µg of cDNA for 4 h (*see* **Note 4**). Imaging studies are preformed 2–4 d after transfection.

3.5. Microperifusion

Precise control of the contents, temperature, and volume of the external bathing solutions is absolutely necessary to ensure reproducible experiments. Most investigators employ temperature-regulated, microperifusion chambers mounted on the specimen stage of inverted and upright fluorescent microscopes. Our experience indicates that continuous superfusion of cells with different solutions, rather than by direct, manual application of chemical reagents into the chamber and onto cells using micropipets, enables more precise control and reproducibility in altering experimental conditions.

In our studies, cells are constantly superfused by defined buffer solutions administered 2–5 mL/min; in a 1-mL microperifusion chamber, higher flow rates (>5 mL/min) could dislodge cells, whereas slower fluid flow (<2 mL/min) does not enable sufficiently rapid solution changes. Fluid flow is driven by peristaltic pumps or gravity flow. Temperature of the superfusate is maintained at 37°C. A steady rate of fluid removal from the chamber is maintained by a small plastic tube immersed in the chamber and attached to two 500-mL Erlenmeyer vacuum flasks connected in series. In one flask, the effluent is collected, whereas the other flask, proximal to the vacuum source, contains a desiccant that prevents contamination of the vacuum pump; negative pressure is provided by a fish tank or Eppendorf pipettor electric pump running in reverse mode. Alternatively, one could use a peristaltic pump with both perifusion chamber inflow and outflow matched for constant fluid levels. Experimental conditions are altered by changing the perifusate (*see* **Note 5**).

3.6. Measuring $[Ca^{2+}]_i$ Using Wide-Field Fluorescence Imaging

We utilize an inverted microscope equipped for multiparameter digitized video fluorescence imaging. Two perifusion systems mounted onto the specimen stage of Nikon TE-2000U microscopes are used in our laboratory: Harvard/Medical Systems PDMI-2 Micro-Incubator and Warner Instruments Series 20 Open Bath Recording/Imaging Chambers. The chambers are connected to a TC-202A Bipolar Temperature Controller and dual-channel TC-344B Chamber System Heater Controller, respectively. The PDMI-2 system is used for experiments involving perifusion of bicarbonate-buffered solutions via a peristaltic pump, whereas the Warner chambers are used in studies in which the solutions are perifused by gravity flow and rapid laminar flow fluid exchange is required. Glass cover slips (25-mm circular) with cells loaded with synthetic Ca^{2+} indicators or expressing biosynthetic Ca^{2+} sensors are placed into a Teflon glass cover-slip holder. Two chambers are available for use with the PDMI-2 system. The MSC-TD (Harvard Apparatus AH 65-0051) has an optical window of 19 mm × 31.5 mm and 1 mL volume. The MSC-PTD has a small rectangular optical window 9.5 mm wide and 19 mm long, and volume less than 1 mL for fast fluid transfer. The modular design of the Warner Series 20 chamber system is very flexible, enabling a wide choice of chamber sizes and configurations. We use closed and open bath Series 20 imaging chambers. For small-volume chambers (36–70 μL), cells are cultured on 15-mm circular cover slips. Larger chambers (>100 μL) use 25-mm circular cover slips. Superfusion of cells is initiated immediately after placing them into the imaging chamber and maintained throughout the experiment.

The cells are visualized with a ×40 or ×60 oil-immersion fluorescence objective. Larger tissues such as single intact islets of Langerhans are studied using a ×10 or ×20 fluorescence objective. Objectives with a high numerical aperture and long working distance are preferred with inverted microscopes. If you use an upright microscope, a ×60 water-immersion objective is highly recommended for experiments with living cells.

A 75-W xenon or mercury bulb is used for conventional wide-field fluorescence imaging. Excitation light is attenuated with neutral-density filters and the excitation wavelength is controlled by optical filters. Currently, we employ a Sutter or Prior filter wheel to set excitation wavelengths. The filter wheel is attached to a filter/shutter control device connected to a computer. Image acquisition and analytical software controls filter wheel settings and shutter activity. A separate emission filter wheel is utilized to change emission light wavelengths and to perform multiparameter emission microphotometry.

Fluorescent signals from cells are captured with an intensified video camera or a digital CCD. There are many cameras available for fluorescence microscopy. Because of space limitations, we will not review the wide range of digital cameras here. The choice of camera depends on resolution and acquisition speed requirements as well as spectral properties. The reader is encouraged to investigate individual specifications of the digital cameras currently available from manufacturers' websites for additional information. A video camera is suitable for Fura and some single-wavelength dyes, but the limited 8-bit digitization range makes it ill-suited for FRET probes as well as dyes with high dynamic response, such as Fluo-3 and Fluo-4.

3.6.1. Measuring $[Ca^{2+}]_i$ Using Fura-2

1. Make all experimental solutions fresh daily for imaging studies. Pre-equilibrate solutions to required temperature, and if bicarbonate-buffered perfusates are used, then bubble solutions with $O_2 : CO_2$ gas 15–30 min before experiments.
2. Remove cover slips with monolayers of cells from multiwell tissue culture plates and place them into the microperifusion chamber.
3. Load cells with Fura-2AM. We limit exposure of Fura-2 to bright light in the laboratory, and we load cells with Fura-2 either in a cell incubator with bicarbonate buffered solutions or, if loading at room temperature with HEPES-buffered solutions, use a light-tight box. See above comments about loading conditions. Do not use bicarbonate buffers in room air.
4. During Fura-2 loading, set the controller for the desired chamber temperature. Make certain that all fluids flow freely in all tubing. Degassing of solutions in tubing can form bubbles that restrict/occlude flow. Adjust and set fluid inflow rate. Set acquisition parameters using image acquisition and analysis software. Set excitation filter wavelengths (using the image acquisition software). For Fura-2, use 340-nm and 380-nm excitation filters, 400DCLP dichroic mirror, and 510/40-nm emission filter. For typical experiments, we record image pairs (exposure time of 50–250 ms for each excitation wavelength) at 0.20- to 10-s intervals. Set the shutter to automatically close between image pair exposures to reduce the amount of time cells are exposed to UV light. The duration of our experiments varies from 5 min to 2 h. It is important to completely plan the acquisition profile prior to running an experiment, and some software packages allow users to program and store protocols for experiments. This can greatly simplify the procedures during an experiment and aid in repeating experimental protocols with precision.
5. Mount chamber onto the microscope stage and immediately begin rinsing cells for 10–15 min with standard external solution to remove excess Fura-2AM. Adjust and set fluid efflux rate to match influx rate. This prevents large fluctuations in perifusate volume in the chamber. Make certain to maintain a constant vigil on the status of the fluid efflux from the chamber in order to prevent overflow accidents, which can severely damage an inverted microscope and the objectives.
6. Prior to exposing cells to excitation light, preliminary focus can be accomplished using bright-field light and the binocular eyepieces. Be certain to focus the objective with great caution. Approach the cover slip with the objective slowly; once the oil has contacted the surface of the cover glass, make adjustments only with the fine-focus controller. Do not extend the objective too far. Impacting the cover glass with the metal surface of the objective will break the cover glass; the cover slips are extremely fragile and remarkably little force is required to shatter them with the objective. Not only will this cause loss of a potentially important biological

sample, but it also risks severe (and quite costly!) damage to the internal components of the microscope because of contamination with saline solutions. For most cells, we set the focal plane about halfway through the cells.

7. Insert a neutral-density filter (Chroma UV filters, 0.3–2.0) into the excitation light path before opening the excitation light shutter (*see* **Note 6**).
8. Open the shutter and visualize cells using fluorescent excitation light. Choose an appropriate field-of-view. Use the acquisition software to refocus using either the 340-nm or 380-nm excitation light. In general, we use the 380-nm illuminated cells for focus because this generally is a brighter image than the 340 nm. Next, set the camera gain. This can be accomplished using either a peripheral camera illumination controller or the camera gain controls in the image acquisition software (*see* **Note 7**). Once set, the camera gain is not readjusted during the experiment. It is important to be consistent here because arbitrary gain changes can affect conversion of ratio values to ion concentration unless in vivo calibrations are used. Finally, use the acquisition software to delineate regions of interest (ROIs) in which imaging data are collected and displayed during the experiment. These can consist of individual cells, groups of cells, and/or subcellular regions in one or multiple cells.
9. Recheck the experimental protocol, acquisition settings, perifusion solutions, fluid flow, and image focus. Begin the experiment, changing experimental conditions as required. During each experiment, we simultaneously display the 340-nm, 380-nm, and ratio images and graphical depictions of the time-dependent changes in ratio values.
10. A cell-free or in vitro calibration method can be performed to convert the F_{340}/F_{380} ratio into molar $[Ca^{2+}]$. A calibration curve plotting the F_{340}/F_{380} ratio as a function of $[Ca^{2+}]$ can be used to provide an estimate of $[Ca^{2+}]_i$. To construct the curve, the F_{340}/F_{380} ratio is measured in a series of buffered solutions with defined $[Ca^{2+}]$ and the potassium salt of Fura-2 (usually 1–5 µ*M*). This calibration method, however, does not take into account possible effects of intracellular environment on Fura-2 ***(6)***. At the end of each experiment, we perform an in vivo (or *in situ*) calibration. To determine the fluorescence intensity values at saturating $[Ca^{2+}]$, the cells are exposed to 10–20 µ*M* ionomycin in the presence of extracellular Ca^{2+}. Under these conditions, the fluorescence maximum value for the intensity ratio is reached in about 30–60 s. After this, ionomycin, in solutions containing 10–20 m*M* EGTA and no added Ca^{2+}, is applied. The amount of time required to reach the minimum fluorescence intensity for the ratios varies and can take 5–60 min before reaching a steady-state value. Some labs prefer to use 4-Br-A23187 as the ionophore because it aids in setting R_{min}. Do not use the nonbromonated form of this ionophore because its intrinsic fluorescence will interfere with the calibration of Fura-2. To convert the fluorescence intensity ratio into molar $[Ca^{2+}]$, we use the following equation ***(1)***:

$$[Ca^{2+}] = K'_d\, \beta\, [(R - R_{min})/(R_{max} - R)]$$

where K'_d is the relative Ca^{2+} dissociation constant of Fura-2 (150–220 n*M*), β is the ratio of the 380-nm fluorescence intensity recorded at 0 and saturating $[Ca^{2+}]$, *R* is the ratio of Fura-2 fluorescence intensities at 340 nm and 380 nm, R_{max} is the fluorescence ratio at saturating $[Ca^{2+}]$, and R_{min} is the fluorescence ratio at 0 $[Ca^{2+}]$.

11. We save all images to the computer hard drive and program the software to automatically export and save the intensity data to spread sheets. After each experiment, the data and images are archived onto CD-R or DVD-R storage media for offline analysis at computer workstations in the main laboratory. Also, the computers in the imaging suite are networked to the workstations to facilitate downloading experimental data to offline sites. At the end of each session, we suggest deleting imaging files or transferring them to CD-R or DVD-R storage media because they can rapidly fill most computer hard drives with 60–80 gigabytes storage capacity.

3.7. Measuring $[Ca^{2+}]_i$ Using Laser Scanning Confocal Microscopy

Confocal laser scanning microscopy (CLSM) allows visualization of subcellular $[Ca^{2+}]$ gradients with high spatial and temporal resolution. The primary advantage of CLSM over standard, wide-field epifluorescence imaging is the reduction of fluorescence emission that originates from outside the plane of focus. The contribution of out-of-focus fluorescence reduces image contrast and resolution of fine microanatomical structures, confounds interpretation of quantitative imaging experiments, and hinders visualization of signaling events occurring within specific subcellular locations. This can be especially problematic in thick biological samples, such as intact single pancreatic islets of Langerhans (which are spherical clusters of cells 100–300 μ*M* in diameter consisting of 1000–3000 cells) or in tissue slices. An example of this artifact is illustrated in **Fig. 1** (see Color Plate 1, following p. 274), which shows the distribution of Fura-2 fluorescence in mouse islets of Langerhans in vitro; Fura-2 fluorescence appears to be present throughout the islet cells. CLSM images from a consecutive series of optical *z*-axis sections through a Fura-2-loaded mouse islet however clearly reveal that Fura-2 fluorescence is confined to the outer cell layers of the islet and not present within the core of the islet (*see* **Fig. 2**).

Single-wavelength excitation and ratiometric dual-wavelength emission imaging can be conducted with CLSM. Although a wide range of fluorescent Ca^{2+} indicators and genetically targeted Ca^{2+} biosensors are available for CLSM, most studies have utilized single-wavelength excitation Ca^{2+}-sensitive dyes such as Fluo-3 or Fluo-4. The choice of a specific Ca^{2+} indicator depends on the laser excitation wavelengths available. Many CLSM systems utilize an argon–krypton laser light source, which produces three discrete excitation lines: 488 nm, 568 nm, and 647 nm. Alternatively, many confocal systems use argon and He–Ne lasers (488- and 543- or 633-nm lines). Until recently, UV laser sources were uncommon because the lasers required high power and suffered from short life. Solid-state diode lasers are now available that could soon expand the wavelengths available in laser-based confocal systems.

It is reasonable to question the need for ratiometric imaging for quantitative measurements of $[Ca^{2+}]$ in confocal systems because the in-focus volume, which can be many voxels (three-dimensional pixels), is a constant and much

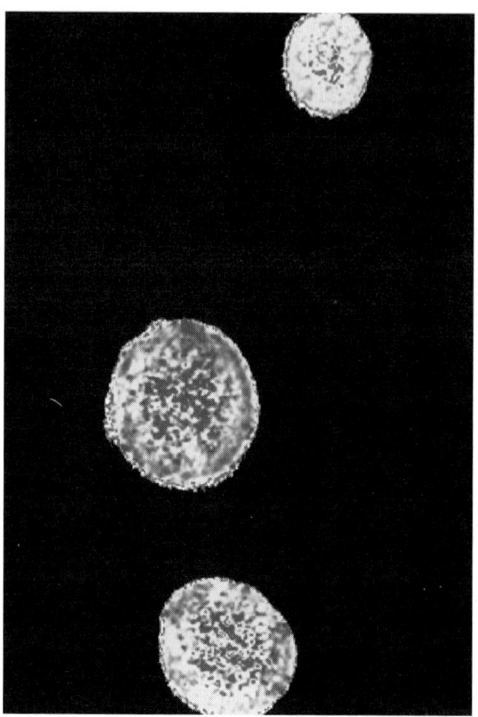

Fig. 1. Pseudocolored image of mouse islets loaded with Fura-2. The islets were imaged by conventional wide-field epifluorescence using a ×10 objective. The apparent distribution of the Fura-2 fluorescence throughout the islets is an artifact resulting from the contribution of fluorescence from above and below the focal plane. (*See* Color Plate 1, following p . 274.)

less than the volume of a cell. The main rationale is to account for dye dilution caused by changes in cell volume during measurements ***(25,28–30)***. Most scanning confocal systems contain at least two emission detector channels that permit emission ratio-based measurements. Emission ratio dyes like Indo-1 can be used in confocal imaging. Also, one can perform ratiometric imaging of cells coloaded with dyes that increase and decrease fluorescence intensity with $[Ca^{2+}]$, such as Fluo-3 and Fura red, respectively. With appropriate sets of dichroic and emission filters, simultaneous recordings of fluorescence emission intensities (stimulated by a single wavelength of excitation light) from cells loaded with multiple Ca^{2+}-sensitive dyes (e.g., Fluo-3 and Fura red) have been used to perform ratiometric dual-wavelength emission measurements of $[Ca^{2+}]$ ***(28–30)***. Excitation-ratio dyes like Fura-2 are not used because appropriate UV lasers are not commonplace and fast excitation line selection and switching requires expensive hardware.

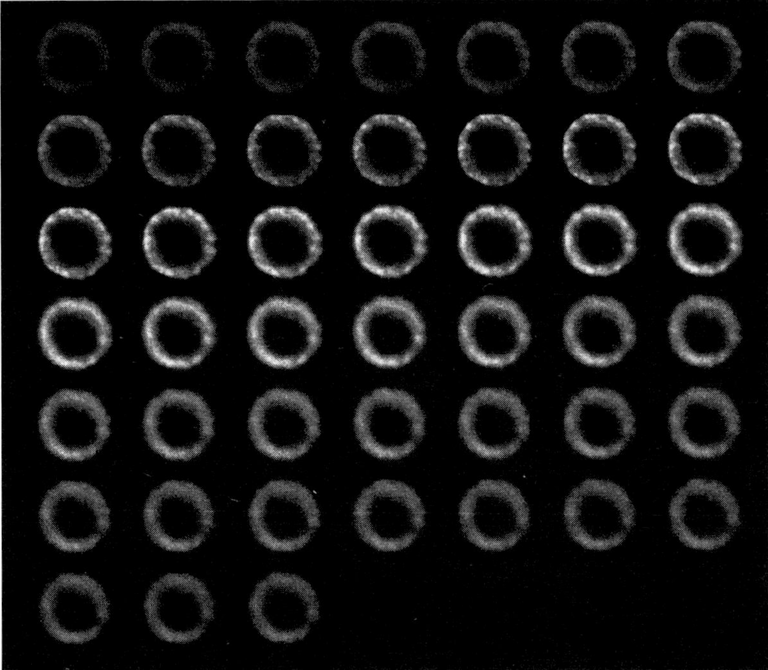

Fig. 2. Serial optical *z*-sections of a mouse islet loaded with Fura-2. The islet was imaged using a UV confocal scanning laser microscope. Note the annular fluorescence and the absence of fluorescence within the central regions of the islet. Spacing between each section was 1 μm.

Ca^{2+} signals can be studied using the confocal microscope operating in one of three possible imaging modes. The single line-scanning mode (*xt*-scanning, where *x* represents a line one pixel wide and *t* is time in seconds) provides the highest temporal resolution. Depending on scanner design, line sample rates range from about 667 to 4000 lines/s. In this mode, the laser beam rapidly and repeatedly scans along the line across the same region of an optical field. This approach has greatly facilitated the study of elementary Ca^{2+} signals and the initiation and propagation of intracellular Ca^{2+} waves ***(27)***. More recently, new CLSM systems allow for multiple lines (and different shapes) to be employed in *xt*-scanning. It is important to minimize laser intensity because bleaching and photodamage are especially high as a result of the energy being confined to tiny cell volumes. A common solution might be to decrease confocality by opening the system pinhole(s).

Analogous to the wide-field epifluorescence method described earlier, spatio-temporal [Ca^{2+}] gradients in a thin optical section of cells can be imaged using the *xyt*-scanning mode. Although the temporal resolution using *xyt*-scanning is slower than line-scanning, more spatial information can be collected. Confocal imaging systems allow the operator to adjust the scanning rate, sampling density, and size of the optical area for *xyt* imaging of biological samples. Temporal and spatial resolution is dependent on the laser scanning speed: the slower the scan rate, the lower the temporal resolution, the brighter the image (up to the point where all dye molecules have been excited), and the higher the spatial resolution (increased pixel dwell time decreases the impact of detector noise). Excessively slow scans will rapidly bleach the probes and will cause photodamage. Because the images in the *xyt*-scanning mode are confocal, contribution of fluorescence from outside the focal plane is minimized; this important optical property greatly facilitates the identification of Ca^{2+} signals originating from different subcellular compartments with precision. Requirements for spatial and temporal resolution will dictate the type of confocal to be used for *xyt*-scanning. If high temporal resolution is required (such as that necessary to visualize initialization and propagation of Ca^{2+} waves through the entire cell or a collection of cells), then high frame or image acquisition rates should be employed. Standard galvanometer mirror-scanning confocal microscopy systems can scan a full frame (512 × 512 pixels) in approx 1 s and are capable of 10- to 30-Hz scanning small *xy* regions (e.g., 64 × 64 and 128 × 128 pixels). However, the image intensity and contrast are low and spatial resolution is limited. A better solution for high-speed, high-resolution imaging is to use a spinning disk confocal microscope and image capture using high-sensitivity, high-speed CCD digital cameras. These microscopes are capable of data acquisition rates between 30 and120 Hz with excellent full-frame spatial resolution. One disadvantage is that these systems typically are designed to acquire a single emission wavelength. Acquisition of multiple emission images either requires multiple CCD detectors or a device to record all wavelengths on different portions of the same camera chip.

The third mode, *xyzt*-scanning, allows resolution of spatio-temporal $[Ca^{2+}]$ gradients in the volume of the cell. Dynamic three-dimensional cell imaging involves repeated and rapid acquisition of *xy* images in multiple *z*-axis optical sections over time. The changes in *z*-axis steps can be accomplished by (a) a stage-stepping motor attached to the focus knob that precisely and reproducibly alters the position of the microscope stage, (b) an intrinsic/internal motorized focus (on some automated microscopes), (c) a high-speed piezoelectric device that controls the position of the viewing objective without moving the microscope stage, or (d) a galvanometer stage plate that can accurately displace the stage up to approximately 100 μ*M*.

3.7.1. Measuring $[Ca^{2+}]_i$ by CLSM in the Line-Scanning Mode

1. Grow cells on cover slips and load with single-wavelength excitation Ca^{2+}-sensitive fluorescent dyes (e.g., Fluo-3 or Fluo-4). Place the coverslip into a microperifusion chamber mounted on the specimen stage of the confocal microscope

system and superfuse cells with buffers for 15–30 min to remove excess indicator. Initial focus of the cells should be accomplished using condenser light.

2. Use *xy* scans to identify the cell or cells to be studied. Refocus the image under fluorescent laser excitation light (for Fluo-3, excitation line 488 nm from the Ar/Kr laser and emission wavelength detected at 510–530 nm). Set the gain (30–60% pixel saturation), black levels, and pinhole/slit width. Use minimal laser power settings (1–10%) to avoid photobleaching and phototoxicity. Configure the line-scanning acquisition in accordance with the CLSM system you are using; adjust position of the line (or lines) as required.
3. Begin the experiment. For rapid solution exchanges, we suggest using a gravity-fed perifusion system or microspritzer devices.
4. Archive the data for offline analysis. A wide choice of imaging software is available for offline analysis of line scan data. Data can be expressed as a ratio of the pixel fluorescence intensity (F) along the line scan measured at each time-point relative to baseline (F_0). The pseudoratio value (F/F_0) represents the foldchange in F relative to F_0 and can be converted to $[Ca^{2+}]$ by the equation ***(1,31)***:

$$[Ca^{2+}] = K'_d\,[(\beta F/F_0) - (1/\alpha)]/(1 - \beta F/F_0)]$$

where K'_d is the apparent Ca^{2+} dissociation constant of the dye, α is the ratio of F_{max}/F_{min}, and β is the ratio of F_0/F_{max}. F_{max} is the maximum fluorescence intensity value (following application of 10 μM ionomycin) and F_{min} is the minimum fluorescence intensity in the absence of Ca^{2+} (10 μ*M* ionomycin, 10–20 m*M* EGTA, and no added Ca^{2+}). The data can be displayed as an image consisting of a combined collection of the consecutive line scans showing the changes in F, $[Ca^{2+}]$, or R (depicted in gray scales or pseudocolored, respectively) or graphically, expressing the pseudoratio, R, or $[Ca^{2+}]$ as a function of time.

3.7.2. Single-Wavelength Excitation Measurements of $[Ca^{2+}]_i$ Using CLSM

1. Grow cells on glass cover slips and load with single-wavelength Ca^{2+} indicator. Like the line-scanning experiments, the fluorophore of choice for *xyt*-scanning is Fluo-3 or Fluo-4, although many others are commercially available. Fluo-3 fluorescence intensity is very low at resting baseline $[Ca^{2+}]_i$ (50–100 n*M*) and increases 4-fold to 10-fold at saturating $[Ca^{2+}]$. We typically load cells in KRB or KRBH containing 1–5 μ*M* Fluo-3/AM for 15–20 min at room temperature or at 37°C.
2. After the dye loading, wash cells with extracellular solutions for 10–20 min by continuous superfusion.
3. Set up the confocal excitation and emission wavelengths, and adjust photomultiplier gain, slit width, and black levels in accordance with the confocal system operator's guide. Identify a field-of-view with sufficient number of cells and a cell-free area for background ROI. Use acquisition software to create ROIs that can consists of single cells, cell clusters, or single or multiple subcellular regions. Set up the acquisition program (duration of experiment, number of images to be acquired, and online graphical plotting formats, if available).

4. Perform the experiment and, afterward, conduct an in vivo calibration. Be sure to carefully monitor fluid exchange in the microperifusion chamber to avoid accidental flooding and exposure of microscope to saline solutions. If an accident does occur, immediately cease the experiment. Dry the microscope with absorbent paper, lightly cleaning all affected surfaces with a pad of absorbent paper moistened with distilled water. If saline solution has entered the body of the microscope or the interior of an objective, do not hesitate to contact the microscope technician for immediate cleaning and repair. Objectives used on confocal microscopes can easily cost over $10,000 each and can be seriously damaged if saline solution is wicked inside them. Let the system administrator determine whether fluid has entered any part of the system.
5. Imaging data are stored and analyzed as earlier.

3.7.3. Dual-Wavelength Ratiometric Emission Measurements of $[Ca^{2+}]_i$ Using CLSM

1. Culture cells on cover slips. Load cells with Fluo-3/AM and Fura red/AM for 15–20 min at room temperature or at 37 °C. Fluo-3 and Fluo-4 produces a much brighter signal than Fura red. For loading rat hepatoma cells, we used a 1:3 ratio of Fluo-3/AM and Fura red/AM For other cell types (mouse insulinoma cells), we used a 1:5 ratio.
2. Rinse cells 15–30 min with perifusate by continuous superfusion. During this time, perform the initial focus, identify a field-of-view, and set up the confocal system and data acquisition; identify ROIs and remember that the ROI used for background subtraction should be located in a cell-free area. Both dyes are excited by the 488-nm line of an Ar or Ar/Kr laser, and emission intensities are monitored simultaneously at 535 nm (Fluo-3) and 622 nm (Fura Red) (*see* **Note 8**). Recheck focus before performing the experiment.
3. Conduct the experiment, monitoring the status of the perifusion system throughout.
4. Save image and data files to CD-R or DVD-R storage media. Background-subtracted data can be expressed as the time-dependent change in the ratio of Fluo-3 (FI_{FL3}) and Fura Red (FI_{FR}) fluorescence intensity: As $[Ca^{2+}]$ increases, FI_{FL3} and FI_{FR} increase and decrease, respectively. Converting FI_{FL3}/FI_{FR} into $[Ca^{2+}]$ can be accomplished using in vivo calibration ***(28,29)***.

3.8. Measuring $[Ca^{2+}]$ in Organelles

Ca^{2+} signaling is regulated by localized control of Ca^{2+} fluxes. To fully define and understand subcellular Ca^{2+} signaling, the contribution of the $[Ca^{2+}]$ within organelles must be directly studied. $[Ca^{2+}]$ gradients in mitochondria and endoplasmic reticulum (ER) have been imaged using synthetic Ca^{2+} indicators Rhod-2 and MagFura-2, respectively, and are loaded into cells using AM derivatives of the dyes. Cationic Rhod-2 partitions into mitochondria because of inner mitochondrial membrane hyperpolarization. Mag-Fura-2 (or Furaptra) has a low affinity for Ca^{2+}, thus making it a suitable indicator of Ca^{2+} fluxes in regions of high [Ca^{2+}]. This indicator, however, distributes throughout the cytosol and organelles, depending on loading conditions (low temperature favors dye compartmentalization because esterase activity is lowered while diffusion rates are not appreciable affected). Measurements of ER Ca^{2+} fluxes with Mag-Fura-2 are generally conducted in permeabilized cells in order to eliminate the contribution of fluorescence originating from the dye distributed in the cytosol. Alternatively, cells are incubated at 37°C for as long as overnight to allow transporters to clear the cytosol of indicator. Mag-Fura-2 has been used to detect ER Ca^{2+} signaling in intact cells. This involves loading cells with the AM form of the indicator, followed by an overnight incubation in growth medium (without Mag-Fura-2/AM), during which time cytosolic but not organelle-sequestered Mag-Fura-2 is removed by plasma membrane anion transporters.

Genetically targeted Ca^{2+} biosensors represent the most recent approach to directly image intraorganelle Ca^{2+} signals. Recombinant forms of aequorin and fluorescent fusion protein chimeras have been expressed in a wide range of eukaryotic cells. By including a targeting sequence in the expression vector, the Ca^{2+} biosensors can be expressed selectively in specific subcellular compartments such as the ER, mitochondria, Golgi, and nucleus. Single-wavelength excitation, dual-wavelength excitation, and dual-wavelength emission microspectrofluorometry can be used to quantitatively image subcellular Ca^{2+} signaling with genetically targeted fluorescent biosynthetic Ca^{2+} sensors. Although wide-field and confocal imaging can be used, we recommend CLSM or spinning disk optical confocal imaging because these approaches allow for the more reliable identification of the location of the fluorescent signal than standard wide-field imaging. This is especially important if the goal of a study is to record mitochondrial flux vs cytosolic (or ER) fluxes (*see* **Note 9**).

3.8.1. Quantitative Dual-Wavelength Ratiometric Imaging of $[Ca^{2+}]_{organelle}$ Using Cameleon and Ratiometric Pericam Ca^{2+} Biosensors

1. Culture cells onto glass coverslips the day before transfecting with the biosensors. In 6-well culture plates, we seed each well (containing a 25-mm circular glass coverslip) with 0.5-1 × 10^6 cells.
2. Day 2, transfect cells with 1–2 μg DNA per well. There are several choices for gene transfer and the most optimal reagent or method will depend on cell type. We have used liposomal methods for transiently transfecting insulin-secreting cell lines, hepatoma cells, neuroendocrine cell lines, and primary vascular smooth muscle cells; the transfection efficiency ranged between 5 and 30%.
3. We perform imaging experiments with the transfected cells 48–96 h after transfection using conventional wide-field or single-spinning disk confocal microfluorometry to measure spatio-temporal [Ca^{2+}] gradients in the cytoplasm, ER,

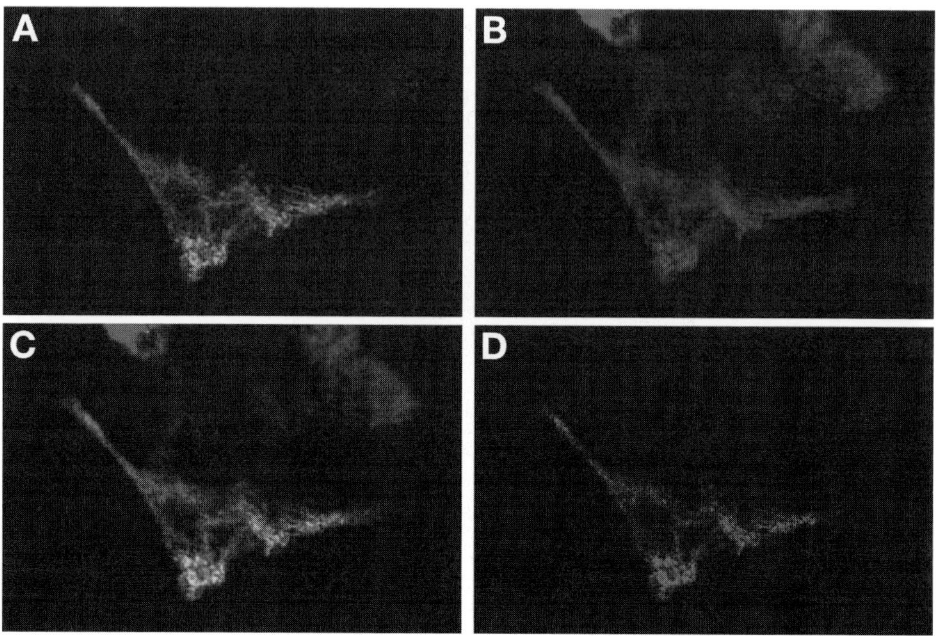

Fig. 3. Expression of mitochondrially targeted ratiometric pericam (RPC-mt) in neuroendocrine cells. Laser scanning confocal images of AtT20 cells colabeled with **(A)** RPC-mt (green) and **(B)** MitoTracker Red. **(C)** An overlay image constructed using Adobe Photoshop. The yellow color indicates colocalization of the two dyes. Note that some cells in the field-of-view do not express RPC-mt and remain red in the overlay image. **(D)** Map of colocalization using Bio-Rad confocal image processing software. Blue indicates regions of overlap between green and red channel fluorescence. (*see* Color Plate 2, following p. 274.)

mitochondria, or nucleus. The subcellular distribution of biosensors targeted to the mitochondria (*see* **Fig. 3;** Color Plate 2, following p. 274) and ER (*see* **Fig. 4**) can be easily visualized with confocal imaging. An example of real-time measurements of [Ca^{2+}] oscillations in the cytoplasm of an insulin-secreting βTC3 cell expressing YC2.1 is shown in **Fig. 5**. Note the inverse relationship between the intensities of the FRET donor (*see* **Fig. 5**, top panel) and FRET acceptor (*see* **Fig. 5**, middle panel). For imaging cameleons, we recommend 436–440 nm excitation, 455DCLP dichroic, and 485 nm and 535 nm emission. Preferred filters for the ratiometric pericams are 410–415 nm and 480–485 nm excitation, 505DRLP-XR dichroic, and 535 nm emission. *See* **Table 3** for filter sets. To minimize photobleaching (and phototoxicty), we insert neutral-density filters into the excitation light path; attenuation of light between 50 and 99% seems to work best for the biosensors. We choose cells with intermediate brightness to study; in our experience, brightly fluorescent cells tend to be unresponsive, possibly reflecting an adverse effect of overexpression of the calmodulin-containing biosensor and buffering of Ca^{2+} transients ***(16)***. A

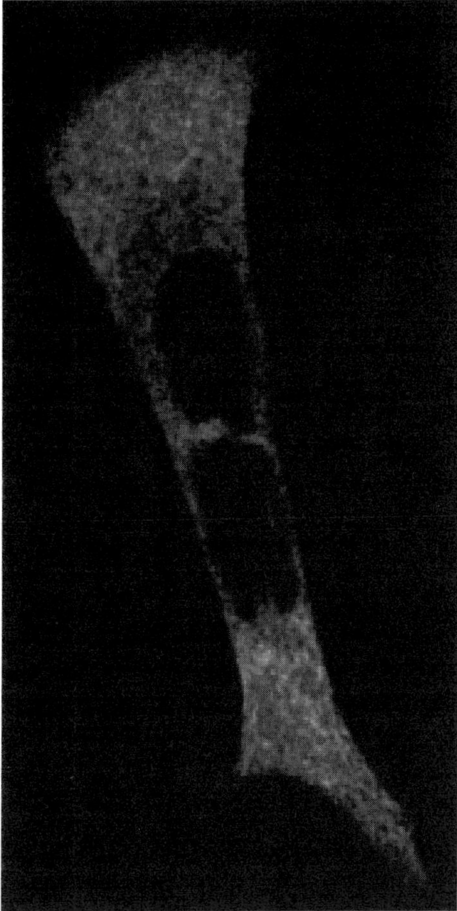

Fig. 4. Confocal micrograph of AtT20 cells expressing YC4er. Cells were transfected with YC4er using lipofectamine and imaged 2 d after transfection. Note the reticular pattern of distribution of the biosensor throughout the cells and absence of expression in the nuclei. Image obtained with a Bio-Rad 1024 MRC confocal laser scanning system mounted on an upright fluorescence microscope and a 60× water immersion objective using 488-nm excitation and 520-nm emission.

protocol to estimate the intracellular concentration of the biosensor has been described by Miyawaki and his colleagues ***(16)***.

4. We convert the ratio data into a molar value of $[Ca^{2+}]$ by in vivo calibration using 10 μ*M* ionomycin in the presence and absence of Ca^{2+} (plus EGTA). For the ER cameleon calibration, we add ionomycin in the presence of 10–20 m*M* $[Ca^{2+}]$ in the extracellular solution to evoke maximum ratio, then for minimum ratio, we administer an extracellular solution containing 10–20 m*M* EGTA and no added Ca^{2+}. The data are converted to molar $[Ca^{2+}]$ using the following equation ***(16)***:

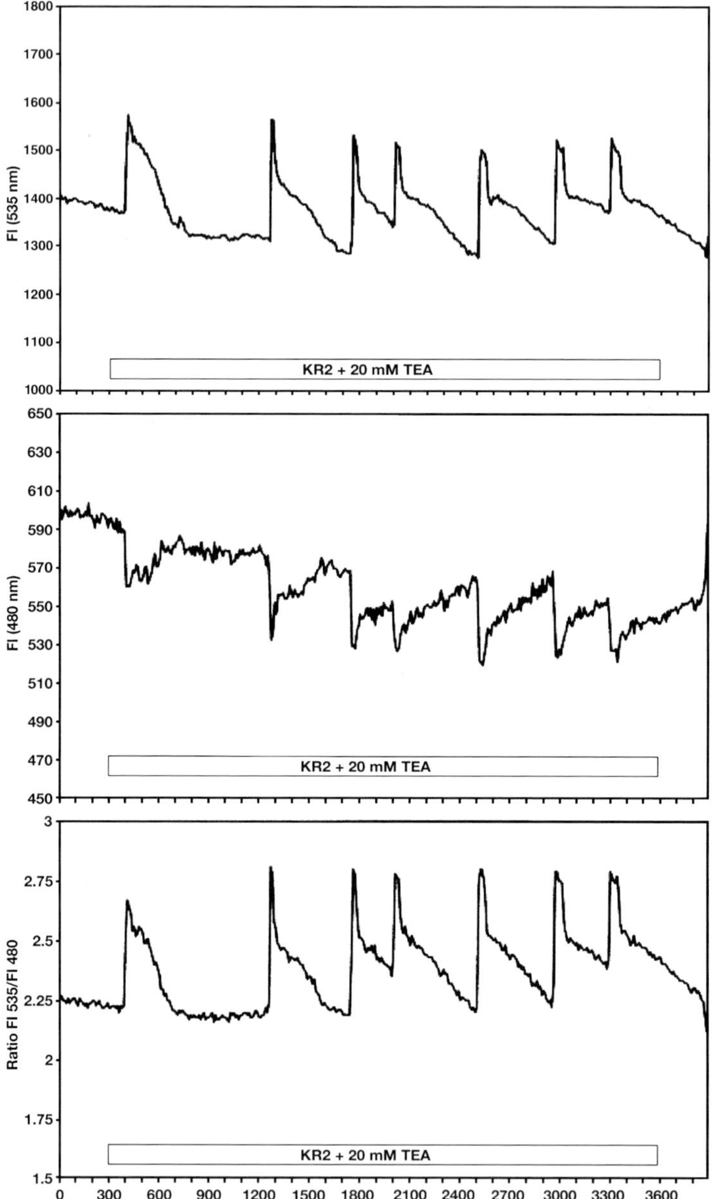

Fig. 5. [Ca^{2+}] oscillations in the cytoplasm of a βTC3 cell expressing YC2.1. The oscillatory changes in fluorescence intensity of the FRET acceptor (upper panel; FI535), FRET donor (middle panel; FI480), and ratio (lower panel; Ratio FI 535/480) induced by glucose (KR2; 2 m*M*) and tetraethylammonium (TEA) (open bar) were measured using conventional wide-field fluorescence microscopy. YC2.1 excitation was 440 nm; FRET donor and acceptor emissions were recorded at 480 nm and 535 nm, respectively. Time (s) is indicated on the *x*-axis.

Table 3
Filters for Imaging Fluorescence Calcium Biosensors

Biosensor	X_1	X_2	DC	M_1	M_2
Yellow Cameleon	440DF10		455DRLP	480DF30	535DF25
Red Cameleon					
YRC2	480DF10		505DRLP	535DF25	565EFLP
CRC2	440DF20		455DRLP	480DF30	565EFLP
SapRC2	400DF15		455DRLP	510WB40	565EFLP
Camgaroo	480DF30		505DCLP	535DF25	
Pericam	480DF10	410DF10	505DRLP-XR	535DF25	

Note: Our suggestions for excitation (X_1 and X_2), dichroic (DC), and emission (M_1 and M_2) filters for biosynthetic Ca^{2+} biosensors. Consult filter manufacturer for product details and recommended alternatives.

$$[Ca^{2+}] = K'_d \, [(R - R_{min})/(R_{max} - R)]^{(1/n)}$$

where K'_d is the apparent dissociation constant and n is the Hill coefficient. Published values for biosensor K'_d and n are summarized in **Table 2**.

4. Notes

1. Prior to seeding cells, cover slips are cleaned and sterilized in absolute ethanol, air-dried while exposed to UV light in a tissue culture hood for 1 h. Coating cover glasses with substrates, such as poly-L-lysine or collagen, is not required for attachment and growth of the cells we study. However, this is not the case for all cell types and use of coated glass cover slips might be necessary. The size and shape of the cover slips depends on the specimen chamber that will house the samples on the microscope stage of the imaging system. We use Harvard Apparatus and Warner Instruments microperifusion chambers. For the Harvard MP-4 system, the microperifusion chamber employs 25-mm circular glass cover slips, whereas 15-mm circular cover glass is used in the Warner PC-21 microperifusion system.
2. Manufacturers provide detailed instructions regarding storage, stability, reconstitution conditions, and handling precautions for the indicators. Unless instructed otherwise, we recommend that immediately upon receipt, the AM form of the dye be frozen at –20°C in a desiccator until ready for resuspension. Acetoxymethyl ester dye derivatives are shipped in multiple vials or a single vial containing 50 μg or 1 mg, respectively, and are soluble in dimethylsulfoxide (DMSO), ethanol, or methanol. Our common practice is to obtain multiple vials of the dye (50 μg per vial) and reconstitute the dye on the day of the experiments. We use water-free DMSO (stored at room temperature in a desiccator) to dissolve the membrane-permeable form of the dye to yield a final concentration of 1 mM (e.g., 50 μl DMSO/50 μg Fura-2AM; M_r = 1002). This volume and concentration provides enough indicator to load cells on 10–50 cover slips. The reconstituted AM form of the dye can be frozen and reused. Although repeated

freeze–thawing cycles do not seem to affect Fura-2, Fluo-3, and Fura red loading or fluorescent properties, we advise using newly reconstituted AM batches of the indicator at least once per week.

3. Pluronic is a nonionic detergent that helps to evenly disperse the AM form of the dye in an aqueous solution. Alternatively, bovine serum albumin can be used to keep the AM esters in solution; AM has limited aqueous solubility and precipitated dye will not enter cells, causing poor loading. Thus, increasing the dye concentration thus can decrease loading if the dye begins to precipitate. Precipitated dye is typically seen as bright debris over the surface of the cells and cover slip that rapidly bleaches from view. Metabolism of the cleaved AM moiety can produce toxicity in some cell types.
4. We have found that time in culture is a critical factor that determines transfection efficiency when using a lipofectamine method: a 12- to18-h culture period prior to transfection appeared optimal for cell types we have studied, including insulinoma, pituitary, and hepatoma cell lines. Even so, the maximum transfection efficiency we have achieved by the liposomal approach was approx 30% and ranged from 5 to 30%. Transfection of primary cells like pancreatic β-cells and rodent vascular smooth muscle cells with a liposome-based method was problematic, usually yielding an extremely low transfection efficiency. An alternative approach is to transfect cells with the biosensors using a viral-mediated gene shuttle vector. This requires construction of a viral vector because most of the biosensors are currently available as a plasmid cDNA. Notwithstanding this limitation, transfection efficiency is greatly improved (80–90%) and allows expression of the sensors in cell lines and primary cells as well as more complex biological tissues such as brain slices, pancreatic islets of Langerhans, and vascular preparations. Another key advantage of viral transfection is the potential for selective expression of a biosensor in a specific cell type within a multicellular tissue. This is accomplished by infecting tissues with viral vectors whose expression is under the control of a cell-specific promoter. For example, by driving adenovirus- and baculovirus-mediated expression with the rat insulin promoter sequence (RIP1), we have selectively and specifically labeled pancreatic β-cells in intact mouse islets of Langerhans with GFP-labeled fusion proteins. Finally, although it is conceivable that a transgenic approach might be used to express biosensors in cells in vivo, there is little evidence in the literature that this method will be useful in mammalian models. Transgenic expression of YC2 and camgaroo-2 in *Caenorhabditis elegans* and *Drosophila* neurons, respectively, has been reported ***(32,33)***.
5. All solutions, including KRB and KRH buffers, should be freshly prepared on the same day as the imaging experiments. We strongly recommend avoiding direct application of drugs or other reagents into the specimen chamber by micropipetting; this method does not allow precise control of drug concentration and risks mechanical disruption of cells, which can affect Ca^{2+} signaling. Other methods are available for drug or reagent delivery to cells, including microelectrodes attached to a pressurized microspritzer device or a gravity-fed U-tube connected to a computer-controlled manifold, which receives fluid input from one or more reservoirs. These

methods have the advantage of allowing the administration of reagents to highly localized regions or to individual cells and are especially useful when administering drugs that are available in limited amounts.

6. Under no circumstances should you expose the cells (or your eyes) to unattenuated arc lamp light. Doing so risks photobleaching of the indicator and cell photodamage. Make certain that the digital camera gain is not activated before diverting the emission light path from the binocular eyepieces to the camera. Also, turn off room lights and the condenser lamp before activating the camera. Unfortunately, some intensified digital cameras have limited protection circuitry to prevent damage to the intensifier caused by a sudden exposure to bright light. Apart from outright failure, strong light also greatly decreases the life of Gen-3 intensifiers or can "burn-in" patterns.
7. For Fura-2 imaging using the intensified video camera, we adjust the camera gain using the 380-nm image by initially increasing the camera gain to near saturation, then setting the final adjustment 20–30% below saturation. This is important because as [Ca^{2+}] increases and the 380-nm signal decreases, but as the [Ca^{2+}] returns to baseline or overshoots baseline (which can occur in some Ca^{2+} signaling responses), the 380-nm intensity increases. Setting the camera gain to saturation reduces image quality and impairs reliable quantification of the imaging intensity data. Using the CCD cameras, the 340-nm and 380-nm image gains can be set independently.
8. Use laser power setting of 3–10%, although up to 30% can be required depending on the quality of dye loading. Because emission intensity of Fluo-3 is higher than Fura red, gain settings for the red channel (Fura red) will be higher than the green channel (Fluo-3). As a general rule, we set the gain so that cell fluorescence intensity was at least fivefold above background. The optimal pinhole size, determined by objective numerical aperture (NA) and wavelength, should ideally have different diameters, but many systems have only one pinhole. For live cell work, one must typically open the pinhole wide enough to get adequate signal intensity while keeping the laser intensity low to prevent photobleaching and phototoxicity.
9. Targeted Ca^{2+} biosensors seem to offer an advantage in compartment selectivity; however, compartment physical characteristics, low expression levels, general phototoxicity, and perhaps chaperone content might still seriously hamper measurements. Cameleon probes are not well suited for CLSM using the 488-nm laser because CFP is poorly excited and YFP is directly excited. Some Ar lasers provide a 457-nm line that has been used to excite cameleon probes; however, this line is typically very weak and directly excited YFP. Recent utilization of 405-nm diode lasers in some confocal systems permits strong excitation of CFP with minimal YFP excitation and offers promise that targeted cameleon probes will become more widely used for confocal measurements of compartmental Ca^{2+} fluxes in single cells, tissue slices, or multicellular preparations.

Acknowledgments

This work was supported by an American Diabetes Association Research Award (MWR), NIH DK64162 (MWR), and NIH DK63493 (LHP and MWR).

References

1. Grynkiewicz, G., Poenie, M., and Tsien, R. Y. (1985) A new generation of Ca^{2+} indicators with greatly improved fluorescence properties. *J. Biol. Chem.* **260,** 3440–3450.
2. Kao, J. P. Y., Harootunian, A. T., and Tsien, R. Y. (1989) Photochemically generated cytosolic calcium pulses and their detection by fluo-3. *J. Biol. Chem.* **264,** 8179–8184.
3. Minta, A., Kao, J. P. Y., and Tsien, R. Y. (1989) Fluorescent indicators for cytosolic calcium based on rhodamine and fluorescein chromophores. *J. Biol. Chem.* **264,** 8171–8178.
4. Tsien, R. Y. (1981) A non-disruptive technique for loading calcium buffers and indicators into cells. *Nature* **290,** 527–528.
5. Tsien, R. Y., Pozzan, T., and Rink, T. G. (1982) Calcium homeostasis in intact lymphocytes: cytoplasmic free calcium monitored with a new, intracellularly trapped fluorescent indicator. *J. Cell Biol.* **94,** 325–334.
6. Roe, M. W., Lemasters, J. J., and Herman, B. (1990) Assessment of Fura-2 for measurements of cytosolic free calcium. *Cell Calcium* **11,** 63–73.
7. Takahashi, A., Camacho, P., Lechleiter, J. D., and Herman, B. (1999) Measurement of intracellular calcium. *Physiol. Rev.* **79,** 1089–1125.
8. Baubet, V., Le Mouellic, H., Campbell, A. K., Lucas-Meunier, E., Fossier, P., and Brulet, P. (2000) Chimeric green fluorescent protein-aequorin as bioluminescent Ca^{2+} reporters at the single-cell level. *Proc. Natl. Acad. Sci. USA* **97,** 7260–7265.
9. Miyawaki, A., Llopis, J., Heim, R., et al. (1997) Fluorescent indicators for Ca^{2+} based on green fluorescent proteins and calmodulin. *Nature* **388,** 882–887.
10. Rizzuto, R., Simpson, A. W., Brini, M., and Pozzan, T. (1992) Rapid changes of mitochondrial Ca^{2+} revealed by specifically targeted recombinant aequorin. *Nature* **358,** 325–327.
11. Emmanouilidou, E., Teschemacher, A. G., Pouli, A. E., Nicholls, L. I., Seward, E. P., and Rutter, G. A. (1999) Imaging Ca^{2+} concentration changes at the secretory vesicle surface with a recombinant targeted cameleon. *Curr. Biol.* **9,** 915–918.
12. Foyouzi-Youssefi, R., Arnaudeau, S., Borner, C., et al. (2000) Bcl-2 decreases the free Ca^{2+} concentration within the endoplasmic reticulum. *Proc. Natl. Acad. Sci. USA* **97,** 5723–5728.
13. Griesbeck, O., Baird, G. S., Campbell, R. E., Zacharias, D. A., and Tsien, R. Y. (2001) Reducing the environmental sensitivity of yellow fluorescent protein: mechanisms and applications. *Proc. Natl. Acad. Sci. USA* **276,** 29,188–29,194.
14. Isshiki, M., Ying, Y. -S., Fujita, T., and Anderson, R. G. W. (2002) A molecular sensor detects signal transduction from caveolae in living cells. *J. Biol. Chem.* **277,** 43,389–43,398.
15. Miyawaki, A., Griesbeck, O., Heim, R., and Tsien, R. Y. (1999) Dynamic and quantitative Ca^{2+} measurements using improved cameleons. *Proc. Natl. Acad. Sci. USA* **96,** 2135–2140.

16. Miyawaki, A., Mizuno, H., Llopis, J., Tsien, R. Y., and Jalink, K. (2001) Cameleons as cytosolic and intra-organellar calcium probes, in *Calcium Signalling*, 2nd ed. (Tepikin, A.V., ed.), Oxford University Press, Oxford, pp. 3–16.
17. Mizuno, H., Sawano, A., Eli, P., Hama, H., and Miyawaki, A. (2001) Red fluorescent protein from *Discosoma* as a fusion tag and pardner for fluorescence energy transfer. *Biochemistry* **40,** 2502–2510.
18. Nagai, T., Sawano, A., Park, E. S., and Miyawaki, A. (2001) Circularly permuted green fluorescent proteins engineered to sense Ca^{2+}. *Proc. Natl. Acad. Sci. USA* **98,** 3197–3202.
19. Nagai, T., Ibata, K., Park, E. S., Kubota, M., Mikoshiba, K., and Miyawaki, A. (2002) A variant of yellow fluorescent protein with fast and efficient maturation for cell-biological applications. *Nature Biotechnol.* **20,** 87–90
20. Nakai, J., Ohkura, M., and Imoto, K. (2001) A high signal-to-noise Ca^{2+} probe composed of a single green fluorescent protein. *Nat. Biotechnol.* **19,** 137–141.
21. Pinton, P., Tsuboi, T., Ainscow, E. K., Pozzan, T., Rizzuto, R., and Rutter, G. A. (2002) Dynamics of glucose-induced membrane recruitment of protein kinase C βII in living pancreatic islet β-cells. *J. Biol. Chem.* **277,** 37,702–37,710.
22. Truong, K., Sawano, A., Mizuno, H., et al. (2001) FRET-based *in vivo* Ca^{2+} imaging by a new calmodulin-GFP fusion molecule. *Nature Struct. Biol.* **8,** 1069–1073.
23: Zhang, J., Campbell, R. E., Ting, A. Y., and Tsien, R. Y. (2002) Creating new fluorescent probes for cell biology. *Nature Rev. Cell Biol.* **3,** 906–918.
24. Baird, G. S., Zacharias, D. A., and Tsien, R. Y. (1999) Circular permutations and receptor insertion within green fluorescent proteins. *Proc. Natl. Acad. Sci. USA* **96,** 11,241–11,246.
25. Dustin, L. B. (2000) Ratiometric analysis of calcium mobilization. *Clin. Appl. Immunol. Rev.* **1,** 5–15.
26. Hallett, M. B., Hodges, R., Cadman, M., et al. (1999) Techniques for measuring and manipulating free Ca^{2+} in the cytosol and organelles of neutrophils. *J. Immunol. Methods* **232,** 77–88.
27. Thomas, D., Tovey, S. C., Collins, T. J., Bootman, M. D., Berridge, M. J., and Lipp, P. (2000) A comparison of fluorescent Ca^{2+} indicator properties and their use in measuring elementary and global Ca^{2+} signals. *Cell Calcium* **28,** 213–223.
28. Lipp, P. and Niggli, E. (1993) Ratiometric confocal Ca^{2+} measurements with visible wavelength indicators in isolated cardiac myocytes. *Cell Calcium* **14,** 359–372.
29. Lipp, P., Luscher, C., and Niggli, E. (1996) Photolysis of caged compounds characterized by ratiometric confocal microscopy: a new approach to homogeneously control and measure the calcium concentration in cardiac myocytes. *Cell Calcium* **19,** 255–266.
30. Roe, M. W., Moore, A. L., and Lidofsky, S. D. (2001) Purinergic-independent calcium signaling mediates recovery from hepatocellular swelling: implications for volume regulation. *J. Biol. Chem.* **276,** 30,871–30,877.
31. David, G., Talbot, J., and Barrett, E. F. (2003) Quantitative estimate of mitochondrial [Ca^{2+}] in stimulated motor neurons. *Cell Calcium* **33,** 197–206.

32. Kerr, R., Lev-Ram, V., Baird, G., Vincent, P., Tsien, R. Y., and Schafer, W. R. (2000) Optical imaging of calcium transients in neurons and pharyngeal muscle of *C. elegans*. *Neuron* **26,** 583–594.
33. Yu, D., Baird, G. S., Tsien, R. Y., and Davis, R. L. (2003) Detection of calcium transients in *Drosophila* mushroom body neurons with camgaroo reporters. *J. Neurosci.* **23,** 64–72.
34. Arnaudeau, S., Kelley, W. L., Walsh, J. V., Jr., and Demaurex, N. (2001) Mitochondria recycle Ca^{2+} to the endoplasmic reticulum and prevent the depletion of neighboring endoplasmic reticulum regions. *J. Biol. Chem.* **276,** 29,430–29,439.

4

Multifluorescence Labeling Techniques and Confocal Laser Scanning Microscopy on Lung Tissue

Maria Stern, Douglas J. Taatjes, and Brooke T. Mossman

Summary

Lung tissue consists of more than 40 individual cell types that might interact to produce adverse pathologies. After injury, a number of signaling proteins expressed in various epithelial and other cell types have been linked to the advent of apoptosis, compensatory proliferation, and adaptation to stress. We describe here the use of immunochemistry and multifluorescence approaches using confocal laser scanning microscopy to define the signaling pathways (protein kinases C and mitogen-activated protein kinases) activated by asbestos fibers after inhalation. Using these approaches, we are able to localize signaling events in distinct cell types of the lung and determine their status in the cell cycle (resting or nonresting). Moreover, we are able to determine whether various signaling proteins colocalize in cells and the sites affected by asbestos fibers.

Key Words: Immunofluorescence; confocal laser scanning microscopy; asbestos; extracellular signal-regulated kinases (ERKs); PKCδ; proliferation marker; Ki-67.

1. Introduction

Confocal laser scanning microscopy (CLSM) is a laser-based imaging technology widely used in pathology and cell biology research. This approach can be used to identify lesions and affected cell types. CLSM provides increased resolution over conventional wide-field microscopy and has the ability to reject out-of-focus fluorescence. It also allows a decrease in autofluorescence signal caused by collagen deposits in some tissues, such as lung, by the use of fluorescent probes that excite at higher wavelengths *(1)*. It is a powerful technique for studying cell signaling by environmental agents in the initiation and pathology of lung disease and/or repair and adaptation. Inhaled environmental agents such as asbestos might elicit cell signaling pathways at the cell membrane. Using antibodies specific to phosphorylated (i.e., activated) signaling proteins,

From: *Methods in Molecular Biology, vol. 319: Cell Imaging Techniques: Methods and Protocols*
Edited by: D. J. Taatjes and B. T. Mossman © Humana Press Inc., Totowa, NJ

related signaling events and gene transactivation by signaling proteins can be documented in vivo. The mitogen-activated protein kinases (MAPKs), including extracellular signal-regulated kinases (ERKs) as well as some isotypes of protein kinase C (i.e., PKCδ) have been linked to both cell proliferation and the development of apoptosis in response to toxic agents such as oxidants and asbestos fibers. In this chapter, we describe multifluorescence techniques to document nuclear and membrane translocation and increased activation of PKCδ and ERK 1/2 in lung tissue after inhalation of asbestos *(2,3)*. Use of a Ki-67 antibody specific for nonresting epithelial cells indicates that these protein kinases are often colocalized in proliferating cells *(4)*.

2. Materials

2.1. Processing and Sectioning

Mice are administered a lethal dose of sodium pentobarbital (Abbot Laboratories, Chicago, IL) before the chest cavity is opened, a polyurethane catheter is inserted into the trachea, and the lungs are instilled with 1X phosphate-buffered saline (PBS) at a pressure of 25 cm water. The unfixed lungs are separated by suturing, removed from the chest, snap-frozen by plunging into liquid-nitrogen-cooled isopentane, placed in OCT embedding compound (Tissue Tek, Torrance, CA), and frozen at –80°C. Sections are cut at –23°C in a cryostat using disposable knives and then retrieved onto Superfrost +/+ slides. Sections are quickly examined under the light microscope to assure that the tissue structure has not been damaged during the sectioning process. Slides are stored in a slide box at –80°C until use.

2.2. Reagents

1. 10X PBS (for 1 L): Add 2.76 g of sodium phosphate monobasic and 14.1 g of sodium phosphate dibasic (anhydrous) to 500 mL of deionized water (dH_2O). Add 90 g of sodium chloride and bring volume to 1L, followed by adjustment of the pH to 7.4. For 1X PBS, dilute 1 : 10 with distilled water (dH_2O) and adjust pH to 7.4 if necessary. Both solutions are stable at room temperature (RT).
2. PBS-PFA fixative: 3.7% paraformaldehyde (PFA), 1X PBS. PFA should be handled under a fume hood and glove protection is required. Weigh 3.7 g of PFA and add to 100 mL of 1X PBS. Heat stirring solution at 60°C until PFA dissolves. When solution becomes clear, remove flask from the heating plate, cool to RT, and filter through Whatman filter paper. This solution is made fresh each time and chilled prior to use.
3. Sodium dodecyl sulfate (SDS): 0.5% SDS, 1X PBS. Procedure should be handled with respiratory mask and gloves. Weigh 0.05 g of SDS and add to 10 mL of 1X PBS. Shake the tube until mix goes into solution. This solution is stable at RT.
4. 1% Bovine serum albumin (BSA)/PBS: 1 mg/mL BSA, 1X PBS, 0.02% sodium azide. Store solution at 4°C. This solution is stable until evidence of bacterial growth.

5. 10% Normal goat serum (from secondary antibody host) in 1X PBS. Solution is made fresh prior to each use.
6. 5% Blocking reagent from mouse-on-mouse kit (M.O.M. kit; Vector Laboratories, Burlingame, CA) in 1X PBS. Solution is made fresh prior to each use.
7. Mounting medium: AquaPolyMount (Polysciences Inc., Warrington, PA) should be stored at 4°C and warmed to RT for a few minutes before each use.

2.3. Equipment

A humid environment is necessary to prevent evaporation of reagents during prolonged incubations. Inexpensive humidity chambers can be created in the lab. For instance, a large (6-in. diameter) glass Petri dish with a moistened filter paper on the bottom can be used as a humid chamber. Because slides should not touch wet paper directly, a piece of parafilm elevated 2–4 mm above the filter paper on which place the slides can be used *(5)*.

2.4. Controls

Staining controls omitting primary antibodies are required for each of the secondary antibodies for every staining. Negative controls using isotype control antibodies or nonimmune serum from the same species as the primary antibody are also helpful for validating the specificity of staining results.

2.5. Antibody Sources

1. PKCδ: rabbit polyclonal nPKC δ (C-20) antibody (Santa Cruz, cat. no. sc-937).
2. p-ERK: rabbit polyclonal Phospho-p44/42 MAP Kinase (Thr202/Tyr204) (Antibody, Cell Signaling Inc., cat. no. 9101).
3. p-JNK: rabbit polyclonal JNK (SAPK) [pTpY183/185] (Biosource, cat. no. 44-682).
4. Cytokeratin: pan antibody produced in mouse (Sigma, cat. no. C2562).
5. MAC3: rat anti-mouse antibody (BD Biosciences Pharmingen, cat. no. 553322).
6. proSP-C: rabbit anti-human prosurfactant protein C polyclonal antibody (Chemicon, cat. no. AB3786).
7. Ki-67: monoclonal rat anti-mouse, clone TEC3 antibody (DAKO, cat. no. M7249).

3. Methods

3.1. Single Immunofluorescence Labeling

3.1.1. MAPK and PKC

Because anti-p-ERK and anti-p-JNK antibodies used in our laboratory are all derived from rabbits, they can be detected using the same secondary antibody. Therefore, an identical procedure can be used to successfully obtain each labeling. The anti-PKCδ antibody is a mouse monoclonal; thus, it should be used in conjunction with the mouse-on-mouse kit.

3.1.1.1. p-ERK, p-JNK Antibodies, Rabbit Polyclonal

1. For fixation, slides are placed into a Coplin jar filled with fresh 3.7% PFA for 10-min at RT followed by two 5-min washes in 1X PBS (*see* **Note 1**).
2. Wash slides in 1X PBS twice for 5 min.
3. Place slides into –20°C methanol for 10 min for permeabilization. Permeabilization is advised to ensure free access of the antibody to its antigen. Storage of methanol as well as incubation of the tissue should take place at –20°C.
4. Wash slides in 1X PBS twice for 5 min.
5. Circumscribe the tissue section with a Pap-pen. Using a Q-tip, dry around the section and circumscribe the tissue section with a Pap-pen to create a hydrophobic border. This hydrophobic border allows the use of small volumes of solutions to cover the entire section without wasting unreasonably large amounts of expensive reagents.
6. Place the slides in a humid chamber.
7. For antigen retrieval, sections can be treated with 1% SDS in 1X PBS for 5 min at RT to enhance antibody staining ***(6,7)***. Apply 1% SDS for 5 min on each section.
8. Wash sections twice for 5 min in 1X PBS in a Coplin jar.
9. Block in 10% normal goat serum in 1X PBS for 1 h at RT. Serum used for the blocking step has to be from the animal species in which the secondary antibody was raised in to prevent nonspecific binding. Therefore, for primary antibodies raised in other animal species than mouse, 10% normal serum diluted in 1X PBS should be used for 1-h incubation.
10. Wash twice in 1X PBS for 5 min.
11. Apply rabbit polyclonal antibody diluted in 1% BSA/1X PBS (PKCδ, 3 μg/mL; p-ERK, 1 : 250; p-JNK, 2 μg/mL). In most instances, we recommend applying primary antibodies overnight at 4°C to enhance staining intensities while lowering nonspecific background staining.
12. Wash slides twice in 1X PBS for 5 min.
13. Cover sections with biotinylated goat anti-rabbit IgG (Vector Laboratories) for 1 h at RT (*see* **Notes 2–4**).
14. Wash slides twice in 1X PBS for 5 min.
15. Incubate sections in strepavidin Alexa 568 conjugate at 1 : 400 dilution in 1X PBS for 1 h at RT in the dark. During the incubation with fluorophore-conjugated secondary antibodies, the humid chamber should be covered with aluminum foil to avoid exposure to room light and potential fluorophore photobleaching. Secondary antibodies are chosen with the desired conjugated fluorophore and to bind to the primary antibody.
16. Rinse slides in 1X PBS for 5 min at RT.
17. Rinse twice with dH_2O for 1 min.
18. Mount section with cover slip using AquaPolyMount. Use no. 1-½ cover slips. Apply a drop of AquaPolyMount to cover the section and attach a cover slip. Using forceps, apply minimal pressure to the top of the cover slip to remove air bubbles from underneath and to create a tight seal (*see* **Notes 6,7**).
19. Store slides in a lightproof slide box at 4°C until analysis.

3.1.1.2. PKCδ Antibody, Mouse Monoclonal

1. Take slides out of –80°C and immediately place them in 4% PFA for 10 min at RT.
2. Wash slides in 1X PBS twice for 5 min.
3. Place slides into –20°C methanol for 10 min.
4. Wash slides in 1X PBS twice for 5 min.
5. Circumscribe the tissue section with a Pap-pen.
6. Place the slides in a humid chamber.
7. For antigen retrieval, apply 1% SDS for 5 min on each section.
8. Wash sections twice for 5 min in 1X PBS in a Coplin jar.
9. To minimize nonspecific antibody binding of mouse monoclonal antibodies on murine tissue, slides are incubated in 5% blocking reagent from a mouse-on-mouse kit (Vector Laboratories, Burlingame, CA) in 1X PBS for 1 h.
10. Wash twice in 1X PBS for 5 min.
11. Pre-incubate sections in 10% concentrated protein from a mouse-on-mouse kit in 1X PBS for 5 min.
12. Apply 1 : 250 dilution of PKCδ antibody in 10% concentrated protein from a mouse-on-mouse kit in 1X PBS overnight at 4°C.
13. Wash twice in 1X PBS for 5 min.
14. Apply 1 : 400 dilution of goat anti-mouse IgG conjugated to Alexa 647 in 1% BSA/1X PBS for 1 h at RT in the dark.
15. Rinse slides in 1X PBS for 5 min at RT.
16. Rinse with dH_2O for twice for 1 min.
17. Mount section with cover slip using AquaPolyMount.

3.1.2. Cell Type Markers

To establish whether the protein of interest is specific to bronchiolar and alveolar epithelial cells vs macrophages that infiltrate lung after an injury, lung sections can be colabeled with cell-type-specific markers.

3.1.2.1. Anti-Cytokeratin; Epithelial Cell Marker; Mouse Monoclonal Pan (Mixture) Antibody

1. Take slides out of –80°C and immediately place them in 4% PFA for 10 min at RT.
2. Wash slides in 1X PBS twice for 5 min.
3. Place slides into –20°C methanol for 10 min.
4. Wash slides in 1X PBS twice for 5 min.
5. Circumscribe the tissue section with a Pap-pen.
6. Place the slides in a humid chamber.
7. For antigen retrieval, apply 1% SDS for 5 min on each section.
8. Wash sections twice for 5 min in 1X PBS in a Coplin jar.
9. Block in 5% blocking reagent from a mouse-on-mouse kit in 1X PBS for 1 h at RT.
10. Wash twice in 1X PBS for 5 min.
11. Pre-incubate sections in 10% concentrated protein from a mouse-on-mouse kit in 1X PBS for 5 min.

12. Apply 1:100 dilution of anti-cytokeratin in 10% concentrated protein from a mouse-on-mouse kit in 1X PBS overnight at 4°C.
13. Wash twice in 1X PBS for 5 min.
14. Apply 1:400 dilution of goat anti-mouse IgG conjugated to Alexa 647 in 1% BSA/1X PBS for 1 h at RT in the dark.
15. Rinse slides in 1X PBS for 5 min at RT.
16. Rinse twice with dH_2O for 1 min.
17. Mount section with cover slip using AquaPolyMount.

3.1.2.2. MAC-3; Rat Anti-Mouse M3/84 Antibody That Reacts With MAC-3 Antigen Expressed on Mouse Mononuclear Phagocytes

1. Take slides out of –80°C and immediately place them in 4% PFA for 10 min at RT.
2. Wash slides in 1X PBS twice for 5 min.
3. Place slides into –20°C methanol for 10 min.
4. Wash slides in 1X PBS twice for 5 min.
5. Circumscribe the tissue section with a Pap-pen.
6. Place the slides in a humid chamber.
7. For antigen retrieval, apply 1% SDS for 5 min on each section.
8. Wash sections twice for 5 min in 1X PBS in a Coplin jar.
9. Incubate sections in 10% normal goat serum in a humid chamber on orbital shaker for 1 h at RT for blocking.
10. Wash twice in 1X PBS for 5 min.
11. Apply 20 µg/mL solution of MAC-3 antibody in 1% BSA/PBS overnight at 4°C.
12. Wash twice in 1X PBS for 5 min.
13. Apply 1:400 dilution of goat anti-rat IgG conjugated to Alexa 647 in 1% BSA/PBS for 30 min at RT in the dark.
14. Rinse slides in 1X PBS for 5 min at RT.
15. Rinse twice with dH_2O for 1 min.
16. Mount section with cover slip using AquaPolyMount.

3.1.2.3. Anti-SP-C; Rabbit Prosurfactant Protein C Polyclonal Antibody, Detects Alveolar Type II Epithelial Cells in Mouse Lung Tissue

1. Take slides out of –80°C and immediately place them in 4% PFA for 10 min at RT.
2. Wash slides in 1X PBS twice for 5 min.
3. Place slides into –20°C methanol for 10 min.
4. Wash slides in 1X PBS twice for 5 min.
5. Circumscribe the tissue section with a Pap-pen.
6. Place the slides in a humid chamber.
7. For antigen retrieval, apply 1% SDS for 5 min on each section.
8. Wash sections twice for 5 min in 1X PBS in a Coplin jar.
9. Incubate sections in 10% normal goat serum in a humid chamber on orbital shaker for 1 h at RT for blocking.
10. Wash twice in 1X PBS for 5 min.
11. Apply 3 µg/mL Solution of anti-SP-C antibody in 1% BSA/PBS overnight at 4°C.
12. Wash twice in 1X PBS for 5 min.

13. Apply 1:400 dilution of goat anti-rabbit IgG conjugated to Alexa 647 in 1% BSA/1X PBS for 30 min at RT in the dark.
14. Rinse slides in 1X PBS for 5 min at RT.
15. Rinse twice with dH_2O for 1 min.
16. Mount section with cover slip using AquaPolyMount.

3.1.3. Proliferation Marker: Ki-67; Clone TEC-3, Cell Proliferation Marker, Rat Anti-Mouse Monoclonal

1. Take slides out of –80°C and immediately place them in 4% PFA for 10 min at RT.
2. Wash slides in 1X PBS twice for 5 min.
3. Place slides into –20°C methanol for 10 min.
4. Wash slides in 1X PBS twice for 5 min.
5. Circumscribe the tissue section with a Pap-pen.
6. Place the slides in a humid chamber.
7. For antigen retrieval, apply 1% SDS for 5 min on each section.
8. Wash sections twice for 5 min in 1X PBS in a Coplin jar.
9. Incubate sections in 10% normal goat serum in a humid chamber on orbital shaker for 1 h at RT for blocking.
10. Wash twice in 1X PBS for 5 min.
11. Apply 1:25 dilution of Ki-67 antibody in 1X PBS overnight at 4°C.
12. Wash twice in 1X PBS for 5 min.
13. Apply 1:300 dilution of goat anti-rat IgG conjugated to Alexa 647 in 1% BSA/PBS for 30 min at RT in the dark.
14. Rinse slides in 1X PBS for 5 min at RT.
15. Rinse twice with dH_2O for 1 min.
16. Mount section with cover slip using AquaPolyMount.

3.2. Double and/or Triple Labeling

To simultaneously detect two or three primary antibodies raised in different host animal species (e.g., rabbit polyclonal p-ERK, mouse monoclonal PKCδ, and rat monoclonal Ki-67 (*see* **Fig. 1**; *see* Color Plate 3, following p. 274), antibodies are combined together and applied on preblocked slides (*see* **Notes 1,5,** and **8**) overnight at 4°C in a humid chamber. For detection of primary antibodies, a cocktail of two or three fluorophore-conjugated secondary antibodies can be applied to the sections or, alternatively, a sequential incubation with each of the secondary antibodies separately (*see* **Notes 3,9**). For examples of labeling techniques described above, *see* **refs.** ***2*** and ***3***.

3.3. Fluorescent Detection of Cell Nuclei

There are a number of fluorescent dyes for cell nuclei, including SYTOX Green and propidium iodide (PI). SYTOX Green, a green nucleic acid-binding fluorescent dye, is also useful to demonstrate the general cellularity of a tissue. Diluted in 1X PBS to 5 μ*M*, it is applied directly on a tissue section for 15 min

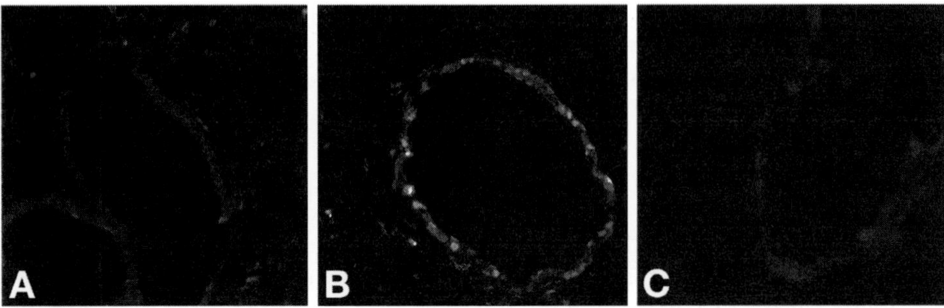

Fig. 1. Triple-labeling using mouse monoclonal PKCδ (blue), rabbit polyclonal p-ERK (red), and rat monoclonal Ki-67 (green) antibodies on sham animals **(A)** and animals exposed to crocidolite asbestos for 4 d **(B,C)**. White pixels appear at the areas of colocalization of all three antibodies. Colocalization of blue and red colors demonstrate colocalization of p-ERK and PKCδ. A negative control omitting primary antibody is also provided and shows no fluorescence signal **(C)**. (*See* Color Plate 3, following p. 274.)

after the excess secondary antibody has been washed off. The sections should then be washed at least two times for 10–15 min in 1X PBS. SYTOX Green is excited by 488-nm wavelength light (*see* **Note 10**), leading to a potential overlap with tissue autofluorescence. However, because the signal emission of SYTOX Green is quite intense, autofluorescence is not a major problem. In addition, this leaves two wavelengths free for staining with antibodies against proteins of interest. (*See* **Note 11.**)

Propidium iodide (PI), a nuclear stain for cells in the G1 phase of the cell cycle, can also be used as a nuclear marker of cellularity. The working solution consists of 10 μg/mL of PI and 10 μg/mL of RNase A diluted in 1X PBS. Tissue sections should be incubated for 30 min at RT in PI, followed by two 10-min washes in 1X PBS. PI is excited by yellow light, resulting in red fluorescence emission.

4. Notes

1. Depending on what staining procedures the sample is going to be used for, the tissue might receive PFA fixation prior to OCT. If this is the case, the PFA fixation step can be withdrawn from the staining procedure. No other changes are needed.
2. Always optimize antibody dilutions using recommended concentrations as a guideline. Every system is slightly different, and trial-and-error experiments might be required to optimize the antibody or a technique for your specific system. Try doing a dilution series for all primary and secondary antibodies.
3. There are a few approaches to overnight incubation with a primary antibody. Incubations can be done at 4°C in a refrigerator or in a cold room on an orbital shaker to assure even application of the antibody. If an orbital shaker is used, cover the section with a piece of parafilm to ensure even distribution and minimal

evaporation of the antibody solution. As a rule, most primary antibody incubations can also be performed for 1 h at RT on an orbital shaker. However, it is not generally recommended because of the possible loss of quality in labeling and resulting unevenness of binding.

4. The background signal can also be decreased by adding a second blocking step using the same blocking reagent after washing away the primary antibody and prior to incubation with the secondary antibody.
5. To amplify the signal, a biotin/strepavidin system can be employed. Biotin has a very specific affinity for strepavidin/avidin that allows a decrease in background fluorescence as a result of nonspecific binding of a secondary antibody. This system is especially beneficial when doing multiple labeling to match the signal intensities of different antibodies.
6. Remember to prewarm the AquaPolyMount to RT prior to use to avoid bubbles when applying to a tissue section. Other mounting media containing antifade reagents are widely available from commercial sources.
7. After the cover slip is mounted, remove excess medium around the cover slip on the slide and/or on the cover slip itself by gently blotting with a bibulous paper.
8. In the case of double or triple labeling, the possibility exists that different fixation conditions are required for preservation of the various antigens. Try to perform the entire staining procedure using a fixative that is specifically required for one of the antibodies. If a required technique is much harsher, attempt to do a sequence labeling by adding a second blocking step to decrease the fluorescent background and to even out the balance between signaling intensities.
9. When blocking sections for multilabeling using an antibody that requires specific reagents (e.g., colabeling of a rabbit polyclonal antibody with a mouse monoclonal antibody), label both probes using only reagents from a mouse-on-mouse kit needed for mouse antibody to simplify the procedure. If binding of a nonmurine-raised antibody fails, perform a sequence labeling using different blocking reagents and antibody diluents for each probe.
10. SYTOX Green can emit at the higher wavelength and be detected by the red channel. This leads to detection of nuclear labeling using a wide-band green filter. However, it does not cause any problems while imaging with the confocal microscope using the sequential scanning mode.
11. To minimize or avoid autofluorescence, use those probes that fluoresce at higher wavelengths (i.e., 568 nm and 647 nm rather than 488 nm).

References

1. Taatjes, D. J., Palmer, C. J., Pantano, C., Hoffmann, S. B., Cummins, A., and Mossman, B. T. (2001) Laser-based microscopic approaches: application to cell signaling in environmental lung disease. *Biotechniques* **31(4),** 880–894.
2. Cummins, A. B., Palmer, C., Mossman, B. T., and Taatjes, D. J. (2003) Persistent localization of activated extracellular signal-regulated kinases (ERK1/2) is epithelial cell-specific in an inhalation model of asbestosis. *Am. J. Pathol.* **162(3)**, 713–720.

3. Lounsbury, K. M., Stern, M., Taatjes, D., Jaken, S., and Mossman, B. T. (2002) Increased localization and substrate activation of protein kinase C delta in lung epithelial cells following exposure to asbestos. *Am. J. Pathol.* **160(6)**, 1991–2000.
4. Schmidt, M. H., Broll, R., Bruch, H. P., Bogler, O., and Duchrow, M. (2003) The proliferation marker pKi-67 organizes the nucleolus during the cell cycle depending on Ran and cyclin B. *J. Pathol.* **199(1)**, 18–27.
5. Carson, F. L. (1996) *Histotechnology: A Self-Instructional Text*, ASCP Chicago, IL.
6. Brown, D., Lydon, J., McLaughlin, M., Stuart-Tilley, A., Tyszkowski, R., and Alper, S. (1996) Antigen retrieval in cryostat tissue sections and cultured cells by treatment with sodium dodecyl sulfate (SDS). *Histochem. Cell. Biol.* **105(4),** 261–267.
7. Wilson, D. M., 3rd and Bianchi, C. (1999) Improved immunodetection of nuclear antigens after sodium dodecyl sulfate treatment of formaldehyde-fixed cells. *J. Histochem. Cytochem.* **47(8),** 1095–1100.

5

Evaluation of Confocal Microscopy System Performance

Robert M. Zucker

Summary

The confocal laser scanning microscope (CLSM) has enormous potential in many biological fields. When tests are made to evaluate the performance of a CLSM, the usual subjective assessment is accomplished by using a histological test slide to create a "pretty picture." Without the use of functional tests, many of the machines could be working at suboptimal performance levels, delivering suboptimum performance and possibly misleading data. To replace the subjectivity in evaluating a confocal microscope, tests were derived or perfected that measure field illumination, lens clarity, laser power, laser stability, dichroic functionality, spectral registration, axial resolution, scanning stability, photomultiplier tube quality, overall machine stability, and system noise. These tests will help serve as a guide for other investigators to ensure that their machines are working correctly to provide data that are accurate with the necessary resolution, sensitivity, and precision. Utilization of this proposed testing approach will help eliminate the subjective nature of assessing the CLSM and allow different machines to be compared. These tests are essential if one is to make intensity measurements.

Key Words: Confocal microscope; lasers; coefficient of variation; photomultiplier tubes; field illumination; axial resolution; spectral registration; laser stability; beads; microscope lenses; quality assurance; quantification; spectroscopy

1. Introduction

The confocal laser scanning microscope (CLSM) has been evaluated by using the following biological test samples: beads, spores, pollens, diatoms, fluorescent plastic slides, fluorescence dye slides, silicone chips, and histological slides from plants or animals ***(1–10)***. In most cases, the test sample is of biological origin and is sometimes used in the course of research in the individual's research laboratory. This is a testing procedure recommended by the manufacturers of most CLSM equipment. In our opinion, this is too arbitrary a test when applications (i.e., intensity measurements or colocalization studies) other than "pretty pictures" are needed. Unfortunately, this technology differs from flow

From: *Methods in Molecular Biology, vol. 319: Cell Imaging Techniques: Methods and Protocols*
Edited by: D. J. Taatjes and B. T. Mossman © Humana Press Inc., Totowa, NJ

cytometry and does not have a universal standard with which to evaluate the CLSM or the image quality derived from a CLSM. It would be advantageous to have better methods to measure system performance and image quality.

The CLSM consists of a standard high-end microscope with very good objectives, different lasers to excite the sample, fiber optics to deliver the laser light to the stage, acoustical transmission optical filters (AOTFs) to regulate the laser light onto the stage, barrier filters, dichroics, and pinholes to control the light, electronic scanning devices (galvanometers), detection devices to measure photons (i.e., photomultiplier tubes [PMTs]) and various other electronic components. For this system to operate correctly, it is important for it to be properly aligned and to have all of the components function correctly. Instrument performance tests that have been devised include the following: laser power, laser stability, field illumination, spectral registration, lateral resolution, axial *Z* resolution, lens cleanliness, lens functionality, and *Z*-drive reproducibility ***(1–10)***. This list is not inclusive and additional parameters might be needed to assess if the CLSM is working properly.

Because a confocal microscope can provide spectacular three-dimensional (3D) data of biological structures, one can have a tendency to overlook many of the quality assurance (QA) parameters that might be necessary as controls. The CLSM could function at suboptimum conditions for long periods of time, delivering inferior data, with the problems being resolved only when the investigator cannot achieve the desired images or there is a hard failure of the system necessitating a service personnel visit. Inferior performance of a CLSM could be attributed to sample preparation. However, the image data might also be incorrectly interpreted if the CLSM is not working correctly. Because all CLSM images are digital and made with sophisticated optical equipment, it is now possible to derive tests that can evaluate some of the components installed in the machine. The CLSMs should not be evaluated by using only a subjective "pretty" image on a histological slide. QA on the CLSM is essential to ensure that it is performing properly and delivering accurate and reproducible data. This review attempts to incorporate QA procedures into the operation/maintenance of confocal microscopes. This review also emphasizes that scientists need to evaluate their CLSM system performance to ensure that it is working properly.

2. Materials

2.1. Field Illumination: Fluorescent Slides

The field illumination test slides consisted of three fluorescent plastic slides (Delta; Applied Precision Inc, Issaquah, Washington) that had excitation peak wavelengths of 408 nm (blue), 488 nm (orange), and 590 nm (red) and emission peak wavelengths of 440 nm, 519 nm, and 650 nm, respectively. The orange slides (488 nm) were used to test for visible field illumination and alignment. The blue slides (408 nm) were used for ultraviolet (UV) field illumination and alignment. Field illumination can also be measured using four Fluor-ref slides (Microscopy Education, Springfield, MA) or four Chroma slides (Chroma, Brattleboro, VT). Although extensive tests were not made with these slides, in preliminary studies they appear to work well if the surface is clean and free of debris.

2.2. Power Meter

The power meter used to measure light on the microscope stage was either a Fieldmaster or a Lasermate Q (Coherent, Auburn, CA) with visible (LN36) and UV detectors (L818). A power meter (1830C) from Newport Corporation with an SL 818 visible wand detector can also be used for power measurements. A remote control box for the Coherent UV Enterprise laser was used to regulate UV laser power (model 0163-662-00; Coherent, Santa Clara, CA). On most confocal systems, there is a 10× lens: Zeiss uses a 10× Plan Neofluar (numerical aperture [NA] of 0.3) and a Leica has a 10× Plan Fluorotar (NA of 0.3) or 10× Plan Apo (NA = 0.4). The dry 10× lens was used to take power measurements. The machinist's plans for building the power detector holder are available by e-mailing the author.

2.3. Beads

Various bead types are useful as test particles to access machine functionality. The beads were obtained from Spherotech (Libertyville, IL), Molecular Probes (Eugene, OR) or Coulter electronics (Hialeah, FL).

Spherotech beads that were used included the 10-µm Rainbow (EX 365, 488, 568) fluorescent particles (FPS-10057). The following beads were used for preliminary field illumination tests: yellow beads (5.5 µm, FPS-5052) EX 488 for visible field illumination; UV beads (5.5 µm, FPS-5040) EX 365 for UV field illumination; blue beads (5.5 µm FPS-5070, EX 647) for 647 nm field illumination. The 6.2-µm Rainbow beads with three different intensities (FPS-6057-3) were used for early statistical PMT tests. The following PSF Rainbow beads were used: 0.16 µm, FP-02557-2s; 0.5 µm, FP 0857-2; and 1.0 µm, FP-0557-2. The polystyrene 10-µm beads (refractive index [RI] = 1.59) were mounted with optical cement (RI=1.56) on a slide using a no. 1.5 size cover glass. The Leica immersion oil has a refractive index of 1.518.

The following Molecular Probes beads were used: TetraSpeck™ beads (T7282 0.5 or 1 µm, EX 365, 488, 568, 647) were used for spectral registration tests and point spread functions (PSF); PS-Speck Microscope Point Source Kit, consisting of beads (175 nm, P-7220) of different wavelengths, were used for acute deconvolution PSF measurements; 15-µm FocalCheck microspheres (F24634 kit) consisting of an orange ring and blue throughout (F7236) for UV and visible colocalization or green, orange, and dark red ring stains (F7235) for visible colocalization of the 488, 543, and 633 laser lines. Fluorospheres (10 µm, Fullbright

Green II; Coulter, Hialeah, FL; EX 488) were used for preliminary statistical PMT tests. The 10-μm Spherotech Rainbow beads were substituted for these beads in later experiments.

Bead slides were made by dropping 3–5 μL of diluted beads onto a slide, allowing the liquid to dry and then covering the spot with Permount, glycerol, water, or oil and sealing it with a 1.5 cover glass. Antifade from Vector (Vectoshield H-1000) or from Molecular Probes (Slowfade light S-7461) is useful to decrease bleaching.

2.4. Biological Test Slides

FluoCells 1 (F-14780; Molecular Probes, Eugene, OR) were stained with three fluorochomes (Mitotracker Red CMXRos, BODIPY FL phalloidin, DAPI) and used as biological test slides. Additional slides were made in our laboratory with cells grown on cover slips, fixed with paraformaldehyde, and stained with DAPI for UV excitation or other suitable fluorochromes for visible excitation.

2.5. Axial (Z) Resolution Test

The axial resolution of the CLSM is tested using a single reflecting mirror obtained from Leica or Edmonds Scientific. A 21-mm square (#31008; Edmonds Scientific, Philadelphia, PA) was glued onto a microscope slide and a cover glass (#1.5; Fischer, Pittsburgh, PA) was placed on top of the slide with a drop of immersion oil (Leica Immersion oil, n = 1.518) The cover slip is placed firmly onto the mirror to remove all excessive oil. This type of standard test slide can also be obtained from a confocal manufacturer or Spherotech (Libertyville, IL).

2.6. Square Sampling Galvanometer Check

It is important to ascertain whether there was square sampling or rectangular sampling in an image. A computer chip was glued onto a glass slide and used as a test substrate. A commercial grid product can also be obtained from MicroBrightField (Williston, VT) or Geller MicroScientific (Topsfield, MA). A digital tagged image file format (TIFF) image was obtained using a dry 20× objective and the amount of small boxes observed was counted by eye in the vertical and horizontal directions. If there is the same number of boxes per inch in the vertical and horizontal directions, then it can be assumed that the sampling of pixels is square. If they are not equivalent, then the sampling of pixels is rectangular, which is not desired. Problems with the galvanometer can also be detected with this grid.

2.7. Laser Beam Shape

The UV laser beam can be checked using an inexpensive lens (12 mm outer diameter [od]; B1099; Melles Griot) held in a lens holder (13 mm inner diameter [id]; H1089,) and focusing the beam onto a white piece of paper to show its configuration mode.

2.8. Confocal Microscope

The majority of data presented in this chapter was derived on either a Leica TCS-SP1 or a Leica TCS4D (Heidelberg, Germany) confocal microscope system. These systems contained an argon–krypton laser (Melles Griot; Omnichrome) emitting 488-, 568-, and 647-nm lines and a Coherent Enterprise UV laser emitting 351- and 365-nm lines. The system contains an AOTF and the following three dichroics for visible light applications: single dichroic (RSP500); double dichroic (DD); and triple dichroic (TD). The Leica-derived tests were shown to be applicable to other point scanning systems that contain different types of laser, objective, or other hardware configuration. For comparison purposes, similar tests were made on two different Zeiss 510 units containing three lasers (argon 488 [25 mW], HENE 543 [1 mW], and HENE 633 [5 mW]) with a merge module and an AOTF. In addition, for comparison purposes, similar tests were made on a Leica SP2 unit that contained three lasers (argon 488 [50 mW], HENE 543 [1.2 mW], and HENE 633 [10 mW]) with a merge module and an AOTF.

2.9. Software Analysis

The analysis of the images was made on workstations that contained Leica, Zeiss, and Bitplane (Zurich Switzerland) software packages. If necessary, the TIFF images were imported into Image Pro Plus (Media Cybernetics, Silver Springs, MD) or Image J (NIH) for more intensive measurements and analysis.

2.10. PMT Spectral Check

The PMT spectral response was measured over a large spectral region using an inexpensive Pariss fluorescence calibration lamp consisting of a defined mixed-ion gas (model 816025; LightForm Inc, Hillsborough, NJ).

3. Methods

3.1. Field Illumination

The fluorescent slide was placed on the stage and the maximum intensity was found on the surface of the slide. It is important to measure the field illumination at a specific depth in the plastic slide, as the intensity distribution might change from the surface to the interior of the slide. The depth of focus was adjusted between 30 and 100 μm, depending on the objective that was used: (5× [100 μm], 10× [75 μm], 20× [50 μm], 40× [40 μm], 63× [30 μm], 100× [30 μm]). Investigators should also be careful not to observe an illumination field deep within the plastic slide samples, as it will usually yield a better field illumination than regions closer to the surface because of various optical distortion factors.

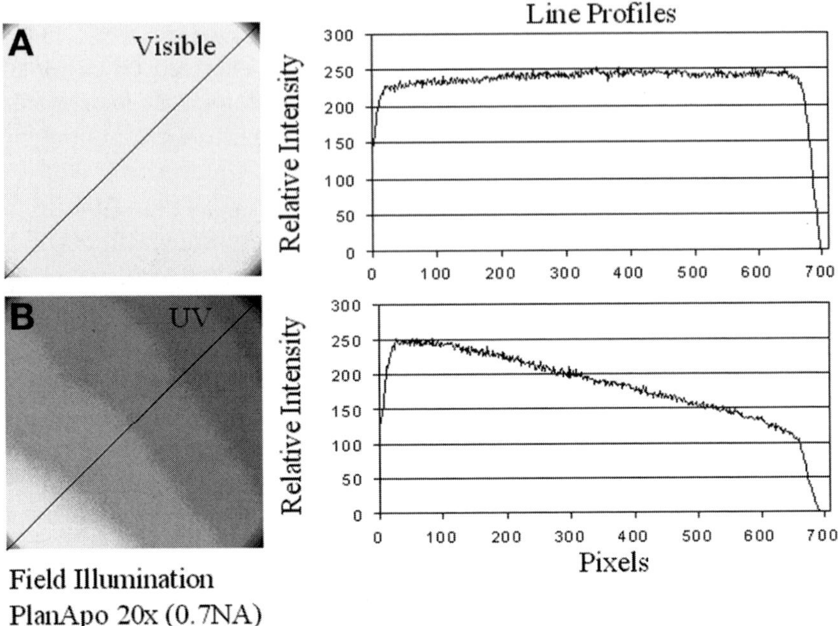

Fig. 1. Field illumination. Field illumination pattern of visible **(A)** and UV **(B)** excitation using a 20× (PlanApo; NA = 0.7) lens. The visible field illumination shows uniform illumination, with the brightest intensity being in the center of the objective. The line running diagonally in panels A and B measures the histogram intensity of the field illumination graphically represented in panels **C** and **D**. The variation in intensity from the left to right side of the field is less than 10% for visible excitation and more than 150% for UV excitation. Acceptable field illumination has the brightest intensity in the center of the objective, decreasing less than 25% across the field. The intensity regions were prepared by using Image Pro Plus to divide the grey scale value into 10 equal regions and a median filter was used for additional processing.

Data derived from a 20× PlanApo lens (NA = 0.7) zoomed to a factor of 1.2 is used to illustrate good visible field illumination (488 nm), and a misaligned UV (365 nm) system is used to yield a poor field illumination (*see* **Fig. 1**). The images were obtained with either a UV plastic slide (excitation [ex] 365 nm; emission [em] 440–480 nm) or a visible plastic slide (ex 488 nm; em 505–550 nm) located securely on the stage. The original images were contoured into 10 intensity ranges using Image Pro Plus software. The line running diagonally in **Fig. 1A,B** measures the histogram intensity of the field that is represented in the graphs in **Fig. 1C,D**.

The maximum intensity should be in the center of the objective and decreasing less than 25% across the field in all directions, as shown with visible 488-nm excitation in **Fig. 1A**. It should not be in the bottom corner, as illustrated with

UV illumination in **Fig. 1B.** As shown in **Fig. 1**, the visible light (**Fig. 1C**) had less than a 10% decease in intensity across the field, whereas the UV light (**Fig. 1D**) had a 150% decrease across the field. If the maximum light intensity is not located in the center of the field, there is an alignment problem that needs to be addressed. The nonuniform pattern shown in **Fig. 1** with UV illumination clearly illustrates a field illumination problem, which will affect intensity measurements in an image. Although **Fig. 1** was obtained with UV optics, it represents the type of field illumination that can also occur with visible excitation. This pattern is unacceptable with any CLSM optical system, as the maximum intensity should be in the center of the objective and not in a corner. Each laser line must be checked to ensure that they are aligned properly, as they use different dichroics to ensure that the beams are colocalized. In addition, the field illumination of one lens is not necessary identical to the field illumination of the other lenses, necessitating that each lens be checked with the suitable dichroic that will be used in the experiment.

In our Leica system, the three visible wavelengths of light are derived from one Omnichrome argon–krypton laser. This enables the field illumination to be tested at one wavelength (488 nm) and allows one to assume that it is equivalent to testing field illumination with the other wavelengths. Because the UV line is derived from a different laser (Enterprise, Coherent), it is essential to check all objectives for proper field illumination (**Fig. 1**) at the 365-nm excitation in addition to the 488-nm excitation. Newer designed confocal systems (Leica SP2, Leica AOBS, Zeiss 510, Zeiss 510 Meta, Nikon C1, Olympus FluoView FV1000, and Bio-Rad Radiance 2100) use three individual lasers with a merge module, which requires that all laser wavelengths have correctly aligned beams emitted from the merge modules. In these systems, all three lines have to be individually tested for field illumination. One laser line might be perfectly aligned, yielding acceptable field illumination, whereas the other laser lines might be misaligned, yielding intensity values for which the brightest region is not in the center of the field, as illustrated in **Fig. 1**.

3.2. Power Meter

The equipment used to acquire power readings include a detector, a machine-shop-built detector holder, and a portable power meter, which are illustrated in **Fig. 2.** To measure the power output of the different wavelengths, either a UV or visible probe (Coherent probe detectors [L818, LN36]) or Newport corporation wand visible probe detector (SL 818) is secured in a special holder that fits onto the microscope stage during the measurement of either UV or visible laser light (*see* **Fig. 2**). The test should be done with a dry objective (2.5× to 20×; preferably 10×) at a fixed position, usually at the top of its movable tract.

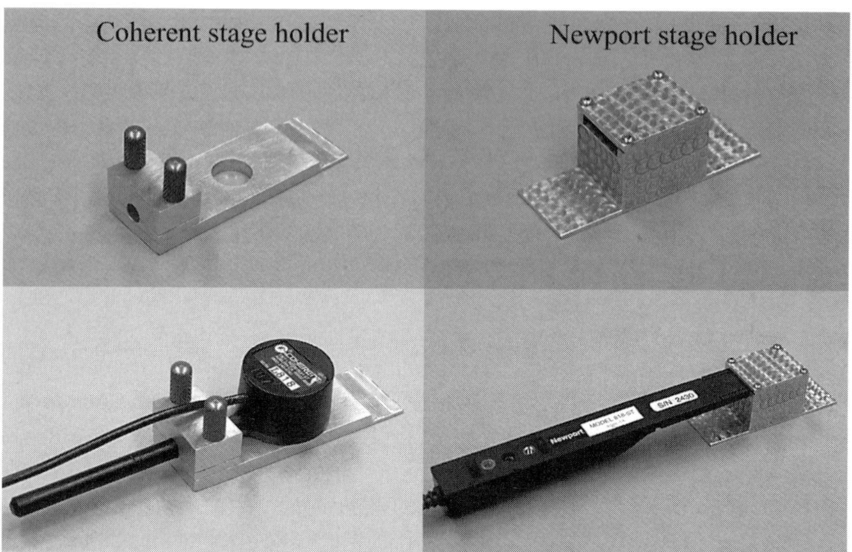

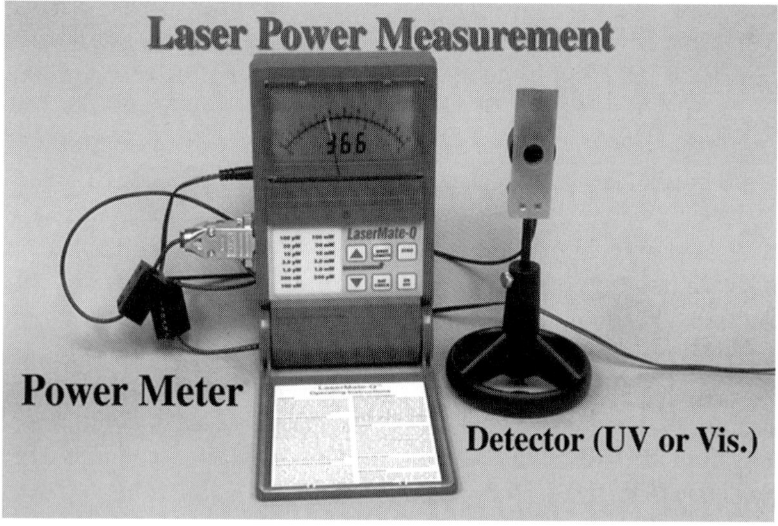

Fig. 2. Power meter and detector holders. A holder to secure detectors on the microscope stage was made in the machine shop in a size similar to a 3×1 microscope slide. The detector holders for the Coherent (LN 36) or Newport wand (SL818) are represented in the photographs. With small objective circumferences (i.e., Leica 10× PlanApo (NA = 0.4), the Coherent detectors can be placed on top of the objective. A Coherent power meter and detector are used to measure laser power on the microscope stage.

Currently, most systems will have a dry 10× lens (NA = 0.3 or 0.4) that can be used to access power using a power meter detector. The use of a lens that has a different lens design, magnification, or NA will affect the laser power transmission and measurement.

The test was made using a 10× (NA = 0.3) or a 10× (NA = 0.4) objective in the following manner. The lens is raised to its maximum specified height. The detector is secured on the stage and grossly centered using either laser light or mercury fluorescent light. The detector position is then more accurately adjusted to achieve maximum signal intensity by using the microscope's *x*/*y* joystick. The power meter is adjusted to the specific wavelength (365, 488, 568, or 647 nm) and the maximum power of laser light is read on the digital scale. The CLSM zoom factor is set from 8 to 32 to reduce the beam scan and to focus it into the "sweet spot" of the detector. The scanner is set at bidirectional slow speed to reduce the time period that the power meter is reading 0. The maximum digital reading from the power meter was recorded. The power derived from this measurement is dependent on the magnification and NA of the lens used. Each lens will have a unique set of values, which is dependent on the objective's NA and other transmission factors. The power meter diode was not reliable and could only be used as a crude estimate of the functioning of the laser. This might change in the future with better designed systems. The Coherent probe detectors LN818 and LN36 can also be placed directly on top of a 10× objective without a holder.

A comparison of the maximum power output derived from different lasers and different optical systems was made on a Zeiss 510, a Leica TCS-SP1, and a Leica TCS-SP2. The maximum power was measured with a Coherent power meter using two different 10× (NA = 0.3, SP1 and Z510) lenses and one 10× (NA = 0.4, SP2) lens on three different types of CLSM system: Zeiss 510, Leica TCS-SP2, and Leica TCS-SP1. The Leica TCS-SP1 system has one 75-mW argon–krypton laser (model 643) emitting three laser wavelength lines. The newer CLSM systems from all manufacturers are designed with three lasers that use different dichroic components to merge the different laser wavelengths. The Zeiss 510 contained three different lasers (30-mW argon, 1 mW HeNe [543 nm], and 5-mW HeNe [633 nm]) with the multiple wavelengths aligned with a merge module. The Leica SP2 contained three different lasers (50-mW argon, 1.2-mW HeNe [543 nm], and 10-mW HeNe [633 nm]) with the multiple wavelengths aligned with a merge module.

Table 1 shows a comparison of power with three different systems from Leica and Zeiss using a triple dichroic to reflect different wavelengths of laser light to the stage. The power meter values shown in **Table 1** can be used to determine both the maximum power output and dichroic reflectance and, thus, functionality in the system. The values reported in **Table 1** were acquired from functioning systems that were subjectively assessed to be problem-free, aligned correctly, and functioning properly. There are no assurances that these machines were producing ideal power readings or were perfectly aligned when these tests were made. However, we believe that the values demonstrate the type of data that can be achieved with this test. The acquired data can be used to compare

Table 1
Comparison of Relative Power on Microscope Stage

	Leica SP1	Leica SP2	Zeiss 510
Laser	Ar–krypton	(3-Laser)	(3-Laser)
488 nm	1.1	4.5	3.13
543–568 nm	1.45	0.22	0.26
633–647 nm	1.6	2.8	1.21

Note: The maximum power was measured in milliwatts with a LN 36 detector and a Coherent Lasermate power meter adjusted to the specific excitation wavelengths. The power was measured on the stage of a Zeiss 510, a Leica SP2, and a Leica SP1 system. This test can be used to determine if the system has acceptable laser power by measuring the power on the microscope stage and not opening up the scan head.

similar machines and internally control an individual machine for proper performance over a period of time.

3.3. UV Power Test

The test was carried out in a manner similar to that described for the visible power measurement. A power meter (Lasermate/Q with UV detector [L818]; Coherent, Santa Clara, CA) was used to measure the light emitted from a UV Enterprise laser. A Coherent UV, 60-mW, 3-yr-old Enterprise laser delivered normal power output at the laser head (over 40 mW of laser power), but only about 500 μW of maximum power through a Plan Fluor 10× (NA = 0.3). However, it should be noted that when our system had insufficient output under these conditions (approx 500 μW through a 10× Leica [NA = 0.3] lens), we also had insufficient light for many UV experiments using higher magnification objectives (40×, 63×, and 100×).

3.4. UV Beam Shape

In addition to power requirements of a UV system, it is important that the UV beam have the correct mode. The beam should be radially symmetric with a Gaussian intensity distribution. The beam should have a TEM00 configuration (transverse excitation mode or Gaussian mode). The UV laser beam can be checked using an inexpensive lens (12 mm od; B1099; Melles Griot) held in a lens holder (13 mm id; H1089) and focusing the beam onto a white piece of paper to show its configuration mode.

3.5. Bead or Histological Power Meter

The crude power of the system can also be assessed by recording the PMT voltage necessary to acquire an image at almost saturation values by using standard histological samples like the FluoCell 1 slide (F-14780; Molecular Probes,

Eugene, OR), beads like the 10-µm Spherotech beads (FPS-10057-100), or Applied Precision fluorescent plastic slides. If conditions are identical between machines, this PMT value can be used as a reference value to compare CLSM units and to establish their acceptable performance levels. Leica technicians routinely use a 40× lens to measure the fluorescence saturation of a histological plant sample. If it saturates in the PMT range between 600 and 700 units in PMT 1, the system is passed as having adequate power.

Because of the optical limitations of the stage, the power meter detector cannot be used with higher power optics (40×, 63×, 100×). Thus, it will be useful to use a test sample (fluorescence histological slide, colored plastic, or large beads) to assess UV and visible power with these higher-magnification lenses. Using maximum UV power, it was found that the 10-µm Spherotech bead saturates PMT 1 (low-noise PMT, R6857) at a voltage setting of 650 using a 100× PlanApo lens (NA = 1.4). When saturation occurred at the higher values, it indicated less power throughput in the CLSM system, and when saturation occurred at relatively lower PMT values, it indicated greater power throughput. The saturation values on other substrates or slides (biological or histological specimens) can also be used and have to be defined by the user's laboratory, as the degree of staining can be different.

3.6. Dichroic Functionality

The dichroic comparison test was accomplished in the following manner. An API fluorescent plastic slide (orange) was placed on the stage using either 488-nm or 568-nm excitation light. After the dichroic was switched into position, the PMT was kept constant and the mean gray scale value (GSV) of a region of interest (ROI) in the image was determined for each acquisition condition. The values for the three dichroics can be compared to determine which one has the highest reflectivity (*see* **Table 2**). The maximum values are normalized to 1 and the other values are reported as a percentage of the maximum GSV. It is important to use a bandpass or barrier filter to collect the desired light. Light at different regions of the spectra may have unique information and it is sometimes advisable to use a narrower bandpass with higher reflection to get the desired fluorescent information.

3.7. Axial (Z) Resolution (Mirror)

The axial resolution test is considered the "gold standard" of resolution in confocal microscopy (***1,3,7,8***). Although it is not the only criterion for a good image, the axial resolution of the system should be maximized to yield a minimal axial Z-resolution value. The reflected surface of the mirror can be found in either *xy* or *xz* scanning by observing the brighter spot with an open aperture. Initially, axial resolution is tested in the reflection mode with the 100× objective

Table 2
Comparison of Relative Dichroic Reflectance

Wavelength	Dichroic	10× (0.3/NA)	63× (1.2/NA)
488	SD	0.95	0.97
	DD	1.00	1.00
	TD	0.84	0.80
568	SD	0.05	0.06
	DD	0.62	0.72
	TD	1.00	1.00

Note: The relative laser power was measured with the 488-nm and 568-nm wavelengths using six different magnification objectives and three dichroics. The 10× and 63× data are shown for clarity. The table demonstrates the relative reflectivity of the dichroics in the system. The test was accomplished by measuring the intensity in an ROI of an image using either 488-nm or 568-nm excitation light, a fluorescent plastic slide, and one of three specific dichroic and maintaining the PMT at a constant voltage. An ROI of the image yielded the mean value for each acquisition condition (wavelength, objective, and dichroic). The GSV of the two images are divided to yield a ratio that is expressed in the table as a fraction. The value of 1.00 is the maximum dichroic refection and is expressed as a bold number. The dichroic with the maximum reflection should be used when only one fluorochrome is required. Unexpectedly, the double dichroic (DD) yielded the best reflectivity with 488-nm wavelength light with most lenses and the triple dichroic (TD) yielded the best reflectively with the 568 nm wavelength light (30% more light reflected than the DD) with most objectives. This table can be used to choose the dichroic that should be used with each excitation wavelength for optimized reflection.

(NA = 1.4; PlanApo lens), zoom of 16× to 24×, a large pinhole diameter opening, and minimum laser power. After the reflected surface is found by scanning in the *xz* mode, the pinhole aperture is reduced to a minimum size. The reflected image is then obtained with a frame averaging of 2–4, and the intensity profile across the reflected surface is determined as shown in **Fig. 3.** The maximum of the peak is determined and then the half-maximum intensity value of the profile is obtained to determine the full width at half-maximum (FWHM) distance to determine the axial resolution. The data can be observed graphically or it can be transferred into Excel to measure the peak and the half-maximum values. The specification for axial registration in a Leica TCS-SP system is below 350 nm. The pattern of the axial resolution curve is also important. One looks for a symmetrical large peak and smaller peaks and valleys to the left of it, indicating diffraction patterns of an acceptable lens (**Fig. 3B**).

The axial *Z*-resolution of three different lenses on an aligned Leica TCS-SP1 system was the following: a 40× (Fluor, NA = 1.0) was 610 nm; a 63× water-immersion lens (PlanApo, NA = 1.2) was 390 nm; and a 63× oil-immersion lens (PlanApo, NA = 1.32) was 315 nm. These are good values for high-resolution work on any confocal microscope. These axial registration values will change depending on lens quality and system alignment.

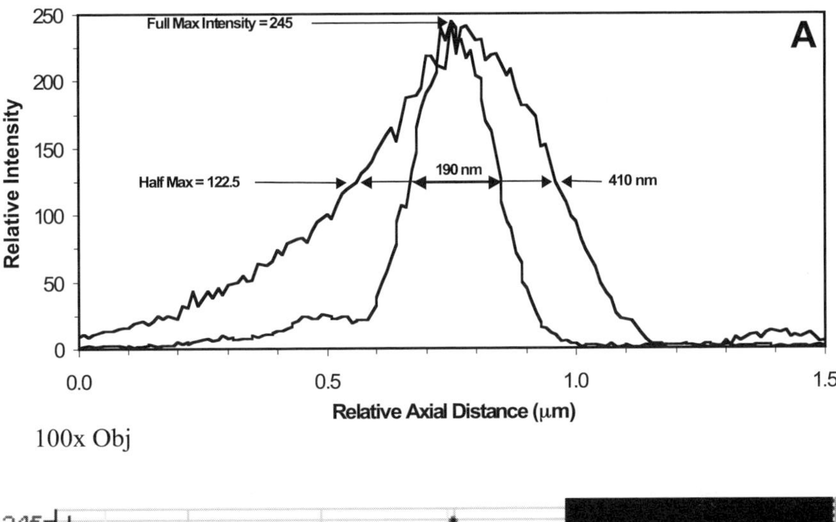

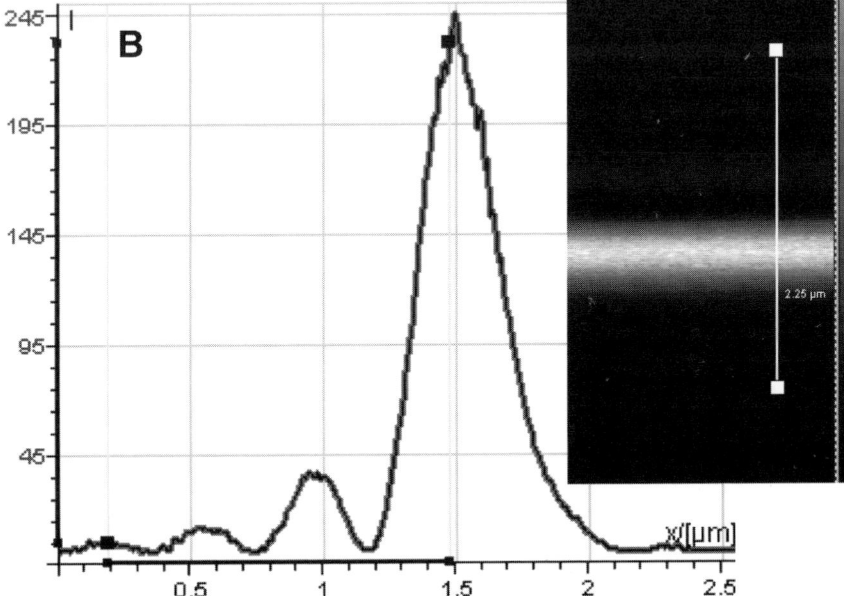

Fig. 3. **(A)** Axial resolution. The axial resolution was made with two 100× lenses (NA = 1.4) on the same Leica TCS-SP1 confocal system. The peak intensity of the histogram is 245 and the half-maximum intensity is measured at 122.5. One lens gave an excellent full width at half-maximum (FWHM) of 190 nm, whereas the other lens yielded a poor value of 410 nm. The system was aligned properly in both cases. **(B)** The axial resolution of a PlanApo 63× (NA = 1.32, 330 nm) showing a symmetrical major peak and a diffraction pattern consisting of smaller peaks and valleys. This pattern is suggestive of an excellent-quality lens.

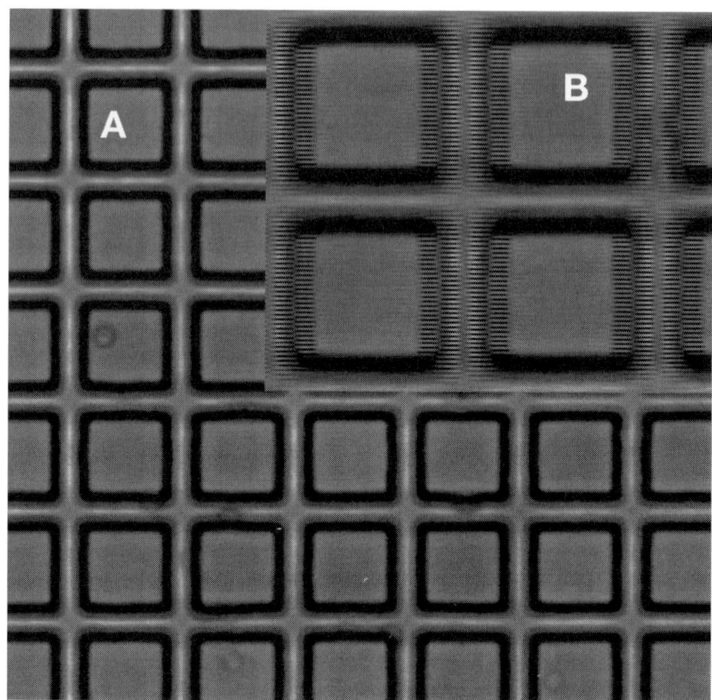

Fig. 4. Field alignment. A computer chip that is mounted on a slide shows 25 boxes per inch. The *x* and *y* difference should be equivalent, yielding square boxes and not rectangles. To align the system for bidirectional scanning, the scanning lines should show a smooth transition **(A)** and not the individual scan lines in the inset **(B)**.

3.8. Axial (Z) Registration (Beads)

One-micrometer beads from Molecular Probes (Tetraspec, T7284) or Spherotech (Rainbow, FP–0857-2) are first located in the *xy* direction and then scanned in the *xz* direction. The power is adjusted for saturation and then they are zoomed approx 8× and averaged: 4×. The size of the bead in the horizontal is compared to the vertical size. The difference between the two numbers will yield the axial resolution of the lens. This method is slightly more subjective than the gold standard axial mirror test, but it does yield similar values. Sometimes the values for unknown reasons are better and sometimes they are worse.

3.9. Square Pixels, Phase Alignment, and Galvanometer Evaluation

The pixel size and symmetry in *XY* directional field scanning can be checked by using a computer chip attached to a glass slide or a slide obtained from Microbright Field or Geller MicroScientific (*see* **Fig. 4**). The confocal should be set up in the reflective mode using a 10× or 20× dry lens. The small boxes in the vertical and horizontal direction should be compared by counting or by a measured standard line. If there are the same number of boxes per inch in the vertical and horizontal directions, then it can be assumed that the sampling of pixels is a square. If they are not equivalent, then the sampling of pixels will be rectangular, which is not desired. This test ensures that the scanning in the *X* and *Y* directions yields a perfect square and the information will be registered correctly. This test can be used to assure that alignment exists in bidirectional scanning. This test can also be used to evaluate galvanometer function. The vertical lines should be straight without wiggles. One should be equivalent to multiple averaged scans.

3.10. Spectral Registration With Beads (1 μm) (UV and Visible)

The 1 μm multiple wavelength fluorescent beads (Tetraspec, T7284 Molecular Probes, or Rainbow beads, FP–0857-2 Spherotech) were used to monitor the visible spectral registration of lenses (100× PlanApo, NA = 1.4; 63× PlanApo, NA = 1.2; PlanFluor 40×, NA = 1.0) or the registration between multiple beams (UV and 568 nm in our case). The bead was located at low zoom in *xy* and the gain and offset were adjusted to their respective optimum image quality levels. Next, an *xz* scan was obtained at the proper zoom magnification. Care was made to make adjustments at the lower power levels to reduce possible bleaching effects. By balancing laser light intensity with the AOTF, the laser crossover between the detection channels was minimized. To check for laser crosstalk, one laser light line is closed with the AOTF and the signal is observed in the other channels with the PMT setting that is necessary to acquire the proper signal with the respective excitation lines. Subjectively, it can be ascertained how much crossover occurs. If too much crossover exists, the test will be invalidated. The bead was imaged (*xy* and *xz* scans) with an 8× to 24× zoom, a slow to medium scanning rate, and frame averaged four to eight times. The registration of bead fluorescence images between the 365-nm UV wavelengths and the 568-nm visible wavelengths in an aligned system was almost superimposable (*see* **Fig. 5A**). **Figure 5B** shows the *xz* registration between the 365-nm UV line and the 568-nm visible line in a misaligned system. The data for **Fig. 5B** were acquired at a different time using the same 100× lens, suggesting that the system is misaligned, as the difference between the peaks was 650 nm (acceptable difference is 210 nm). The 568-nm line was chosen instead of the 488-nm line to minimize the crossover fluorescence between the visible and UV wavelengths.

3.11. Focal Check Beads (UV and Visible Lines)

Molecular Probes produces a series of beads (Focal Check) that can be used to assess the different visible wavelengths from multiple lasers in a confocal system. These beads have different fluorescent excitation rings and core colors that can be used to assess colocalization of laser light from multiple lasers. We used these beads

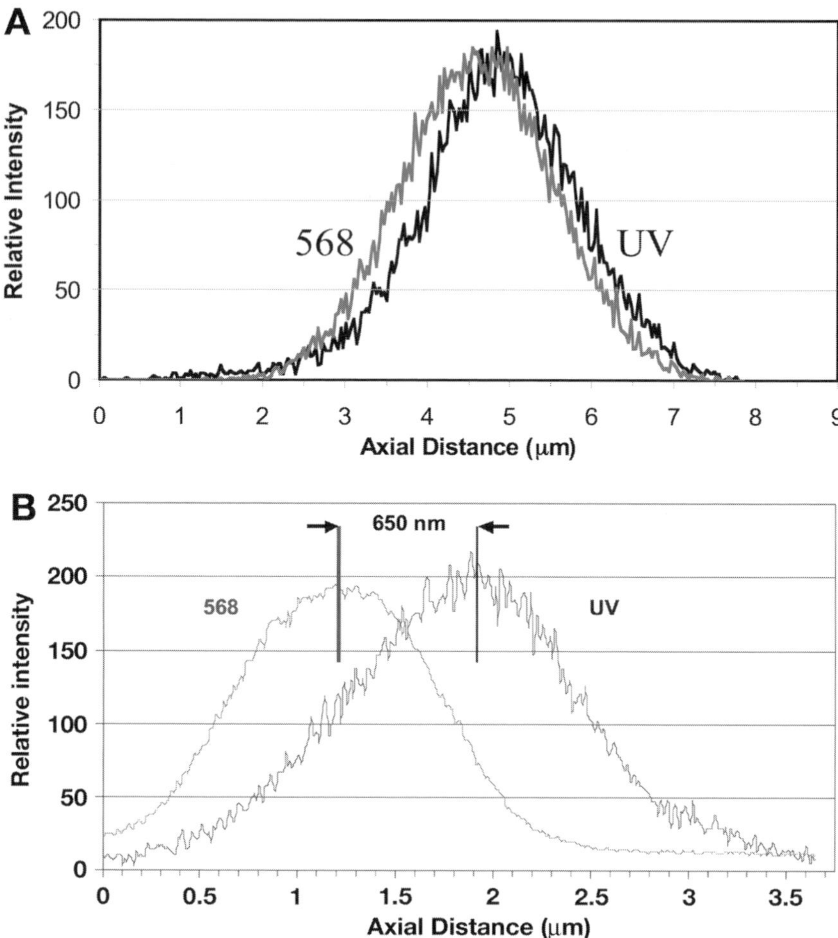

Fig. 5. Spectral registration (UV and visible). The *xz* spectral colocalization of this UV (365 nm) and visible wavelength (568nm) was evaluated with a 100× PlanApo NA = 1.4 lens using a 1-μm multiple wavelength fluorescent bead (TetraSpec T7284; Molecular Probes). An aligned system has a FWHM of less than 210 nm **(A)**, whereas a misaligned system has a FWHM difference of 650 nm **(B)**. The bead was imaged using *xz* scans with a 24× zoom, a slow scanning rate, and averaged eight times. The 568-nm line was chosen instead of the 488-nm line to minimize the crossover between the visible and UV wavelengths.

to examine the UV (365 nm) and visible lines (568 nm) in our TCS-SP1 confocal system that had an argon–krypton laser emitting three lines and a UV Enterprise laser. A 15-μm bead with UV interior and orange fluorescence ring exterior (F7236; Molecular Probes) was used to show that the UV and 568-nm lines were aligned (*see* **Fig. 6;** *see* Color Plate 4, following p. 274). In a similar manner, the F7237 bead consists of a UV blue core and a green ring and this can be used to show

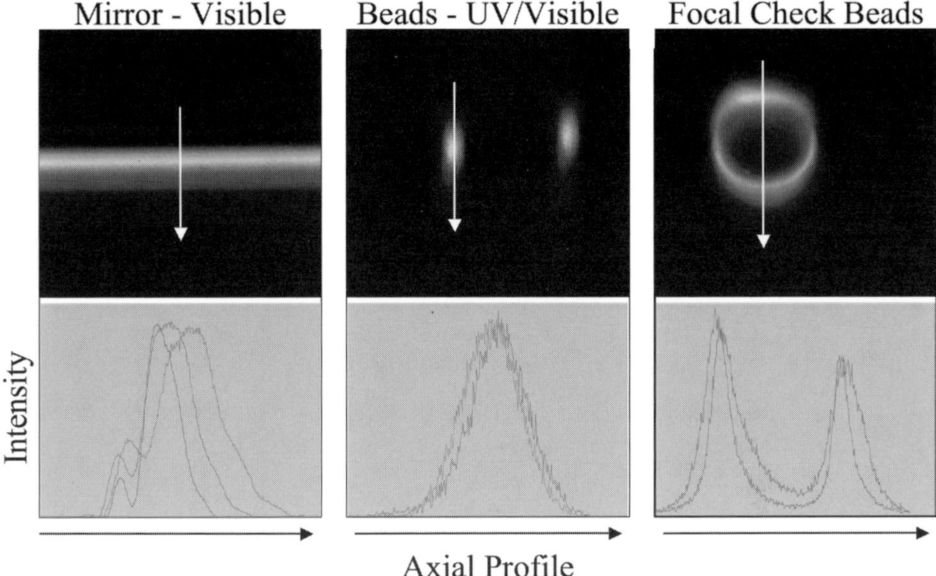

Fig. 6. Spectral registration methods. Three methods are used to check spectral registration of different laser lines. A mirror can be used for visible wavelengths on a Leica TCS-SP system (**left**). A bead scanned in the *xz* directions can be used on all confocal systems (**middle**) to show alignment of laser beams. Focal check beads (**right**) can be used to show alignment of multiple laser beams. In all cases, it is useful to make measurements on the peak height position to determine the spectral registration. (*See* Color Plate 4, following p. 274.)

colocalization of the 488-nm and 365-nm laser lines. This bead might have slightly more crosstalk than the blue core with a red ring (F7236). Normally, the bead should reveal concentric fluorescent rings that have maximum values in the same focal plane with either an *xy* or *xz* scan. The laser power and AOTF should be adjusted to reduce crosstalk between the emitted fluorescence. Any deviation between the concentric localization of the rings or the maximum diameter of the rings and core bead size suggests misalignment. In the newer confocal systems that have three lasers and a merge module, it is recommended to test for colocalization with a focal check bead (F7235) that has three rings representing green, orange, and red or a focal check bead (F7239) that has a red ring and a green core. Individual pinholes in a Zeiss system might have to be realigned monthly and this test can indicate if the pinhole needs adjustments. Smaller beads should yield more accuracy.

3.12. Lens or System Spectral Registration (Beads or Reflective Mirror)

This spectral registration test demonstrates the ability of the CLSM to colocalize different wavelengths of fluorescence in the same plane. To evaluate the spectral registration of the 365-, 488-, 568-, and 647-nm lines, either a 1-μm

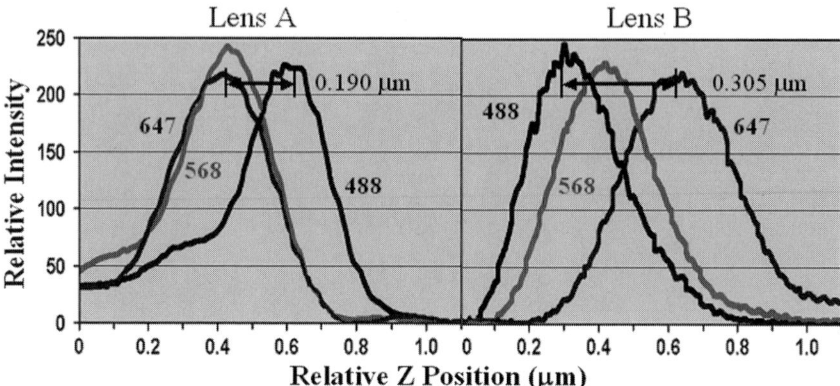

Fig. 7. Spectral registration (visible). The visible spectral registration of a 100× Plan Apo NA = 1.4 objective was evaluated using a front-surface, single-reflection mirror with the same lens at different times. A 10-nm slit was put over each wavelength and the reflection of each line was measured sequentially. The AOTF and PMT intensity was adjusted so that the maximum intensity of each line was 250 GSV. Lens B was sent back to the factory, as it did not meet the following: (1) spectral registration for UV (365 nm) and visible (568 nm); (2) spectral registration for the three visible lines; and (3) axial resolution specifications. The refurbished Lens A showed excellent registration among the three visible lines with the difference being less than 220 µm. Refurbished Lens A also had an axial registration below 350 nm. This single-reflection mirror test will yield slightly better spectral registration than 1-µm bead data for the 647-nm excitation line, as the fluorescence emission occurs in the far-red range (>660 nm) and many lenses have difficulty colocalizing this far-red emitted light with the fluorescence emitted from the 488-nm and 568-nm wavelength excitation.

multicolored bead (Tetraspec or Rainbow) or a front-surface, single-reflective mirror (*see* **Fig. 7**) was used. The front-surface, single-reflective mirror can be used to check visible spectral registration in a Leica TCS-SP system, in a similar manner to what was described for **Fig. 3** for axial *Z* registration. In the Leica SP system, a 10-nm refection bandwidth is put over each excitation wavelength and the reflection is measured sequentially. By tweaking the AOTF and PMT voltage adjustments, the intensity of each reflected line was adjusted so that the maximum intensity of the image was approx 250 GSV. This mirror test is more accurate than the bead tests, but the data obtained skew the results slightly toward better values in normal operating conditions. The emission from either specimens or beads are normally recorded at least 10–40 nm above the excitation wavelengths and not at the excitation wavelength. Many lenses have difficulty in colocalizing far-red fluorescence with the blue and green fluorescence, so measuring the emission at 647 ± 10 nm will yield better resolution than measuring the emission at 660–700 nm.

Figure 7 represents a 100× lens measured over a time period of approx 6 mo on the same CLSM system. On testing Lens B, problems in axial resolution and spectral registration existed. The separation between the 488-nm and 647-nm line was 305 nm in lens B and the axial registration was 410 nm. This lens (B) was returned to the factory to correct this spectral registration problem in visible light and a spectral registration problem between UV and visible light. Upon repair, lens A showed perfect colocalization between the 488-nm and 647-nm lines and acceptable registration among the 488-, 568-, and 647-nm lines. The UV and 568 nm (*see* **Fig. 5**) also showed acceptable registration after repair.

Depending on the laser configuration, this test can reveal characteristics of the lens' spectral registration or system alignment. In an argon–krypton laser system, because all three lines are derived simultaneously, the test will reveal the lens characteristics. If a three-laser system with a merge module is used, it will reveal the combination of both lens characteristics and laser alignment. It is useful to make this test with more than one objective, as it will be rare that a system will contain multiple lenses with axial and spectral registration problems.

A 1-μm bead can be used to test visible wavelengths for colocalizing in a similar manner to that described for colocalizing UV and visible light. Using the bead test, it is necessary to reduce the laser light to reduce bead bleaching and then adjust the laser light with the AOTF and PMT voltage so fluorescence crosstalk between the emitted wavelengths from different laser lines is minimized. It is also useful to suspend the beads in an antifade medium to reduce bleaching for this test, as they will be zoomed at high magnification, which increases photobleaching.

3.13. Laser Stability (Visible, Long Term [Hours])

Laser stability measurements were made over hours to evaluate the possible fluctuations in power (*1,11*). The laser power fluctuations were initially measured both in PMT 1 (blue light sensitive, low noise, R6857) and in the transmission detector, using a fluorescent plastic slide with very low laser power that was reduced with either neutral-density filters or with the AOTF adjustments. The transmission optical system without a slide showed results similar to PMT 1 with fluorescent plastic slides, and this was the desired optical system to perform this test if it exists on the system, as it eliminated any possibility of bleaching or laser interaction with the substrate.

To measure laser stability using the transmission optical system, the microscope is first aligned for Kohler illumination with a histological slide, which is then removed from the optical path. The image intensity is measured using the transmission detector for the three wavelengths of the argon–krypton laser by sequentially measuring the laser light with the 488-, 568-, and 647-nm wavelengths, with the power being adjusted by the AOTF transmission control so that the transmission detector voltage remains constant for all three wavelengths. The test usually takes a few hundred scans separated by 10–30 s over a period of 2 h. The intensity of a large ROI of the field is averaged and plotted

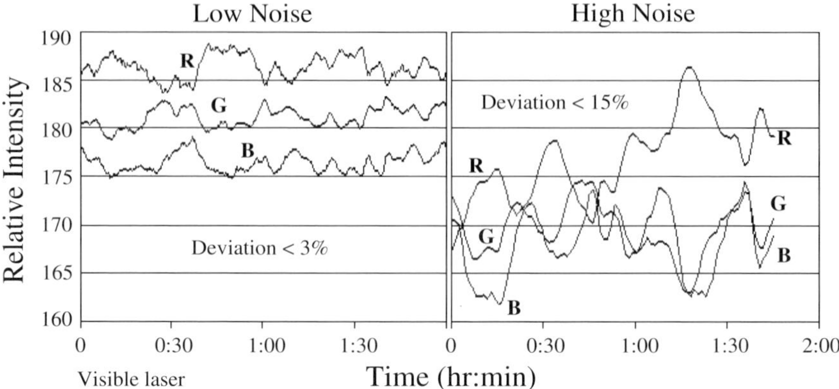

Fig. 8. Laser noise. The periodic change in laser power was measured using transmitted interference optics without a slide in the optical path. The two panels show a visible system delivering low-power fluctuations and one delivering high-laser-power fluctuations over time. The 488-nm and 568-nm lines are cycling periodically directly opposite the 647-nm line. Variations in this power intensity occur over hours and never seem to stabilize. To measure laser stability over time, the PMT was kept constant and the laser power of the three lines was adjusted with the AOTF. Next, over 200 samples were sequentially measured every 15 s. After the test was complete, the intensity of a large ROI was evaluated and plotted over time. The three lines are designated the following R (red, 647 nm), G (green, 568 nm), and B (blue, 488 nm). The laser power instability might be the result of either laser light entering a fiber with an incorrect laser polarization, thermal instability in the AOTF, or a poorly aligned system. The reason for the source of power instability is not known.

over time for the three wavelengths. The goal of this test is to have a straight line with no variations in power to ensure accuracy in the intensity measurements. It is not necessary to save the scan, but it is useful to save the data measurements in an Excel file, text file, or equivalent.

The type of laser power stability data that can be achieved with this test is represented in **Fig. 8**. It shows a relatively stable visible system yielding low-power fluctuations (**Fig. 8A**; <3%) and an unstable visible system (**Fig. 8B**; >15%), yielding high-laser-power fluctuations (source of noise is not identified). The goal of this test is to achieve a flat stable line. There is periodic noise in the laser system that exceeds the manufacturer's (Ominichrome) laser stability fluctuations specifications of less that 0.5% over a 2-h time period. The 488-nm and 568-nm lines have a periodic cycle that is directly opposite the 647-nm lines, and stability is never achieved (**Fig. 8**). The AOTF may be reponsible for these power fluctuations.

The fiberoptics might influence artifacts resulting from deterioration with time or improper polarization alignment. Fiberoptical problems will attenuate the laser power and necessitate using higher laser power or higher PMT settings in the operation of the CLSM. One test procedure recommended by the manufacturer to

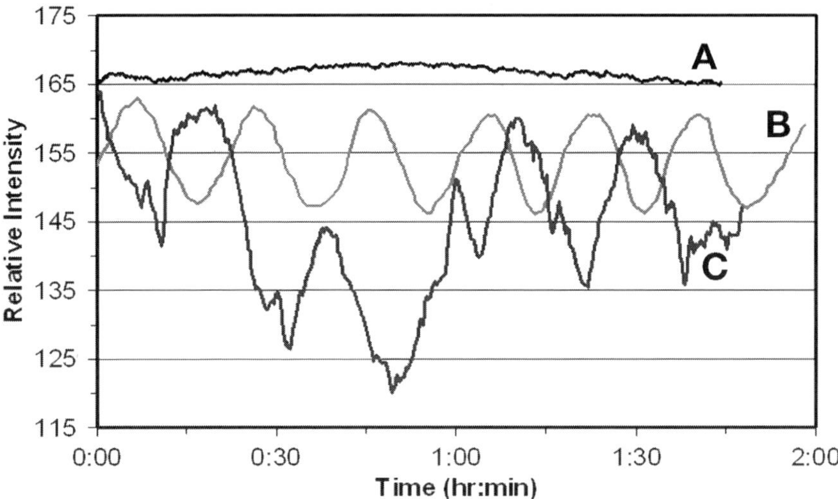

Fig. 9. Laser Stability of a Coherent Enterprise UV laser. The Coherent Enterprise UV laser delivers less than 1% peak-to-peak noise. The laser was connected to an LP 20 water–water cooler, which should be set at least 10°C above the cooling water of the building. It should also be set above the ambient temperature of the room. Improper set points for laser cooling resulted in poor thermal regulation of the laser (trace B). Improper fiber alignment resulted in additional laser intensity variations (trace C). The elimination of the temperature and polarization issues resulted in proper laser stability (trace A, 3% power variation over time). The test was conducted by measuring the laser power stability in PMT 1 using a fluorescent plastic slide. Neutral-density filters were used to reduce the power and, thus, minimize slide bleaching. The transmission detector optics gave similar results to the UV fluorescent plastic slide and was also used to measure UV laser stability.

ensure that polarization is correct after the alignment procedures is to wiggle the fiberoptic and see if the image returns to the same intensity, suggesting that the polarization is correct. This is a fairly crude test, but it will demonstrate whether the system fiberoptic needs further polarization alignment. Better procedures are being tested by the manufacturers to eliminate this polarization alignment problem.

3.14. Laser Power Stability (UV Long Term)

The Coherent Enterprise laser delivers less than 1% peak-to-peak noise and is considered a very quiet and stable laser. The Coherent laser was tested in a manner similar to that described for visible lasers using the transmitted optical system or the PMT system with blue-colored fluorescent plastic slides and very low laser power. A relatively stable line showing minimal fluctuations should be obtained (*see* **Fig. 9,** trace A). The temperature of the cooling system must be regulated properly or power fluctuations will occur (**Fig. 9,** traces B and C). One source of power instability appears to be way the laser is cooled and how the laser heat is dissipated. This was illustrated with our Coherent Enterprise

UV laser that was connected to a Coherent LP 20 water–water exchange cooler. This cooler should be set at least 10°C above the circulating cooling water of the building and it should be set above the ambient temperature of the room. Improper set points for the LP 20 cooler resulted in temperature regulation problems of the circulating cooling water in the laser, which, in turn, resulted in the improper regulation of the laser power (**Fig. 9,** trace B).

3.15. Laser Power Stability (Short Term [Seconds])

To measure laser stability using the transmission optical system, the microscope is first aligned for Kohler illumination with a histological slide, which is then removed from the optical path. The image intensity is measured using the transmission detector for the three wavelengths of the argon–krypton laser by sequentially measuring the laser light with the 488-, 568-, and 647-nm wavelengths, with the power being adjusted by the AOTF transmission control so that the transmission detector voltage remains constant for all three wavelengths. In contrast to the long-term stability test, the short-term test uses only one scan for each wavelength. The intensity of a single-line scan is averaged and the mean, standard deviation, and coefficient of variation (CV) are determined (*see* **Fig. 10**). The stability of the laser and the system noise while scanning is determined by this test. This test will measure the noise in the laser and the system and how much averaging should be necessary to improve image quality at the PMT setting that is used. The goal of this test is to have a straight line with no variations in power. If power fluctuations exist, they can be averaged to increase image quality. Averaging will reduce the image CV. If transmission optics are unavailable on the system, this test can also be done with fluorescent plastic using very low laser power to reduce bleaching.

3.16. CV Principle

One of the major elements of a poor image in a confocal system is related to using the PMT at high voltage values. This can be the result of insufficient sample staining, a misaligned system, or a failing hardware component or attenuation of light through a fiberoptic. If the unit could be operated at lower PMT values, then the image quality would be improved and the PMT noise decreased. The noise present in the system can be evaluated using a large (10-µm) bead (Spherotech) or an Applied Precision fluorescent substrate slide. The 10-µm bead consisting of nearly uniform size and intensity (CV = 5% by flow cytometry) is zoomed 4× to increase the number of pixels contained in the ROI with either a 100× objective (PlanApo, NA = 1.4) or a 63× (water [NA = 1.2] or oil [NA = 1.32]) lens. The bead is located in the center of the field and the image of the bead is obtained at the center cross-section of the bead, which relates to its maximum diameter. This large bead suspended in an antifade solution permitted repeated measurements with minimal bleaching of the bead sample. A fixed ROI was

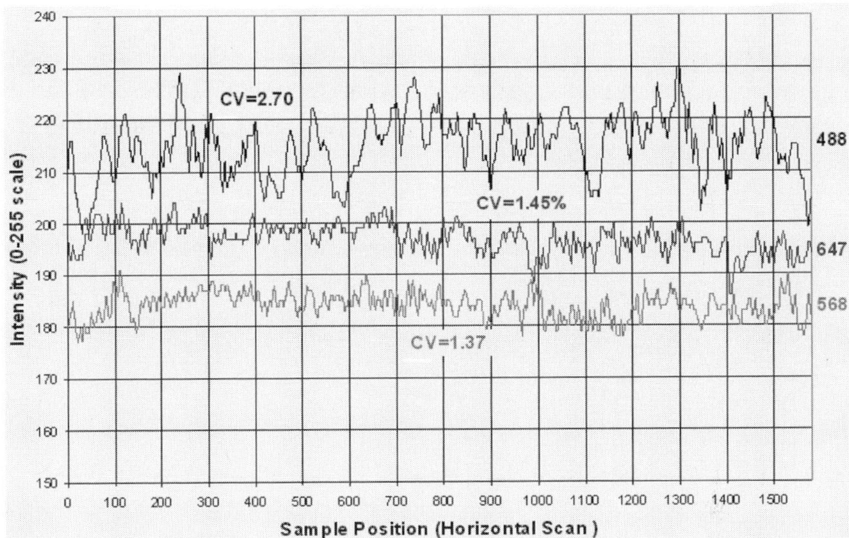

Fig. 10. Laser noise (short time). The transmission PMT was set to a constant value and the intensity of the laser lines was adjusted with the AOTF to bring the intensity to approximately the same point for graphic representation. The intensities of the three lines were made from one individual scan at each wavelength and it demonstrated that 488 nm (CV = 2.70), 568 nm (CV = 1.45), and 647 nm (CV = 1.37) had variations in intensity at very low PMT settings. The blue line (488 nm) was twice as noisy as the green (568 nm) or red (647 nm) lines. The more intensity variations, the more averaging that will be necessary to create a good image.

determined in the bead image that consisted of approximately half of the bead's area. The mean and standard deviation of the pixel intensities in the ROI of the bead were determined to yield the CV. This can also be done with a clean fluorescent substrate slide measured at a specific depth, so the intensity is reproducible.

The CV is defined as the standard deviation (σ)/mean (μ). It is important to maintain the machine variables (pinhole = size Airy disk, PMT voltage, averaging, etc.) at reproducible values for these studies. The laser power was set at a constant value that allowed the mean intensity level of the bead to be approx 150 (out of 255) for each PMT setting. The noise associated with the various settings can be evaluated by varying PMT settings, frame averaging, scan speed, image size, and laser power ***(1–3)***. The CV that we are measuring is actually the variation of pixel intensity within the bead, as opposed to the variation of intensity among a population of beads, as measured on a flow cytometer. An increase in CV might imply that there is either a decrease in laser power or a system alignment problem that results in higher PMT values. A noisy laser can also be detected by this test.

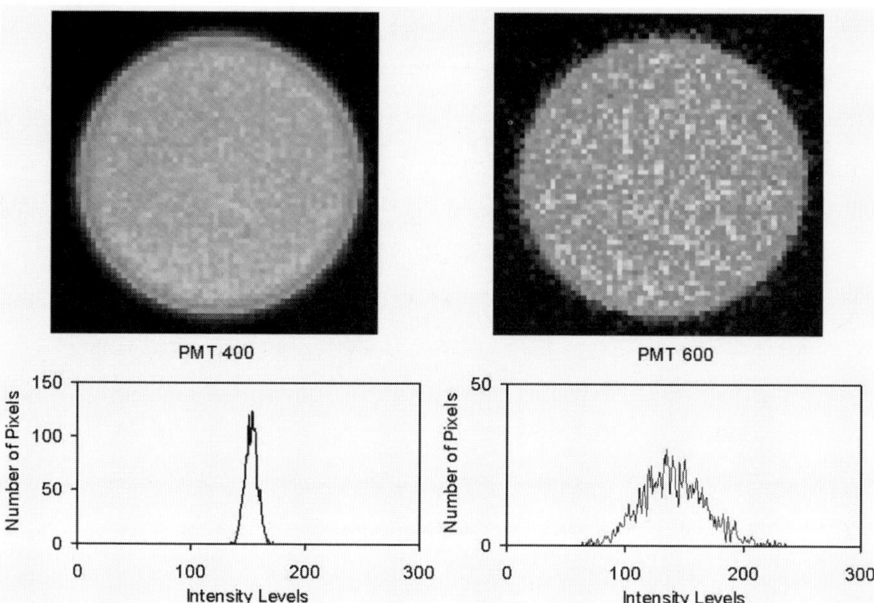

Fig. 11. Bead pixel variations. TIFF images of a 10-µm Spherotech bead were obtained with two different PMT settings (PMT = 400, PMT = 600) with a zoom of 4 and no frame averaging using a 100× PlanApo lens (NA = 1.4). A ROI was drawn in the interior of the bead and the histogram of the population of pixel intensities is displayed in the bottom panels. The mean pixel intensity in both images was approx 150 intensity levels and was obtained by keeping the PMT at 400 or 600 and adjusting the laser power with the AOTF.

Figure 11 illustrates the pixel distribution of a 10-µm bead that was measured with a PMT voltage setting of 400 and 600 and a zoom of 4× and an Airy disk of 1. These two bead images were obtained in the following manner. The mean intensity value in the ROI within the bead was set at channel no. 150 by adjusting the AOTF manipulation, instead of actually lowering/raising the laser power. The higher PMT voltage yielded a broader histogram, which translated into more pixel intensity variations. Because the CV (σ/μ) is defined as the standard deviation (σ) divided by the mean (μ), one can compare the quality of images using this technique. As the quality (less bead noise) of the images increases, the CV of the population of pixel intensities within the bead decreases. To compare image quality between different confocal microscopes, it is critical that as many variables as possible be kept constant ***(1,2)***.

The relationship between PMT voltages and frame averaging influenced the CV value and image quality. If the laser power was kept constant and the system power was adjusted with the AOTF, it was found that either increased frame averaging or lower PMT voltages could decrease the bead CV value and, thus, improve image quality. The relationship shown in **Fig. 12** demonstrates that at higher PMT set-

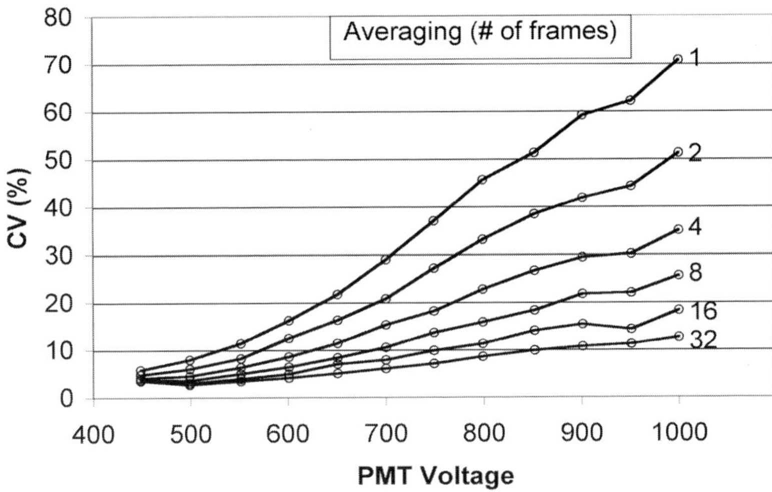

Fig. 12. Relationship among averaging, PMT, and CV. The noise present in the system was evaluated using a 10-μm bead with a 100× PlanApo objective (NA = 1.4). If the mean is assumed to be constant (channel no. 150), the histogram distributions generated by averaging can be produced and the CV and standard deviation (SD) calculated. This test was made by decreasing the PMT value and adjusting the laser power with the AOTF to ensure that the mean pixel intensity was at a value of 150. The higher PMT values were taken first to minimize possible bleaching. Images corresponding to 1, 2, 4, 8, 16, and 32 were obtained at each PMT setting. The CV is defined as the standard deviation (σ) of pixels in a bead divided by the mean (μ) intensity ($CV = \sigma/\mu$). The noise at a specific setting can be reduced if frame averaging is increased. The averaging decreases the pixel variation, which lowers the SD and decreases the CV. Theoretically, by averaging the image *n* times, the CV and SD are decreased by the square root of *n*.

tings, it is necessary to frame average more to reduce the CV and increase image quality. The same relationship exists in biological samples (*see* **Fig. 14**).

3.17. PMT Function

The PMT is the detecting system and there exists different quality PMTs of the same type in a confocal system. This CV bead test method can be used to access the operation and quality of the PMTs in the confocal system. The use of the Leica SP system easily allowed for switching of PMTs and pairing them with different excitation wavelengths. In effect, any PMT could be used in conjunction with any of the four excitation wavelengths. Although the PMT position will affect the CV, it is not considered to be a major contributor, and in this assessment, all of the PMTs were considered equivalent. PMT 1 and PMT 3 have an extra mirror present to reflect the light, whereas, PMT 2 collects the light directly after the prism. There are two types of PMT used in the Leica system: PMT 1 is considered low

Table 3
PMT Comparison and Noise

Excitation	Emission	PMT#	PMT Voltage	CV(%)	Relative CV
488 nm	505–555 nm	1	474	6.06	100
		2	428	6.58	108.65
		3	425	6.23	102.86
	555–600 nm	1	471	6.02	100
		2	432	7.00	116.25
		3	421	6.46	107.17
568 nm	580–630 nm	1	439	4.00	100
		2	411	4.88	122
		3	393	4.49	112.11
647 nm	665–765 nm	1	802	20.30	100
		2	732	22.70	111.68
		3	675	20.30	100.12

Note: The noise of the system was evaluated using a 10-μm bead (Spherotech) and a 100× PlanApo (NA = 1.4) objective. The intensity of a 10-μm bead was determined at a constant laser power, a zoom of 4, and no averaging using various PMT settings. The emitted light was measured in each of three PMTs. The pixels in each ROI were set to a mean of approx 150 and the standard deviation of pixel distribution was measured to determine the CV. The CV of the pixel intensity within the bead was measured at each PMT setting. PMT 1 is low noise and blue sensitive, whereas PMT 2 and PMT 3 are far-red sensitive. The quality and part of the performance of each PMT can be measured with this test.

noise (R6357) and PMT 2 and PMT 3 (R6358) have high efficiency and sensitivity in the far-red wavelength regions. Zeiss also has different types of PMT in their system. The system was set up with a triple dichroic (TD) using 488-nm, 568-nm, and 647-nm wavelength excitation. The three PMTs were adjusted to allow the mean pixel intensities to be at channel 150. The relative intensities were measured with the three PMTs for all conditions (*see* **Table 3**). Lower CVs will translate into better image quality and will require less frame averaging to produce the equivalent image quality. This is a test that is useful to determine system quality and identify a possible problem in PMT performance prior to a hard failure. The test can be done with dichroics and barrier filters to assess the efficiency of each filter in the light path. It is important to define the bandpass region that is being evaluated, as dichroics will eliminate specific regions in the transmission. A similar type of test can be made on filter systems to evaluate the PMTs.

3.18. Spectral Scanning: PMT Comparison

The PMT performance was measured using the Leica spectral scan feature. This feature allows for a sequential scan across the whole range of 400–800 nm in units as small as 5 nm. The test used an inexpensive, defined mixed-ion light source, (LightForm Inc., Hillsborough, NJ) to measure the spectral response of

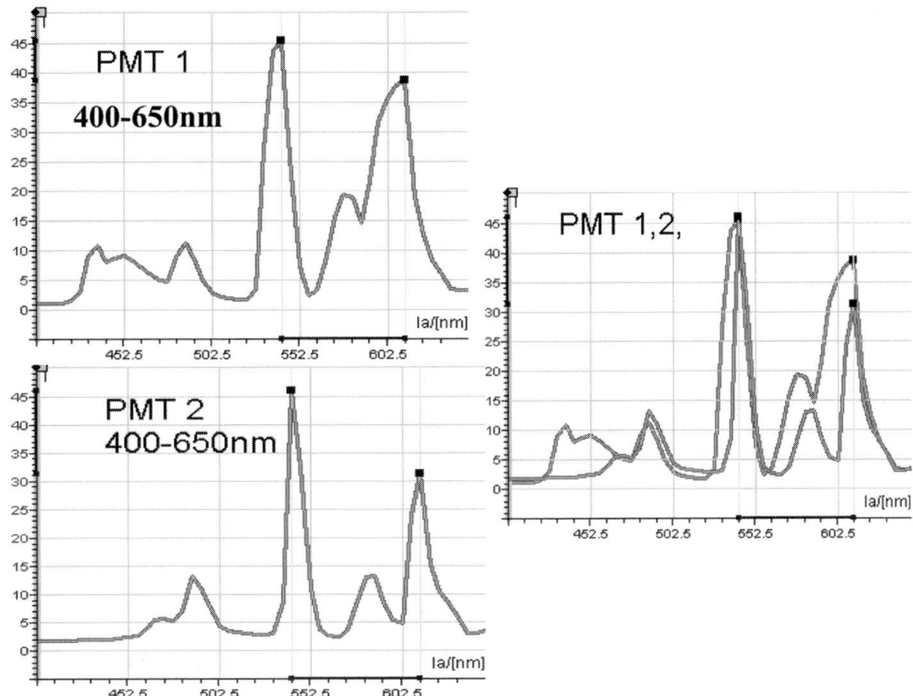

Fig. 13. Spectral scanning. The PMT performance was measured across the 400- to 650-nm spectral range using the Leica spectral scan feature. The slide holder was removed and the light source (LightForm Inc., Hillsborough, NJ) was positioned on the stage. A spectral scan consisting of 50 increments of 5 nm each was made between 400 and 650 nm using a 10× (NA = 0.4) PlanApo lens without averaging and an Airy disk of 1. A large ROI was used to analyze the intensity of each 5-nm scan and the data are displayed as an intensity curve between 400 and 650 nm. Note the broadness in the peaks acquired from PMT 1 relative to the tighter and smaller CVs in PMT 2. PMT 3 is similar to PMT 1 and is not shown for figure clarity. This scan shows a suppression of PMT 2 wavelengths below 450 nm. It also demonstrates a lack of resolution of the PMT 1 peaks relative to PMT 2 peaks ***(24)***.

a PMT across a large spectral region. The slide holder was removed and the light source positioned on the stage. A spectral scan consisting of 50 increments of 5 nm each was made between 400 and 650 nm using a 10× (NA = 0.4) PlanApo lens without averaging and an Airy disk of 1. The efficiency of the light-collection system was low, necessitating that a large PMT voltage setting be used. The intensity of each 5-nm scan was calculated as the mean from a large ROI and the data were displayed as an intensity graph between 400 and 650 nm (*see* **Fig. 13**). A similar test can be made on a Zeiss 510 Meta using a 10.7 nm scan size. If necessary, the pinhole can be opened to let in more light.

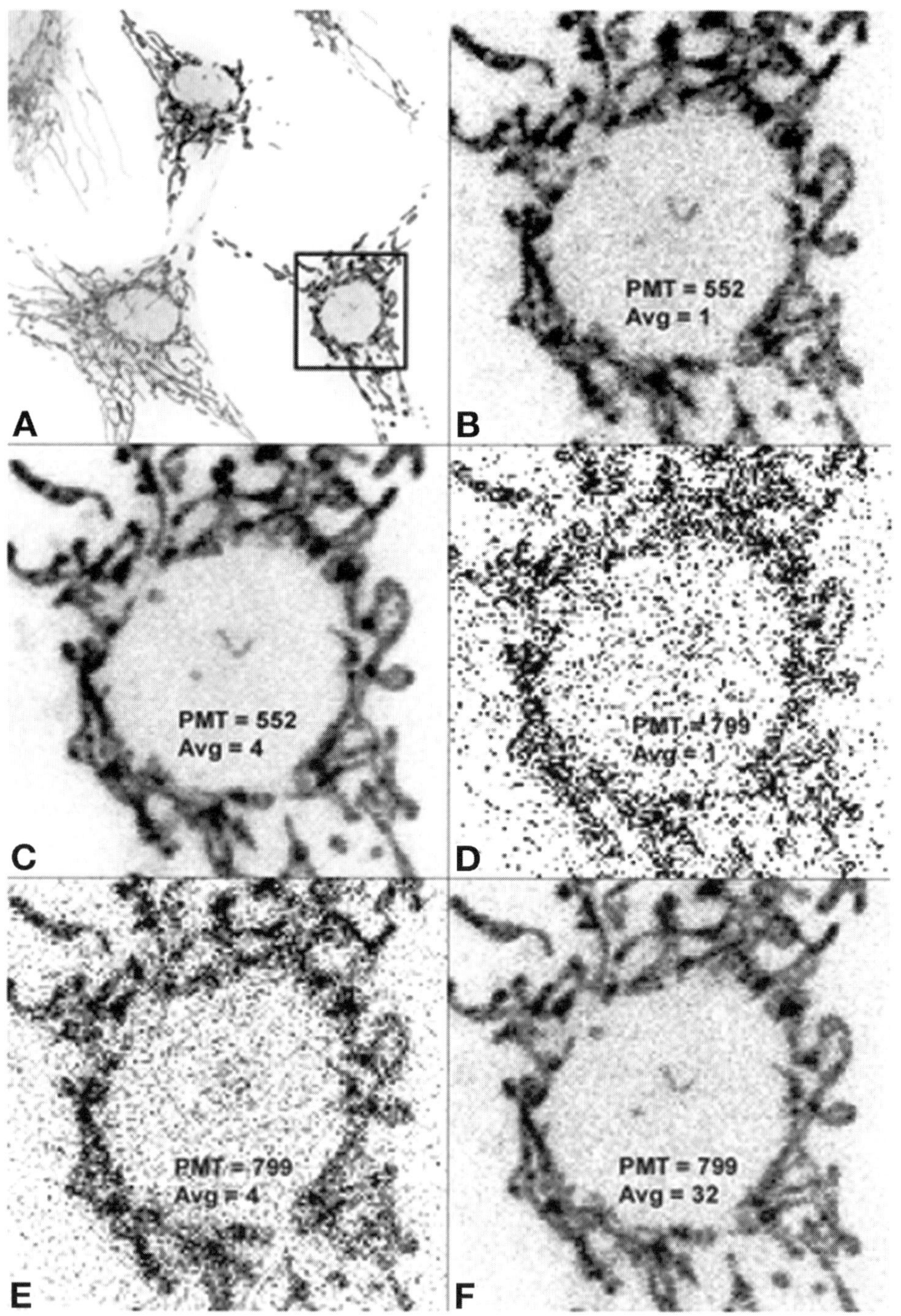
A
B
PMT = 552
Avg = 1
C
PMT = 552
Avg = 4
D
PMT = 799
Avg = 1
E
PMT = 799
Avg = 4
F
PMT = 799
Avg = 32

3.19. Biological Samples

The CV technique developed on beads was applied to biological specimens (MP FluoCells; chicken cells stained with Acridine Orange) to observe if the same relationship between PMT voltage and averaging described on beads are maintained on biological cells. FluoCells® (F-14780, Molecular Probes) were excited with a 568-nm laser line and detected with a 580 to 630-nm bandpass filter in PMT 2. The difference in image quality illustrated by averaging 1, 4, or 32 times at 2 different PMT settings (552 or 799) is shown in **Fig. 14**. Figure 13B–F shows images zoomed 4× using Image Pro Plus to clearly illustrate the individual pixels. The CVs of a selected ROI in the nucleus varied with the number of frames averaged and the PMT voltage. The best image quality (low CV) consisted of either low PMT voltages (**Fig. 14B,C**) with minimal amounts of frame averaging, or high PMT voltages with 32 frames averaged (**Fig. 14F**). High PMT settings (**Fig. 14D,E**) with minimal amounts of frame averaging (one or four) demonstrated high CVs and poor image quality. In all cases, the increase in averaging resulted in a decrease in the CV and a corresponding increase in image quality. In contrast, raising the PMT voltages increased the CV and decreased image quality. The higher PMT settings necessitated the use of greater frame averaging to increase image quality. **Figure 14** shows the relationship among PMT voltage, frame averaging, and CV on image quality on cells, which was similar to that described with beads in **Fig. 11.** The noise in this figure is also reduced as the square root of the frames averaged ***(12,13)***. The CVs of a ROI in the nucleus of the various panels of **Fig. 14** was the following: B, 49%; C, 40%; D, 212%; E, 109%; F, 49%.

3.20. Sensitivity

The sensitivity of a confocal microscope is an important parameter to determine, as its value influences the acquisition settings of the PMT voltage, laser power, and frame that are used to acquire images. The values derived will relate

---

Fig. 14. PMT and averaging of cells. FluoCells (F-14780; Molecular Probes) were excited with a 568-nm laser line and detected with a 580- to 630-nm bandpass filter in PMT 2. Averaging 1 measured the resolution, 4 or 32 times at two PMT settings (552 or 799). **(A)** shows the distribution of three cells at normal magnification, whereas **(B)–(F)** show one cell located in the box in **(A)** that was zoomed 4× with Image Pro Plus. The settings in the different panels were the following: (A) control (PMT = 552, AV = 1); (B) (PMT = 552, AV = 1), (C) (PMT = 552, AV = 4), (D) (PMT = 799 AV = 1), (E) (PMT = 799, AV = 4), and (F) (PMT = 799, AV = 32). Note the difference in pixel variations in the six panels of the same cell acquired at different PMT/averaging settings. The CVs of a ROI in the nucleus of the various panels were the following: (B) 49%; (C) 40%; (D) 212%; (E) 109%; (F) 49%.

Table 4
Confocal Laser scanning Microscope Sensitivity: Relationship Between Laser Power and CV

Laser type	Wavelength mW	Power mW	CV-Bead% SD/Mean
	Fixed Power Comparison		
Argon Krypton	488	1	4
(75mW, Leica)	568	0.2	4.6
Argon 25mW	488	1	1.3
HeNe 1mW	543	0.2	1.9
(Zeiss)			
	Maximum Power Comparison		
Argon Krypton	488	1.1	3.8
(75mW, Leica)	568	1.45	2.6
Argon 25mW	488	3.2	1
HeNe 1mW	543	0.23	1.9

Note: The sensitivity from a Leica TCS-SP1 containing an argon–krypton laser emitting three laser lines and a Zeiss 510 containing three individual lasers and a merge module are represented. The CVs were obtained from a 10-μm bead using a 100× PlanApo objective (NA = 1.4) The laser power was derived by using a 10× (NA = 0.3) objective and a power meter situated on the stage. By setting the power to a fixed value of either 1 mW of 488-nm laser light or 0.2 mW laser of 543-nm laser light on the stage, the sensitivity of two machines was measured. The CV of the bead was observed to be almost three times lower with the 488-nm laser lines using the Zeiss 510 system compared to the Leica TCS-SP1 system. By increasing the lasers to their maximum power, the CV values were decreased. Sometime maximum power measurements are useful to indicate alignment of the system and functionality of different components. There are many explanations for this difference.

to alignment, equipment functionality, and general performance of the CLSM. **Table 4** represents two confocal microscopes that have different configurations: a Leica TCS-SP1 containing one argon–krypton laser emitting three laser lines, and a Zeiss 510 system that contains three individual lasers with a merge module. The test particle was a 10-μm Spherotech bead and measurements were made using a 100× PlanApo objective (NA = 1.4) with a zoom factor of 4 with 488-nm excitation in both systems. The laser power in both systems was measured on the stage using a 10× (NA = 0.3) objective and a power meter detector secured on the stage. It is essential that the acquisition parameters be as equivalent as possible when trying to compare machines from the same or different manufacturer. Because there are considerable differences between the designs of different machines, extreme care must be made when interpreting the results of data obtained from different manufacturers' machines. The variables that must be considered included specific wavelength, laser type, objective lens, optical components, pinhole size, scan speed, zoom, and pixel size. Other variables not mentioned could also affect the measurement. It should be emphasized again that it is extremely difficult to compare different manufacturers' machines because of the number of variables that must be kept constant, and conclusions drawn from this head-to-head test can only be made if all the variables are nearly equivalent.

To illustrate the test, we used a Leica machine that has an older Omnichrome argon–krypton laser emitting three laser lines and a new Zeiss 510 CLSM containing three relatively stable lasers. It would be expected that by using different lasers, the tests would reveal different sensitivity values (*see* **Table 4**). The sensitivity of two machines was made by maintaining the laser power at a constant value of 1 mW for 488-nm wavelengths and 0.2 mW for 568/543-nm wavelengths. The data from this test revealed that 1 mW of 488-nm power measured on the scan head yielded a CV value of 4% with a Leica TCS-SP1 and a CV value of 1.3% with a Zeiss 510. Comparable power readings on both systems showed the CV to be almost three times lower with the 488-nm and 568-nm lines with the Zeiss 510 system as with the Leica TCS-SP system containing an Omnichrome argon–krypton laser. The significance of a higher CV value means that the samples will have to be frame averaged to increase the image quality. These CV values will change if the laser power and PMT values are changed. The quality of lasers will also affect the CV values, as shown by the data in this test. Finally, increasing the laser power to a maximum value resulted in the CV being lowered with both the Zeiss 510 and Leica TCS-SP1 systems. However, it is not recommended to operate lasers at maximum power because of sample bleaching considerations and laser lifetime considerations.

This CV sensitivity data could be considered an initial reference point that can be used by other investigators to compare their CLSM performance with similar machines. It seems to be possible with this approach that the sensitivity of systems in different laboratories can be compared. Because of the number of variables that must be kept constant, extreme care must be made when interpreting this test in the comparison of different manufacturers' machines. However, there is no reason why similar machines from the same manufacturer cannot be compared if they are run under nearly identical conditions.

4. Notes

4.1. System Performance Tests

Image quality is an important parameter in the evaluation of a confocal microscope performance. Unfortunately, image quality is too often used as the "gold standard" to evaluate confocal microscope functionality and performance. Variables that effect image quality should be assessed to ensure that the system is delivering its optimum performance. In the cases in which intensity measurements are required, it is essential that the machine be stable to deliver reproducible data. A series of tests were either adapted from the literature or devised in our laboratory to measure the system performance of the confocal microscope ***(1–8)***. These tests include the following: laser power measured at the stage, field illumination, laser stability, dichroic performance, PMT performance, system linearity, axial resolution, spectral registration, sensitivity, and lens quality. This list is not inclusive but represents what can be tested and interpreted to ensure that the machine is operating properly.

4.1.1. Field Illumination

Field illumination is one of the easiest and most important tests to make on a confocal microscope. Many CLSM units in laboratories that have been checked for field illumination using a plastic fluorescent slide (Applied Precision) have demonstrated unacceptable field illumination patterns. This test should be made with all objectives and all wavelengths of visible and UV light to ensure that the machine is delivering proper field illumination. Field illumination should be relatively uniform, with the maximum intensity being in the center of the objective and decreasing less than 25% across the field according to one manufacturer. The light should come in the center of the objective and it should decrease in all directions in a similar manner. The decrease is dependent on the characteristics of the objective and its magnification. Most alignment procedures are made using high-magnification objectives (40×). However, this does not always translate into good performance with different magnification objectives.

The illumination intensity across the observation field can be measured with different types of test specimen in order to ensure that a homogeneous field illumination exists. The following test substrates have been used: concentrated fluorescent dye suspended in a hanging-drop well slide, small concentrated fluorescent beads (1–3 µm) or large concentrated fluorescent beads (10 µm) (Spherotech), fluorescent specimens, uranyl glass slides, Altuglas, or plastic fluorescent slides (Applied Precision), a piece of tissue paper stained with fluorescent dye or fluorescent dye solution (Fluorescein [F-7505] or Rhodamine B [R-6626]; Sigma, St Louis, MO) and mixed with immersion oil (Leica Immersion oil, $n = 1.518$) ***(1)***. A histological sample derived from plant or animal, which is usually the choice of service field engineers ***(1,3,8)***. The plastic slides (Applied Precision) were found to be the most consistent sample to test field illumination. We use the blue slide for UV excitation and the orange slide for 488-nm and 568-nm excitation. The red slide was found to bleach rapidly with 568-nm excitation; therefore, it is preferable to use the orange slide for this wavelength. The Chroma red slide was chosen for the 365-nm, 488-nm, 543-nm, and 568-nm excitations. The green slide can also be used for 488-nm excitation.

Initially, the surface of the slide is determined as it is the region that emits the most intense fluorescence in the *z* axial direction. It is important to measure the field illumination at a specific depth in the plastic slide, as the intensity distribution might change from the surface to the interior of the slide. Investigators should also be careful not to observe illumination fields deep within the plastic slide samples, as it will usually yield a better field illumination than regions closer to the surface because of various optical distortion factors ***(14)***. It is also important that the plastic slide be placed on a firm surface to eliminate any possibility of substrate flex. Although it has been proposed to use this fluorescent slide to take daily measurements for system stability, the factor of reproducible depth in addition to instability of laser power makes this application questionable.

There are many test substrates that have been used to measure field illumination. Some criticisms of these tests are as follows. Uranyl glass has previously been used to check field illumination, but it is currently difficult to obtain and we have observed that plastic slides have higher fluorescent efficiency than the uranyl glass at all wavelengths. A field of small or large beads on a slide (Spherotech) can be used, but it is essential that all of these beads be located at the same plane or the image will be inaccurate. To eliminate this potential error, a stack of images can be obtained from the beads, followed by a maximum projection of the stack to obtain an image of the beads that represents field intensity. However, the downside of this method is that it is very time-consuming. Histological samples derived from plant or animal can also be used to measure field illumination, and these are usually the choice of service field engineers. Histological samples will show the illumination pattern indicating proper or uneven illumination. In our experience, histological samples are not sensitive enough to properly measure field illumination. It usually will yield a sense of false security for the investigator. Solutions of fluoresce fluid are unstable and may shift with time. If there is a discrepancy between the plastic substrate test slide and other test slides in measuring field illumination, it might be the result of a greater sensitivity on the plastic substrate.

In our system, the three visible wavelengths of light are derived from one Omnichrome argon–krypton laser, which allows us to test the field illumination at one wavelength (488 nm) and assume that it is equivalent to testing field illumination with the other wavelengths. Because the UV line is derived from a different laser (Enterprise, Coherent) it is essential to check all objectives for proper field illumination at the 365-nm excitation in addition to the visible 488-nm excitation. Newer designed confocal systems use three individual lasers with merge modules, requiring that all laser wavelengths have correctly aligned beams emitted from the merge modules. In these systems, each of the three lines has to be individually tested. One laser line might be perfectly aligned, yielding acceptable field illumination, whereas the other laser lines might be misaligned, yielding intensity values in which the brightest region is not in the center of the field, as illustrated in **Fig. 1** with UV illumination.

4.1.1.1. Objectives

The lenses are the engines that drive this technology and their selection is very critical for optimum performance. Some of the factors to consider are the following: high numerical aperture (NA) relative to the specific magnification, flat field objectives, long working distance (WD) relative to NA, good fluorescence transmission, and good achromatic correction at desired wavelengths ***(12)***. Currently, microscope manufacturers make lenses that go though extratight QA procedures and have been classified as confocal objectives. These should be bought, as they are believed to be of a higher quality than ordinary fine microscope objectives. As a general rule, one should use the smallest magnification and the largest NA lens to acquire images ***(13)***. These lenses offer a larger field-of-view and better light transmission. Although it is critical to reject out-of-focus light for confocality, it is also necessary to have sufficient laser light entering the system, and lenses should be chosen that have good fluorescent transmission, characteristics. Opening the pinhole can increase transmission, but this should be done only if insufficient light is present, as this parameter decreases the confocality of the system. The choice of a good fluorescent transmission lens is sometimes chosen over PlanApo and spectral precision.

It is suggested that a confocal core lab should have a full range of high-quality microscope objectives consisting of air, water, multi-immersion, and oil lenses (typically planapochromat of the highest numerical apertures available) to satisfy the multiple applications in a core laboratory. It is necessary to match low-magnification oil and water objectives with higher-power water and oil objectives. Water-immersion objectives increase the working distance of the higher-power objectives. A complete set might be the following: 5× dry, 10× dry, 20× dry—used for low magnification imaging of macroscopic specimens (e.g., whole-mount mouse brains and fetuses), a 10× oil, 20×, 40× oil, 63× oil—used for conventional microscopy and 10× and 20× multi-immersion lenses combined with a 63× water lens and a 63× oil lens. A 100× lens has problems with alignment and low light yield and UV incompatibility. It is suggested that a 63× lens be used and zoomed slightly to address the 100× magnification range.

Objectives have unique characteristics and should be chosen for the specific applications accordingly. For instance, the Leica 100× (**Fig. 2;** NA = 1.4) is not recommended for UV applications, as it has a bull's-eye pattern with UV excitation. Because of the design of the 100× objective, it is recommended to use a zoom of 2× when using UV light in order to achieve a drop off of less than 25% across the field. Leica recommends that the PlanApo 63× (NA = 1.32) be used for UV work, as it has more uniform UV field illumination and better UV light transmission.

Other objectives that are useful in conventional microscopy might have a fluorescence bull's-eye pattern, making these lenses unsuitable of confocal applications *(1,3)*. It is important to acquire lenses that are compatible for confocal microscopy applications and test the lenses for field illumination accuracy using UV and visible excitation wavelengths. This field illumination test allows for a system evaluation consisting of both the objective properties and the confocal microscope laser alignment. This bull's-eye intensity profile has been obtained with different magnification objectives using all manufacturers' systems (Biorad, Nikon, Leica, Zeiss). The incompatibility of different lenses with confocal microscope systems can increase this bull's-eye effect and this parameter should be considered in choosing lenses. The problem might be the result of the lasers under filling the objective, which results in operating a lens under suboptimum conditions, resulting in the field illumination problems. One recommended solution to poor field illumination or bull's-eye illumination is to increase the zoom factor. However, this enlarges the illumination center and pushes the lower intensities off the field-of-view. Increasing zoom also increases the magnification and bleaching rate of the sample and this might defeat the purpose of using a low-magnification objective to observe a large field-of-view or might rapidly bleach the sample. This field illumination effect has to be monitored with each laser wavelength and each objective, as the alignment, wavelength, and lens design can influence the field illumination pattern. In summary, to eliminate field illumination problems, the system should be aligned correctly, with the brightest light being focused into the center of the field and decreasing less that 25% in all directions equally from the center with most lenses.

Not all problems with the field illumination test are the result of poor alignment, lens design/quality, or incompatibility of a lens with specific wavelengths of light. A dirty lens will yield both poor field illumination and poor resolution. If a lens is dirty or covered with dried oil, then it would yield a nonuniform pattern *(3)*. In one example, the intensity of the field from a 20× (NA = 0.6) dirty lens varied by as much as 70%, with the maximum intensity being off center on the right side of the image. After cleaning the lens to remove oil and other particles, an acceptable illumination pattern was obtained, with the maximum intensity being located in the center of the image and decreasing in intensity by less than 10% from the center maximum *(3)*.

4.2. Power Meter Readings

This power test appears to be one of the most useful tests because it quickly evaluates both the system alignment and performance. For the adequate operation of a CLSM, a sufficient laser power must excite the specimen. If a system is misaligned or functioning suboptimally, it can be assessed by a test that measures laser power. The power test can indicate if the system is aligned properly up to the plane of excitation on the stage or if the machine has a defective component (i.e., a dying laser or a defective fiber). It should be emphasized that this test is performed at the microscope stage prior to the light reflecting the dichroics for a second time and penetrating through the emission pinhole and the emission barrier filters or prisms (if they exists in the system) and into the PMT. In our experience, without sufficient power throughput in the system, the voltages will have to be increased to high values to visualize fluorescence derived from specimens, which will introduce PMT noise. In addition, the cause of the decreased laser power might result in other problems (i.e., laser instability, loss of axial resolution, increased laser noise, increased PMT noise, fiber polarization, broken fibers, AOTF malfunction).

The laser power measurements are listed in **Table 1** and are useful to illustrate the maximum power that can be obtained from a CLSM having these different laser configurations. These measurements serve as a valuable reference for an individual lab to QA their system over time and as a comparison with other similar CLSM for adequate performance. In a three-laser system, it appears that there is sufficient power with the 633-nm and 488-nm lasers, but because the maximum power of the 543-nm laser is so low and the attenuation of the 543-nm laser is so great, at least 0.2 mW of power are necessary from the 1.2-mW HeNe 543-nm laser to provide sufficient light to excite the samples. If the system is not aligned properly, the laser output will be decreased. With insufficient power, the PMTs will need to be operated at high-voltage settings, which increases the system noise and produces unacceptable images. Less power throughput in the confocal system suggests a problematic laser, a fiber polarization problem, defective AOTF, or just a poorly aligned system. The maximum laser power is dependent on the laser, optical configuration, and the specific objective used. Using a 10× objective (NA = 0.3), it is desirable to have at least 1 mW of power on the microscope stage for each laser line derived from a 75-mW Omnichrome 643 argon–krypton laser. Other confocal systems with different laser configurations will naturally have different power values, as indicated in **Table 1**.

It is important to measure the power output to evaluate system performance for all three lines after installation to make sure that the system is aligned properly and the laser is functioning correctly. These power values will not only serve as a reference to ensure the system is performing properly but can be useful to notify the confocal manufacturer of deviations from acceptable values, which will mean either laser failure or misalignment. A new Omnichrome 75-mW argon–krypton mixed-gas laser delivered the following power outputs: 488 nm, 1.10 mW; 568 nm, 2.68 mW; 647 nm, 1.60 mW. After time and proper laser alignment, almost 3 mW for each line have been achieved in this system. However, on another day the maximum power with a Leica TCS-SP1 system using a Plan 10× Fluor (NA = 0.3) was the following: 488 nm (1.1 mW), 568 nm (1.45 mW), and 647 nm (1.65 mW). Similarly, the maximum power measured on installation with a Leica SP2 system using a Plan 10× Fluor (NA = 0.3) was the following: 4.6 mW (TD, 488 nm), 6.5 mW (DD, 514 nm), 0.22 mW (TD, 543 nm), and 2.8 mW (TD, 633 nm). The values on the same machine taken a few months later were the following: 2.87 mW (TD, 488 nm), 3.9 mW (DD, 514 nm), 0.093 mW (TD, 543 nm), and 1.45 mW (TD, 633 nm). The reason for the fluctuations is unknown, but it appears that they might be attributed to fiber polarization problems, and AOTF instability problems.

4.3. Bead/Histological Power Meter Sample

If a power meter is not available, the crude power of the system can be assessed using standard histological samples like the FluoCells slide (F-14780; Molecular Probes) or beads like the 10-µm Spherotech beads (FPS-10057-100) and recording the PMT value that is necessary to acquire an image at almost saturation values If conditions are identical between machines. This PMT value can be used as a crude reference value to compare CLSM units and to establish their acceptable performance levels. Scientists desiring a more accurate method to test performance will find major problems with this type of testing because of the large range of acceptable PMT voltage values. Another reason to doubt the data from using a histological sample is that individual PMTs and samples can vary greatly in quality. Two similar PMTs on our machine had almost 100 PMT unit differences in the 700 PMT range. What is even more troubling with this test is that the PMT voltage is expressed as a logarithmic relationship relative to an intensity increase, which means that the difference of only a few PMT units can be translated into a huge difference in intensity.

In our opinion, these tests are very crude and subjective because of the acceptability of such a large range of PMT amplification values, the variations in staining between different plant samples, variations in PMT characteristics, and the logarithmic relationship between PMT and intensity ***(1)***. It does, however, yield a rough estimate to determine if there is sufficient power in the system. It is useful for each laboratory to have a reference slide to determine PMT saturation values with different laser wavelengths to determine if there is adequate power in the system and if the CLSM is adequately performing.

The comparison of two lenses to transmit light at an Airy disk of 1 can be relatively compared using a saturation test with almost any uniformly stained sample. On a Leica system, the 100× (NA = 1.4) does not transmit as much light as a 63× (NA = 1.32) lens; thus, for low-light applications, this 100× lens should not be used when there is insufficient staining.

4.4. Laser Adjustments

Argon, UV, and argon–krypton lasers need to be aligned and adjusted on a regular basis, as they occasionally go out of alignment. The investigator can easily measure laser power over time using a power meter positioned on the stage (*see* **Table 1**). Either the loss of laser power or poor optical alignment will reduce the laser power in the system, necessitating an increase in PMT to compensate for lack of laser light intensity. If minor adjustments are made to the mirrors with the horizontal and vertical knobs located on the back of these lasers, the laser power sometimes can be increased. However, these visible lasers are usually enclosed in a box, with the rear knobs being inaccessible for adjustment by the investigator. In fact, most confocal manufactures do not allow the user to adjust these controls, as it is the responsibility of the manufacturer on a service contract to ensure that the lasers and system are functioning properly. For example, in our system it is possible to tweak the Coherent UV Enterprise laser, but it is not possible to adjust the argon–krypton laser, as it is enclosed in a box that the manufacturer required not be open or the service contract would be invalidated. The investigator usually will not notice a problem with laser power or alignment, but will continually have to increase the laser power to compensate for the reduced system laser power. This use of increased laser power will not only shorten the life of the laser, but will not correct the CLSM system problems that are indicated by reduced power. The reduced laser power might result in a system yielding poor resolution and system noise.

If there is insufficient light entering the system, a careful realignment of the laser beam might be required (a separate procedure done by qualified personnel) to increase the laser output. If this alignment does not solve the insufficient system power values (similar to data shown in **Table 1**), then the laser might need to be replaced. Knowing specifications of laser power output on a stage is a critical parameter to assess system performance. In the future, it is suggested that the different manufacturers should specify these power values obtained on the microscope stage for different laser configurations, thus allowing investigators to determine their CLSM performance in their laboratories.

4.5. UV Power Test

One of the major problems that occur with UV confocal systems is that there is insufficient UV power output. The UV power transmission can be decreased from a number of factors, which include misalignment, aging fiberoptic, polarization mismatch between fiber and laser, unfocused collimator lens, and dying laser. This measurement of UV power helps assess the system performance and determines if adequate UV power is being transmitted through the system and if the fiber is in good condition. Although it is recommended to take the measurements at the back objective aperture region to eliminate the characteristics of the lens from influencing the test, we were not able to mount our detector probes in a sufficiently stable manner to allow for repeatable measurements.

An objective designed with good UV transmission characteristics should be used to increase the power throughput. Attenuation of the laser light through the low-power optics of the system will still occur and the power values obtained are relative values that are highly dependent on the specific type of objective used. Because our power detector does not work with higher-power optics (40×, 63×, 100×) because of optical limitations of the stage, it will be useful to use a histological test slide sample, fluorescence slides, or bead sample to assess UV power with these higher-power lenses. Experiments can also be done with histological test slides or fluorescent colored glass to approximate the laser output with higher-power objectives. For instance, by using maximum UV laser power, it was found that the 10-µm Spherotech bead saturates PMT 1 (low-noise PMT) at a setting of 650 using a 100× PlanApo lens (NA = 1.4). Leica technicians routinely use a 40× lens to measure the fluorescence saturation of a histological plant sample. If it saturates in the PMT range between 600 and 700 units in PMT 1, the system is passed as having adequate power. In our opinion, this test is very crude and subjective because of the acceptability of such a large range of PMT amplification values, the variations in staining between different plant samples, variations in PMT characteristics and the logarithmic relationship between PMT and intensity ***(1,3)***. It does, however, yield a rough estimate to determine if there is sufficient laser power in the CLSM. It would be useful to measure these values on a user histological test slide or bead when the machine is first installed and deemed working acceptably by the manufacturer's service and sales representative.

4.6. Dichroic Reflectance (Single-Wavelength Excitation)

Dichroic filters are made to reflect or reject specific wavelengths of light and pass the desired excitation/emission wavelengths of light. Dichroics do not always perform the way they were designed to perform in a CLSM (*see* **Table 2**). Dichroic tests should be made to determine the optimum system efficiency and ascertain the performance of individual dichroics with a variety of objectives and wavelengths. In certain images, it is important to have narrow bandpasses and not broad long passes in acquiring the fluorescent emission. It is important to use the dichroics that reflect the maximum amount of light at the desired wavelengths to increase the efficiency of the CLSM ***(1,3,6,7)***. By placing a fluorescent slide on the stage and measuring the relative intensity of an image, the efficiency to reflect light in a confocal system of the specific dichroic filter can be evaluated. The double dichroics (DD) and triple dichroics (TD) are more complicated dichroics than the single dichroics (SD) and, in theory, should reflect less light, as they were made to reject more light and pass fewer specific wavelengths of light. The placing of either a single, double, or triple dichroics in the light path should reflect successively less light using the 488-nm excitation line. Using a Leica confocal filter system, normally the 488-nm line should use the SD (RSP500), the 568-nm line should use the DD (488/568), and the 647-nm line should use the TD (488/568/647). However, the DD reflected the 488-nm light best and the TD reflected the 568-nm light best with all objectives. In principle, the better the reflection, the more efficient the dichroic. Using 568-nm excitation, the TD reflected 30% more light than the recommended DD dichroic. Comparing the RSP500 and DD dichroics with 488-nm excitation shows the efficiency difference between the low-power and high-power objectives, necessitating the need to test all of the objectives. From these data, it can be surmised that when using single-wavelength excitation, the DD should be used in preference to the RSP for 488-nm excitation for all lenses, and the TD should be used instead of the DD for 568-nm excitation for all lenses. This is a QA test to determine the efficiency of the dichroics in CLSM and help in determining which dichroic should be used in a single-wavelength excitation experiment. This dichroic test is used with single-wavelength excitation and application of the data allows the system to be run at lower PMT values, which translates into less noise and better performance. With multiple excitation wavelengths, the dichroics have to be chosen to balance the power of the emitted fluorescence from each fluorochrome, but this test is helpful in making the decision. A lambda scan of the dichroics will indicate their actual spectra in the system and whether there are alignment problems.

4.7. Axial (Z) Resolution

The axial resolution test is considered the "gold standard" of resolution in confocal microscopy ***(1,3,7,8)***. The axial resolution test is made using a 100× PlanApo (NA = 1.4) objective and has yielded below 350 nm. It should be emphasized that this is the only performance specification in 2006 that a company has said it will guarantee on a confocal microscope. Normally in a functioning system, values between 280 nm and 350 nm with a 100× (NA = 1.4) lens were obtained. A 63× PlanApo (NA = 1.32) lens should meet the specification of 400 nm, although Leica does not currently guarantee this value on a TCS-SP system. If the laboratory does not have a 100× PlanApo (NA = 1.4) objective and it is not possible to borrow one for comparison purposes from another confocal facility, it is useful to have other lenses as a reference point. The axial *Z* resolution of three different lenses was the following: a 40× (Fluor, NA = 1.0) was 610 nm; a 63× water-immersion lens (PlanApo, NA = 1.2) was 390 nm; and a 63× oil-immersion (PlanApo, NA = 1.32) was 315 nm. The excellent resolution that was obtained with the 40× and 63× lenses on our aligned system can serve as a system standard for axial resolution in a correctly aligned machine for other investigators using Leica TCS-SP equipment. It is important that the lenses achieve good values or the resolution in the system will be inadequate. It is also important that the pattern of the axial resolution be symmetrical with suitable diffraction regions (peaks and valleys) to the left of the major peak (*see* **Fig. 3B**). Normally, the axial registration does not change over time assuming the laser lines are stable. However if alterations are made in the scan head (i.e., galvanometer replaced) or when the lasers in the system are replaced, it will be necessary to realign the system and measure the axial resolution again. The quality of the lens by this test will relate to the quality of the biological image and that is why it is called the "gold standard."

It is important to compare the user-determined test slide with that of the service technician's slide to ensure that both specimens are yielding the same value. It should be emphasized that not all lenses are created equal and some will yield better resolution than others, as clearly illustrated in **Fig. 3**. If possible, lenses should be chosen from the manufacturer for excellent quality. Currently, there is a grade of lenses defined as confocal grade by one manufacturer. These lenses should be acquired, as these lenses undergo higher QA procedures in the factory and they are guaranteed to show excellent axial resolution, spectral registration, and other excellent lens characteristics. Other manufacturers should let you evaluate the lenses that are purchased prior to acceptance of the CLSM system. It is very important to have the best quality objectives on a CLSM.

4.8. Axial Registration (Beads)

This method is slightly more subjective than the axial z mirror test, but it does yield similar values most of the time. For unknown reasons, the values might be better or worse than the mirror-derived values. The mirror test is more accurate and should be used if available.

4.9. Square Pixels

The changing of a galvanometer stage will require that the symmetry in *XY* directional field scanning be measured. Measurements of objects will be inaccurate if the *xy* scanning does not yield a perfect square. This value should stay constant, but it must be checked and adjusted periodically. The phase adjustment in bidirectional scanning can also be checked and adjusted by this test. A lack of adequate phase alignment will result in a decrease in resolution. In cases that require the highest-resolution image, unidirectional scanning should be done, knowing that the scanning time will double. The galvanometer stability can also be checked using the square pixel test

4.10. Spectral Registration (UV and Visible)

The 1-μm multiple wavelength fluorescent beads (Tetraspec, T7284, Molecular Probes; or Rainbow beads, Spherotech) were used to monitor the UV and visible colocalization. The registration of bead fluorescence images between the 365-nm UV wavelengths and the 568-nm visible wavelengths in an aligned system was almost superimposable (*see* **Fig. 5A**), whereas in a misaligned system (*see* **Fig. 5B**), the difference between the peaks was 650 nm (acceptable difference is only 210 nm). The 568-nm line was chosen instead of the 488-nm line to minimize the crossover fluorescence between the visible and UV wavelengths. If the wavelengths are not aligned, then colocalization and fluorescence resonance energy transfer (FRET) studies cannot be effectively made on the machine. The image data must also be expressed as a maximum projection to eliminate the spectral mismatch. It appears for an unknown reason that if there is proper spectral registration between UV and visible wavelengths, then the UV field illumination might not be uniform, and vice versa. Both field illumination and spectral registration parameters must be checked. In addition, the confocal machines have separate collimator lenses that are used to align the UV light for different magnification lenses. It is very difficult to get all of the objectives to show proper spectral registration between UV and visible wavelengths.

The spectral registration of the 365-, 488-, 568-, and 647-nm lines can be made with either a small (0.5- or 1.0-μm multicolored bead) or a front-surface, single-reflective mirror (*see* **Fig. 7**). The spectral registration with the mirror on a Leica system is a superior test to the bead, as the laser light can be measured sequentially or simultaneously to eliminate any crosstalk between adjacent emission wavelengths. In addition, no bleaching occurs at high zoom magnifications with the mirror.

With new confocal systems that contain three visible lasers, the spectral registration test measures both the lens spectral registration and the laser spectral registration. It would be useful to measure a few different objectives to determine if the spectral registration of the lasers is correct or if a pattern of misalignment occurs. It is highly unlikely that different lenses will show the same spectral mismatch and, thus, the pattern observed should indicate if there are potential problems with either the lasers or with the objectives.

Molecular Probes produces a series of different-sized beads (Focal Check 1 μm, 6 μm, 15 μm) with different fluorescent rings and core bodies to assess colocalization from multiple lasers. These data should reveal if the laser lines are aligned correctly. The smaller bead should be more accurate but slightly harder to use.

4.11. Laser Power Stability

Power stability in a CLSM can be influenced by many factors, which include the lasers, PMTs, electronics, electronic component failure, fiberoptics transmission, fiberoptical polarizations incompatibility, AOTF thermal regulation, thermal heat dissipation, optical components, and galvanometers. The data obtained from this power stability test alert the investigator to possible errors that might exist in the acquisition of intensity measurements in biological and physiological experiments ***(1,11)***. The reason for such high variations in CLSM systems is unknown. For an investigator, it is not initially important to know where the source of instability is being generated, but only that it exists. Once the problem is identified, trained microscope service personnel will be able to troubleshoot the system and hopefully remove the source of the power instability.

Laser stability measurements should be made on the CLSM for the investigator to have confidence that the CLSM is not introducing artifacts in experiments requiring intensity measurements or time-dependent physiological experiments. In our system, there was periodic noise in the laser power tested by the transmitted light detectors that exceeds the manufacturer's (Ominichrome) laser stability fluctuations specifications of less that 0.5% over a 2-h time period. The 488-nm and 568-nm lines have a periodic cycle that is directly opposite the 647-nm line (*see* **Fig. 9**). Argon lasers and helium–neon lasers (543 nm or 633 nm) are considerably quieter than the argon–krypton laser and are definitely preferable.

In a confocal microscope, there are different ways to measure power stability over time (hours). These include the following: (1) manufacturer-installed pin diodes; (2) laser meters on the microscope stage connected to a readout device; (3) fluorescent emission intensity from a plastic slide detected by a PMT; (4) transmission optical system detection. The pin diode test was not stable and should only be used as a subjective assessment of power. A laser power meter can be connected to a UV or visible (VIS) wavelength detector situated on the microscope stage and then be continuously used to monitor the power output with either a chart recorder or equivalent computer software (Coherent).

Simultaneous comparison of the measurements using a pin diode in the Leica SP CLSM and either a power meter on the stage or the transmission average intensity (not shown) demonstrated that the pin diode has unstable power readings, whereas the other two measurements (transmission optics detection and power meter detection) were relatively stable over time. The pin diode should not be used as an absolute indicator for power or stability, as the power derived from it can vary in intensity over time. It can, however, be used as a subjective assessment of the laser performance and system alignment.

A UV or VIS detector situated on the microscope stage and connected to a suitable power meter can monitor the CLSM laser power. The power output intensity is continuously monitored with either a chart recorder or equivalent computer software. Manual measurements are deemed not accurate enough and are very time-consuming. If transmission optics is not available on the system, a power test can be made that uses a fluorescent slide sample placed in the light path. However, the investigator must be aware that repeated samplings of a fluorescence slide could bleach the sample, which will decrease the fluorescence intensity and increase the transmission intensity. Therefore, the laser power should be decreased with the AOTF to minimum values to help reduce slide bleaching, as decreasing the laser power with the power supply might result in laser instability. In addition, one must be aware of possible energy excitation of the fluorochrome in the slide.

In our experience, the most reliable method to measure the laser power stability consisted of using the transmission optics of the CLSM without a fluorescence slide in the optical path. If transmission optics is not present, a fluorescent-colored slide can be used, but extreme care must be made to reduce the laser power with an AOTF to reduce possible interaction of the laser beam on the sample. Sometimes the output fluorescence intensity might increase as a result of repetitive additive excitation or decrease resulting from bleaching.

Lasers used in flow cytometry or confocal microscopy equipment should be stable with low peak-to-peak noise and minimal power fluctuations over hours. Laser noise can originate from different sources, which include the AOTF, laser polarization mismatch, heat dissipation, and power supplies. One of the most likely causes is fiber polarization, which might be mismatched with the polarization of the laser. One of the most likely sources is a poor power-supply regulation that results in light output fluctuations at the frequency of line current used to run the power-supply ***(15)***. Noise in a helium–neon laser might be found at frequencies of a few hundred kilohertz as a result of either radio-frequency energies used to pump the laser medium or of fluctuations in the medium itself ***(15)***. The DC power supply used should be the correct type (Omnichrome power supply 171B or 176B with Omnichrome argon–krypton laser) to produce low noise and should be operated at "Light mode" (constant power) and not constant current mode ***(15)***. The 171B power supply had transformers and heating problems and has been replaced by the 176B model with rectifiers that regulate heat better. Typically, a Coherent Enterprise laser, Coherent 90-5 or Coherent 70-4, will have less than 1% peak-to-peak noise ***(15)*** and the power will not fluctuate over time. The air-cooled argon laser from Uniphase or Spectra Physics used in benchtop flow cytometers or confocal microscopes will also have less than 1% peak-to-peak noise according to the manufacturer's specifications. The argon–krypton (Melles Girot, Omnichrome 643) laser contains three simultaneous laser lines that yield power intensity fluctuations of less than 0.5% for 2 h (personal communication and website). If this is true, where do the fluctuations in power intensity in excess of 10% as shown in **Figs. 8** and **9** come from if the laser is not generating it? Is it the AOTF or fiber polarization? Lasers can be checked with power meters in front of the beam or by special electronic boards that connect to the power supply to determine laser stability. However, this testing might not be possible by the investigator, as the laser is in a sealed compartment and the investigator is not allowed into this compartment or the service contract will be canceled.

4.12. Laser Power Stability (UV)

The Coherent Enterprise laser delivers less than 1% peak-to-peak noise and is considered a very stable laser. The argon air-cooled lasers, HeNe lasers, and Spectrophysics argon–krypton laser are all rated at less than 1% peak-to-peak noise. However, even with the Coherent UV Enterprise laser or a HeNe laser (543 nm), periodic noise and large power fluctuations were observed. One source of power stability appears connected to the way the laser is cooled and how the laser heat that is generated is dissipated. This was illustrated with our Coherent Enterprise UV laser that was connected to a Coherent LP 20 water–water exchange cooler. This cooler should be set at least 10°C above the circulating cooling water of the building and it should be set above the ambient temperature of the room. Improper set points for the LP 20 cooler resulted in temperature-regulation problems of the circulating cooling water in the laser, which, in turn, resulted in the improper regulation of the laser power (*see* **Fig. 9,** trace B). In addition, problems with proper fiber alignment also appeared to occur with the UV system, resulting in power fluctuations (*see* **Fig. 9,** trace C). The elimination of these temperature and polarization issues resulted in proper laser cooling and laser stability (**Fig. 9**, trace A; <3% noise). The water exchanger appears to have better thermal regulation than the noisier air–air (LP5) exchanges, but as illustrated in **Fig. 9**, the cooler must be set properly or thermal power instability might occur.

4.13. Laser Configurations (Fiberoptic Polarization)

This argon–krypton laser has been incorporated into the older-designed systems from which most of these measurements have been made. These argon–krypton lasers will deteriorate with time because of the gas escaping, resulting in a continuously reduced power output over time. The new systems from Leica, Zeiss, Nikon, Olympus, and Bio-Rad all have merge modules and individual lasers. The merge module design allows for the incorporation of multiple lasers in a confocal microscope that are less noisy and more stable than the Omnichrome argon–krypton laser supplied by Melles Girot. It is recommended to acquire a stable low-noise laser. One major problem with the design of the current version of confocal microscopes that use merge modules is that the laser lines are directed into the scan head with fiberoptics. This is in contrast to the older versions of confocal microscopes, which used direct coupling with only dichroics to deliver the light to the microscope's scan head. The use of the fiberoptics makes it critical to align the polarization in the fiberoptic with that of the laser's polarization value *(1,15)*. Failure to do this could result in laser power instability in the CLSM system. The fiberoptics also deteriorate with time, which will attenuate the laser power and necessitate running the machine with higher laser power or higher PMT settings. One test procedure to ensure that polarization is correct after the alignment procedures is to wiggle the fiberoptic and see if the image returns to the same intensity, suggesting that the polarization is correct. This is a fairly crude test, but it will demonstrate whether the system needs further alignment. Confocal microscope manufacturers are supposedly devising more reliable tests to check for polarization and ensure laser power stability within the CLSM.

4.14. Heat Dissipation

The dissipation of heat is a very important variable to consider in measuring laser stability. Improper heat dissipation with the air-cooled lasers will result in laser power fluctuations. In our case, we have observed fluctuations with an argon–krypton air-cooled laser that (1) had a restrictor in the exhaust line and (2) used a smaller exhaust duct (4 in. instead of 5 in.) to remove heat. In both cases, the heat was not dissipated correctly and the laser power in the CLSM fluctuated above 20%. Removal of these problems reduced the laser calculations to less than 5%. All lasers have to dissipate heat properly and their thermoregulators must be set correctly or the power to the lasers will fluctuate as illustrated using the Coherent Enterprise laser in **Fig. 9**.

4.15. AOTF

The AOTF could also introduce power fluctuations in the CLSM system by improper thermal regulation. The AOTF is a birefringence crystal capable of rapid and precise wavelength selection. Earlier CLSM systems used dichroics, barrier filters, and neutral-density filters to regulate the proper intensity of laser light that illuminates the samples and did not use the AOTF. However, the AOTF is enormously useful in operating a confocal microscope, as wavelength selection can be very easily and accurately controlled. It is also used to control the power that illuminates a sample, which effectively acts to reduce the crosstalk between detectors. The original AOTFs that were installed in older CLSM models were not temperature regulated and could have introduced some power instability into the confocal microscope systems. This problem is very disturbing to investigators expecting to make comparative intensity measurements on biological samples. The laser has 1% fluctuations, but there is over 10% observed on the stage. The AOTF is suspected of introducing these power fluctuations.

In summary, the laser power instability might be attributed to the power supply, AOTF thermal regulation, improper thermal heat dissipation, electronic component failure, or fiberoptical polarization incompatibility, but the definitive source of laser power fluctuations has not been identified at this time.

4.16. Standard Bead and Fluorescent Plastic Slide Tests

In flow cytometry, alignment beads, linearity beads, and chicken red blood cells (CRBCs) are used to ensure that the machine is functioning properly ***(16,17)***. It would be useful if a suitable bead or fluorescence slide test sample were used for a similar assessment on the CLSM. Fluorescent-colored plastic slides can be used as a reference standard; however, the issue of the depth that the measurement is taken at and possible surface irregularities will both affect the measurements. If these two variables are controlled, the plastic slide can be effectively used. Most of the description will be on a bead, but the data are applicable to uniform plastic slides at a specific depth that are lying flat on a microscope stage.

4.17. Biological Test Slides

It is important to have a reliable sample that can be used to test the machine performance and image quality. The most useful histological test slide in our laboratory was a FluoCell slide (Molecular Probes; F-14780) stained with three fluorchromes. This slide allows for proper evaluation of resolution, crosstalk between detector channels, and observation of the emission from multiple excitation wavelengths (UV, 488 nm, and 568 nm). The resolution of biological structures such as mitochondria (Mitotracker red), nucleus (DAPI), and tubules (Alexa 488) can be assessed with the slide. The pollen or diatom slides (Carolina Biological) have also been used to demonstrate fine structure at various magnifications with different excitation wavelengths. Leica service engineers use a fluorescent plant tissue that can be excited using all of the wavelengths. The histological plant test sample has been used for a combination of power output, field illumination, resolution, and overall assessment of the machine. Most sales and service personnel use their favorite histological slide as their gold standard to determine if the system is functioning properly. Although a trained confocal person can evaluate many parameters on a confocal microscope, these observations on histological slides are subjective and cannot be confirmed, and the machine performance should not be totally based on this type of sample as a performance standard, as illustrated by the data described in this chapter.

A biological sample is a very subjective method of addressing total system performance. Similar to beads, biological test samples (FluoCells, F-14780, Molecular Probes, Acridine Orange-stained CRBCs) show a relationship between frame averaging and PMT voltage. The higher the voltage, the more averaging that is necessary to produce a desirable image. Because the relationship between PMT and CV (image quality) is logarithmic, very small changes in PMT values can indicate major changes in system performance ***(1)***. In our opinion, the use of an acceptable range of hundreds of PMT units is far too large to effectively assess instrument performance.

4.18. General Confocal Principle

Generally, as the PMT is used at higher values, the sample needs to be averaged more to reduce the noise in the image. How much averaging is done is dependent on the amount of bleaching in the sample and the amount of time available to average. The more averaging, the better the image, but after a few averages, there is a limiting return, with factors of bleaching affecting image quality. We have found that averaging between three and six times dependent on PMT value yields good images. **Figure 12** shows the relationship between CV averaging and PMT value. In certain cases, the pinhole size might need to be increased to lower the PMT voltage and decrease the noise. This will sacrifice some confocality in the system.

4.19. CLSM Comparison

It is possible to compare similar and different systems using this CV test. This is done by first measuring a specific fixed power value on the stage, keeping all staining parameters constant and then measuring the noise of a single scan line. The amount of fluctuation that occurs in a line scan may indicate the instability of the system to produce a good image. It will be directly related to PMTool talk. The more line fluctuations that occur during scanning, the more averaging that will have to be used to create a desirable image.

4.20. Image Size

A sample should be adjusted using medium speed. A slower scan speed is equivalent to averaging more, as the laser spends a longer time on each pixel. Larger-sized images, (e.g., 1024 × 1024 or 2048 × 2048) give only slightly better detail than 512 × 512 but allow for extra zooming, which could be useful in visualization of structures. Compared to the 512 images the 1024 images are 4 times larger and the 2048 images are 16 times larger. They will take 4 and 16 times, respectively, longer to acquire, and it will need 4 and 16 times more disk space, respectively, to store the data. Most of our routine work is done at 512 × 512 resolution *(18)*.

4.21. PMT Factors

The PMT is the detection unit of a confocal microscope and its quality and function is extremely important to a confocal microscope. Not all PMTs are equal; therefore, the PMTs should be checked to determine if the unit in the confocal microscope has good sensitivity. The bead tests describe a procedure that can be used to measure PMT sensitivity. The PMT should have a large dynamic range, be linear, and show good sensitivity in the wavelengths measured *(7,19)*. Often, the confocal user will set the PMT at high voltages to observe an image and be unaware how this setting influences image quality.

Operating a system with high PMT values will generate poor image quality because of reduced signal and excessive noise. This noise will have to be eliminated by averaging to yield a good image. Frame averaging or high laser power will increase image quality, but bleaching usually occurs by repeated scans over the same sample or with high laser power. It would be useful to have the best PMT with the least noise and greatest dynamic range installed in the machine.

4.22. Spectral Scanning: PMT Comparisons

The PMT performance was also measured using the Leica spectral scan feature. This feature allows for a sequential scan supposedly across the whole range of 400–800 nm in units as small as 5 nm. Using the light form lamp, a spectral scan consisting of 50 increments of 5 nm each was made between 400 and 650 nm using a 10× (NA = 0.4) PlanApo lens or with an Airy disk of 1 and without averaging ***(24)***. The efficiency of the system was low, necessitating a large PMT voltage setting. It is essential to use a reflecting mirror and not a dichroic to measure the spectral scans so that no attenuation of light will be introduced across the spectral range. In our system, PMT 2 has a direct path to the detector, whereas PMT 1 and PMT 3 need to be reflected off an additional mirror prior to entering the detector. PMT 2 showed sharper bands than PMT 1 and PMT 3. However, PMT had less response in both the lower blue regions and far-red regions. This might be the result part of the light path, quality of PMT, PMT alignment, or type of PMT. The PMT pattern with the tightest peaks and the largest valleys define the PMT with the best spectral response, which is located in the second position in our machine. PMT 2 should thus be used for spectral imaging experiments. This test has been adapted to the Zeiss 510 Meta system.

4.23. Interference Contrast and Confocal

Interference contrast is a very useful parameter in microscopy and it can be combined with fluorescence. However, because the microscope system was designed for light to traverse through two interference filters, when this optical system is applied to a confocal microscope there is distortion in the fluorescence signals. The fluorescent light traverses the interference contrast filter and excites the sample, and then the emitted fluorescence travels back down through the same interference contrast filter and back through the scan head. The resulting image shows a duplication of very small particles (0.17 μm, PSF beads) and a distortion of larger particles. PSF beads show two spots and 0.5 μm beads show an egg shaped image instead of a round image. The same distortion that is observed on beads will occur on biological structures in cells (*see* **Fig. 15**). For optimum resolution of data that will be deconvoluted later, it is recommended to remove the interference filters when acquiring an image.

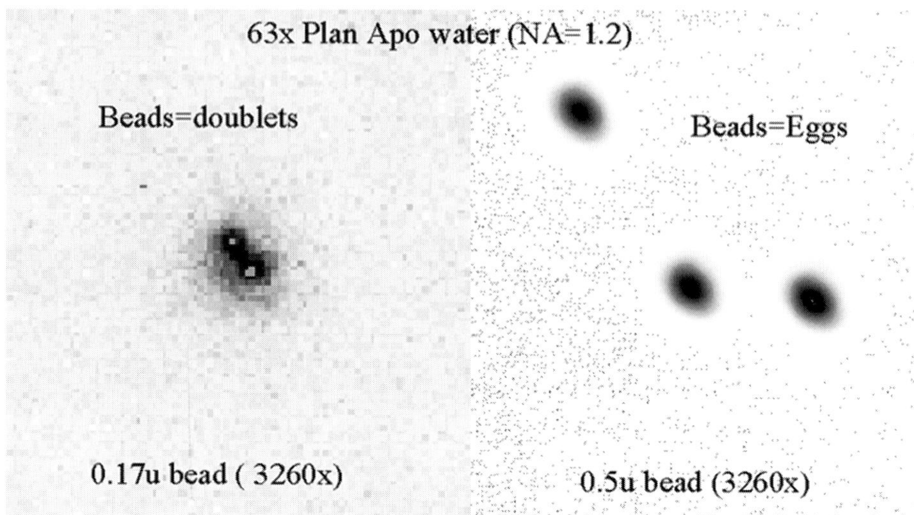

Fig. 15. Incompatibility of interference and CLSM. The Normarski interference contrast light path consists of an analyzer, a polarizer, and two interference filters. In acquiring an interference image simultaneously with a fluorescent image, the interference filters interfere with the fluorescent image, resulting in doublets using PSF beads (0.17 μm) and egg-shaped objects with 0.5-μm beads. All objectives show aberrations with the interference contrast transmission optics in place and it is assumed that the same aberrations will occur with biological structures.

4.24. Good Images: Take-Home Lessons

To obtain a good confocal image, it is necessary to balance laser power, PMT voltage, frame averaging, pinhole aperture size, scan speed, and zoom magnification ***(1–6)***. For a novice user, it might be difficult to know how to balance these factors to produce an ideal image. Although all of these factors affect image quality, we have found that the zoom greatly bleaches samples and high PMT settings will introduce excessive noise that can only be eliminated through averaging, which, in turn, can also bleach the sample. The zoom factor increases the bleaching by the square of the zoom factor. If possible, magnification should be obtained with objectives, not with the zoom magnification ***(13)***. The PMT values should be kept as low as possible, even to the extent of increasing the pinhole diameter, decreasing the scan speed, and increasing the laser power. Although the pinhole diameter determines the confocality of the system and should ideally be set at the size of the first Airy disk, the pinhole diameter can be increased if insufficient fluorescent laser light is emitted from the sample, making visualization noisy and difficult. In a filter system, the detection bandpass barrier filter can also be replaced with a Schott long-pass filter to increase the amount of detected light, or in a Leica spectrometer system (TCS-SP), the slit can be adjusted to wider values. However, the quality of the image can also be affected by the spectra of wavelengths that are recorded. A slow scan speed can be used, which increases the time the laser beam resides at each pixel. These acquisition factors should be regarded as starting points for using the CLSM and not as an absolute rule, as every sample might have unique qualities that must be considered prior to imaging.

The aim of acquiring a good image is to lower the CV of the image. The challenge for the investigator is to create this image using a standard pinhole (Airy disk = 1, eliminates light) without bleaching the sample and without using excessive laser power or long exposure times. At a given PMT setting, increased frame averaging produces a greater reduction of CV (decreased noise). The use of lower PMT settings allows for less averaging because of the decrease in noise in the system (*1,3*) **(Figs. 11 and 12)**. Antifade compounds are useful to reduce the bleaching, allowing for increased averaging with fixed samples. Another way to reduce bleaching is to reduce the size of the sample image field (e.g., from 512 × 512 to 128 × 128) acquired. This will result in shorter exposure times with less bleaching while keeping the PMT and laser power settings constant. However, this also results in a smaller sample size and less resolution. For performance tests, the loss of minor details as a result of decreased resolution is not a significant issue, but for biological samples, structural details are very important and good images should be taken at pixel values of 512 × 512 or higher.

In summary, to visualize any fluorescent specimen, a sufficient amount of light has to be delivered onto the sample located on the stage. If the sample cannot be adequately visualized, then it will be necessary to increase the PMT voltage. As the PMT voltage is increased, the image noise also increases. At higher PMT settings, it is necessary to use more frame averaging in order to reduce the noise in the image. However this leads to longer exposure times and possible sample bleaching.

4.25. Laser Choices

The new configuration of CLSM uses the quiet helium–neon lasers (543 nm and 633 nm) that will yield lower power values at the microscope stage compared with the noisy 75-mW argon–krypton laser. These low-noise lasers should deliver better signal-to-noise performance, as they have less peak-to-peak noise than the argon–krypton laser (*15*). The advantages of these small air-cooled lasers are clear: They are more stable, less noisy, last longer, and create fewer problems for the field engineers. Lasers should be chosen dependent on power, dye excitation noise, and frequency of service. Currently, higher-power argon lasers (30, 50, or 100 mW) are being installed to allow one to use the following excitation lines: 458, 478, 488, 514, and 532 nm. The HeNe 543-nm laser is 1–1.2 mW and HeNe 633-nm laser is either 5 or 10 mW.

4.26. CV Principle

One aim of this research was to apply similar statistical procedures, used for many years in flow cytometry, as a standard to evaluate CLSM images and system performance ***(16,17,20)***. One method to assess flow cytometry system performance is to use a population of uniform fluorescent beads and measure the fluorescence and light scatter of approx 10,000 beads. The measurement of 10,000 beads yields a distribution of fluorescent intensities and sizes, which correlates to the particle variation and system performance factors ***(17,18)***. From this distribution, the CV can be measured. We have applied the principle of CV of a population of beads to evaluate confocal system noise, image quality, and system performance. Instead of using thousands of beads to produce a fluorescent histogram, this novel technique uses thousands of pixels from a single bead to generate a population distribution. From these pixel-intensity values, the population means and standard deviation, and thus the CV, can be determined. Given the impracticality of imaging tens of thousands of beads to get a distribution of fluorescence intensities or particle sizes with a CLSM, we have analyzed a large bead consisting of many pixels. The intensity deviation of these pixels represents the noise in the scanned image of the bead. If the beads are uniform in intensity and size, then they can represent a standard for the evaluation of image quality and the performance delivered by a specific manufacturer's system. Generally, it is assumed that the smaller CV represents a system that is properly aligned, stable, and yielding good resolution and system performance. The inherent sources of noise were analyzed by measuring the pixels contained within a large ROI in one large bead. By zooming the bead 4×, problems with the field illumination, focal excitation plane, and the presence of unequal illumination at the edges of the bead were eliminated.

Although the CV Spherotech bead noise test might be useful to evaluate the reproducibility of the system over weeks and months, its applicability in our laboratory was found to be limited as a daily/monthly test. The laser power and stability in the CLSM system were extremely variable for unknown reasons. Perhaps this can be attributed to the quality of the argon–krypton laser alignment instability, variable temperature dissipation, or the AOTF temperature instability.

4.27. CV Test

The CV that we are measuring in a confocal microscope is actually the variation of pixel intensity within the bead, as opposed to flow cytometry, where the variation of intensity is among a population of beads. The 10-μm bead appeared to be the correct size using an Airy disk of 1, as the image that was captured contained relatively homogenous pixels throughout the bead area using two different microscope systems (Zeiss 510 and Leica TCS-SP1). The area inside the bead was of sufficient size to allow a uniform ROI to be defined within it. It was helpful to zoom the bead four times to increase the quantity of pixels contained within the ROI. Repeated sampling of the same bead resulted in minimal bleaching and the CV did not change significantly during subsequent scans. It is important to maintain the machine variables (pinhole size, PMT voltage, averaging, etc.) at reproducible values for all studies. The laser power was set at a constant value that allowed the mean intensity level of the bead to be approx 150 (out of 255) for each PMT setting.

The relationship between the PMT and calculated CV is the following: As the PMT voltage is raised, the distribution of the pixels in the bead increases, resulting in an increased CV. The value gives the reference of each PMT relative to each other and suggests which one should be used for optimum performance. In a Leica system, this information is useful to evaluate the PMTs and, thus, determine the best PMT that should be used to acquire signal at the lowest voltage settings and possibly how much averaging will be necessary to achieve a desired image quality ***(1–3)***. By using the most efficient PMT, the image CV is decreased as lower PMT values are used. An increase in CV values might imply that there is a decrease in laser power after the stage measurement from misalignment, resulting in higher PMT values. This CV test on beads can detect a noisy laser. With the argon–krypton laser, fast scans actually yielded less noise than slower scans, which was interpreted as when the pixel dwell time was longer, the laser fluctuations were more detectable.

There are numerous factors that effect the CV measurement. These include the fluorescence of the bead at the emission and excitation wavelengths, optical components and efficiency of the system, maximum system laser power obtainable, and functionality of optical components in the system and electronic components (PMTs). In addition, many acquisition parameters such as scan speed, pinhole setting, and objectives will affect the CV value. When all of these factors are considered, the CV will essentially be a measure of the system's relative sensitivity. Lower CVs will translate into better image quality with less image averaging and less bleaching. Because PMTs will deteriorate with time, it is important to measure the initial CV (bead) and then to periodically measure how the CV (image quality) changes over time as a reference point for possible replacement (*see* **Table 3**).

4.28. Sensitivity Test

The sensitivity of any fluorescence optical system depends on the intensity of the light source, the efficiency of the optical system, and the quality of the detection system ***(7,19,21)***. For confocal microscopes specifically, the sensitivity comprises variables that include PMT noise, laser noise, alignment, and system efficiency. It would be extremely useful if there was a test that could assess sensitivity in this optical equipment. We believe that we have developed a fluorescent bead test that can be used to measure sensitivity over time, so that an assessment can be made on how the machine is performing over time. The test can also be used to compare the sensitivity of two machines from one manufacturer or compare the machines from different manufacturers with regard to sensitivity and performance. Extreme care must be made to ensure that all variables are equivalent when undertaking this exercise. We have shown that the CV bead test confirms principles of noise reduction by averaging sequential frames. The noise is reduced inversely as the square root of the number of frames averaged ***(18,22)***.

The voltage setting of the PMT was the primary determinant of image quality and bead noise. An increase in PMT value was always accompanied by an increase in image noise and pixel CV distributions. This test can be used to assess the sensitivity of an individual machine, and if the acquisition variables are rigorously controlled, it could measure the sensitivity between different machines. For example, the best CV on a Leica TCS-SP1 with UV excitation that was obtainable using a 10-μm Spherotech bead was 19%, whereas the best CV obtainable for 488-nm excitation was 3.8%, and for 568-nm excitation, it was 2.5%. These percentages are invaluable to determine the sensitivity of our TCS-SP and compare its sensitivity to other CLSMs. The values will represent how well the system is aligned and how well each system is functioning regarding this sensitivity parameter. These values are dependent on the maximum power throughput in a system.

The CV bead test has been used to monitor sensitivity in a CLSM system. This test is important not only to detect defective lasers and system alignment problems but also to set some rational basis to determine how much averaging will be necessary to remove the noise from an image. To compare the sensitivity of CLSM system lasers, one can measure a defined milliwatt power on the stage and then conduct the bead noise test as described previously **(Subheading 3.)**. Because the power on the stage is the same under different systems and configurations, the CV should be an indicator of system sensitivity that includes efficiency, laser noise, and PMT detector quality. Using this test, we found that 1 mW of power yielded a CV of 4% on a Leica TCS-SP and 1.3% on a Zeiss 510 system. With the 568-nm line, using 0.2 mW of power yielded 4% on a Leica TCS-SP system and 2% on a Zeiss 510 system. This test reveals, as expected, that the noisy argon–krypton laser yields a higher CV than the quieter three individual-laser systems.

The comparison of the CV from an argon–krypton laser system (Leica TCS-SP1) and air-cooled argon/HeNe system (Zeiss 510) at the same milliwatt ranges were made and it was determined that the system with individual lasers was functioning better than the system with only a single Omnichrome argon–krypton laser emitting three wavelengths. The data can be attributed to the different style lasers. It should be emphasized that there are many variables that can affect these CV measurements and they should be used as approximate values, not absolute values. However, the approach appears to be very useful to compare like machines where there can be no argument that the test was done incorrectly. It should be emphasized that reducing the light to increase confocality will also decease the sensitivity of the system, as fewer photons hit the PMT and the PMT voltage is raised. It is important to do these tests at a uniform pinhole setting of one Airy disk.

The comparison of maximum power UV (365 nm) and visible (568 nm) excitation on a 10-μm bead revealed the following PMT and CV values: UV (PMT = 679 V, CV = 19%) and 568-nm (PMT = 382 V, CV = 5%) excitation. Explanations for the difference in CV between the two excitation wavelengths include the following: (1) The bead might not be excited as well in an UV system as in a visible system; (2) the optical system might be more efficient and less attenuated with visible excitation compared to UV excitation; or (3) there is insufficient UV laser power necessitating the increase of the PMT value to observe the bead at channel 150 (pixel mean). This bead test appears to be a reproducible test with both wavelengths that can be used to compare the sensitivity in a CLSM machine and could also be used to compare different machines. However, in our experience, because of some unknown laser power fluctuations, daily tests on a confocal microscope do not appear to be as useful, simple, or efficient as they are on a flow cytometer.

4.29. AOTF and Spectrophotometer Sliders and Functional Wavelengths

The functioning of the AOTF and spectrophotometer sliders can be checked by the following two procedures. (1) Set the machine up in the reflection mode similar to that for an axial *Z*-resolution test. Positioning the 5-nm sliders over the individual laser lines indicates the approximate relative position of the laser line. By moving the 5-nm slider above or below the laser line, the laser light should eventually be attenuated. The laser light should be reflected from regions 2–3 nm below and above the desired laser line of choice. If light is reflected below or above the laser line being checked, it suggests that the sliders or AOTF might not be selecting the correct lines. The 488, 568, and 647 nm laser lines had 5 nm, 8 nm, and 13 nm of reflected bandpass, respectively. A lambda scan between 470 and 670 nm can be made to show the width of the reflected laser lines; (2) Fluorescent plastic can be placed on the stage and the sliders can be placed close to the excitation band to determine where the emission lines are and where the excitation lines are. This test will determine the acceptable detection wavelengths above and below the excitation wavelength. It should be less than 10 nm below the excitation wavelength. It will be more difficult to test the functionality of the AOTF or proper functioning of excitation wavelengths in a filter containing the confocal system. However, by measuring the emission wavelengths of the laser line through different barrier filters from PMT 1, PMT 2, and PMT 3, a pattern of acceptable fluorescence can be determined.

4.30. Purchase Considerations

The purchase of a new confocal machine is a very difficult and complex decision *(23)*. Too often it is decided on subjective criteria, such as whether a specific machine can observe a phenomenon on a slide or generate a "pretty picture" during the demonstration. The pretty picture might be primarily determined by lens characteristics or the functionality of the demonstration unit. It is critical when comparing machines from different manufacturers that they be set up in a similar fashion. The laser power, objectives, scans speed, illumination and detection pinhole size, and other hardware components are factors that affect image quality. Other critical factors entering into the buying decision should include service and company support after the machine is delivered. QA, organizational, and service issues should not be overlooked in choosing which manufacturer's machine to buy, as the buyer will have a long-lasting relationship with the vendor and it should be based on trust and knowledge that the machine is operating at its top performance levels. The ability of a manufacturer to guarantee specifications on their machine and address the issues of QA would rate very high on the list of criteria for choosing which CLSM machine to buy. It does the investigator little good if he/she has the best designed machine that does not perform to those advertised high standards in his/her laboratory.

Microscopes connected to a confocal scanning apparatus can also be combined with digital cameras if a suitable port exists. The confocal microscope should contain excellent objectives and the microscope should be ordered with a 100-mW bulb light housing instead of a 50-mW housing so that the microscope can be used also for conventional fluorescence applications. Very high-quality lenses consisting of water, multi-immersion, and oil should be ordered. These should have the largest NA and greatest working distance possible. Units are sometimes ordered with lower-quality objectives because of budgetary constraints. This is one of the biggest mistakes in operating a confocal facility. The lenses are the engines that drive this technology and they should always be the best quality available.

4.31. Summary

We have described the following tests to measure system performance in a CLSM: field illumination for individual lenses, laser power indicators, dichroic efficiency, chromatic lens aberration, axial resolution, spherical registration, bead noise, PMT performance, sensitivity, laser stability, and noise analysis ***(1)***. The work of the field engineer also has to be checked carefully, as the field engineer might not be able to accomplish all of these tests during an installation or a preventative maintenance visit. Many sales and service representatives might have different levels of machine understanding, and without specifications provided by the manufacturers, the level of a correctly aligned and functional machine is open to question and debate. Unfortunately, even sales/service representatives from confocal companies can make mistakes in the judgment of what constitutes a correctly aligned machine; thus, it becomes necessary to use these tests to ensure that the machines are working correctly in the scientist's laboratories. Unfortunately, the CLSM might function at suboptimum conditions during its operation and problems are resolved only when the investigator cannot achieve the desired images or there is a hard failure of the system necessitating a service personnel visit. It is the responsibility of the core director in each lab to ensure that the machine is working at acceptable levels of performance.

Acknowledgments

The authors thank Jeff Wang of Spherotech for providing a 10-μm bead that was uniform in size and fluorescence intensity and did not readily bleach with repeated samplings. The authors also thank Jeremy Lerner for providing the light source to evaluate spectral imaging. Thanks also to Earl Puckett of the US EPA instrument shop for building a stage adapter for Newport or Coherent power meter detectors and to Keith Tarpley of CES Corporation for providing excellent technical assistance in the preparation of the illustrations. Figures 1, 3A, 5B, 7–9, 12, and 14 and Tables 2 and 3 have been previously published in *Cytometry* (vol. 44, 2001) and the journal has allowed these figures to be reproduced in this book chapter. The research described in this article has been reviewed and approved for publication as an EPA document. Approval does not necessarily signify that the contents reflects the views and policies of the Agency, nor does mention of trade names or commercial products constitute endorsement or recommendation for use.

References

1. Zucker, R. M. and Price, O. T. (2001) Evaluation of confocal system performance. *Cytometry* **44,** 273–294.
2. Zucker, R. M. and Price O. T. (2001) Statistical evaluation of confocal microscopy images. *Cytometry* **44,** 295–308.
3. Zucker, R. M. and Price, O. T. (1999) Practical confocal microscopy and the evaluation of system performance. *Methods* **18,** 447–458.

4. Centroze, V. and Pawley J. (1995) Tutorial on practical confocal microscopy and use of the confocal test specimen, in *Handbook of Biological Confocal Microscopy,* 2nd ed. (Pawley, J., ed.), Plenum, New York, pp. 559–567.
5. Centroze, V. and Pawley, J. (1998) Practical laser scanning confocal light microscopy: obtaining optimum performance from your instrument, in *Cell Biology*, 2nd ed. (Celis, J., ed.), Academic, New York, Vol. 3, pp. 149–169.
6. Sheppard, C. J. R. and Shotton, D. M. (1997) *Confocal Laser Scanning Microscopy*, Bios Scientific, New York.
7. Marjlof, L. and Forsgren, P. O. (1993) Accurate imaging in confocal microscopy, in *Methods of Cell Biology* (Matsumoto, B., ed.), Academic, San Diego, CA. pp. 79–95.
8. Carter, D. (1999) Practical considerations for collecting confocal images, in *Confocal Microscopy Methods and Protocols* (Paddock, S., ed.), Methods in Molecular Biology Vol. 122, Humana, Totowa, NJ, pp. 35–57.
9. Pawley, J. (1995) Fundamental limits in confocal microscopy in *Handbook of Biological Confocal Microscopy,* 2nd ed. (Pawley, J., ed.), Plenum, New York, pp. 19–36.
10. Pawley, J. (2000) The 39 steps: a cautionary tale of quantitative 3-D fluorescence microscopy. *Biotechniques* **28(5),** 884–886.
11. Swedlow, J. R., Hu, D., Andrews, P. D., Roos, D. S., and Murray, J. M. (2002) Measuring tubulin content in *Toxoplasma gondii*: a comparison of laser scanning confocal microscopy and wide field fluorescence microscopy. *PNAS* **99,** 2014–2019.
12. White, N. S., Errington, R. J., Fricker, M. D., and Wood, J. L. Aberration control in quantitative imaging of botanical specimens by multi dimensional fluorescence microscopy. *J. Microsc* **181,** 99–116.
13. Piston, D. W. (1998) Choosing objective lenses: the importance of numerical aperture and magnification in digital optical microscopy. *Biol. Bull.* **195(1),** 1–4.
14. Czader, M., Liljeborg, A., Auer, G., and Porwit, A. (1996) Confocal 3-dimensional DNA image cytometry in thick tissue sections. *Cytometry* **25(3),** 246–253.
15. Shapiro, H. (1995) *Practical Flow Cytometry,* 3rd ed., Wiley–Liss, New York.
16. Watson, J. V. (1991) *Introduction to Flow Cytometry*, Cambridge University Press, Cambridge.
17. Muirhead, K. (1993) Quality control for clinical flow cytometry, in *Clinical Flow Cytometry Principles and Applications* (Bauer, K. D., Duque, R. E., and Shankey, T. V., eds.), Williams and Wilkins, Baltimore, MD.
18. Russ, J. C. (1998) *Image Processing Handbook,* 3rd ed., CRC, Boca Raton, FL.
19. Pawley, J. B. (1994) Sources of noise in three dimensional microscope data sets, in *Three Dimensional Confocal Microscopy: Volume Investigations of Biological Specimens,* Academic, New York.
20. Watson, J. V. (1992) *Flow Cytometry Data Analysis: Basic Concepts and Statistics*, Cambridge University Press, Cambridge.
21. Art, J. (1995) Photon detectors for confocal microscopy, in *Handbook of Biological Confocal Microscopy,* 2nd ed. (Pawley, J., ed), Plenum, New York, pp. 183–195.

22. Cardullo, R. A. and Alm, E. J. (1998) Introduction to image processing, in *Methods in Cell Biology* (Sluder, G., and Wolf, D. E., ed.), Academic, New York. Vol **56,** 99–115.
23. Steyger, P. (1999) Assessing confocal microscopy systems for purchase. *Methods* **18(4),** 435–446.
24. Lerner, J. L. and Zucher, R. M. (2004) Calibration and validation of spectroscopic imaging. *Cytometry* **62,** 8–34.

6

Quantitative Analysis of Atherosclerotic Lesion Composition in Mice

Marilyn P. Wadsworth, Burton E. Sobel, David J. Schneider, Wendy Tra, Hans van Hirtum, and Douglas J. Taatjes

Summary

Comparative quantitation has become an increasingly desirable tool in determining compositional differences of aortic plaque lesion in transgenically altered mice. To this end, methodology has been developed to identify lipid, cellularity, collagen, and elastin components using traditional bright-field microscopy, fluorescence, and polarized light microscopy, employing both confocal and wide-field imaging systems. Subsequent imaging processing and analysis on the digitally captured images reveals differences in compositional components as influenced by diet, age, and gender. This method can be expanded to employ a rich variety of histochemical and immunohistochemical staining protocols.

Key Words: Atherosclerotic plaques; composition; lesion; quantitation; lipid; cellularity; collagen; fluorescence microscopy; polarized light microscopy; confocal scanning laser microscopy; image analysis.

1. Introduction

With the advent of digital image acquisition, image analysis has advanced from a mostly qualitative to a "semi"-quantitative science *(1)*. A variety of commercially produced software packages are available to perform sophisticated analyses on digital format images. For instance, such analyses might be useful in determining the morphometrical effect of a specific treatment or a transgene on cells and tissues. Indeed, we have been interested in developing computer-assisted image analysis methods for the determination of atherosclerotic lesion composition in transgenic mice *(2–4)*. A paradigm developed in the last 15 yr suggests that two types of atherosclerotic plaque exist: (1) highly cellular, stable plaques that attenuate the vessel lumen but are usually benign clinically and (2) plaques that contain a paucity of cells but are rich in lipid. These plaques (also

From: *Methods in Molecular Biology, vol. 319: Cell Imaging Techniques: Methods and Protocols*
Edited by: D. J. Taatjes and B. T. Mossman © Humana Press Inc., Totowa, NJ

referred to as "vulnerable" plaques) are prone to rupture, especially in the thin shoulder regions, resulting in clinical sequlae such as myocardial infarction and stroke ***(5–9)***. To address the molecular and biochemical mechanisms involved in plaque formation, sophisticated imaging methods must be developed to accurately assess lesion composition. The methods we are now developing are highly dynamic, and we continue to extend and optimize them.

The methods described in this chapter have been developed to specifically characterize the composition of atherosclerotic plaques in mouse aortic lesions. Mouse models, particularly the apolipoprotein E-knockout mouse ($ApoE^{-/-}$) are used extensively in cardiovascular research ***(10,11)***. We have demonstrated that plaques show compositional changes based on diet, age, and gender in $ApoE^{-/-}$ mice compared with control C57BL6 mice ***(3)***. The lesion compositional elements we are interested in studying include lipid, total cellularity, collagen, elastin, smooth muscle cells, and macrophages. These techniques employ traditional histochemical staining and immunostaining visualized with confocal scanning laser microscopy, wide-field transmitted, fluorescent, and polarized light digital image capture, and computer-assisted image processing and analysis. Thus, they might be of interest to those working in other fields requiring quantitative analysis of tissue composition.

2. Materials

2.1. General Equipment and Supplies

2.1.1. Fixation and Processing

1. Scale with glassine weigh paper.
2. Weighing spatula.
3. 200-mL Glass beaker with stir bar.
4. Glass vials with plastic snap top.
5. Fine-tip forceps.
6. Razor blade.
7. Plastic tissue-processing capsules.
8. Stirrer/hot plate.
9. Disposable Tissue-Tek embedding cryomolds 10 mm × 10 mm × 5 mm.
10. Liquid nitrogen (LN_2) in a Dewer flask (Cole Parmer Instrument Company, Vernon Hills, IL) (*see* **Note 1**).
11. 2-Methyl-butane.
12. Plastic cup with attached strings for suspension in LN_2 (to hold the 2-methyl-butane) (*see* **Note 2**).
13. Dissecting microscope.
14. OCT (optimal cutting temperature) embedding medium; Tissue-Tek; Sakura Finetek USA, Torrance, CA.
15. Freezer bag or aluminum foil for storage.
16. Solvent-proof pen.

2.1.2. Cryostat Sectioning

1. Cryostat chuck.
2. OCT embedding medium.
3. Disposable cryostat knives.
4. Small paintbrush.
5. Fisher Superfrost Plus-coated slides or equivalent (Fisher Scientific).
6. Slide box.
7. Large dissecting pin.

2.1.3. Staining

1. Scale with glassine weigh paper.
2. 100-mL glass beaker with stir bar.
3. Coplin glass staining jars.
4. 50 mm × 15 mm Glass Petri dishes.
5. No.1 filter paper.
6. Weighing spatula.
7. Wooden applicator sticks.
8. Forceps.
9. Oven.
10. Graduated cylinder.
11. Size 1-½ glass cover slips.
12. Kimwipes EX-L low-lint cleanroom wipes (Fisher Scientific).
13. Slide tray.
14. Plastic transfer pipets.
15. Aluminum foil.

2.2. Reagents and Solutions

2.2.1. Fixation

1. Phosphate-buffered saline (PBS) 10X stock solution.

   Solution A: 2.76 g sodium phosphate monobasic in 100 mL of distilled water (DH_2O).

   Solution B: 14.1 g sodium phosphate dibasic (anhydrous) in 500 mL of DH_2O.

   Mix: Solution A: 95 mL
   Solution B: 405 mL

   Add 90 g of sodium chloride and increase volume to 1 L. The pH of the solution is then adjusted to 7.2–7.4.
   PBS (1X working solution): 0.01 *M* phosphate buffer, 0.15 *M* NaCl. Dilute one part of the 10X stock solution with nine parts of DH_2O and adjust pH to 7.4, if necessary.
2. 3% Paraformaldehyde (PFA): 3 g PFA, 100 mL PBS, stirred and heated to 60°C (do not boil) until the PFA dissolves completely into solution. Cool to room temperature before using. This fixative should be freshly prepared as required. For

light microscopy studies, we store this fixative for up to 1 wk at 4°C. Always prepare formaldehyde solutions in a fume hood and wear protective gloves.

2.2.2. Tissue Processing

1. 5% Gelatin: 5 g gelatin Type A from porcine skin (Sigma Chemical Co., St. Louis, MO), 100 mL of DH_2O, dissolve with low heat.
2. 10% Gelatin: 10 g Gelatin Type A from porcine skin, 100 mL of DH_2O, dissolve with low heat.

2.2.3. Histologic Stains

1. Oil red O (lipid stain).

   Oil red O stock solution: 2.5 g Oil red O (CI 26125; *see* **Note 3**) (Polysciences, Inc, Warrington, PA), 500 mL of 98% isopropanol, mix well.
   Oil red O working solution: 24 mL Oil red O stock solution, 16 mL of DH_2O. Mix well and let stand for 10 min before filtering. Filtered solution is stable for several hours.
2. Aldehyde Fuchsin (elastic stain).

   Alcoholic Basic Fuchsin, 0.5% stock solution: 2.5 g Basic Fuchsin (CI 42500; *see* **Note 4**) (Sigma Chemical Co., St. Louis, MO), 500 mL of 70% ethyl alcohol. This solution is stable for several months when stored at room temperature.
   Aldehyde Fuchsin working solution: 50 mL 0.5% alcoholic Basic Fuchsin, 2.5 mL acetaldehyde (Aldrich Chemical Co., Milwaukee, WI) (*see* **Note 5**), 1.0 mL hydrochloric acid, concentrated. Mix the three solutions, cover, and allow to sit overnight at room temperature. Filter, cover tightly, and store at 4°C. This solution is stable for 3 wk. Warm to room temperature before using. Use suitable precautions for flammability, ventilation, and body protection.
   Light Green solution.
   Light Green stock solution: 0.2 g Light Green SF yellowish (CI 42095) (Fisher Scientific, Fair Lawn, NJ), 100 mL DH_2O, 0.2 mL glacial acetic acid, mix well.
   Light Green working solution: 10 mL Light Green stock solution, 50 mL of DH_2O.
3. Picrosirius red (collagen stain). Picrosirius red working solution: 0.1% Sirius red F3BA (CI 35780) (Pfaltz and Bauer, Inc., Waterbury, CT) in saturated aqueous picric acid, pH 2.0. Solution is stable at room temperature for 1 yr (*see* **Note 6**).
4. 0.01 *N* HCl.
5. 70% ETOH.

2.2.4. Fluorescent Stains

1. PBS.
2. PBS/1.0% bovine serum albumin (BSA, Fraction V; Sigma Chemical Co.)/0.1% Triton X (10 μL in 10 mL PBS).
3. PBS/1.0% BSA.
4. 10% Normal serum (from secondary antibody host) diluted in PBS/1.0% BSA.
5. Glycerol-based mounting media containing antifade reagent.

6. Primary antibody: α-smooth muscle actin, asm-1 mouse monoclonal antibody (Chemicon International, Temecula, CA), 5 μg/mL final concentration diluted in PBS/1%BSA.
7. Secondary antibody: Alexa 568 goat anti-mouse IgG (Molecular Probes, Eugene, Oregon), 10 μg/mL final concentration diluted in PBS/1%BSA.

3. Methods

3.1. Fixation/Processing Methods

Mouse heart and aorta are dissected out and immersion fixed overnight in 3% PFA/PBS at 4°C as previously described ***(2,3)***. During fixation, the hearts are bisected with a cut parallel to both atria. Residual external fat coating the aortic region is carefully removed with forceps under a dissecting microscope. Following fixation, the hearts are removed from the fixative and placed into plastic processing cassettes and rinsed for 1 h under running water. The container can be covered with gauze and a rubber band to prevent the processing cassettes from washing out. Retain the fixative for use later in the procedure. Capsules are immersed in 5% gelatin for 2 h at 42°C and then transferred into 10% gelatin and infiltrated overnight at 42°C. The following morning, the capsules are refrigerated at 4°C for 3 h to solidify the gelatin. The gelatin-encased hearts are removed from the cassettes and trimmed into a rectangle parallel to the primary atria/heart cut and immersed into 3% PFA/PBS for 2 h to harden the gelatin. The capsules are then transferred into PBS to rinse for 30 min before freezing. The hearts are embedded with OCT medium in labeled embedding molds (cut side up so that the first sections taken will reveal the sinus area) and snap-frozen in 2-methyl butane cooled in LN_2 and stored at –80°C until the time of cryostat sectioning.

3.2. Sectioning

This method is based on the technique of Paigen and co-workers ***(12)***. All sectioning is done in a cryostat using disposable knives. OCT-embedded tissue blocks are carefully mounted so that the first sections taken are from the side showing the bisected heart face. Sections are examined at step intervals until the aortic sinus area is revealed. The aortic sinus area is characterized by the presence of valves and a bulging shape. Orientation adjustments are made as needed until sections show the end of the aortic sinus area, identified by the thinning of the connecting valves and prominence of three valve cusps. Ten-micrometer sections are collected onto Fisher Superfrost Plus-coated slides as follows. Six slides each hold five step sections collected over a well-defined 320-μm area of interest. This area is defined by three prominent valve cusps at the juncture of the aortic sinus region to the end of the valve region, when the valves disappear and the aorta becomes more rounded in appearance. The first slide holds sections 1, 7, 13, 19, and 25. The second slide holds sections 2, 8,14, 20, and 26; the third through sixth slides continue in the same fashion. Several additional single-section slides are collected at the end of the 300 μm area to act as controls for subsequent staining procedures. Because this technique requires absolute collection of serial sections, compressed or wrinkled sections must be corrected by immediately covering with a drop of water and straightened using a dissecting microscope and teasing needle. The sections on the slides are finally air-dried for 30 min to ensure proper adhesion before being stored in a slide box at –80°C.

3.3. Staining for Lipid and Cellularity (DNA Binding Dyes: DAPI, SYTOX Green, and Propidium Iodide)

Oil red O method for neutral lipids: Pearse method (*see* **Note 7**)

1. Cut 10-μm frozen sections, place on coated slides, and allow to air-dry at room temperature for 30 min before staining (*see* **Note 8**).
2. Stain sections for 10 min in a Coplin jar containing filtered Oil red O working dilution.
3. Gently wash in distilled water.
4. Dry the slide area around the tissue section and incubate with 5 μg/mL DAPI for 15 min in a moist, light-protected chamber (easily created by lining a large Petri dish with moistened filter paper, and inserting two orange sticks on which rest incubating slides). Cover the top lid with aluminum foil to protect the slides from light. DAPI can be replaced with a DNA-binding stain of choice depending on the wavelength desired (*see* **Note 9**).
5. Rinse 2X 5 min in PBS.
6. Mount in glycerol-based mounting medium containing antifade reagent (*see* **Note 10**).

Results: Lipid is stained bright red (*see* **Fig. 1A,B**). Other tissue components are stained faint pink to pale yellow. The DNA component is fluorescent in the desired wavelength (*see* **Fig. 2A;** *see* Color Plate 5, following p. 274) (*see* **Note 11**).

3.4. Elastic Stain

Aldehyde fuchsin elastic stain (method)

1. Air-dry slides for at least 30 min.
2. Rinse the slides in distilled water 1X 5 min.
3. Rinse in 50% ethanol, 5 min.
4. Rinse in 70% ethanol, 5 min.
5. Place sections in Aldehyde Fuchsin solution for 15 min at room temperature.
6. Rinse the slides in 70% ethanol for 10 dips and check differentiation with the microscope. Elastic fibers should be deep blue to purple.
7. Put the slides in distilled water for 5 min to stop the differentiation. At this point, the staining can be continued with the Picrosirius red stain for collagen, counterstained with a Light Green working solution for 1–2 min, or dehydrated and cover-slipped for permanent mounting.

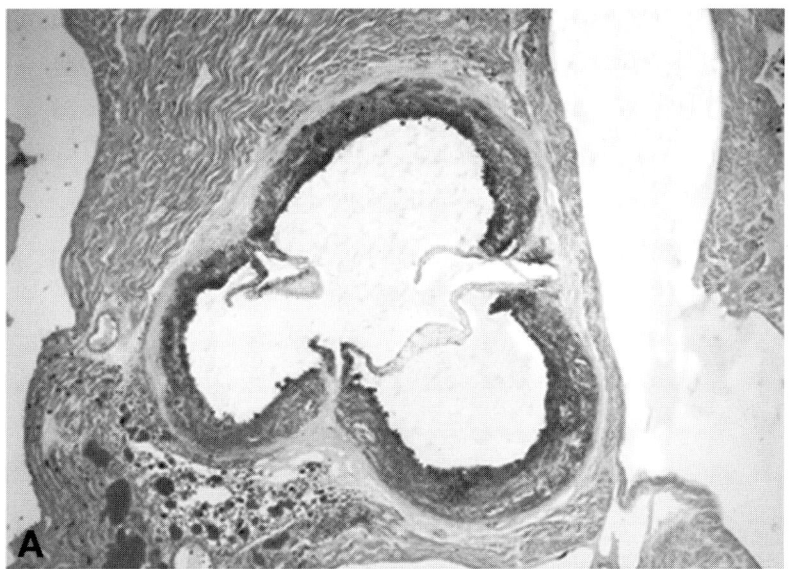

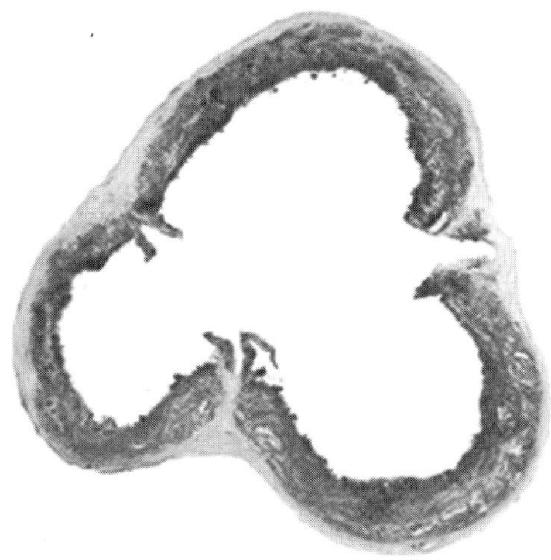

Fig. 1. **(A)** Cryostat section from mouse aorta stained with Oil red O and imaged with a 4X objective lens on a wide-field light microscope. **(B)** Final appearance of cropped image isolating the aorta from surrounding tissues for quantitative compositional analysis. Scale bar = 500 μm.

8. Dehydrate with 10–15 dips in two changes each of 95% alcohol, 100% alcohol, and three changes of xylene and mount with a synthetic mounting medium.

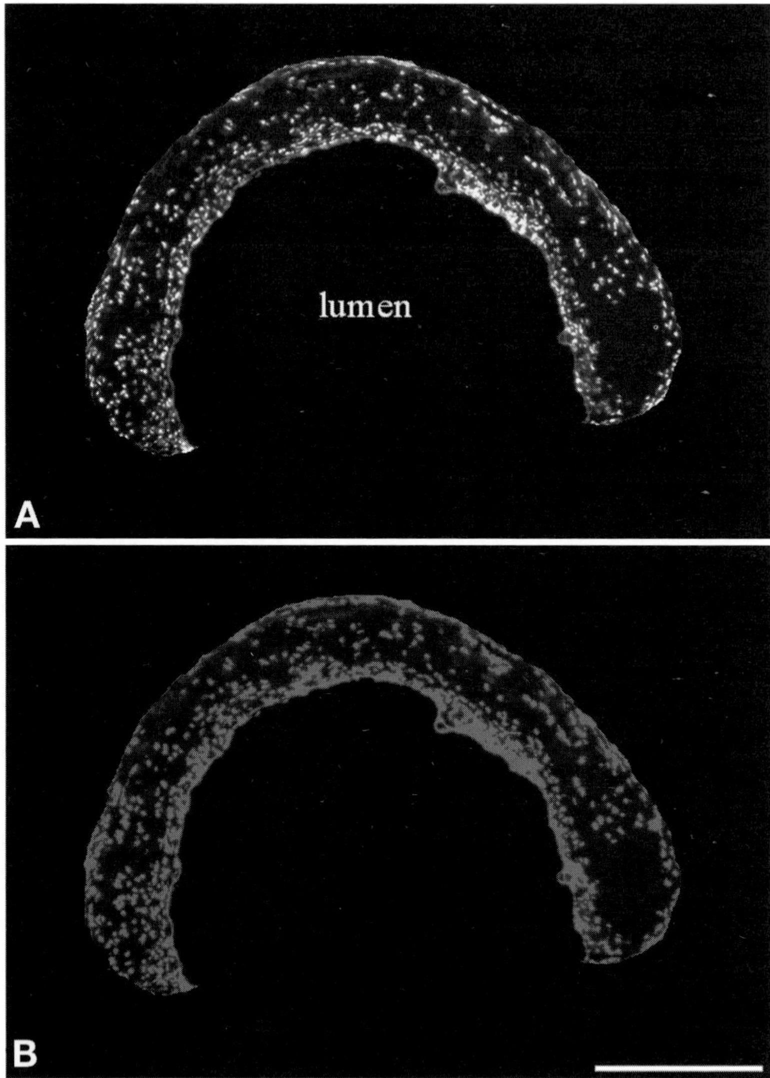

Fig. 2. **(A)** Gray-scale fluorescent capture of DAPI-stained nuclei in one portion of the lesion. The media has been cropped away, revealing the lesion cellularity only. **(B)** Red pseudocolor overlay of the image in **(A)** indicating all pixel intensities included in the threshold. Scale bar = 250 μm. (*See* Color Plate 5, following p. 274.)

Results: Elastic fibers should be purple to deep purple and other tissue elements green if Light Green counterstain is used. If combined with Picrosirius red stain, collagen fibers will be red and can also be viewed with polarized light.

3.5. Collagen

Picrosirius red stain: Dolber and Spach Method ***(13)***

1. Air-dry cryostat sections for at least 30 min prior to staining.
2. Hydrate sections in distilled water for 2 min.
3. Stain for 90 min in filtered 0.1% Sirius red F3BA/saturated picric acid at room temperature.
4. Wash in 0.01 N HCl for 1 min.
5. Rinse in 70% ETOH.
6. Dehydrate with 10–15 dips in two changes each of 95% alcohol, 100% alcohol, and three changes of xylene and mount with a synthetic mounting medium.

Results: In bright field, collagen is stained red and cytoplasm is stained yellow (*see* **Note 12**). For more accurate determination of collagen using this stain, the sections are viewed by polarized light microscopy ***(14)***.

3.6. Immunostaining

1. PBS/1.0%BSA/0.1% Triton X-100, 15 min.
2. PBS/1.0% BSA, 10 min.
3. 10% Normal serum directed against the host of the secondary antibody, diluted in PBS/1%BSA, 30 min.
4. Primary antibody, diluted in PBS/1% BSA overnight at 4°C.
5. Rinse with PBS/1% BSA, 2X 10 min at room temperature.
6. Secondary antibody diluted with PBS/1% BSA, 60 min.
7. PBS/1%BSA 2X 5 min.
8. Nuclear stain, 15 min.
9. PBS/1% BSA 2X 5 min.
10. Mount in glycerol-based mounting medium containing antifade reagent (*see* **Note 13**).

3.7. Imaging Procedures and Digital Capture

3.7.1. Confocal Scanning Laser Microscopy

Digital images are captured using a Bio-Rad 1024 confocal laser scanning microscope using Lasersharp 4.2 software (Bio-Rad Laboratories, Hercules, CA). The laser power, iris, gain, and offset for each channel were set using the *Set col* overlay as an intensity guide (*see* **Note 14**). Images are captured using a 1024 × 1024 box size for increased resolution and in sequential capture mode to avoid channel-to-channel bleed-through. Images are Kalman filtered three times on a final capture for possible noise reduction. The Bio-Rad .pic format is converted to .tif format using a Photoshop plug-in, or by using the export function available in Lasersharp software.

3.7.2. Wide-Field Bright-Field Microscopy

Digital images are captured using an Olympus BX50 upright light microscope (Olympus America, Inc., Lake Success, NY) with an attached Optronics MagnaFire digital camera (Optical Analysis Corp., Nashua, NH) and associated MagnaFire software (version 2.0). The captured images are displayed in 1280 × 1024 pixel RGB format. The objective lenses used are first cleaned and the microscope aligned for Koehler Illumination. The condenser aperture is set at one-half open, and the LBD, ND6, and ND25 filters are engaged. The microscope illumination potentiometer is set at 5 and the beam splitter is pulled out fully to deliver 100% of the signal to the camera. Exposure is set appropriately using a clip detect mode (*see* **Note 15**), an infrared (IR) cut filter is selected to filter out infrared light from the subject, and a white balance is done on a blank area of the slide.

3.7.3. Wide-Field Fluorescence Microscopy

Fluorescence images are captured using the Olympus BX50 microscope as described above. The exposure is set using the clip detect mode, and the white balance preset is set to "fluorescence." The IR cut filter is chosen and images are captured in 1280 × 1024 8-bit gray-scale format (*see* **Fig. 2A**).

3.7.4. Polarized Light Microscopy (see *Fig. 3*)

Polarized images are captured in 1280 × 1024 8-bit gray-scale format and combined with an additional gray-scale unpolarized image captured by turning the polarizer to an unpolarized position using the clip detect function to avoid oversaturation and inserting a neutral density filter. An RGB image is created from the gray-scale images using the *color/merge* function. The unpolarized image is dropped into the green and blue channels and the polarized image is cropped into the red channel (*see* **Fig. 3A;** Color Plate 6, following p. 274). Dropping the unpolarized image into two channels considerably lightens the final image and assists in more accurately discriminating lesion borders during the subsequent image cropping (*see* **Fig. 3B**). Combining the Picrosirius red stain (imaged with polarized light) with the Aldehyde Fuchsin stain (imaged unpolarized in bright field) provides definition to the elastic lamina. After cropping, the green and blue channels are discarded, leaving an 8-bit gray-scale image of the polarized red channel (*see* **Fig. 3C**).

3.8. Computer-Assisted Image Processing and Analysis

3.8.1. Image Processing

Image processing is easily accomplished with Adobe Photoshop (Adobe Systems Inc., San Jose, CA). Images are opened and the *lasso* and *magic wand* tools are used to selectively highlight and fill unwanted areas with black.

Cropping can be assisted with the addition of a WACOM tablet and Intuos 2 stylus (CDWG, Vernon Hill, IL), which is a more elegant drawing tool that replaces the traditional mouse and mouse pad. Using this tool, the artery (*see* **Fig. 1B**) and, subsequently, the lesion are isolated from the surrounding tissue for analysis (*see* **Note 16**).

3.8.2. Image Analysis

Although we originally performed the image analysis in Adobe Photoshop using a modified version of the protocols published by Lehr and co-workers ***(15,16)***, it is now done using MetaMorph (Universal Imaging Corp., Downingtown, PA) software. Cropped digital images are opened and the appropriate (precalibrated) objective calibration is set by choosing *Measure/Calibrate Distance/Apply*, followed by minimizing this window. This will allow for area measurements to be expressed in calibrated square microns. Pixel values will be displayed at the bottom of the screen.

3.8.2.1. Measurements of the Artery and Lesion Using an RGB Threshold

Color thresholding is applied by choosing the following software options:

1. Color threshold "HSI."
2. Hue range "inclusive."
3. Hue 0–255.
4. Saturation 0–255.
5. Intensity 1–250.
6. State "inclusive."

3.8.2.2. Measurements for Lipid

1. Color threshold "HSI."
2. Hue range "exclusive."
3. Hue 10–240.
4. Saturation 55–255.
5. Intensity 0–255.
6. State "inclusive."

3.8.2.3. Measurements for Cellularity Using Gray-Scale Threshold (**Fig. 2B**)

1. Gray-scale 90–255.
2. State "inclusive."

Once the image is thresholded, the *Integrated Morphometry* feature will measure the thresholded area and log the area values onto a Microsoft Excel spreadsheet (Microsoft Corp., Redmond, WA). The logged values are then converted into an Excel chart for presentation (*see* **Fig. 4**). Animal comparisons are calculated and expressed as both area and percent values.

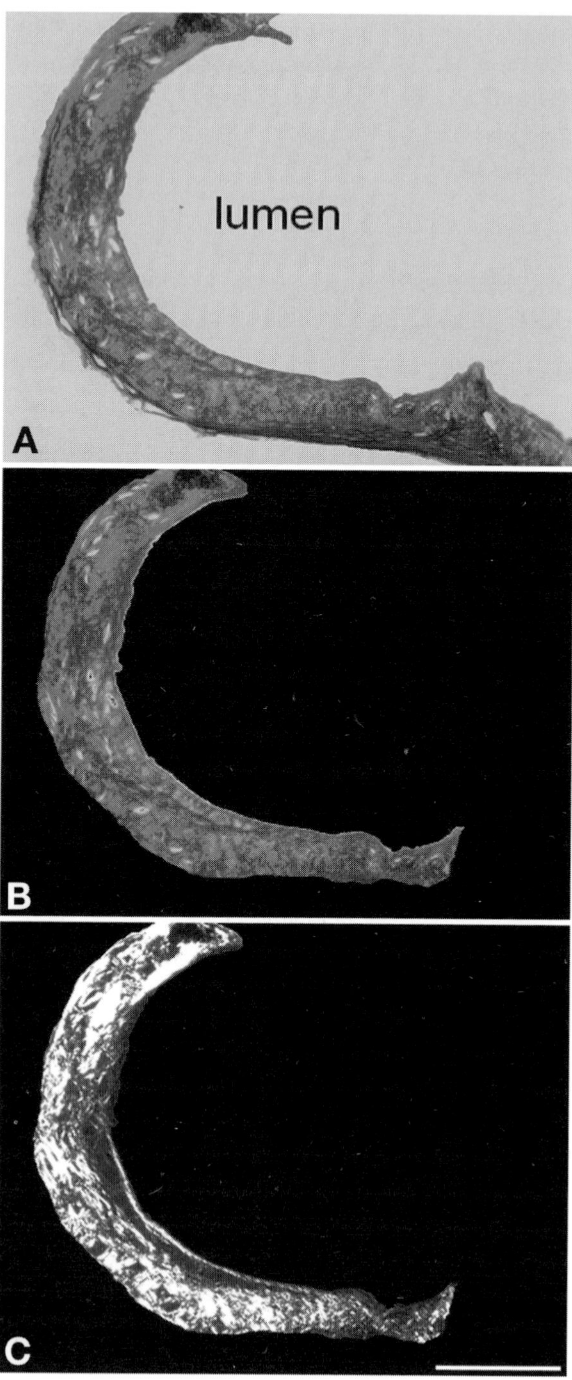

Fig. 3.

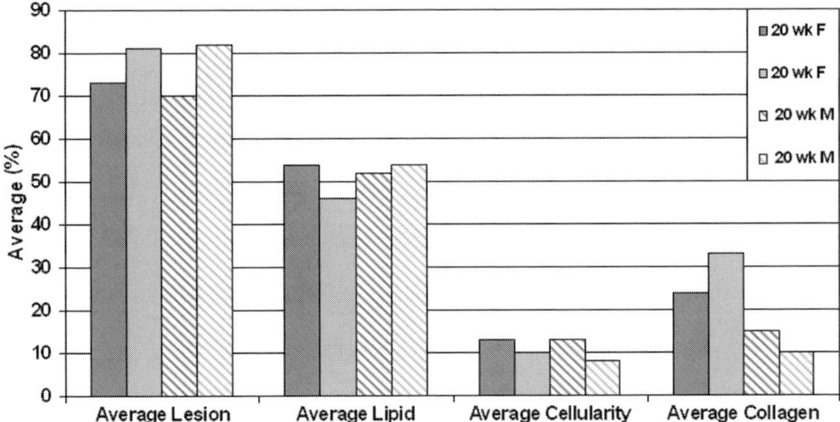

Fig. 4. Microsoft Excel chart displaying the quantitative evaluation of atherosclerotic lesion composition in 20-wk-old male and female Apo $E^{-/-}$ mice fed a high-fat diet for 16 wk. Note the large increase in lesion collagen content in the female mice as compared to their male counterparts.

4. Notes

1. There are a wide variety of insulated flasks available. For safety, any flask used must be able to withstand the low temperature of liquid nitrogen. For our use, a 1-L-capacity flask works well.
2. We made this by punching a hole and attaching a long string on either side of a 150-mL plastic cup. The cup is gently suspended in the LN_2 and the strings are secured to the carrying handle of the container. One hundred milliliters of the 2-methyl-butane are added very slowly to the cup. The nitrogen will boil vigorously until the liquid is cooled, so use precautions such as eye protection and pour very slowly.
3. Stains can be purchased from a variety of sources and can be identified by their unique color index (CI) number.
4. Basic Fuchsin is also known as Pararosaniline and Basic Red 9.
5. Acetaldehyde is a highly volatile substance and great care must be taken during its use. It is used in this elastic stain as a substitute for paraldehyde, which must be

Fig. 3. **(A)** Picrosirius red stain from the atherosclerotic lesion captured in the gray scale with polarized light and merged into the red channel with a bright-field unpolarized image that was merged into the green and blue channels. These channels are used for cropping purposes for orientation and will be discarded later. **(B)** Cropped image with the media removed revealing only the isolated lesion. The image was cropped by turning off the red and blue channels and using the green channel's bright-field image to identify the elastic lamina as a landmark for cropping purposes. **(C)** Gray-scale image of the red polarized channel after the green and blue channels have been discarded. Scale bar = 250 µm. (*see* Color Plate 6, following p. 274.)

obtained with a DEA (Drug Enforcement Administration) number. Moreover, paraldehyde must be freshly opened, with the remainder of the container discarded.

6. When weighing and using chemicals refer to the Material Safety Data Sheet (MSDS) for that chemical and use the appropriate precautions, including working in a fume hood and wearing protective gloves and mask.
7. Oil red O is a physical method of staining that relies on the greater affinity for solubility of the dye in lipid than the dye solvent. Alcoholic fixatives should be avoided because they act as lipid solvents. A positive control is usually not needed because most tissue contains some fat.
8. Oil red O staining must be performed on cryostat sectioned frozen tissue. Lipids are solubilized by the organic solvents used in traditional paraffin embedding.
9. The excitation (Ex)/emission (Em) spectra for some of the DNA-binding dyes we have used are listed as follows: DAPI (4′ ,6-diamidino-2-phenylindole dihydrochloride), 359 nm Ex/461 nm Em; SYTOX Green, 504 nm Ex/523 nm Em; propidium iodide (containing RNase), 535 nm Ex/617 nm Em. All of these dyes can be excited by a conventional mercury arc lamp on a wide-field fluorescence microscope. However, for the confocal microscopic excitation of DAPI, an ultraviolet laser is required.
10. Aqueous mounting media must be used for cover-slipping, because resin-based mounting media contain lipid-extracting solvents. When adding a cover slip to the slide, care must be taken to not apply pressure on the cover slip, which might displace the lipid droplets. To optimize resolution, match the cover slip thickness being used with the appropriate corrected objective.
11. Oil red O can be viewed with transmitted light, and it has fluorescent properties when excited by far-red light. This stain is compatible with other fluorescent staining procedures, such as DNA labeling and immunostaining for cellular identification. Thus, with a confocal microscope with three excitation wavelengths, multiple components can be examined. We found fluorescent comparisons are best made with sections stained and imaged in a single batch. Fluorescent staining and imaging, done over time, can introduce slight variability in intensity as a result of fluorophore age, bulb life, imaging conditions, and other variables, which can influence intensity comparisons.
12. Picrosirius red staining can be performed on either fixed paraffin or cryostat sections. Possibly because of the highly acid nature of the Picrosirius red stain, we have been unsuccessful in combining it with immunofluorescent dyes. However, we have successfully used it in combination with hematoxylin stains and the Aldehyde Fuchsin stain for elastic fibers. Picrosirius red can be viewed with transmitted, fluorescent (yellow light excitation), or polarized light (preferred).
13. Primary antibodies of interest can be used in cocktail combinations and combine nicely with a DNA-specific fluorophore such as DAPI, SYTOX Green, propidium iodide, or TOTO. Note that for widefield imaging, the bleedthrough of SYTOX Green nuclear stain can also be detected with a wide-band green-filter cube.
14. The Bio-Rad *Set col* is a look-up table (LUT) that assigns specific colors to pixel intensity ranges on an 8-bit scale. In this case, pixel intensities from 0 to 5 are

depicted as green, 6 to 249 are depicted as gray, and 250 to 255 are depicted as red. This LUT is a useful guide to ensure that a full dynamic range of gray scale is depicted in the image. An optimized image would contain a few green and red pixels, with the majority represented by gray. This method produces the same effect as adjusting a histogram of intensity values on an image.

15. The "clip detect" feature in the MagnaFire software is similar in function to the Bio-Rad *Set col* LUT described in **Note 14**. Essentially, this feature represents saturated pixels by the color red. The exposure time can then be adjusted to allow for only a few red pixels (saturated intensity) in the field.
16. Cropping of the images can also be done in MetaMorph by using the WACOM tablet and stylus and the MetaMorph *drawing tool* option to draw around unwanted areas, followed by the *Display/Graphics/Paint outside* (or inside) region to fill the unwanted areas. The highlighted area must then be reselected and deleted.

References

1. Oberholzer, M., Ostreicher, M., Christen, H., and Bruhlmann, M. (1996) Methods in quantitative image analysis. *Histochem. Cell Biol.* **105**, 333–355.
2. Taatjes, D. J., Wadsworth, M. P., Schneider, D. J., and Sobel, B. E. (2000) Improved quantitative characterization of atherosclerotic plaque composition with immunohistochemistry, confocal fluorescence microscopy, and computer-assisted image analysis. *Histochem. Cell Biol.* **113**, 161–173.
3. Wadsworth, M. P., Sobel, B. E., Schneider, D. J., and Taatjes, D. J. (2002) Delineation of the evolution of compositional changes in atheroma. *Histochem. Cell Biol.* **118**, 59–68.
4. Taatjes, D. J., Schneider, D. J., Bovill, E. G., and Sobel, B. E. (2001) Microscopy-based imaging of the pathogenesis of cardiovascular disease. *Microsc. Anal.* **50**, 21–23.
5. Falk, E. (1985) Unstable angina with fatal outcome: dynamic coronary thrombosis leading to infarction and/or sudden death: autopsy evidence of recurrent mural thrombosis with peripheral embolization culminating in total vascular occlusion. *Circulation* **71**, 699–708.
6. Davies, M. J., Bland, M. J., Hangartner, W. R., Angelini, A., and Thomas, A. C. (1989) Factors influencing the presence of acute coronary thrombi in sudden ischemic death. *Eur. Heart J.* **10**, 203–208.
7. Sobel, B. E. (1999) Increased plasminogen activator inhibitor-1 and vasculopathy: a reconcilable paradox. *Circulation* **99**, 2496–2498.
8. Sobel, B. E. (1999) The potential influence of insulin and plasminogen activator inhibitor type 1 on the formation of vulnerable atherosclerotic plaques associated with type 2 diabetes. *Proc. Assoc. Am. Physicians* **111**, 313–318.
9. Newby, A. C., Libby, P., and van der Wal, A. C. (1999) Plaque instability—the real challenge for atherosclerosis research in the next decade? *Cardiovasc. Res.* **41**, 321–322.
10. Breslow, J. L. (1996) Mouse models of atherosclerosis. *Science* **272**, 685–688.

11. Warden, C. H., Hedrick, C. C., Qiao, J. -H., Castellani, L. W., and Lusis, A. J. (1993) Atherosclerosis in transgenic mice overexpressing apolipoprotein A-II. *Science* **261**, 469–472.
12. Paigen, B., Morrow, A., Holmes, P. A., Mitchell, D., and Williams, R. A. (1987) Quantitative assessment of atherosclerotic lesions in mice. *Atherosclerosis* **68**, 231–240.
13. Dolber, P. C. and Spach, M. S. (1993) Conventional and confocal fluorescence microscopy of collagen fibers in the heart. *J. Histochem. Cytochem.* **41**, 465–469.
14. Junqueira, L. C. U., Bignolas, G., and Brentani, R. R. (1979) Picrosirius staining plus polarization microscopy, a specific method for collagen detection in tissue sections. *Histochem. J.* **11**, 447–455.
15. Lehr, H. -A., Mankoff, D. A., Corwin, D., Santeusanio, G., and Gown, A. M. (1997) Application of Photoshop-based image analysis for quantification of hormone receptor expression in breast cancer. *J. Histochem. Cytochem.* **45**, 1559–1565.
16. Lehr, H. -A., van der Loos, C. M., Teeling, P., and Gown, A. M. (1999) Complete chromogen separation and analysis in double immunohistochemical stains using Photoshop-based image analysis. *J. Histochem. Cytochem.* **47**, 119–125.

7

Applications of Microscopy to Genetic Therapy of Cystic Fibrosis and Other Human Diseases

Thomas O. Moninger, Randy A. Nessler, and Kenneth C. Moore

Summary

Gene therapy has become an extremely important and active field of biomedical research. Microscopy is an integral component of this effort. This chapter presents an overview of imaging techniques used in our facility in support of cystic fibrosis gene therapy research. Instrumentation used in these studies includes light and confocal microscopy, transmission electron microscopy, and scanning electron microscopy. Techniques outlined include negative staining, cryo-electron microscopy, three-dimentional reconstruction, enzyme cytochemistry, immunocytochemistry, and fluorescence imaging.

Key Words: Gene therapy; gene transfer vectors; light microscopy; confocal microscopy; electron microscopy; cystic fibrosis.

1. Introduction

Over the past 15 yr researchers at the University of Iowa have been working to identify the specific gene and cell structure that is compromised in cystic fibrosis (CF) patients. Significant effort has been expended to identify effective viral and other vectors for transfer of the normal gene construct. Light and electron microscopy are essential tools for developing strategies for genetic therapy in the treatment of disease. Negative staining and cryo-electron microscopy are critical in understanding the structure of native and hybrid vectors. Enzyme cytochemistry is used for the detection of markers that aid in determining the efficiency of gene transfer to cells. Immunocytochemistry and fluorescent proteins are employed to identify gene products, locate them within tissues and cells, and track transfected cells over time. Light, confocal, and electron microscopy are used to evaluate pathogenesis, inflammation, and cellular responses to gene transfer. This chapter will highlight some of these techniques as they have been applied in our facility to gene therapy research.

From: *Methods in Molecular Biology, vol. 319: Cell Imaging Techniques: Methods and Protocols*
Edited by: D. J. Taatjes and B. T. Mossman © Humana Press Inc., Totowa, NJ

2. Materials

The following is a partial list of microscopy-related vendors:

2.1. Transmission Electron Microscope Vendors

1. FEI Inc., Hillsboro, OR 97124, USA; (503) 640-7500.
2. Hitachi High Technologies America, San Jose, CA 95134, USA; (800) 548-9001, (408)432-0520.
3. JEOL USA Inc., Peabody, MA 01960, USA; (508) 535-5900.
4. LEO Electron Microscopy Inc., Thornwood, NY 10594, USA; (800) 356-1090.

2.2. Scanning Electron Microscope Vendors

1. FEI Inc., Hillsboro, OR 97124, USA; (503) 640-7500.
2. Hitachi High Technologies America, San Jose, CA 95134, USA; (800) 548-9001, (408)432-0520.
3. JEOL USA Inc., Peabody, MA 01960 USA; (508) 535-5900.
4. LEO Electron Microscopy Inc., Thornwood, NY 10594, USA; (800) 356-1090.
5. RJ Lee Instruments Inc., Trafford PA 15085, USA; (724) 744-0100.

2.3. Research-Grade Light Microscope Vendors

1. Leica Microsystems Inc., Bannockburn IL 60015, USA; (847) 405-0123.
2. Nikon Inc., Melville, NY, USA; (516) 547-8500.
3. Olympus America Inc., Melville, NY, 11747, USA; (516) 844-5039.
4. Carl Zeiss MicroImaging Inc., Thornwood, NY 10594, USA; (914) 747-1800.

2.4. Spot-Scanning Confocal Microscope Vendors

1. Leica Microsystems Inc., Exton, PA 19341, USA; (610) 321-0460.
2. Nikon Inc., Melville, NY, USA; (516) 547-8500.
3. Olympus America Inc., Melville, NY, 11747, USA; (516) 844-5039.
4. Carl Zeiss MicroImaging Inc., Thornwood, NY 10594, USA; (914) 747-1800.

2.5. Multiphoton Microscope Vendors

1. Leica Microsystems Inc., Exton, PA 19341, USA; (610) 321-0460.
2. Carl Zeiss MicroImaging Inc., Thornwood, NY 10594, USA; (914) 747-1800.

2.6. Microscopy Chemicals and Supplies Vendors

1. Electron Microscopy Sciences, Fort Washington, PA 19034, USA; (215) 646-1566.
2. Energy Beam Sciences Inc., Agawam, MA 01001, USA; (413) 786-9322.
3. Polysciences Inc., Warrington, PA 18976, USA; (215) 343-6484.
4. Ted Pella Inc., Redding, CA 96049, USA; (530) 243-2200.
5. Structure Probe Inc., West Chester, PA 19381, USA; (610) 436-5400.

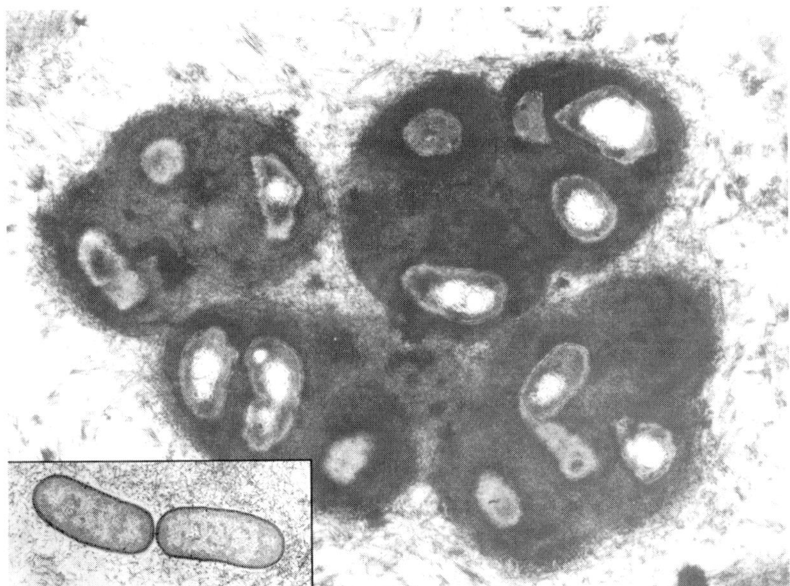

Fig. 1. Transmission electron microscope image of *Pseudomonas* bacteria in sputum from a CF patient. The main image was fixed using OsO_4 dissolved in perfluorocarbon (PFC), dehydrated in 100% ethanol and embedded in Eponate 12 resin, avoiding all aqueous solutions in order to preserve the biofilm matrix surrounding the cells. This technique also works well in preserving the mucus layer on epithelia. The inset shows the same sample processed using a conventional aqueous method, with fixation in gluteraldehyde and OsO_4. Note the loss of the matrix integrity compared to the PFC-processed sample. Inset bacteria are 0.5 μm in diameter.

3. Methods

For much of the past two decades, we have been involved in research to identify the specific gene, gene product, and cell structure that is compromised in CF patients and to develop an efficacious gene therapy treatment. Significant effort has been expended to identify effective viral and other vectors for transfer of the normal gene construct into defective airway cells. In addition to CF, we are also involved with investigations for genetic therapy of hemophilia, hypertension, neurological problems, cancer, and skin disorders. Light and electron microscopy has been instrumental to the advancement of these projects.

Cystic fibrosis is the most common lethal genetic disease in Caucasians, with a carrier rate of 5% in the population. It is caused by mutations in a cell membrane chloride channel called the Cystic Fibrosis Transmembrane Conductance Regulator (CFTR) *(2)*. This gene is expressed in several epithelia, including airway, sweat glands, and the pancreas. Pathology associated with the lung is the major reason for

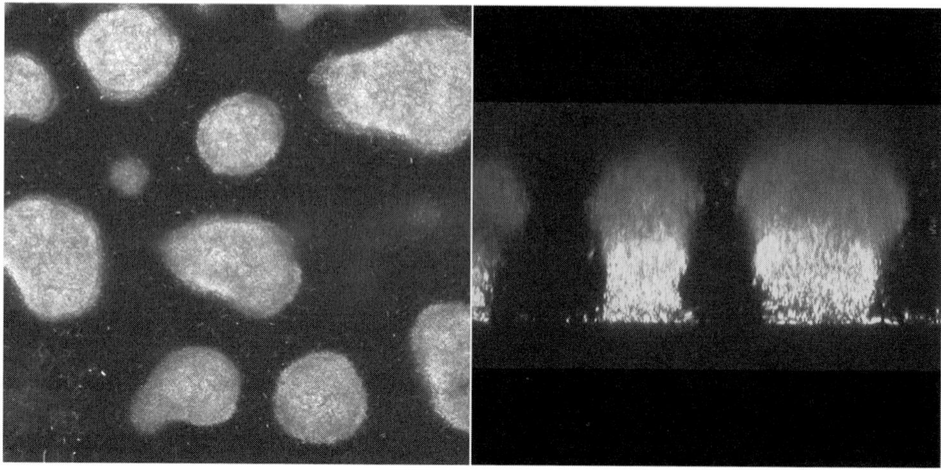

Fig. 2. Confocal images of a green fluorescent protein (GFP) expressing *P. aeruginosa* PAO1 bacterial biofilm grown in a chamber designed to facilitate observation in a confocal microscope. The image on the left is an X-Y optical section through the middle of a field of biofilm columns. The image on the right is an X-Z vertical section through the biofilm columns. The columns are approx 200 µm in height.

the morbidity caused by this disease. These defective ion channels result in high concentrations of chloride in the surface airway liquid, which becomes thick and difficult to expectorate. Consequently, naturally occurring antimicrobial agents are inactivated resulting in high concentrations of bacteria, most commonly *Pseudomonas aeruginosa* and *Staphylococcus aureus* ***(3)*** (*see* **Fig. 1**). Once one of these bacteria has colonized a CF patient's lung, that clone can be detected in sputum for up to a decade ***(4)***. It is believed that these bacteria are able to persist because of the formation of biofilms, that are extremely resistant to antimicrobial treatment (**Fig. 2; 5**). This infection and the resulting inflammation cause chronic damage and scaring to the lung ultimately resulting in loss of function and death. In addition to lung airway epithelia being affected, the pancreas, intestine, sweat glands, and other parts of the body can be compromised, although with less severe consequences. Treatment of CF patients involves antibiotics, enzyme supplements, dietary intervention, chest massage, and even organ transplants.

The gene for CFTR was isolated in 1989 ***(6)***. cDNA was administered to human nasal epithelial, with the result that the CF defect was temporarily corrected as measured by cAMP-stimulated, amiloride-sensitive voltage ***(1)***.

Since 1993, various vector systems have been investigated to more naturally introduce the normal CF gene into human airway epithelial cells. Most of these investigations were carried out using a novel primary human airway cell culture

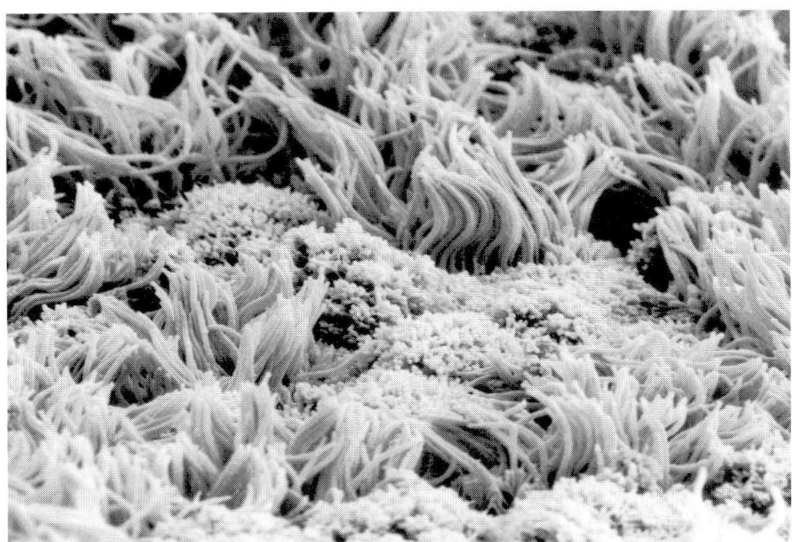

Fig. 3. Scanning electron microscope image of the surface of a 3-wk-old primary human airway culture. The sample was prepared using conventional aqueous methods, thereby removing the mucus covering the culture and allowing the cell surface to be viewed. Note the ciliated cells and the group of microvilli-covered goblet cells in the center of the field. The cilia extend 10 μm from the surface of the cells.

model *(7)*. In this model airway, cells from either CF or non-CF donor lungs are cultured on filter membrane inserts and grown with an air interface on the apical surface. Over the course of 2 wk, the cells differentiate into pseudostratified columnar epithelia containing basal, goblet, and ciliated cells covered by a mucus blanket (*see* **Figs. 3** and **4**). Under sterile conditions, these cultures can be maintained for months.

Initially, CFTR cDNA was complexed with different cationic lipids and polymers (*see* **Figs. 5** and **6**). This combination was effective in stimulating uptake of the DNA and temporarily corrected the CF defect ***(8–11)***. However, effective concentrations of these lipids and polymers tended to be toxic to the cells and the correction was short-lived.

Viral vectors were also considered as a means of transfecting cells. Initial experiments involved adenovirus (*see* **Fig. 7**), which is about 80 nm in diameter, has a double-stranded DNA genome of 36,000 bases, and is able to infect human airway cells. In 1993 an adenovirus expressing CFTR transiently corrected the chloride defect in the nasal epithelium of CF patients ***(12)***. Unfortunately, adenovirus infects many cell types at very low levels and is prone to eliciting an immune response. Protocols were developed that involved mixing adenovirus with lipids (cholesterol and phosphotidyl inositol based),

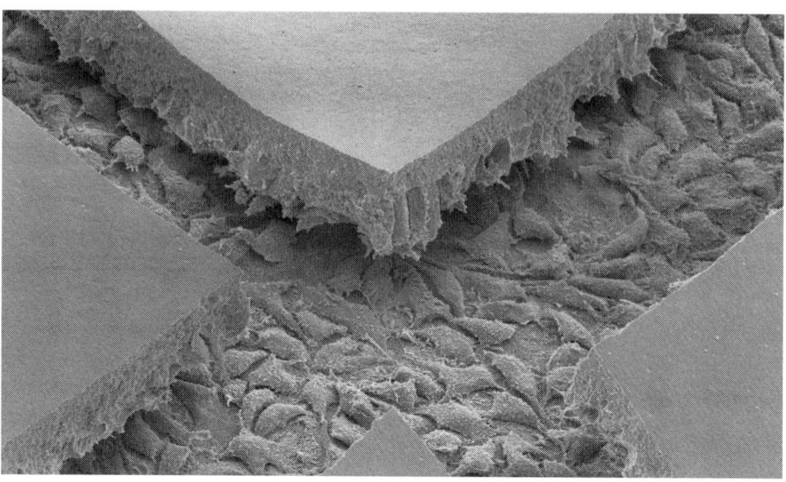

Fig. 4. Scanning electron microscope image of a 3-wk-old primary human airway culture. In contrast to **Fig. 3**, this sample was prepared using the nonaqueous OsO_4/perfluorocarbon method, thereby retaining the thick mucus blanket on the surface of the culture. Note that in this fractured area, the basal cells remained attached to the filter while the goblet and ciliated cells remained with the mucus layer. Total thickness of the cell and mucus layers is 45 µm.

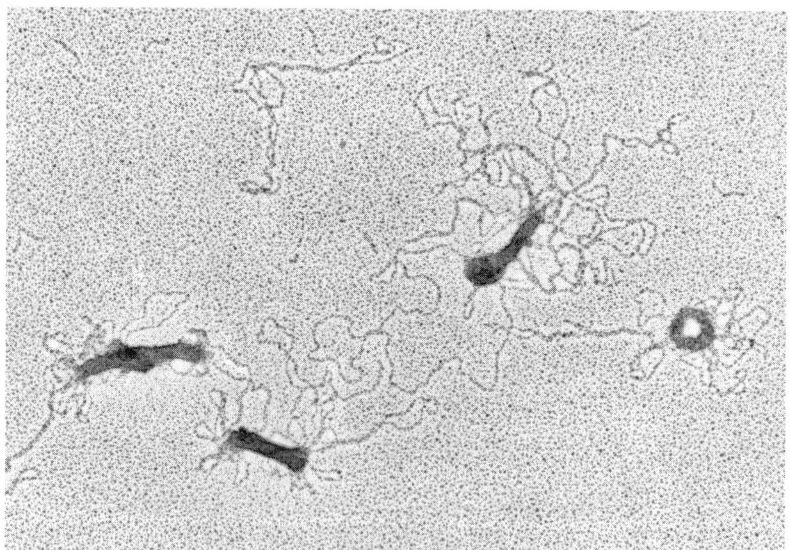

Fig. 5. Combination 1% uranyl acetate negative stain and platinum rotary shadowing of DNA/Poly-L-lysine transfection complexes displaying the typical rod and torus morphologies as well as uncomplexed DNA. The torus on the right is approx 40 nm in diameter.

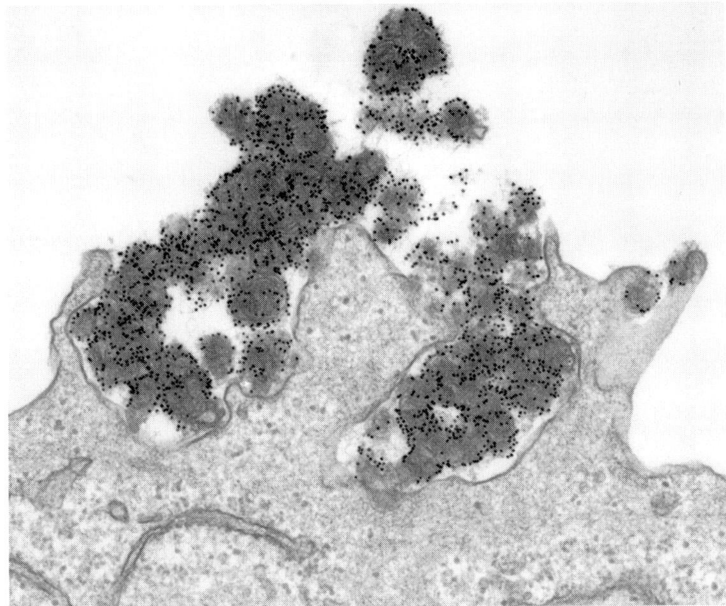

Fig. 6. Transmission electron microscope image of DNA/cationic lipid complexes being engulfed by a COS-7 cell. The DNA has been labeled with 10 nm of gold to facilitate visualization of the fate of the complexes.

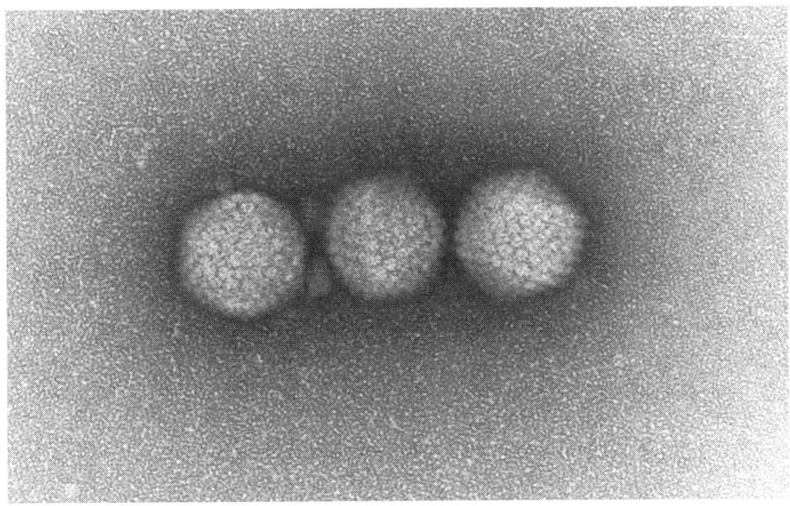

Fig. 7. Negative stain of adenovirus using 1% ammonium molybdate. The particles are 80 nm in diameter.

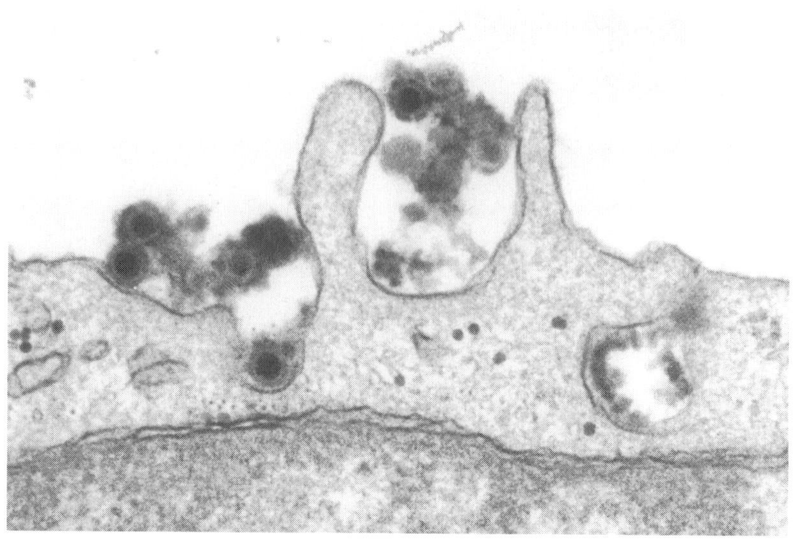

Fig. 8. Transmission electron microscope image of adenovirus/lipid complexes being engulfed by a COS-1 cell.

polymers (poly-L-lysine, dendrimers, polyethylene glycol), or calcium phosphate. These techniques resulted in significant increases in the amounts of virus that were taken into the airway epithelial cells. Analysis indicated that the majority of the viral complexes were taken in by bulk phagocytosis as opposed to a receptor-mediated mechanism (*see* **Fig. 8**). This resulted in significant viral particles being destroyed by lytic activity. Also, as with naked DNA complexes, at effective concentrations the lipids and polymers tended to be toxic. In addition, because adenovirus does not integrate into the cells genome, any correction of the CF defect was transient, necessitating subsequent reapplications. In human subjects, this results in an ever-increasing immune reaction to the adenovirus.

The lentiviruses have several advantages for gene therapy as compared to adenovirus. Prime among these attributes is the ability to transfect differentiated and stem cells. This is a critical advantage because the outer layer in the airway epithelia is made up of differentiated ciliated and mucus-secreting cells that are destined to die and be replaced from offspring of the stem or progenitor cells. Feline immunodeficiency virus (FIV) is 20 nm in size, has a capacity of 10,000 bases, and is able to transfect both differentiated and stem cells. Little if any immunoreaction occurs to FIV. Transfection of airway epithelia with FIV results in a permanent correction of the gene defect. Unfortunately, the apical surface of airway epithelium is naturally very resistant to infection by this virus. Treatment of the airway epithelium with a hypotonic buffer containing EGTA

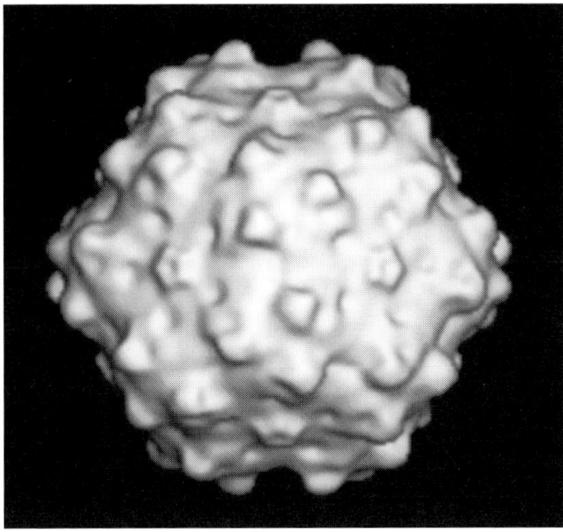

Fig. 9. Adeno-associated virus serotype 5 (AAV5). The 16-Å resolution three-dimentional reconstruction of AAV5 viewed along the twofold axis calculated from cryo-transmission electron microscope images of vitreous-ice-embedded samples.

transiently opens the tight junctions between the cells and allows FIV to transfect the basally occurring stem cells ***(13)***. Alteration of the apical surface of the airway epithelial cells by enzymatic treatment has not resulted in increased uptake of virus into the differentiated cells.

Another promising vector is adeno-associated virus (AAV). AAV is a member of the Parvoviridae family, genus *Dependovirus*, and is 20 nm in size (*see* **Fig. 9**). It has a single-stranded DNA genome and has been shown to be effective in muscle, brain, retina, and liver. It also is able to transduce nondividing cells. Significantly, there is no known pathology associated with the various serotypes of AAV, minimizing the deleterious effects of inflammation. One shortcoming of AAV as a gene therapy vector is the limited length of DNA that is about 4900 bp. Full-length CFTR cDNA (including the promoter, the 5′ untranslated region, the ploy-A-tail, and internal terminal repeats) is 5435 bp. One way past this roadblock is to determine what parts of the cDNA can be deleted and still allow production of a functional channel ***(14)***. Another problem is that recombinant AAV can be difficult to prepare and purify.

Current efforts also involve evaluation of pseudotype virus that incorporates characteristics of two or more types of virus. As a result, vectors can be tailor-made to receptors on specific cells. Proposed experiments involve FIV incorporating glycoproteins from Ross River virus. Alternative approaches involve chemically modifying cell surface receptors, isolating tissue-specific stem cells to transfect and reintroduce into the host, as well as other nongenetic therapies. It should be emphasized that each genetic disease will most likely involve specific and unique approaches to treatment. As in the example of CF, the airway epithelium is naturally resistant to transfection by virus and other vectors. Conversely, the transfection of hemopoetic cells does not present such a resistant barrier.

Because of the nature of this overview, specific techniques and protocols have not been presented. The reader is invited to refer to other chapters in this manual and to the references at the end of the chapter for specific protocols. Inquiries to the authors are also welcome.

4. Note

A wealth of related information is available on the internet. Microscopy Society of America, http://www.msa.microscopy.com/; Microscopy Vendors database, http://www.kaker.com/mvd/vendors.html; and Molecular Expressions Virtual Microscopy, http://micro.magnet.fsu.edu/primer/virtual/virtual.html

References

1. Rich, D. P., Anderson, M. P., Gregory, R. J., et al. (1990) Expression of cystic fibrosis transmembrane conductance regulator corrects defective chloride channel regulation in cystic fibrosis airway epithelial cells. *Nature* **347,** 358–363.
2. Quinton. P. M. (1986) Missing Cl conductance in cystic fibrosis. *Am. J. Physiol.* **251,** C649–C652.
3. Smith, J. J., Travis, S. M., Greenberg, E. P., and Welsh, M. J. (1996) Cystic fibrosis airway epithelia fail to kill bacteria because of abnormal airway surface fluid. *Cell* **85,** 229–236.
4. Romling, U., Fiedler, B., Bosshammer, J., et al. (1994) Epidemiology of chronic *Pseudomonas aeruginosa* infections in cystic fibrosis. *J. Infect. Dis.* **170,** 1616–1621.
5. Costerton, J. W., Stewart, P. S., and Greenberg, E. P. (1999) Bacterial biofilms: a common cause of persistent infections. *Science* **284,** 1318–1322.
6. Riordan, J. R., Rommens, J. M., Kerem, B. S., et al. (1989) Identification of the cystic fibrosis gene: cloning and characterization of complementary DNA. *Science* **245,** 1066–1073.
7. Karp, P. H., Moninger, T. O., Weber, S. P., et al. (2002) An in vitro model of differentiated human airway epithelia. Methods for establishing primary cultures. *Methods Mol. Biol.* **188,** 115–137.
8. Fasbender, A. J., Zabner, J., and Welsh, M. J. (1995) Optimization of cationic lipid-mediated gene transfer to airway epithelia. *Am. J. Physiol.* **269,** L45–L51.
9. Fasbender, A. J., Marshall, J., Moninger, T. O., Grunst, T., Cheng, S., and Welsh, M. J. (1997) Effect of co-lipids in enhancing cationic lipid-mediated gene transfer in vitro and in vivo. *Gene Ther.* **4,** 716–725.
10. Fasbender, A. J., Zabner, J., Chillon, M., et al. (1997) Complexes of adenovirus with polycationic polymers and cationic lipids increase the efficiency of gene transfer in vitro and in vivo. *J. Biol. Chem.* **272,** 6479–6489.

11. Fasbender, A. J., Zabner, J., Zeiher, B. G., and Welsh, M. J. (1997) A low rate of cell proliferation and reduced DNA uptake limit cationic lipid-mediated gene transfer to primary cultures of ciliated airway epithelia. *Gene Ther.* **4,** 1173–1180.
12. Zabner, J., Couture, L. A., Gregory, R. J., Graham, S. M., Smith, A. E., and Welsh, M. J. (1993) Adenovirus-mediated gene transfer transiently corrects the chloride transport defect in nasal epithelia of patients with cystic fibrosis. *Cell* **75,** 207–216.
13. Wang, G., Zabner, J., Deering, C., et al (2000) Increasing epithelial junction permeability enhances gene transfer to airway epithelia In vivo. *Am. J. Respir. Cell Mol. Biol.* **22,** 129–138.
14. Ostedgaard, L. S., Zabner, J., Vermeer, D. W., et al. (2002) CFTR with a partially deleted R domain corrects the cystic fibrosis chloride transport defect in human airway epithelia in vitro and in mouse nasal mucosa in vivo. *Proc. Natl. Acad. Sci. USA* **99,** 3093–3098.

8

Laser Scanning Cytometry

Principles and Applications

Piotr Pozarowski, Elena Holden, and Zbigniew Darzynkiewicz

Summary

The laser scanning cytometer (LSC) is the microscope-based cytofluorometer that offers a plethora of analytical capabilities. Multilaser-excited fluorescence emitted from individual cells is measured at several wavelength ranges, rapidly (up to 5000 cells/min), with high sensitivity and accuracy. The following applications of LSC are reviewed: (1) identification of cells that differ in degree of chromatin condensation (e.g., mitotic or apoptotic cells or lymphocytes vs granulocytes vs monocytes); (2) detection of translocation between cytoplasm vs nucleus or nucleoplasm vs nucleolus of regulatory molecules such as NF-κB, p53, or Bax; (3) semiautomatic scoring of micronuclei in mutagenicity assays; (4) analysis of fluorescence *in situ* hybridization; (5) enumeration and morphometry of nucleoli; (6) analysis of phenotype of progeny of individual cells in clonogenicity assay; (7) cell immunophenotyping; (8) visual examination, imaging, or sequential analysis of the cells measured earlier upon their relocation, using different probes; (9) *in situ* enzyme kinetics and other time-resolved processes; (10) analysis of tissue section architecture; (11) application for hypocellular samples (needle aspirate, spinal fluid, etc.); (12) other clinical applications. Advantages and limitations of LSC are discussed and compared with flow cytometry.

Key Words: Cytometry; fluorescence; cell cycle; apoptosis; nucleus; nucleolus; micronucleus cytoplasm; enzyme kinetics

1. Introduction: Limitations of Flow Cytometry

During the past three decades, flow cytometry (FC)/cell sorting has become commonplace in various disciplines of biology, medicine, and biotechnology. However, because cells are measured while suspended in a stream of liquid and often discarded following the measurement, the analytical capability of FC is limited for many applications, such as the following:

1. Time-resolved events such as enzyme kinetics, drug uptake, or drug efflux cannot be analyzed on individual cells.

From: *Methods in Molecular Biology, vol. 319: Cell Imaging Techniques: Methods and Protocols*
Edited by: D. J. Taatjes and B. T. Mossman © Humana Press Inc., Totowa, NJ

2. The morphology of the measured cell might only be assessed after sorting, which is cumbersome and not always available.
3. Subcellular localization of the fluorochrome cannot be analyzed.
4. The cell once measured cannot be reanalyzed with another probe(s).
5. Analysis of solid tissue requires cell or nucleus isolation that leads to loss of information on tissue architecture.
6. Small-size samples such as fine-needle aspirates or spinal fluid are seldom analyzed by FC because repeated sample centrifugations, which often are required, lead to cell loss that has to be compensated for by starting with a large number of cells per sample.
7. The measured sample is lost and cannot be stored for archival preservation.

The microscope-based laser scanning cytometer (LSC), manufactured since the mid-1990s by CompuCyte Corp. (Cambridge, MA), offers many advantages of FC but has none of the limitations presented above. The analytical capabilities of LSC, therefore, complement those of FC and extend the use of cytometry in many applications ***(1–4)***. This chapter is primarily focused on the applications and capabilities of LSC that are unique to this instrumentation and it updates the previous reviews ***(4,5)***. Clinical applications of LSC were also recently reviewed ***(6)***, as well as the applications of LSC in toxipathology ***(7)***. A separate review that is focused on applications of LSC and other laser-based technologies in cell signaling in environmental lung disease was also recently published ***(8)***.

2. Features of LSC and Parameters That Can Be Measured

The microscope (Olympus Optical Co.) is the key part of the instrument, and it provides essential structural and optical components (*see* **Fig. 1**). The beams from two or three lasers (argon ion and helium–neon; violet laser is optional) spatially merged by dichroic mirrors are directed onto the computer-controlled oscillating (350 Hz) mirror, which reflects them through the epi-illumination port of the microscope and images through the objective lens onto the slide. The mirror oscillations cause the laser beams to sweep the area of microscope slide under the lens. The beam spot size varies depending on the lens magnification, from 2.5 (at 40×) to 10.0 μm (at 10×). The slide, with its *xy* position monitored by sensors, is placed on the computer-controlled motorized microscope stage, which moves at 0.5-μm steps per each laser scan, perpendicular to the scan. Laser light scattered by the cells is imaged by the condenser lens and its intensity is recorded by sensors. The specimen-emitted fluorescence is collected by the objective lens and directed to the scanning mirror. Upon reflection, it passes through a series of dichroic mirrors and optical emission filters to reach one of the four photomultipliers. Each photomultipler records fluorescence at a specific wavelength range, defined by the combination of filters and dichroic mirrors. A light source, additional to the lasers, provides

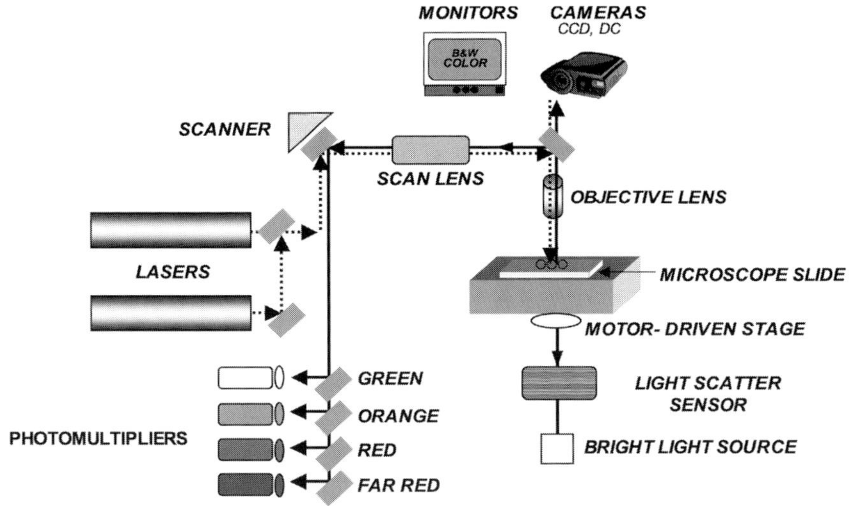

Fig. 1. Schematic representation of the LSC. *See* text for explanation.

transmitted illumination to visualize the objects through an eyepiece or the charge-coupled device (CCD) camera.

The measurement of cell fluorescence (or light scatter) is computer controlled and triggered by a threshold contour set above the background (*see* **Fig. 2**). The following parameters are recorded by LSC for each measured cell/object:

1. *Integrated fluorescence* intensity, representing the sum of intensities of all pixels ("picture elements") within the integration contour area. The latter can be adjusted to a desired width with respect to the threshold contour (*see* **Fig. 2**).
2. The maximal intensity of an individual pixel within this area (*maximal pixel;* "max pixel").
3. The *integration area*, representing the number of pixels within the integration contour.
4. The *perimeter* of the integration contour (in micrometers).
5. The fluorescence intensity integrated over the area of a torus of desired width defined by the *peripheral contour* located around (outside) the primary integration contour. For example, if the integration contour is set for the nucleus, based on red fluorescence (DNA stained by propidium iodide [PI]), then the integrated (or maximal pixel) green fluorescence of fluorescein isothiocyanate (FITC)- stained cytoplasm can be measured separately, within the integration contour (i.e., over the nucleus) and within the peripheral contour (i.e., over the rim of cytoplasm of desired width outside the nucleus). All of the above values of fluorescence (parameters 1, 2, and 5) are automatically corrected for background, which is measured outside the cell, within the *background contour* (*see* **Fig. 2**).

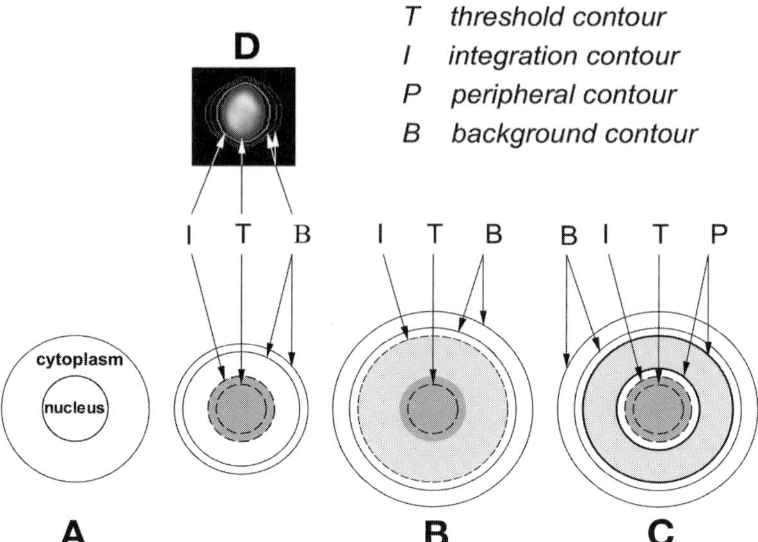

Fig. 2. Different settings for analysis of nuclear, total, and/or cytoplasmic fluorescence by LSC. When nuclear DNA is stained with a red fluorescing dye (e.g., propidium), the threshold contour (T) is set on the red signal to detect the nucleus **(A)**. The integration contour (I) is then set a few pixels outside of T to ensure that all nuclear fluorescence is measured and integrated **(A)**. However, when cytoplasmic fluorescence also is measured, I is set far away from T to ensure that fluorescence emitted from the cytoplasm is integrated as well **(B)**. It also is possible to separately measure nuclear and cytoplasmic fluorescence, as shown in **(C)**. The peripheral contours (P) are then set at the desired number of pixels outside of I and the fluorescence intensities emitted from both areas (viz. within the I boundary and within the P torus) are separately measured and separately integrated. In each case, the background contour is automatically set outside the cell, and the background fluorescence is subtracted from nuclear, cytoplasmic, or total cell fluorescence. The actual cell's contours, as they appear on the monitor, are shown in **(D)**.

6. The *xy* coordinates of maximal pixel locating the measured object on the microscope stage.
7. The computer clock *time* at the moment of measurement.

The software of LSC (WinCyte) allows one to obtain ratios of the respective parameters as a new parameter, and the ratiometric data can be displayed during data analysis. The electronic compensation of fluorescence emission spectral overlap is one of the features of the data analysis. The compensation at the time of data analysis is more convenient than in real time, as it is in most flow cytometers, because it provides an opportunity to test and compare different settings for optimal results.

In addition to the above-listed parameters, the WinCyte software of LSC is also designed to analyze the fluorescence *in situ* hybridization (FISH). Toward this end, the software allows one to establish, within a primary contour representing a nucleus stained with a particular dye (e.g., propidium), a *second set of contours* representing another color (e.g., FITC) fluorescence. Five secondary features are then measured in addition to the major features that were listed above, namely (1) number of secondary contours (i.e., FISH spots); (2) distance between the nearest spots; (3) integrated fluorescence; (4) maximal pixel fluorescence, as well as (5) fluorescence area. The last three parameters (3–5) are measured for each secondary contour.

The measurements by LSC are relatively rapid; having optimal cell density on the slide, up to 5000 cells can be measured per minute. The accuracy and sensitivity of cell fluorescence measurements by LSC are comparable to the advanced flow cytometers ***(1,2)***.

Another methodology that can be utilized to quantify cell constituents measuring fluorescence, in addition to LSC and FC, is fluorescence image analysis (FIA). In FIA, the cell illumination is uniform, provided by a mercury or xenon arc epi-illuminator. A bandpass filter selects fluorescence of a desired wavelength that is imaged at a low depth of focus by a CCD camera. Compared with LSC or FC that utilize photomultipliers, the dynamic range of sensitivity of fluorescence intensity measurement by a CCD is lower in FIA. FIA, thus, cannot provide quantitative analysis of fluorescence intensity that would be on par with FC or LSC. The image, however, can be analyzed by FIA at higher spatial resolution than by LSC.

3. Maximal Pixel of Fluorescence Intensity

Maximal pixel of fluorescence intensity (MPFI) is a useful reporter of local intracellular hypochromicity or hyperchromicity, reflecting a low or high concentration (density) of the fluorescent probe in intracellular compartments. One of the early applications of LSC along this line was to identify cells with condensed chromatin; namely, because DNA in condensed chromatin, such as of mitotic or apoptotic cells, shows increased staining intensity (per unit area of chromatin image, because the same amount of chromatin is compacted at a higher spatial density) with most fluorochromes, the MPFI of these cells is higher than that of cells with the same DNA content but with diffuse chromatin ***(9,10)***. Although mitotic cells can be recognized by FC using a variety of markers (reviewed in **ref. *11***), the advantage of the MPFI approach is that a single fluorochrome is used to distinguish G_1 vs S vs G_2 vs M phase cells. It is possible, thus, to use another color fluorochrome(s) to detect other cell constituents instead of to identify M cells. This approach was used to combine pulse labeling of DNA replicating cells with bromodeoxyuridine (BrdU) (detected with

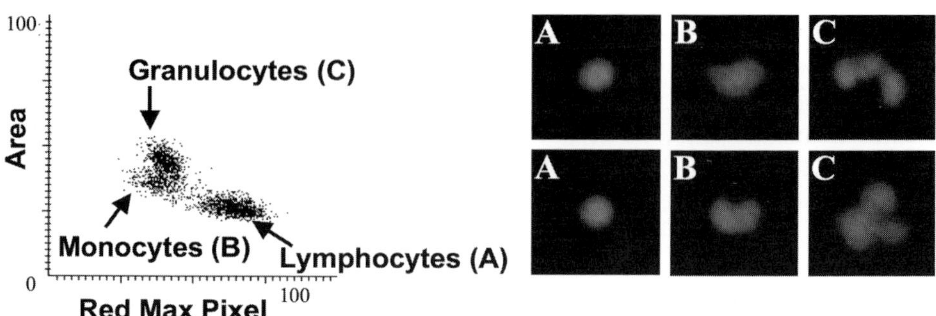

Fig. 3. Distinction of granulocytes, monocytes, and lymphocytes by LSC based on differences in their area vs maximal pixel of red fluorescence, after staining with propidium iodide (*20*).

anti-BrdU mAb) with the identification of mitotic cells to study the cell cycle kinetics by the fraction of labeled mitoses (FLM) method ***(12)***. The FLM assay, initially designed for tritiated thymidine autoradiography ***(13)***, yields a wealth of information on cell cycle kinetics but is cumbersome and time-consuming. Its adaptation to LSC simplifies the procedure and shortens the time of analysis ***(12)***. Similar to mitotic chromosomes ***(8)***, BrdU-labeled meiotic chromosomes were identified by LSC in studies of mutagenesis ***(14)***.

Apoptotic cells having condensed chromatin ***(15,16)*** also can be identified by MPFI of DNA-associated fluorescence ***(17,18)***. It should be cautioned, however, that because both mitotic and apoptotic cells are characterized by a high value of MPFI, the distinction of G_2/M apoptotic cells from mitotic cells is not always possible by this method. This limitation is of particular importance when apoptosis is induced by agents that arrest cells in mitosis and, therefore, the sample contains mitotic cells that die by apoptosis. MPFI has also been found useful to detect localized caspase activity in early apoptotic cells by analysis of local intracellular accumulation of the caspase-cleavage product of the fluorogenic substrate ***(19)***.

The MPFI of DNA-associated fluorescence can also be used to distinguish lymphocytes from monocytes and from granulocytes, the cell types that differ by the degree of chromatin condensation (*see* **Fig. 3**). The fluorescence area, the parameter that reflects nuclear size and is correlated (inversely) with chromatin condensation, also is different for lymphocytes, monocytes, and granulocytes ***(20)***.

Translocation of macromolecules to different subcellular compartments *if it is associated with a change in their local density* (concentration) can also be traced by the measurement of the changes of MPFI. For example, translocation of the proapoptotic regulatory protein Bax and its accumulation in mitochondria, the event that facilitates release of cytochrome-*c* and activation of caspases ***(21,22)***, is detected by LSC as an increase in MPFI of Bax immunofluorescence *(23)*. The analysis of MPFI to detect translocation of other macromolecules might find other applications (e.g., to monitor activation or deactivation of the signal transduction proteins, receptor clustering, accumulation of protein in the Golgi apparatus, etc.).

4. Nuclear vs Cytoplasmic Localization of Fluorescence

Most cellular DNA is localized in the nucleus (the content of mitochondrial DNA is minimal); therefore when DNA is fluorochrome stained, it provides a good marker defining the nuclear boundary. If another color fluorochrome is used to mark other cell constituents, LSC is then able to resolve and separately measure the nuclear and cytoplasmic content of this constituent (*see* **Fig. 2**). This capability can be used to detect translocation of particular proteins from cytoplasm to nucleus, or vice versa (e.g., to monitor the traffic of signal transduction or activation molecules). A classical example of such a protein is nuclear factor kappa B (NF-κB). This ubiquitous factor is involved in regulation of diverse immune and inflammatory responses and also plays a role in the control of cell growth and apoptosis *(24)*. Inactive NF-κB remains in the cytoplasm bound to I-κB protein. Rapid translocation of NF-κB from cytoplasm to the nucleus occurs in response to extracellular signals or DNA damage and is considered to be a hallmark of activation *(24)*. NF-κB was detected immunocytochemically in several leukemic cell lines with FITC-tagged antibody, and its presence in the nucleus vis-a-vis cytoplasm was monitored by LSC measurements of green fluorescence (FITC) integrated over the nucleus vs over the cytoplasm *(25)*. Activation led to a rapid increase in NF-κB-associated fluorescence measured over the nucleus concomitant with a decrease in fluorescence over the cytoplasm, which was reflected by a large increase in the nuclear-to-cytoplasmic fluorescence ratio (*see* **Fig. 4**). One of the virtues of this assay is that NF-κB activation could be correlated with cell morphology, immunophenotype, or cell cycle position *(25)*. The assay of NF-κB translocation by LSC has been found more sensitive than by either of the four alternative methods *(26)*. This application of LSC can be extended to monitor other factors that, upon activation, accumulate in cytoplasm and/or undergo translocation to the nucleus, such as tumor suppressor p53 and signal transduction or cell cycle regulatory molecules. Indeed, the upregulation and translocation of p53 from cytoplasm to the nucleus in response to DNA damage by the topoisomerase I inhibitor camptothecin, detected by LSC, was recently reported *(27)*.

The possibility of measurement of only nuclear-associated fraction of proliferating cell nuclear antigen (PCNA) detected immunocytochemically as offered by LSC provided a good distinction of the S-phase cells in histopathological analysis of renal cell carcinoma *(28)*.

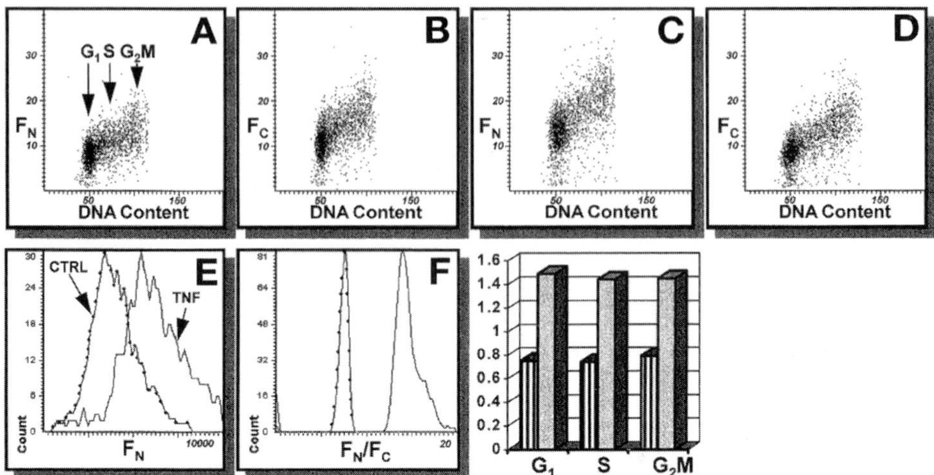

Fig. 4. Changes in intensity of NF-κB immunofluorescence integrated over the cell nucleus (F_N) **(A,C)** and cytoplasm (F_C) **(B,D)** of the U-937 histiomonocytic lymphoma cells, untreated **(A,B)** and treated for 1 h with 10 ng/mL of tumor necrosis factor (TNF)-α **(C,D)**. Note the increase in F_N after the treatment **(E)** and even more pronounced increase in the F_N/F_C ratio **(F)**. Bars indicate F_N/F_C of the cells gated in G_1, S, and G_2/M based on differences in their DNA content as shown in **(A)**; striped bars, prior to TNF-α treatment; shaded bars, after the treatment *(25)*.

It should be stressed, however, that in the case of highly asymmetrically shaped cells (e.g., fibroblasts, neurons) or cells with acentric position of the nucleus (e.g., muscle cells) that grow on slides, the integration of fluorescence from the entire cytoplasm is problematic. A partial solution to the problem might be cell trypsinization followed by their deposition on slides by cytocentrifugation. The cells that grow asymmetrically then become more spherical, with the nucleus centrally located.

5. The Micronucleus Assay/Genotoxicity Studies

The micronucleus assay is widely used to assess the chromosomal or mitotic spindle damage induced by ionizing radiation or mutagenic agents in vivo or in vitro. Because visual scoring of micronuclei is cumbersome, semiautomatic procedures that rely either on flow cytometry or image analysis were developed. The classic image analysis, however, appears to be a complex approach and its practical utility to quantify frequency of micronucleation has been limited to relatively few laboratories with proper expertise and instrumentation. LSC was recently adapted for the analysis of micronucleation induced by genotoxic agents in vivo in mouse erythrocytes *(29)* as well as in vitro in cultured cells *(30)*. The capability of LSC to relocate micronuclei for visual

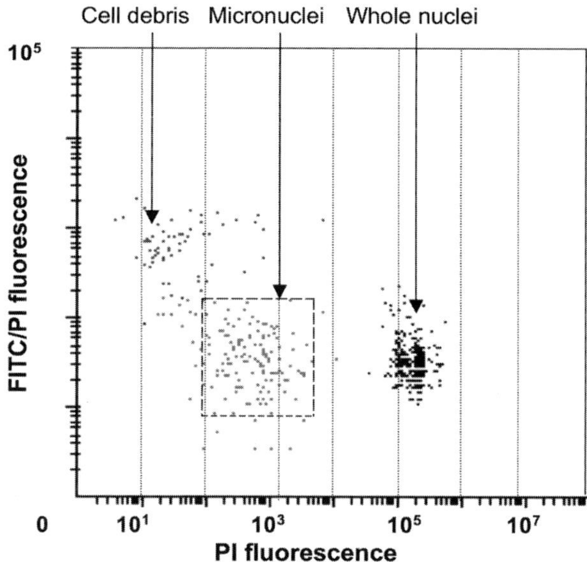

Fig. 5. Detection of micronuclei by LSC. To induce micronuclei, HL-60 cells were treated with mitomycin C, then cytocentrifuged, fixed, and stained with FITC and PI. Micronuclei (located on this scatterplot within the dashed-line gating window where approx 93% of all the events occur) were identified by their low DNA content (PI fluorescence) ranging between 0.1% and 5% of that of the whole G_1 nuclei and the FITC/PI fluorescence ratio that was similar to that of the whole nuclei ***(30)***.

examination was useful to confirm their identification. Multiparameter characterization of micronuclei that took into account their DNA content and protein/DNA ratio (*see* **Fig. 5**) made it possible to establish the gating parameters that excluded objects that were not micronuclei ***(30)***. The percentage of micronuclei assayed by LSC correlated well with that estimated visually by microscopy in both studies ***(29,30)***. LSC, thus, can be used to obtain an unbiased estimate of frequency of micronuclei more rapidly than by conventional examination of the preparations by microscopy. Additionally, unlike flow cytometry, LSC allows one to characterize individual cells with respect to frequency and DNA content of micronuclei in these cells. Furthermore, it can be applied to the cytokinesis-blocked (e.g., by cytocholasin B) micronuclei assay ***(30)***. These analytical capabilities of LSC might be helpful not only to score micronuclei but also to study mechanisms of induction of micronuclei by clastogens or aneugens through their effects on chromosomes or mitotic spindles, respectively.

Assessment of DNA damage is another measure of genotoxicity or can be used as a marker of cell death (apoptosis). LSC has found an application for this purpose as well, specifically for measuring the extent of DNA electrophoretic mobility from individual cells in the "comet" *(31)* or "halo" *(32)* assays.

6. Applications of LSC Utilizing the WinCyte Software Designed for FISH Analysis

6.1. FISH Analysis, Cytogenetic Studies

The semiautomated FISH analysis represents still another LSC application that is based on its ability to spatially resolve the distribution of fluorescent regions within the cell *(2,33)*. As mentioned, the software developed for this application allows one to establish, within a primary contour representing, for example, nucleus stained with a particular dye (e.g., propidium), a second set of contours representing another color (e.g., FITC) fluorescence. An obvious advantage of LSC over visual analysis of FISH is the unbiased selection of the measured cells and their semiautomated, rapid measurement. Furthermore, analysis of the integrated fluorescence intensity of the secondary contours might yield information pertaining to the degree of amplification of particular genome sections. Thus, for example, Kobayashi et al. *(34)*, using the dual color FISH analysis by LSC, revealed an increase in 20q13 chromosomal copy number in several breast cancer cases and correlated it with DNA ploidy and estrogen receptor (ER) or progesterone receptor status. LSC has also found utility in studies employing comparative genomic hybridization (CGH) to reveal cytogenetic aberrations in several types of human cancer *(35–37)* or in studies of DNA ploidy in sperm cells *(38)*. It should be noted, however, that semiautomated FISH measurements by LSC require high-quality technical preparations *(2,33)*.

6.2. Analysis of Nucleoli and Protein Translocations Between Nucleoli and Nucleoplasm

The capacity of the LSC software originally designed for semiautomatic FISH analysis can be applied to other applications. One such application is quantitative analysis of nucleoli and monitoring traffic of molecules between nucleoli and nucleoplasm *(39,40)*. A useful immunocytochemical marker of nucleoli is an antibody to the nucleolar protein nucleolin. Using this antibody, it was possible to estimate the size of individual nucleoli (area and circumference), number of nucleoli per nucleus, total nucleolar area per nucleus, as well as expression of nucleolin separately in nucleoli and nucleoplasm (*see* **Fig. 6**). All of these parameters have been found to strongly correlate with the proliferative status and the cell cycle position of mitogenically stimulated lymphocytes *(40)*. Most interesting, however, was the observation that abundance of nucleolin in nucleoplasm was maximal during the cell transition from the G_0 to G_1 phase of the cycle, which corresponded to the maximal rate of rRNA synthesis

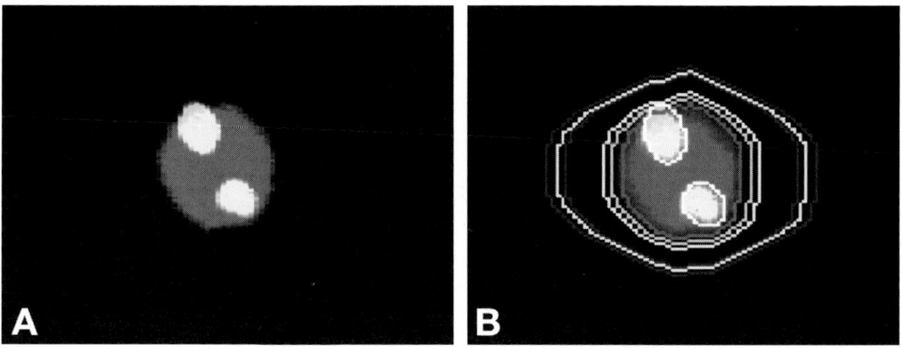

Fig. 6. Analysis of nucleoli in mitogen-stimulated lymphocytes. Nucleolin antibodies were used to immunostain nucleoli in 48 h-phytohemagglutinin-stimulated human lymphocytes; DNA was counterstained with propidium iodide. The software of LSC designed for FISH analysis was used to separately contour the cell nucleus and the nucleoli. The approach allows one to measure number of nucleoli, their size, expression of nucleolin in nucleoli and in nucleoplasm; all in relation to the cell cycle phase ***(40)***.

and its accumulation within the cell. The translocation of nucleolin from nucleoplasm to nucleoli was observed at later stages of lymphocyte stimulation, when the cells were progressing through G_1, S, and G_2/M and when the rate of rRNA accumulation was decreased ***(40)***. Similar application of LSC revealed the cell-cycle-phase-associated nucleoplasm–nucleolar shuttling of cyclin E, which was defective in bladder cancer cells ***(39)***.

6.3. Progeny of Individual Cells/Clonogenicity Assay

Another application of LSC utilizing FISH software was demonstrated in the analysis of progeny (clones) of individual cells ***(41)***. In this application, cellular protein and DNA have been stained with fluorochromes of different colors and the product of tumor suppressor gene *p53* or estrogen receptor was detected immunocytochemically, with still another color fluorescent dye. The threshold contour was set on protein-associated fluorescence, which made it possible to analyze the *whole-cell colony as a single entity* (*see* **Fig. 7**). This approach made it possible to measure a variety of attributes of the progeny of individual cells (phenotype of individual cell colonies), such as colony size (area, circumference, cell number per colony), DNA and protein content per colony, expression of *p53* or estrogen (per colony, per cell, per unit of DNA or protein), the colonies' heterogeneity, and the cell cycle distribution of individual cells within colonies. Such multiparameter analysis provided a wealth of information and was used to study mechanisms by which the cytotoxic RNase–onconase affected proliferative capacity of the cells and induced growth imbalance and

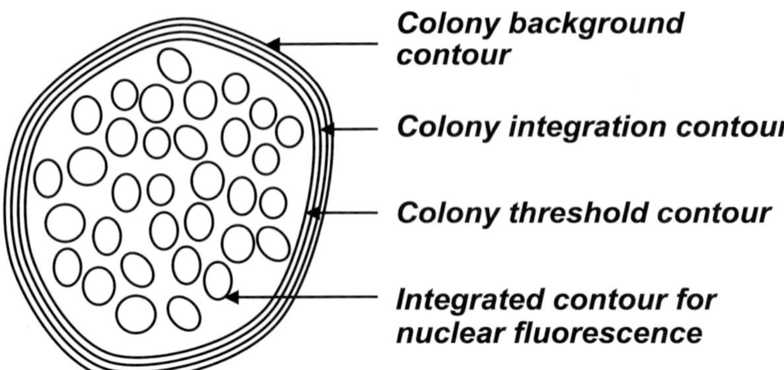

Fig. 7. Strategy used to analyze individual cell colonies by LSC. The MCF-7 breast cancer cell colonies growing on a microscope slide were fixed, and cellular DNA was stained with PI (excited by an argon-ion laser), total protein was stained with BODIPY 630/650X (excited by a neon–helium laser), and individual proteins (tumor suppressor *p53*, or estrogen receptor [ER]) were stained with FITC-tagged Ab (excited by an argon-ion laser). The colony threshold contour and integration contour were set based on BODIPY 620/650X fluorescence. It was possible to estimate total DNA content, content of total protein, and content of *p53* (or ER) per individual colony. Ratiometric analysis revealed ratios of total protein per DNA, *p53* per DNA, or ER per DNA, as well as *p53* (or ER) per total protein for each colony. Furthermore, it was possible to estimate the number of nuclei (cells) per colony and cell cycle distribution within each colony. Such multiparametric analysis of progeny of individual cells provided a sensitive assay to monitor not only the drug-induced changes in cell proliferation but also in their phenotype ***(41)***.

differentiation ***(41)***. Extensions of LSC might make this instrument applicable for automatic analysis of cloning efficiency and multiparameter analysis of cell colonies in soft agar. Such analyses might be useful in studies of mechanisms and effectiveness of antitumor drugs in the field of carcinogenesis and for analysis of primary cultures, assessing tumor prognosis and drug sensitivity. The assay can also be adapted to analysis of microbial colonies.

7. Cell Immunophenotyping

Laser scanning cytometry has been adapted to carry out routine immunophenotyping. Multichamber microscope slides were developed that can be used to automatically screen cells against up to 36 antibodies on a single slide by LSC ***(42–46)***. The chambers are filled with cell suspension by capillary action. In the absence of serum or other proteins in the suspension, the cells strongly attach to the floor of the chambers by electrostatic interactions ***(42,45)***. Various antibody combinations are then introduced into the chambers, the cells are incubated in their presence for 30–60 min, and following a buffer rinse, their fluorescence is measured. The rate of analysis is relatively fast, as it takes approx 20 min to screen the cells distributed in 12 wells labeled with a panel of 36 antibodies (3 antibodies at a time) measuring 3000–5000 cells per well ***(42)***.

Although the rate of measurement by LSC is slower than flow cytometry and the lack of side (90° angle) light-scatter analysis impedes discrimination of lymphocytes from monocytes and granulocytes, certain advantages of LSC might outweigh these deficiencies. Thus, LSC is preferred for hypocellular samples, which cannot tolerate repeated centrifugations that lead to cell loss. It has to be stressed that cell loss during centrifugations, as required for flow cytometry analysis, is not random but preferential to different cell types ***(20)***. LSC is also economical: Because of the small sample size in LSC, the cost of the reagents (mAbs) is reduced by over 80% compared to flow cytometry ***(34)***. Furthermore, LSC provides the possibility to relocate immunophenotyped cells for additional analysis or archival preservation (*see* **Subheading 8.**).

8. The Relocation Feature

8.1. Visual Cell Examination

Because the spatial *xy* position of the measured cell is recorded, it is possible to relocate such a cell for visual examination by microscopy or imaging. The possibility to subject the measured cells to morphological examination is particularly important in studies of apoptosis. This is because apoptosis was originally defined by morphological criteria ***(47)*** and cell morphology still remains the gold standard to identify this mode of cell death ***(15,16)***. Using LSC, it was possible to discriminate between the genuine apoptotic cells and "false-positive" cells in peripheral blood and bone marrow of leukemic patients undergoing chemotherapy ***(18)***. The latter cells were monocytes/macrophages containing apoptotic bodies (probably ingested from the disintegrating apoptotic cells) in their cytoplasm. Although both the genuine apoptotic cells and the "false-positive" cells contained numerous DNA strand breaks and were indistinguishable by FC, analysis of their morphology by LSC made their positive identification made possible ***(18)***. In another study, the eosinophils were identified by LSC as "false-positive" apoptotic cells because of their nonspecific labeling with fluorescein-conjugated reagents ***(48)***. LSC was also helpful to distinguish apoptotic cells from the cells infected by human granulocytic erlichiosis ***(49)***. Based on these observations and other findings, it was concluded that LSC is the instrument of choice for the analysis of apoptosis ***(18,50,51)***.

Several other methods of identification of apoptotic cells, including their recognition by the presence of DNA strand breaks, decreased mitochondrial transmembrane potential, cleavage of poly (ADP-ribose) polymerase, or fractional DNA content (reviewed in **refs.** ***15*** and ***16***), have been successfully adapted to LSC ***(18,52–58)***. More recently, LSC was used to measure activation of caspases during apoptosis by the method utilizing fluorochrome-labeled caspase inhibitors (FLICA) ***(58,59)***, activation of serine-proteases using a similar approach ***(61)***, as well as to detect segregation of RNA from DNA and their separate packaging into apoptotic bodies ***(62)***. In all of these studies, the possibility of cell relocation for visual assessment of apoptosis was a valuable feature of LSC.

The relocation feature of LSC was also useful for the "comet" (single-cell elecrophoresis) assay ***(31)***. In this assay, the cells with damaged DNA were subjected to electrophoresis; then LSC was used to determine the DNA-associated fluorescence of the comets' heads (undamaged DNA), followed by image acquisition of each comet and digital image analysis and computation of tail (damaged DNA) moment and the head-DNA content measurement. A strong inverse correlation was observed between the tail moment and content of undamaged DNA ***(31)***.

8.2. Sequential Analysis of the Same Cells With Different Probes

Laser scanning cytometry allows one to integrate the results of two or more measurements into a single file (the "file-merge" feature). This feature provides the opportunity to measure the same cells more than once, using different settings or probes. For example, it is possible to set the integration contour first on the nucleus and, subsequently, on the whole cell. This approach enables one to separately measure particular constituents in the nucleus and in the whole cell. Such analysis has been applied to reveal translocation of cyclin B1, detected immunocytochemically, from cytoplasm to nucleus during mitosis ***(63)***.

Another application of the file-merge feature of LSC was recently described for analysis of cellular DNA and double-stranded (ds) RNA ***(64)***. The cells were stained with propidium iodide (PI) and measured twice, prior to and after the incubation with RNase. The integrated value of PI intensity of individual cells during the first measurement was proportional to their DNA plus dsRNA content. The PI fluorescence intensity during the subsequent measurement was the result of the dye interaction with DNA only. Thus, when the second measurement was subtracted from the first measurement, the difference ("differential fluorescence" [DF]) represented the RNA-associated PI fluorescence only. DF was then used as a *separate parameter* that was recorded in list mode in the merged file. Cellular protein was also counterstained, but with a fluorochrome of another color of emission than PI. The multiparameter analysis of these data made it possible to correlate, within the same cells, the cellular ds RNA content with DNA content (cell cycle position) or with protein content ***(64)***. This novel approach of using the DF as an additional, discrete parameter for the bivariate or multivariate analysis extends the application of LSC for bivariate or multivariate analysis of other cell constituents, which might be differentially stained with the same wavelength of emission fluorochromes.

Still another application of the file-merge feature and sequential cell staining was to study, within the same cells, a correlation between the supravitally detected cell attributes such as mitochondrial transmembrane potential or induction of oxidative stress, with the attributes to be detected requiring cell fixation, such as the presence of DNA strand breaks or DNA content ***(65)***. This approach revealed that the loss of the transmembrane potential during apoptosis was either very transient or not correlated at all with the activation of caspases and DNA cleavage ***(66)***.

8.3. Enzyme Kinetics and Other Time-Resolved Events

The time of cell measurement by LSC is recorded in the list mode file together with other measured parameters. The relocation feature, in turn, makes it possible to measure the same cell repeatedly. LSC, unlike flow cytometry, thus provides the means to measure kinetic reactions within individual cells in large cell populations. Using the fluorogenic substrate di-(leucyl)-rhodamine 110, the kinetic activity of L-aminopeptidase was measured in several cell types by LSC ***(67)***. The rate of fluorescein di-acetate hydrolysis by esterases as well as the rate of uptake of the lysosomo-trophic fluorochrome acridine orange also was assayed ***(67)***. Several hundred cells per sample were measured with a time resolution of 10–60 s. The kinetic curves constructed for individual cells were matched with the respective cells; the latter were stained with conventional absorption dyes and following the relocation by LSC, they were identified as monocytes, granulocytes, or lymphocytes using bright-field illumination ***(67)***. In a similar manner, the kinetics of dissociation of fluorochromes from nuclear DNA, induced by caffeine, were measured by LSC, and the dissociation plots were constructed ***(68)*** (**Fig. 8**).

Repeated scanning of the same cells causes fluorescence fading. Unfortunately, the fading, which could be extensive when time intervals between scanning are short, imposes a limitation on the time resolution of the kinetic measurement. However, the fading rate as well as the fluorescence recovery rate can be measured in the same cells by LSC, and the results corrected appropriately ***(67)***.

9. LSC in Clinical Pathology

Cytometry still plays a relatively minor role in routine clinical pathology. However, by quantifying key attributes of selected cells in a specimen (tissue section or fine-needle aspirate [FNA]), cytometry can contribute useful prognostic information and help guide therapy. Because little cell loss occurs during sample preparation, LSC is particularly suitable for hypocellular specimens.

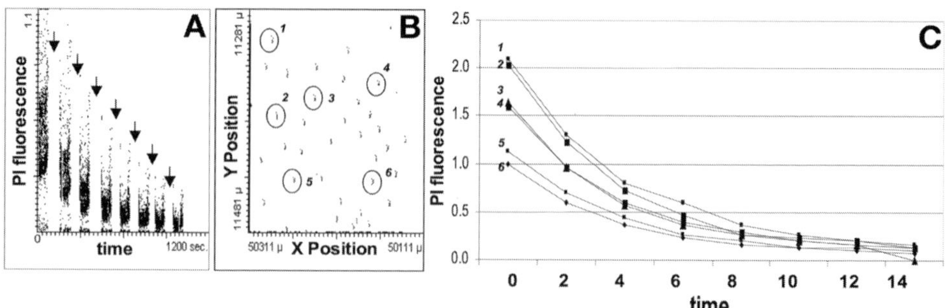

Fig. 8. Application of LSC in analysis of time resolved events; kinetics of removal of the DNA-bound PI from fixed cells by a solution containing caffeine. HL-60 cells attached to the microscope slide were treated with RNase and stained with PI. The cells were then repeatedly [arrows in **(A)**] rinsed with a 50-m*M* solution of caffeine in phosphate-buffered saline and cellular fluorescence was measured at different time-points. **(A)** The scattergram representing time vs PI (red) fluorescence intensity (integrated value) of the rinsed cells; **(B)** the *xy* spatial cell distribution on the slide. Because the same cells were analyzed repeatedly over time and the cell position was recorded in a list mode as one of the parameters, it was possible to select any cell, for example, based on its *xy* coordinates, and, by multiparameter analysis, obtain the kinetic curve for that cell. **(C)** Representative kinetic curves for six such cells marked (circles) in (B). Cells 1 and 2, having at 0 time twice the PI fluorescence intensity of cells 5 and 6, are most likely G_2/M cells; the latter are most likely G_1 cells. Caffeine removes PI from DNA by forming heterologous complexes with this dye in aqueous solutions, thereby altering equilibrium between bound and free PI, resulting in dissociation of the DNA–PI complexes *(68)*.

FNA samples ***(69)***, sputum ***(70)***, bladder washes ***(71)***, neonatal blood samples ***(46)***, and paraffin blocks ***(72)*** each provide an adequate number of cells/nuclei for analysis by LSC. LSC can also be used for analysis of histologic sections. Areas of interest that might be a minor component of the whole section can be selected to exclude extraneous tissues from measurement ***(72)***. The specimens can be destained and restained ***(68)*** to measure additional attributes of the same cells; the relocation feature of LSC allows one to precisely identify each cell by its location on the slide. Several publications describe the usefulness of LSC in analysis of tissue sections, FNA samples, or touch preparations ***(44,73–83)***.

One of the drawbacks inherent in measuring constituents of cells in histologic sections is that most of the cells are transected at different levels. Thus, because only a fraction of a cell or nucleus, unknown in size, is assayed, such measurement provides no information about the quantity of the measured constituent per cell. However, a ratiometric analysis, relating the quantity of the measured nuclear constituent per unit of DNA, normalizes the data and makes them comparable between sections of different thicknesses ***(80)***. Still to be worked out are the computer-assisted analytical methods that will be needed to fully exploit the information in histologic sections. In the case of solid tumors, this includes the relationship between tumor cells and reactive host cells, stroma, proliferating vessels, and so forth, the distribution of proliferating vs apoptotic cells within the tumor, the expression of growth factor receptors in tumor cells according to location and in relation to host cells and blood vessels, and the effect of drug therapies on the functional measurements of the cells. The number of measurable features is increasing, providing new tools to characterize and monitor human tumors in ways not possible by conventional light microscopy.

10. Utility of LSC in Other Applications

The major assets of LSC are the relocation and file-merge capabilities. These features are essential in studies of time-resolved events such as enzyme kinetics, transmembrane transport rates of drugs or metabolites, and other cell functions. Likewise, *in situ* association constants of fluorochrome-conjugated ligands with the respective receptors can easily be assessed for individual cells by LSC by repeatedly measuring ligand binding to the same cells as a function of increasing ligand concentrations. After archival preservation, the same cells can be subjected to further measurements with new probes and the results merged into a single file for multivariate analysis. The same cells can be sequentially studied—first when they are alive (e.g., surface immunophenotyped, subjected to functional assays for a particular organelle, oxidative metabolism, pH, enzyme kinetics, etc.) and then following their fixation (e.g., probed for DNA content to assess DNA ploidy and/or cell cycle distribution, DNA replication, content of an intracellular constituent[s] that can be detected immunocytochemically, etc.). To obtain their cytogenetic profile, cells can be subsequently probed by FISH or *in situ* polymerase chain reaction (PCR). The length of telomere sections of DNA can be conveniently estimated *in situ* by LSC using FISH telomere probes ***(84,85)***. Conventional staining with absorption dyes followed by microscopy can further identify the measured cells and correlate their morphology with any of the measured parameters. If desired, a more sophisticated image analysis of the selected cells can follow. An attachment of LSC to the image analysis system (Kontron KS 100) through standard connections has been described ***(86)***. The images collected by LSC can be digitally processed for further quantitative analysis ***(31)***. LSC can also be combined with a laser-capture microdissection device to obtain histologically homogenous cell populations (e.g., for cytogenetic analysis) ***(87)***.

Laser scanning cytometry has the potential to be used to analyze cell–cell interactions *in situ*. One such application, namely to detect platelet–endothelial cell interactions, has already been demonstrated ***(88)***. This assay can be a sensitive marker predictive of vascular thrombosis. We are currently testing the application of LSC to measure the kinetics of the in vitro "wound healing" in the model of mechanically or thermally damaged monolayer cultures of epithelial cells ***(89)***.

Further applications might require more parameters to be measured. Sequential analyses of the same cells rely on the removal of the fluorescence from the earlier analyses prior to the next measurement. Currently available means of enzymatic or chemical removal of the fluorochrome ***(59)***, or its bleaching, might not always be effective and new methods have to be developed. As mentioned, however, sequential measurements of the cells using the same color fluorescence but employing subtraction of the fluorescence signals to obtain differential fluorescence as a separate parameter ***(64)*** might circumvent, to some extent, a need for more colors. A violet laser, which is already available on new LSC models or can be retrofitted into old models, will further enhance LSC capability for multiparametric measurements with different color probes, alleviating the need for fluorochrome removal. New dye combinations provide additional possibilities to expand analytical capabilities of LSC for multiparameter cytometry ***(90,91)***. Furthermore, a combination of fluorescence and time-delayed luminescence probes that are both color- and time- resolved ***(92)*** and can be adapted to LSC can double the analytical capability of this instrument.

The capability of spatial localization of a fluorochrome within the cell is another feature of LSC that is expected to attract new applications. LSC will also be used for the measurement of the translocation of different factors, such as components of the signal transduction pathway between cytoplasm and nucleus and between nucleolus and nucleoplasm, cytosol and mitochondria, organelle phenotyping, or transorganelle trafficking. LSC can also be used to measure a variety of attributes of live cells, including in vivo trafficking of any protein e.g., when fused with green fluorescent protein [GFP]), in cells cultured in a thermostated, optically transparent culture chamber mounted on the microscope stage.

Analysis of cell-to-cell interactions is another field in which LSC can find further applications. Signal transfer between the cells, cell-to-cell transport of the fluorochrome-tagged molecules, local differences in cell proliferative potential, and apoptosis can be studied on cells growing on slides, live and/or after fixation. Likewise, the role of cytokines or other growth factors released from individual cells on proliferation or apoptosis of the neighboring cells, whether in cell monolayers or in tissue sections, also can be studied by LSC.

By providing quantitative data on FNA, tissue sections, or cytology smears, LSC could also become widely used in clinical pathology. As new diagnostic and prognostic markers are rapidly being developed and their clinical utility becomes more and more evident, the need for quantification of these markers is apparent.

Laser scanning cytometry has all of the required attributes, which, with relatively simple alterations, could be transformed into a three-modular instrument consisting of (1) a static cytometer, (2) a flow cytometer, and (3) a confocal microscope; that is, attachment of a module containing a flow channel to the microscope stage combined with immobilization of the laser beam transforms LSC into a flow cytometer. The existing scanning laser beam, on the other hand, when combined with pinhole apertures incorporated into an optical system of cell illumination and light collection, could transform LSC into a confocal microscope. Such a three-modular cytometer/confocal microscope could offer limitless analytical capabilities.

It was recently proposed that LSC might serve as both microscope and a cytometer in the *Space Station* ***(93)***, where it can be used for monitoring the crew's health and in ongoing biomedical experiments. Toward this end, the "liquidless" staining of specimens, which can be applicable at microgravity conditions, was developed ***(93)***. This futuristic application of LSC could be combined with automated or robotic handling of the specimen and data transmission to Earth.

11. Recent LSC Technology Developments

Since the submission of the original chapter two years ago, a new generation of laser scanning cytometers, the iGeneration instruments, has been released ***(94)***. Like the LSC, iGeneration instruments contain up to three lasers of different wavelengths, now enabling not only precise quantification of fluorescent signal but also laser light loss and imaging of cellular and tissue specimens on a solid substrate ***(94)***. Beams from the lasers are combined into a coincident path and directed to an inverted microscope (a new feature differentiating the iGeneration from the original technology) and onto a focal plane at the specimen.

The iGeneration's inverted format allows analysis of specimens on a variety of media, including microscope slides, microtiter plates, chamber slides, Petri dishes, or user-defined carriers fitting the footprint of a microtiter plate.

Autofocus is integral to iGeneration cytometers, minimizing operator involvement in the analysis. Although high-quality imaging on these systems eliminates the need for an optical microscope, it can still be provided to allow post-scan visualization of any location of interest. An optional robotic arm is also available for large-scale walk-away experiments, allowing automatic loading of up to 45 carriers.

The iGeneration software platform keeps much of the original LSC software functionality intact, as the LSC is considered the "gold standard" in solid-phase quantitative imaging cytometry. Raw scan data can be saved as JPEG or 16-bit image files suitable for data processing, specimen visualization, and quantification by proprietary analytic software. Contours may be generated around cellular events based on the fluorescence intensity, forward scatter, or light absorption, either automatically to select an entire sample area, or user-defined to target specific types of events. Scan images may be assembled into tiled mosaic images, allowing contours to be drawn on tissue sections, cell colonies, and other large events that span multiple scan fields. Once the contours are generated, the software can perform a wide range of analyses and produce output in the form of numerical statistics, scattergrams, histograms, expression maps, or other statistical visualizations. Special mathematical operations allow correction for spectral overlap and tissue autofluorescence, and permit combining different time-lapse images. Analysis protocols are easily "written" by graphically assembling various functional modules to establish the desired analytical workflow. Each module's attributes can be easily modified, and the effect of these changes can be viewed immediately in the scanned images. Customizable protocol templates are provided as frameworks for different assays.

11.1. New Applications

The technology's performance characteristics make it suitable for a number of applications in basic life science research and drug discovery. Many new applications *(93)* are greatly facilitated by iGeneration's inverted format, expanded scatter and light-loss features, and enhanced cytometric analysis software. Space limitations allow only a few to be described.

Live-Cell Studies. iGeneration instruments allow live-cell analysis with minimal sample disruption. In many cases, multiple-dye homogenous staining cocktails can be used, eliminating specimen washing. This is particularly important in apoptosis applications, where fragile, loosely adherent apoptotic cells may be lost during washing, leading to potentially erroneous results. Cell-cycle analysis is performed simultaneously with other markers of plasma and nuclear membrane integrity for comprehensive evaluation of apoptotic pathways. Additionally, live cells can be analyzed and then fixed and restained with new reagents for further analysis. Live-cell toxicological analysis using a multiple-dye homogenous cocktail consisting of cell-cycle and apoptosis related markers is shown in **Fig. 9**. Among the notable drug effects are the abrogation of homotypic adhesion of the Jurkat cells **(Fig. 9A)**. An example of a scan field showing late apoptotic cells exhibiting red (propidium iodide) and green (annexin V staining) and early apoptotic cells hyperstained with Hoechst 33342 (DNA content) and expressing green annexin V are shown in **Fig. 9B**.

Automated Analysis of Tissues and Tissue Microarrays (TMAs). iGeneration platforms are well suited for analysis of samples with preserved tissue architecture, such as tissue imprints, TMAs, and tissue sections *(94–96)*.

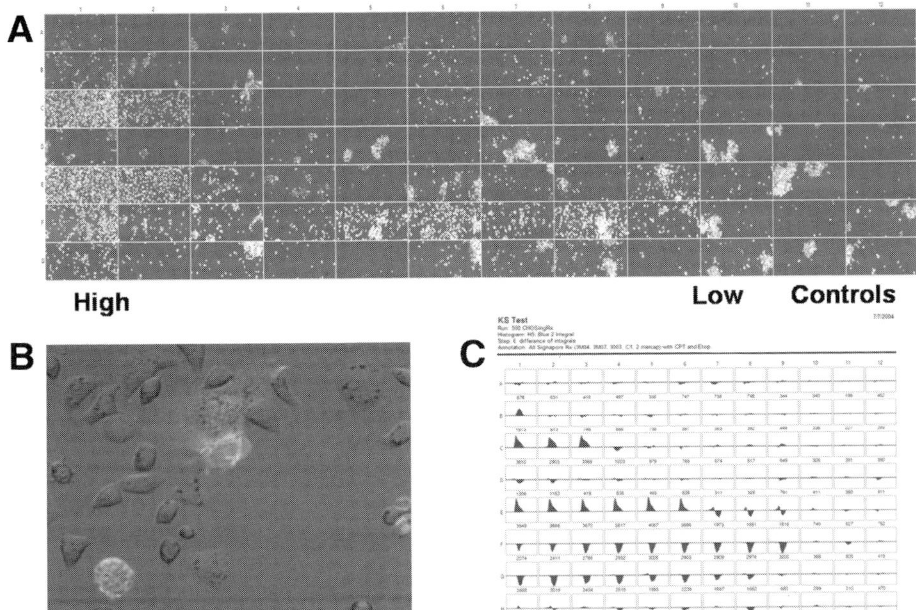

Fig. 9. Live-cell toxicological analysis using a multiple-dye homogenous cocktail. **(A)** Laser scan images of the Jurkat cells treated with a panel of mitochondrial-specific compounds and topoisomerase inhibitors camptothecin and etoposide. **(B)** Portion of a scan field showing early and late apoptotic cells. **(C)** Kolmogorov–Smirnov D-value histograms of DNA staining from adherent cells treated with the same test compounds.

New light-loss modes allow for simultaneous analysis of chromatically and fluorescently stained sections. Multi-scale scanning increases specimen throughput: an initial high-speed scan identifies specific objects or areas of interest for high-resolution re-scanning, which identifies their contents and quantifies constituents of interest. Unique image processing techniques correct for background fluorescence and spectral overlap of both fluorescent dyes and chromophores.

TMA analysis is emerging as the method of choice for biomarker discovery, and requires instrumentation capable of objective analysis and handling of multiple features measured on many tissue elements. An example shown in **Fig. 10** represents a prostate TMA labeled with CD10 developed with the chromophore DAB and counterstained with hematoxylin. Analysis employs blue and red laser light absorption to quantify DAB and hematoxylin, combined with detection of autofluorescence.

Research areas currently benefiting from LSC technology are system biology, early-stage drug discovery (particularly target identification, lead optimization, and predictive and investigative toxicology), biomarker discovery and

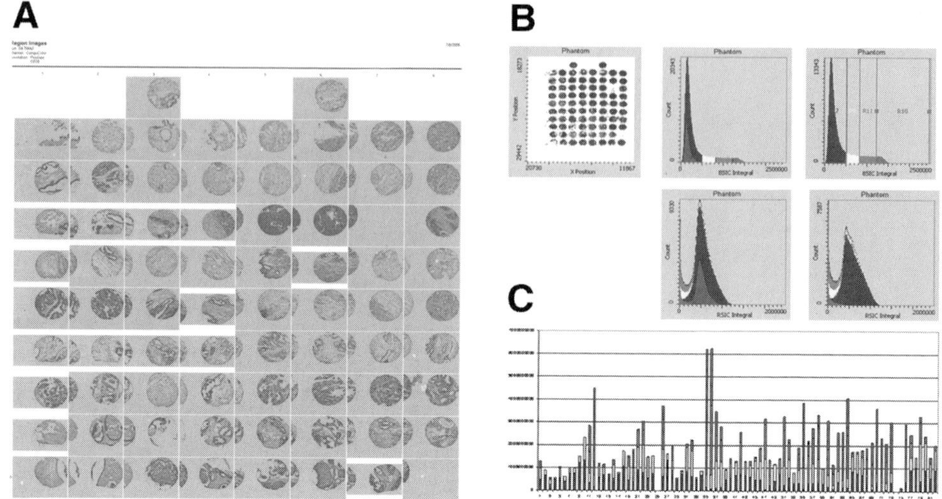

Fig. 10. Tissue microarray analysis of prostate TMA labeled with CD10 developed with the chromophore DAB and counterstained with hematoxylin. **(A)** Mosaic images of the individual core elements. DAB expression is color-coded red, hematoxylin is blue, and autofluorescence is green. **(B)** Quantitative analysis of the core elements using random sampling techniques. **(C)** CD10 expression of the individual elements of the TMA.

validation (with application in phase II and phase III clinical trials), immunophenotyping and immunohistochemistry analysis in clinical research.

Acknowledgments

This work was supported by NCI grant no. CA 28704 and the "This Close" Foundation for Cancer Research.

References

1. Kamentsky, L. A. and Kamentsky, L. D. (1991) Microscope-based multiparameter laser scanning cytometer yielding data comparable to flow cytometry data. *Cytometry* **12,** 81–87.
2. Kamentsky, L. A., Burger, D. E., Gershman, R. J., Kamentsky, L. D., and Luther, E. (1997) Slide-based laser scanning cytometry. *Acta Cytol.* **41,** 123–143.
3. Kamentsky, L. A. (2001) Laser scanning cytometry. *Methods Cell Biol.* **63,** 51–83.
4. Darzynkiewicz, Z., Bedner, E., Li, X., Gorczyca, W., and Melamed, M. R. (1999) Laser-scanning cytometry: a new instrumentation with many applications. *Exp. Cell Res.* **249,** 1–12.
5. Grabarek, J. and Darzynkiewicz, Z. (2002) Versatility of analytical capabilities of laser scanning cytometry (LSC). *Clin. Appl. Immunol. Rev.* **2,** 75–92.
6. Tarnok, A. and Gerstner, O. H. (2002) Clinical applications of laser scanning cytometry. *Cytometry* **50,** 133–143.

7. Roman, D., Greiner, B., Ibrahim, M., Pralet, D., and Germann, P. G. (2002) Laser technologies in toxipathology. *Toxicol. Pathol.* **30,** 11–14.
8. Taatjes, D. J., Palmer, C., Pantano, C., Hoffmann, S. B., Cummins, A., and Mossman, B. T. (2001) Laser-based microscopic approaches: application to cell signaling in environmental lung disease. *Biotechniques* **31,** 880–894.
9. Luther, E. and Kamentsky, L. A. (1996) Resolution of mitotic cells using laser scanning cytometry. *Cytometry* **23,** 272–278.
10. Kawasaki, M., Sasaki, K., Satoh, T., et al. (1997) Laser scanning cytometry (LSC) allows detailed analysis of the cell cycle in PI stained human fibroblasts (TIG-7). *Cell Prolif.* **30,** 139–147.
11. Juan, G., Traganos, F., James, W. M., et al. (1998) Histone H3 phosphorylation and expression of cyclins A and B1 measured in individual cells during their progression through G_2 and mitosis. *Cytometry* **21,** 1–8.
12. Gorczyca, W., Melamed, M. R., and Darzynkiewicz, Z. (1996) Laser scanning cytometer (LSC) analysis of fraction of labeled mitoses (FLM). *Cell Prolif.* **29,** 9–47.
13. Quastler, H. and Sherman, F. G. (1959) Cell population kinetics in the intestinal epithelium of mouse. *Exp. Cell Res.* **24,** 420–438.
14. Schmid, T. E., Attia, S., Baumargartner, A., Nuesse, M., and Adler, I. D. (2001) Effect of chemicals on the duration of male meiosis in mice detected with laser scanning cytometry. *Mutagenesis* **16,** 339–343.
15. Darzynkiewicz, Z., Juan, G., Li, X., Gorczyca, W., Murakami, T., and Traganos, F. (1997) Cytometry in cell necrobiology: Analysis of apoptosis and accidental cell death (necrosis). *Cytometry* **27,** 1–20.
16. Darzynkiewicz, Z., Bruno, S., Del Bino, G., et al. (1992) Features of apoptotic cells measured by flow cytometry. *Cytometry* **13,** 795–808.
17. Furuya, T., Kamada, T., Murakami, T., Kurose, A., and Sasaki, K. (1997) Laser scanning cytometry allows detection of cell death with morphological features of apoptosis in cells stained with PI. *Cytometry* **29,** 173–177.
18. Bedner, E., Li, X., Gorczyca, W., Melamed, M. R., and Darzynkiewicz, Z. (1999) Analysis of apoptosis by laser scanning cytometry. *Cytometry* **35,** 181–195.
19. Telford, W. G., Komoriya, A., and Packard, B. Z. (2002) Detection of localized caspase activity in early apoptotic cells by laser scanning cytometry. *Cytometry* **47,** 81–88.
20. Bedner, E., Burfeind, P., Gorczyca, W., Melamed, M. R., and Darzynkiewicz, Z. (1997) Laser scanning cytometry distinguishes lymphocytes, monocytes and granulocytes by differences in their chromatin structure. *Cytometry* **29,** 191–196.
21. Wolter, K. G., Hsu, I. T., Smith, C. L., Hechusthan, A., Xi, X. G., and Youle, R. J. (1997) Movement of *Bax* from the cytosol to mitochondria during apoptosis. *J. Cell Biol.* **139,** 1281–1292.
22. Zamzani, N., Brenner, C., Marzo, I., Susin, S. A., and Kroemer, G. (1998) Subcellular and submitochondrial mode of action of *Bcl-2*-like oncoproteins. *Oncogene* **16,** 2265–2282.
23. Bedner, E., Li, X., Kunicki, J., and Darzynkiewicz, Z. (2000) Translocation of *Bax* to mitochondria during apoptosis measured by laser scanning cytometry. *Cytometry* **41,** 83–88.

24. Baldwin, A. S., Jr. (1996) The NF-κB and IκB proteins: new discoveries and insights. *Annu. Rev. Immunol.* **14,** 649–681.
25. Deptala, A., Bedner, E., Gorczyca, W., and Darzynkiewicz, Z. (1998) Activation of nuclear factor kappa B (NF-κB) assayed by laser scanning cytometry (LSC). *Cytometry* **33,** 376–382.
26. Mercie, P., Belloc, F., Biblou-Nabera, C., et al. (2000) Comparative methodologic study on NFκB activation in cultured endothelial cells. *J. Lab. Clin. Med.* **136,** 402–411.
27. Deptala, A., Li, X., Bedner, E., Cheng, W., Traganos, F., and Darzynkiewicz, Z. (1999) Differences in induction of p53, p21^{WAF1}, and apoptosis in relation to cell cycle phase of MCF-7 cells treated with camptothecin. *Int. J. Oncol.* **15,** 861–871.
28. Kawamura, K., Kobayashi, Y., Tanaka, T., Ikeda, R., Fujikawa-Yamamoto, K., and Suzuki, K. (2002) Intranuclear localization of proliferating cell nuclear antigen during the cell cycle in renal cell carcinoma. *Anal. Quant. Cytol. Histol.* **22,** 107–113.
29. Styles, J. A., Clark, H., Festing, M. F. W., and Rew, D. A. (2001) Automation of mouse micronucleus genotoxicity assay by laser scanning cytometry. *Cytometry* **44,** 153–155.
30. Smolewski, P., Ruan, Q., Vellon, L., and Darzynkiewicz, Z. (2001) The micronuclei assay by laser scanning cytometry. *Cytometry* **45,** 19–26.
31. Petersen, A. B., Gniadecki, R., and Wulf, H. C. (2000) Laser scanning cytometry for comet assay analysis. *Cytometry* **39,** 10–15.
32. Bacso, Z. and Eliason, J. F. (2001) Measurement of DNA damage associated with apoptosis by laser scanning cytometry. *Cytometry* **45,** 180–186.
33. Kamentsky, L. A., Kamentsky, L. D., Fletcher, J. A., Kurose, A., and Sasaki, K. (1997) Methods for automatic multiparameter analysis of fluorescence *in situ* hybridized specimens with laser scanning cytometer. *Cytometry* **27,** 117–125.
34. Kobayashi, Y., Yesato, K., Oga, A., and Sasaki, K. (2002) Detection of 20q13 gain by dual-color FISH in breast cancers. *Anticancer Res.* **20,** 531–535.
35. Hashimoto, Y., Oga, A., Okami, K., Imate, Y., Yamashita, Y., and Sasaki, K. (2002) Relationship between cytogenetic aberrations by CGH coupled with tissue microdissection and DNA ploidy by laser scanning cytometry in head and neck squamous cell carcinoma. *Cytometry* **40,** 161–166.
36. Harada, K., Nishizaki, T., Ozaki, S., et al. (1999) Cytogenetic alteration in pituitary adenomas detected by comparative genomic hybridization. *Cancer Genet. Cytogenet.* **112,** 38–41.
37. Harada, K., Nishizaki, T., Kubota, H., Harada, K., Suzuki, M., and Sasaki, K. (2001) Distinct primary central nervous system lymphoma defined by comparative genomic hybridization and laser scanning cytometry. *Cancer Genet. Cytogenet.* **125,** 147–150.
38. Baumgartner, A., Schmid, T. E., Maers, H. K., Adler, I. D., Tarnok, A., and Nuesse, M. (2001) Automated evaluation of frequencies of aneuploid sperm by laser-scanning cytometry (LSC). *Cytometry* **44,** 156–160.
39. Juan, G. and Cordon-Cardo, C. (2001) Intranuclear compartmentalization of cyclin E during the cell cycle: disruption of the nucleoplasm–nucleolar shuttling of cyclin E in bladder cancer. *Cancer Res.* **61,** 1220–1226.

40. Gorczyca, W., Smolewski, P., Ardelt, B., Ita, M., Melamed, M. R., and Darzynkiewicz, Z. (2001) Morphometry of nucleoli and expression of nucleolin analyzed by laser scanning cytometry in mitogenically stimulated lymphocytes. *Cytometry* **45,** 206–213.
41. Bedner, E., Ruan, Q., Chen, S., Kamentsky, L. A., and Darzynkiewicz, Z. (2000) Multiparameter analysis of progeny of individual cells by laser scanning cytometry. *Cytometry* **40,** 271–279.
42. Clatch, R. J., Foreman, J. R., and Walloch, J. L. (1998) Simplified immunophenotypic analysis by laser scanning cytometry. *Cytometry* **34,** 3–16.
43. Clatch, R. J. and Foreman, J. R. (1998) Five-color immunophenotyping plus DNA content analysis by laser scanning cytometry. *Cytometry* **34,** 36–38.
44. Clatch, R. J. and Walloch, J. L. (1997) Multiparameter immunophenotypic analysis of fine needle aspiration biopsies and other hematologic specimens by laser scanning cytometry. *Acta Cytol.* **41,** 109–122.
45. Clatch, R. J. (2001) Immunophenotyping of hematological malignancies by laser scanning cytometry. *Methods Cell Biol.* **64,** 313–342.
46. Gerstner, A., Lafler, W., Bootz, F., and Tarnok, A. (2000) Immunophenotyping of peripheral blood by laser scanning cytometry. *J. Immunol. Methods.* **246,** 175–185.
47. Kerr, J. F. R., Wyllie, A. H., and Curie, A. R. (1972) Apoptosis: a basic biological phenomenon with a wide-ranging implications in tissue kinetics. *Br. J. Cancer* **26,** 239–257.
48. Bedner, E., Halicka, H.D., Cheng, W., et al. (1999) High affinity binding of fluorescein isothiocyanate to eosinophils detected by laser scanning cytometry: a potential source of error in analysis of blood samples utilizing fluorescein conjugated reagents in flow cytometry, *Cytometry* **36,** 77–82.
49. Bedner, E., Burfeind, P., Hsieh, T.-C., et al. (1998) Cell cycle effects and induction of apoptosis caused by infection of HL-60 cells with human granulocytic ehrlichiosis (HGE) pathogen measured by flow and laser scanning cytometry (LSC). *Cytometry* **33,** 47–55.
50. Darzynkiewicz, Z., Bedner, E., Traganos, F., and Murakami, T. (1998) Critical aspects in the analysis of apoptosis and necrosis. *Hum. Cell* **11,** 3–12.
51. Darzynkiewicz, Z., Bedner, E., and Traganos, F. (2001) Difficulties and pitfalls in analysis of apoptosis. *Methods Cell Biol.* **63,** 527–546.
52. Li, X., Melamed, M. R., and Darzynkiewicz, Z. (1996) Detection of apoptosis and DNA replication by differential labeling of DNA strand breaks with fluorochromes of different color. *Exp. Cell Res.* **222,** 28–37.
53. Darzynkiewicz, Z. and Bedner, E. (2000) Analysis of apoptotic cells by flow- and laser scanning- cytometry. *Methods Enzymol.* **322,** 18–39.
54. Darzynkiewicz, Z., Bedner, E., Burfeind, P., and Traganos, F. (1998) Analysis of apoptosis by flow and laser scanning cytometry, in *Apoptosis. Detection and Assay Methods* (Zhu L. and Chun J., eds.), BioTechniques Books Series, Eaton, Natick, MA, pp. 75–92.
55. Gorczyca, W., Bedner, E., Burfeind, P., Darzynkiewicz, Z., and Melamed, M. R. (1998) Analysis of apoptosis by laser-scanning cytometry. *Mod. Pathol.* **11,** 1052–1058.

56. Abdel-Moneim, L., Melamed, M. R., Darzynkiewicz, Z., and Gorczyca, W. (2000) Proliferation and apoptosis in solid tumors. Analysis by laser scanning cytometry. *Anal. Quant. Cytol. Histol.* **22,** 393–397.
57. Zabaglo, L., Ormerod, M. G., and Dowsett, M. (2000) Measurement of markers for breast cancer in model system using laser scanning cytometry. *Cytometry* **41,** 166–171.
58. Darzynkiewicz, Z., Li, X., and Bedner, E. (2001) Use of flow and laser-scanning cytometry in analysis of cell death. *Methods Cell Biol.* **66,** 69–109.
59. Bedner, E., Smolewski, P., Amstad, P., and Darzynkiewicz, Z. (2000) Activation of caspases measured in situ by binding of fluorochrome-labeled inhibitors of caspases (FLICA: correlation with DNA fragmentation. *Exp. Cell Res.* **259,** 308–315.
60. Smolewski, P., Bedner, E., Du, L., et al. (2001) Detection of caspases activation by fluorochrome-labeled inhibitors: multiparameter analysis by laser scanning cytometry. *Cytometry* **44,** 73–82.
61. Grabarek, J., Dragan, M., Lee, B. W., Johnson, G. L., and Darzynkiewicz, Z. (2002) Activation of chymotrypsin-like protease(s) during apoptosi detected by affinity-labeling of the enzymatic center with fluoresceinated inhibitor. *Int. J. Oncol.* **20,** 225–233.
62. Halicka, H. D., Bedner, A., and Darzynkiewicz, Z. (2000) Segregation of RNA and separate packaging of DNA and RNA in apoptotic bodies during apoptosis. *Exp. Cell Res.* **260,** 248–256.
63. Kakino, S., Sasaki, K., Kurose, A., and Ito, H. (1996) Intracellular localization of cyclin B1 during cell cycle in gliomas cells. *Cytometry* **24,** 49–54.
64. Smolewski, P., Grabarek, J., Kamentsky, L. A., and Darzynkiewicz, Z. (2001) Bivariate analysis of cellular DNA versus RNA content by laser scanning cytometry using the product of signal subtraction (Differential Fluorescence) as a separate parameter. *Cytometry* **45,** 73–78.
65. Li, X. and Darzynkiewicz, Z. (1999) The Schrödinger's cat quandary in biology: integration of live cell functional assays with measurements of fixed cells in analysis of apoptosis. *Exp.Cell Res.* **249,** 404–412.
66. Li, X., Du, L., and Darzynkiewicz, Z. (2000) During apoptosis of HL-60 and U-937 cells caspases are activated independently of dissipation of mitochondrial electrochemical potential. *Exp. Cell Res.* **257,** 290–297.
67. Bedner, E., Melamed, M. R., and Darzynkiewicz, Z. (1998) Enzyme kinetic reactions and fluorochrome uptake rates measured in individual cells by laser scanning cytometry (LSC). *Cytometry* **33,** 1–9.
68. Bedner, E., Du, L., Traganos, F., and Darzynkiewicz, Z. (2001) Caffeine dissociates complexes between DNA and intercalating dyes: Application for bleaching fluorochrome-stained cells for their subsequent restaining and analysis by laser scanning cytometry. *Cytometry* **43,** 38–45.
69. Clatch, R. J., Walloch, J. L., Foreman, J. R., and Kamentsky, L. A. (1997) Multiparameter analysis of DNA content and cytokeratin expression in breast carcinoma by laser scanning cytometry. *Arch. Pathol. Lab. Med.* **121,** 585–592.
70. Woltmann, G., Ward, R. J., Symon, F. A., Rew, D. A., Pavord, I. D., and Wardlaw, A. J. (1999) Objective quantitative analysis of eosinophils and

bronchial epithelial cells in induced sputum by laser scanning cytometry. *Thorax* **54,** 124–130.
71. Wojcik, E. M., Saraga, S. A., Jin, J. K., and Hendricks, J. B. (2001) Application of laser scanning cytometry for evaluation of DNA ploidy in routine cytologic specimens. *Diagn. Cytopathol.* **24,** 200–205.
72. Kamiya, N., Yokose, T., Kiyomatsu, Y., Fahey, M. T., Kodama, T., and Mukai, K. (1999) Assessment of DNA content in formalin-fixed, paraffin-embeded tissue of lung cancer by laser scanning cytometry. *Pathol. Int.* **49,** 695–701.
73. Grace, M. J., Xie, L., Musco, M. L., et al. (1999) The use of laser scanning cytometry to assess depth of penetration of adenovirus p53 gene therapy in human xenograft biopsies. *Am. J. Pathol.* **155,** 1869–1878.
74. Musco, M. L., Shijun, C., Small, D., Nodelman, M., Sugarman, B., and Grace, M. (1998) Comparison of flow cytometry and laser scanning cytometry for the intracellular evaluation of adenoviral infectivity and p53 protein expression in gene therapy. *Cytometry* **33,** 290–296.
75. Rew, D. A., Reeve, L. J., and Wilson, G. D. (1998) Comparison of flow and laser scanning cytometry for the assay of cell proliferation in human solid tumors. *Cytometry* **33,** 355–361.
76. Gorczyca, W., Darzynkiewicz, Z., and Melamed, M. R. (1997) Laser scanning cytometry in pathology of solid tumors. A review. *Acta Cytol.* **41,** 98–108.
77. Gorczyca, W., Sarode, V., Melamed, M. R., and Darzynkiewicz, Z. (1997) Laser scanning cytometric analysis of cyclin B1 in primary human malignancies. *Mod. Pathol.* **10,** 457–462.
78. Kawamura, K., Tanaka, T., Ikeda, R., Fujikawa-Yamamoto, K., and Suzuki, K. (2000) DNA ploidy analysis in urinary tract epithelial tumors by laser scanning cytometry. *Anal. Quant. Cytol. Histol.* **22,** 26–30.
79. Gorczyca, W., Bedner, E., Burfeind, P., Darzynkiewicz, Z., and Melamed, M. R. (1998) Analysis of apoptosis in solid tumors by laser scanning cytometry. *Mod. Pathol.* **11,** 1–7.
80. Gorczyca, W., Davidian, M., Gherson, J., Ashikari, R., Darzynkiewicz, Z., and Melamed, M. R. (1999) Laser scanning cytometry quantification of estrogen receptors in breast cancer. *Anal. Quant. Cytol. Histol.* **20,** 470–476.
81. Tsukazaki, Y., Numa, Y., Zhao, S., and Kawamoto, K. (2000) Analysis of DNA-ploidy using laser scanning cytometer in brain tumors and its clinical application. *Hum. Cell* **13,** 221–228.
82. Gerstner, A. O., Machlitt, J., Laffers, W., Tarnok, A., and Bootz, F. (2002) Analysis of minimal sample volumes from head and neck cancer by laser scanning cytometry. *Onkologie* **25,** 40–46.
83. Bollman, R., Torks, R., Schmitz, J., Bolman, M., and Mehes, G. (2002) Determination of ploidy and steroid receptor status in breast cancer by laser scanning cytometry. *Cytometry* **50,** 210–215.
84. Kajstura, J., Peroldi, B., Leri, A., et al. (2000) Telomere shortening is an *in vivo* marker of myocyte replication and aging. *Am. J. Pathol.* **156,** 813–819.
85. Izumi, H., Hara, T., Oga, A., et al. (2002) High telomerase activity correlates with the stabilities of genome and DNA ploidy in renal carcinoma. *Neoplasia* **4,** 103–111.

86. Woltmann, G., Wardlaw, A. J., and Rew, D. A. (1997) Image analysis enhancement of the laser scanning cytometer. *Cytometry* **33,** 262–265.
87. Mora, J., Cheung, N. K., Juan, G., et al. (2001) Neuroblastic and Schwannian stromal cells of neuroblastoma are derived from a tumor progenitor cell. *Cancer Res.* **61,** 6892–6898.
88. Claytor, R. B., Li, J. M., Furman, M. I., et al. (2001) Laser scanning cytometry: a novel method for the detection of platelet-endothelial cell adhesion. *Cytometry* **43,** 308–313.
89. Haider, A. S., Grabarek, J., Eng, B., et al. (2003) In vitro wound healing analyzed by laser scanning cytometry. Accelerated healing of epithelial cell monolayers in the presence of hyaluronate. *Cytometry* **53A,** 1–8.
90. Police, A. A., Smith, C. A., Brown, K., Farkas, D. L., Silverman, J. F., and Shackney, S. E. (2002) Multiparameter analysis of human epithelial tumor cell lines by laser scanning cytometry. *Cytometry* **42,** 347–356.
91. Gerstner, A. O., Lenz, D., Laffers, W., et al. (2002) Near-infrared dyes for six-color immunophenotyping by laser scanning cytometry. *Cytometry* **48,** 115–125.
92. Tanke, H. J., De Haas, R. R., Sagner, G., Ganser, M., and van Gijlswijk, R. P. M. (1998) Use of platinum coproporphyrin and delayed luminescence imaging to extend the number of targets FISH karyotyping. *Cytometry* **33,** 453–459.
93. Smolewski, P., Bedner, E., Gorczyca, W., and Darzynkiewicz, Z. (2001) "Liquidless" cell staining by dye diffusion from gels and analysis by laser scanning cytometry: potential application at microgravity conditions in space. *Cytometry* **44,** 355–360.
94. Luther, E., Kamentsky, L., Henriksen, M., and Holden, E. (2004) Next-generation laser scanning cytometry. *Methods Cell Biol.* **75,** 185–218.
95. Megyeri, A., Bacso, Z., Shields, A., and Eliason. J. (2005) Development of a stereological method to measure levels of fluoropyrimidine metabolizing enzymes in tumor sections using laser scanning cytometry. *Cytometry A* **64,** 62–71.
96. Pruimboom-Brees, I., et al. (2005) Using laser scanning cytometry to measure PPAR-mediated peroxisome proliferation and oxidation. *Tox. Path.* **33,** 86–91.
97. Davis, D., Takamori, R., Raut, C. P., et al. (2005) Pharmacodynamic analysis of target inhibition and endothelial cell death in tumors treated with the vascular endothelial growth factor receptor antagonists SU5416 and SU6668. *Clin. Cancer Res.* **11,** 678–689.

9

Near-Clinical Applications of Laser Scanning Cytometry

David A. Rew, Gerrit Woltmann, and Davinder Kaur

Summary

Biological samples from human tissues are characterized by complexity and heterogeneity. The ability to make rapid, reliable, quantitative fluorochromatic measurements on clinical samples allows the development of new and practical assays that could influence diagnosis and treatment in a variety of clinical applications. Laser scanning cytometry (LSC) is a very versatile and adaptable technology that allows for the quantitative analysis of cell samples that are unsuitable for flow cytometry by virtue of their presentation and context. Crucially, it allows the direct visualization of cells and rare events and the correlation of imagery with fluorochromatic measurements. In this chapter, we describe early experiments in the study of cytotoxic drug uptake and resistance in human tumor cells and in the study of sputum cells from asthmatic patients, which harness the specific capabilities of LSC to practical clinical problems.

Key Words: Laser scanning cytometry; drug resistance; cancer; doxorubicin; breast cancer; asthma; sputum; eosinophils; cytospins; bronchial epithelium cells.

1. Introduction

This chapter introduces some applications of laser scanning cytometry (LSC) in the hospital-based clinical laboratory. LSC has proved to be a powerful research tool for cytometric studies with practical clinical applications, which are beyond the capability of flow cytometry (FC).

The key advantage of LSC over FC is that in LSC, the laser beam is scanned over the surface of the sample, constrained on a planar microscope slide or derivative presentational surface *(1)* (*see* **Fig. 1**). In FC, the individual particles move past the beam and are lost to further specific analysis. The LSC allows light excitation of the sample by a 488-nm argon laser, by fluorescence excitation from an Olympus epifluorescence system with a mercury arc lamp, and by normal illumination in conventional or dark-field modes. The direct visualization of cells, tissues, or fluorescent subcellular elements on the constrained

From: *Methods in Molecular Biology, vol. 319: Cell Imaging Techniques: Methods and Protocols*
Edited by: D. J. Taatjes and B. T. Mossman © Humana Press Inc., Totowa, NJ

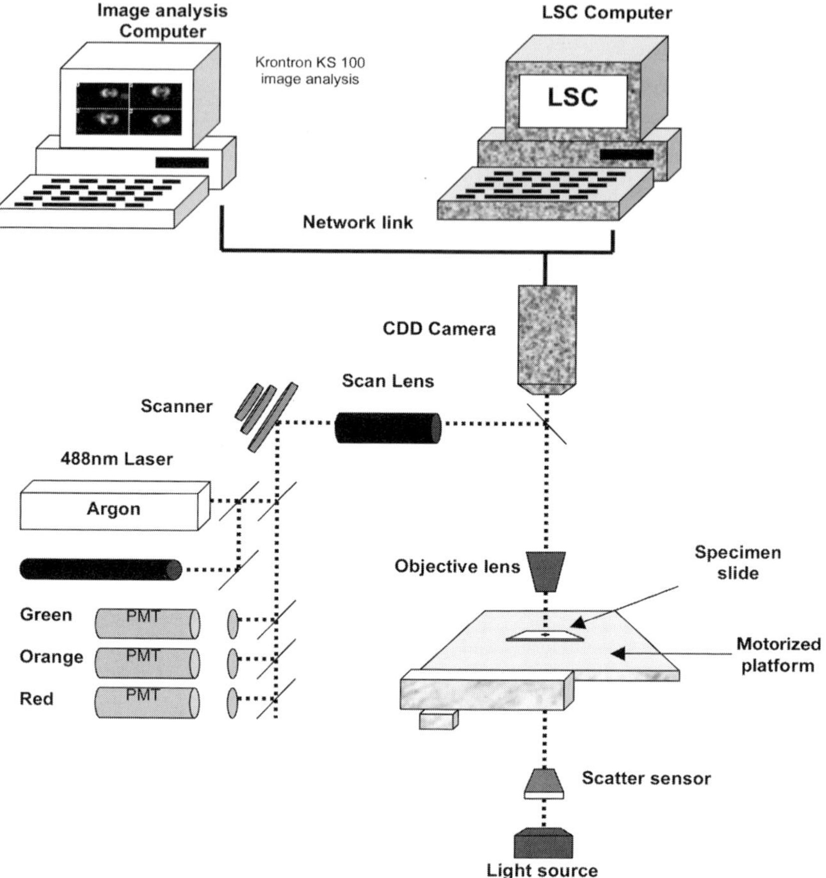

Fig. 1. Setup of the LSC.

surface is thus possible, using either binocular optics or a video digitizing camera.

The microscope slide is moved on the stage in precisely controlled steps by the computer-controlled motor. Each cell image comprises some 100 pixels with a 0.5-μm resolution. Measurements within each array include cell position on the slide, signal area, signal perimeter, peak fluorescence, the time of measurement, and texture. Data are collected and displayed in various graphical forms. Moreover, in LSC, a single cell can be scanned numerous times by the laser beam, with more uniform illumination than is possible with FC and a single light pulse. The speed of scanning of the LSC (1000 cells/min) allows large datasets such as are generated in the study of human tumors and tissues to be collected and analyzed in a realistic time frame.

Use of the microscope slide as the examination plane offers advantages in sample preparation. Whole cells, intact membranes, and normal cell physiology can be studied using chamber slides. The microscope slide allows a broad range of clinical sample types to be studied, including fine-needle aspirates, swab and brushings cytology, smear cytology, viscid and mucoid samples such as sputum, imprint specimens (such as from the cut surface of a carcinoma), and suspensions prepared as for FC. The scanning process allows subcellular fluorescence peaks to be identified and analyzed, such as fluorescence *in situ* hybridisation (FISH) markers ***(2)*** and micronuclei ***(3)***. The instrument is also a powerful tool for immunophenotyping ***(4,5)***.

The LSC only contains one light forward-scatter detector but can contain up to three fluorescence photomultiplier tubes (PMTs): channel 1, 530 ± 15 nm (green); channel 2, 570–630 nm (red/orange); channel 3, >650 nm (far red). The position of the cells can be recorded and the cells of interest can successfully be relocated for reanalysis by storing the information on disk ***(6–9)***. A schematic representation of the LSC is illustrated in **Fig. 1**.

The X-Y coordinates of the cell position on the slide allow recall, visual, or camera inspection of specific cells and the reevaluation and repositioning of cells after further staining. Rare events can be validated within complex cell populations. Cells can be restained and remeasured in studies with additional fluorochromes in multiparameter studies. The sample can be remeasured to generate time-sequenced data using the integral clock and time stamp on each measurement.

More advanced instruments and software now offer up to three coaxial violet, blue and red/infrared lasers, more detectors, and more versatile image analysis software, including the capability for tissue section and FISH spot and multiparameter, high-content analysis.

The methodologies that we describe were developed on a first-generation instrument from Compucyte (Cambridge, MA; www.compucyte.com). This was acquired in 1996 for a small, specialized cytometry research facility within a district hospital, for near-clinical and cytopathological research. The instrument was equipped with a single argon laser (488 nm) and a mercury lamp for excitation, an Olympus microscope, light collection optics, three color PMTs, and forward-scatter detector, and an early version of the Wincyte proprietary analysis software on a Windows-based Pentium computer. Its image analysis capabilities were enhanced through a charge-coupled device (CCD) camera linked to a Kontron image analysis system ***(8)***.

1.1. Specific Applications

Our studies have addressed a range of these capabilities, including single- and dual-parameter DNA content analysis, the measurement of cell proliferation in human tumors following in vivo bromodeoxyuridine labeling ***(9,10)***, viability studies of human tumor cell lines and of cells extracted from human tumors, chamber slide studies, micronucleus genotoxicity assays ***(3)***, rare-event analysis on asthmatic sputum samples, and studies on the subcellular distribution, metabolism, and drug-resistance mechanisms in breast cancer cells ***(11,12)***. In this chapter, we will describe our studies on intracellular cytotoxic drug distribution in encapsulated microspheres and on the detection of dispersed eosinophils in asthmatic sputum.

1.2. Laser Scanning Cytometry Studies of Fluorochromatic Drug Uptake in Human Tumor Cells

Treatment strategies for cancers are often inadequate. We hypothesized that quantitative cytometry and, in particular, LSC can provide useful information about tumor biology and treatment sensitivity not otherwise available to the clinician. Specifically, the decision to use adjuvant chemotherapy for primary, metastatic, or recurrent tumors and the selection of particular cytotoxic agents thereafter is much influenced by empiricism in clinical oncology. This might be depriving some patients of benefits from individually targetted adjuvant chemotherapy. Multidrug-resistance (MDR) mechanisms are a potent cellular source of treatment failure, and other patients might be being treated with ineffectual agents.

Tumor cells in suspension in blood such as leukemias and in malignant ascites are easy to study using FC ***(13)***. White blood cells in particular have specific size, scatter and phenotypic characteristics. Fresh and archival samples of solid tumors pose problems to FC by virtue of their heterogeneity of size, scatter, viability, and biomarker preservation ***(11)***. In contrast, visualization of complex populations and of individual cells by LSC allows verification of features and rare events in complex populations during quantitative cytometry.

The anthracycline and anthraquinone families of cytotoxic agents include adriamycin (doxorubicin), daunorubicin, mitozantrone, epirubicin, and idarubicin. They have a number of potent cytotoxic actions, including intercalation of DNA and inhibition of DNA repair, and are widely used in oncology practice. They possess intrinsic fluorescence and absorb light around 470 nm and emit around 560 nm. They are expelled from the cell by the p170 glycoprotein MDR pump.

Laser cytometry thus offers opportunities to assay the MDR capacity of cancer cells and its modification by MDR blocking agents ***(14)***, to identify those tumors that fail to concentrate the agents and spare their owners from ineffective therapy, to identify those drugs most reliably concentrated in cells from tumor biopsies from a panel of agents, and to study new drug delivery vehicles such as drug-loaded albumin microspheres. The assay of drug uptake alone is insufficient to indicate cell killing. For example, a fluorescent drug might be metabolised, sequestered, or otherwise rendered inert within target cells. Additional assays of drug efficacy are needed to show evidence of induced cell death or irreversible disruption of other vital functions, such as evidenced by mitotic disruption or apoptosis ***(15)***.

We first describe methods to evaluate the binding and internalization of doxorubicin-loaded albumin microspheres to breast cancer cells in vitro using LSC. Doxorubicin microcapsules are a drug delivery system comprising 3-μm human serum albumin (HSA) microcapsules with doxorubicin covalently linked to their outer surface. They are available commercially or they can be produced in the laboratory. In this experiment, we investigate the binding of doxorubicin microcapsules and uptake of doxorubicin in both the wild-type (WT) and doxorubicin-resistant (R) MCF-7 breast cancer cell lines using an in vitro model and LSC.

2. Materials

2.1. Cytometric Reagents

2.1.1. Doxorubicin

We use doxorubicin (Sigma-Aldrich Ltd, Nottingham, UK) as the anthracycline reporter molecule. The drug is used in lyophilized form, obtained in vials containing 10 mg doxorubicin hydrochloride. Each vial is reconstituted under sterile conditions in double-distilled sterile water (ddH_2O) to create stock solutions of 1 mg/mL, which can be stored in polypropylene screw-top vials and frozen at –70°C.

2.1.2. Reagents Used to Bind Doxorubicin to HSA Microcapsules

1. 1-Ethyl-3 (3-dimethyl aminopropyl) carboiimide (EDC) linker, mannitol, pepsin, and Tween 80 (Sigma-Aldrich Ltd).
2. Hydrochloride acid (32%) (Fisher Scientific Ltd, Loughbrough, UK).
3. HSA microspheres.

2.1.3. Selection of Breast Cancer Cell Lines and Culture Conditions

The reagents needed as follows (all purchased from Life Technologies Ltd [Paisley, Scotland]):

1. L-Glutamine.
2. Fetal calf serum (FCS), heat-activated, mycoplasm virus-free.
3. HEPES buffer: 1 *M* L-glutamine (2 m*M*).
4. Phosphate-buffered saline (PBS) without calcium magnesium or sodium biocarbonate.
5. RPMI 1640 medium without L-glutamine or phenol red.

6. Trypsin solution containing 0.25% trypsin.
7. 1 m*M* Ethylenediaminetetraacetic acid (EDTA).

3. Methods

3.1. Source and Production of HSA Microcapsules

The HSA microcapsules have a mean diameter of 3.3 μm. Ours were a gift from Andaris Healthcare Ltd, as a 1 : 2 (w/w) suspension with mannitol (7.5×10^9/mL: number of microcapsules). The microcapsules are washed free of mannitol prior to crosslinking the doxorubicin to the microcapsules as follows.

Microcapsules are suspended at 100 mg HSA/mL in a 1% (v/v) solution of Tween 80 in ddH_2O, vortexed once for 30 s and then left to stand at room temperature for 30 min. The suspension is then centrifuged at 875*g* for 2 min; the supernatant is removed and discarded. The microcapsules are resuspended in 1 mL of ddH_2O (50 mg/mL), vortexed for another 30 s, and centrifuged at 875*g* for 2 min. The supernatant is removed and discarded. This washing step is repeated twice.

3.2. Preparation of Doxorubicin Microcapsules Using 1-Ethyl-3 Carboiimide Linker

All procedures are carried out under sterile conditions. A solution of doxorubicin (3 mg/mL ddH_2O) is added to 100 mg of washed microcapsules; then the mixture is vortexed for 30 s. A solution of EDC (6 mg in 500 μL of ddH_2O) is then added and the mixture is revortexed for 30 s. A magnetic stirrer bar is placed in the reaction vessel and left stirring at 37°C overnight, in the dark. The doxorubicin microcapsules are then collected by centrifugation at 875*g* for 5 min and the supernatant is removed and discarded. Unreacted doxorubicin is removed by resuspending the doxorubicin microcapsules in 5 mL ddH_2O, vortexing, and then centrifuging the suspension at 875*g* for 5 min, with the supernatant then being discarded. This step is repeated until the supernatant appeared colorless to the eye. The drug-loaded albumin microcapsules are then resuspended in 1 mL of ddH_2O and stored at 20°C in the dark.

3.3. Selection and Subculture of MCF-7/WT Cells and MCF-7/R Cells

The MCF-7/WT human breast adenocarcinoma cell line no. 86012803 ***(16)*** can be obtained from the European Collection of Animal Cell Culture (Porton Down, UK). Cells are maintained in culture medium (RPMI 1640 without phenol red, supplemented with 10% [v/v] FCS, 2 m*M* L-glutamine, and 0.04 *M* HEPES buffer) and cultured at 37°C in a humidified incubator containing 5% CO_2.

Confluent cultures of MCF-7/WT cells are subcultured as follows. Cells are washed with PBS once, then trypsin/EDTA 0.1% is added (0.5 mL for a 25-well plate, 1 mL for a 6-well plate, 3 mL for a 25-cm^2 flask [T_{25}], and 5 ml for a 75-cm^2 flask (T_{75}). The culture vessels are incubated in trypsin/EDTA for 3–5 min at 37°C until the cells become detached. The action of trypsin/EDTA is denatured by the addition of an equal volume of culture medium containing 10% FCS. The cells are then aspirated and placed into sterile 15-mL tubes, centrifuged at 450*g* for 8 min, resuspended in 1 mL of culture medium, and then counted using a hemocytometer.

Our MCF-7/R cells *(17)* were kindly provided by Dr. Tim Gant (MRC, Toxicology Unit, University of Leicester). These cells can be cultured in the same way as the MCF-7/WT cells. Additionally, they are maintained throughout in the presence of 0.5 μ*M* doxorubicin. This should be added to the culture medium and filter-sterilized at the time of subculturing.

Both the MCF-7/WT and MCF-7/R cells are normally seeded at 4×10^4 cells/cm^2, equivalent to 1×10^6 cells per T_{25} flask.

3.4. Drug Uptake Studies

The cell lines can then be used to investigate the rate of uptake of free doxorubicin and the binding of doxorubicin microcapsules to the target cells, the time-course of uptake, the patterns of intracellular distribution, the dynamics of drug exclusion and microcapsule metabolism, and the effects on cell viability, cytostasis, and apoptosis over various time periods and using various incubation doses. One can further quantitate the handling of different fluorochromatic drugs and reagents by different cell types. The following subsections illustrate our experiments in this context.

3.5. Measurement the Uptake of Free Doxorubicin and Doxorubicin Microcapsules

Cells are trypsinized (MCF-7/WT, MCF-7/R) and incubated at a cell density of either 2×10^5 or 5×10^5 cell/mL in serum-free culture medium in 5-mL polypropylene tubes at 37°C in a CO_2 incubator for 1 h with 0.5 μ*M* of either free doxorubicin or doxorubicin microcapsules ***(15)***. At 10-min intervals during the incubation, the cell suspensions should be mixed manually. After incubation, they should be washed twice in PBS, resuspended in 500 μL PBS, and then visualized using the LSC.

Time-course experiments can be conducted to establish the rate of uptake of either free doxorubicin or doxorubicin microcapsules after the cells are seeded at 1×10^6 into the T_{25} flask incubated in 3 mL of culture medium at 37°C in CO_2, and incubated for a further 2–96 h.

Following incubation, adherent cells are washed with PBS and detached from the flasks with trypsin. The cell pellet is then resuspended in 1 mL culture medium, cell viability counts are performed, and the cell density are adjusted to

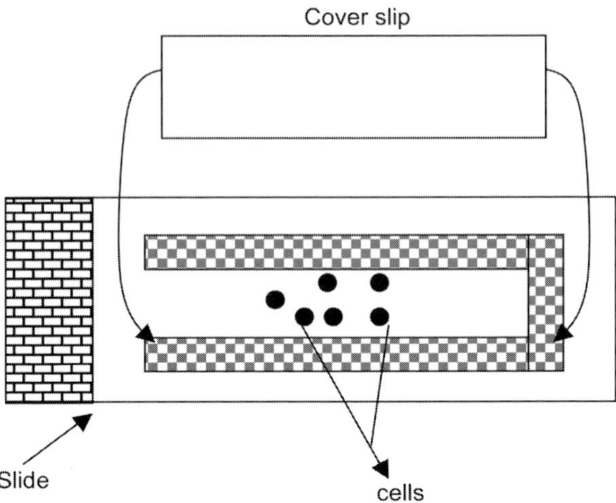

Fig. 2. Construction of a chamber slide.

5×10^5 cells/mL in all cells. Samples are then analyzed and visualized by the LSC to capture the uptake and internalization of the doxorubicin microcapsules by the target cells.

3.6. Slide Preparation for LSC and Fluorescent Microscopy of Labeled Cells

Chamber slides (*see* **Fig. 2**) (*see* **Note 1**) allow the study of spatially fixed but viable cells in fluid media under the microscope of the LSC. This allows a wide range of experimental options in the study of the response of cells to stimulants, toxins, or the binding of additional antibodies and fluorochromes.

3.7. Examination of Samples Using the LSC

The 488-nm argon laser provides excitation. Proprietary Wincyte™ software (Compucyte) is used to control the instrument. The target cells are identified and selected by contouring on laser light scatter and red fluorescence (as determined by the presence of doxorubicin and doxorubicin microcapsules). The integrated laser light scatter and red fluorescence of each cell are displayed in a dot plot of area vs red maximal pixel. The dot plot is very similar in concept and presentation to those familiar to flow cytometrists (*see* **Fig. 3**; Color Plate 7, following p. 274). The signal from each cell comprises some 200 pixels in 0.5-μm increments. The area measures the number of pixels in the field of interest, usually a single cell; thus, area represents the size of the cells or microcapsules. The maximum pixel value is represented the highest fluorescence signal from within the data contour of the individual cell. Each cell can be rescanned

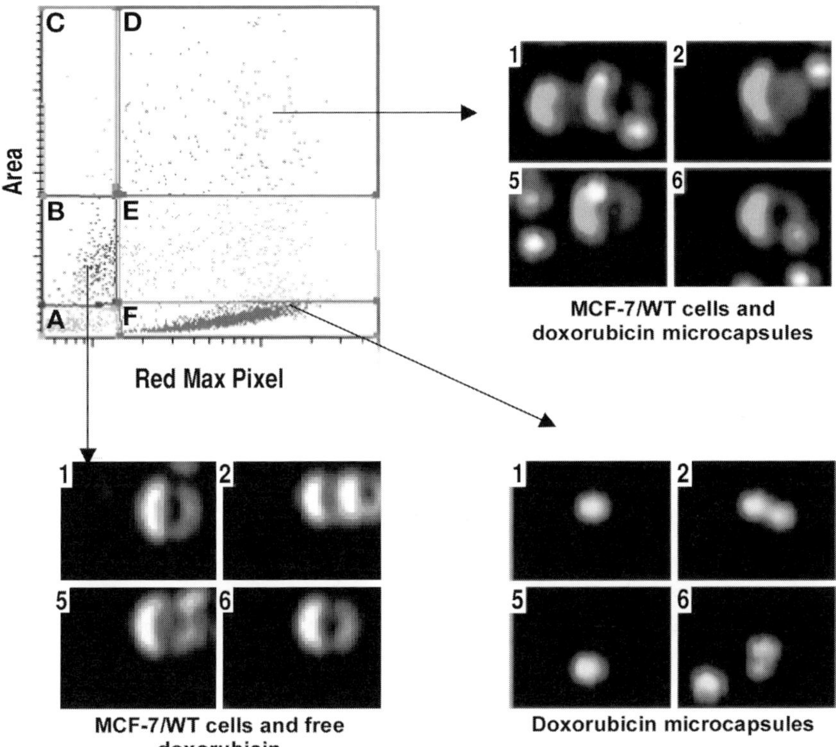

Fig. 3. Assay of cells incubated with either free drug or drug-encapsulated microspheres by LSC. The size of the cells or particles is denoted on the *Y*-axis (Area). The intensity of red, doxorubicin-associated fluorescence, is denoted on the *X*-axis. (*See* Color Plate 7, following p. 274.)

numerous times, because the LSC records the X-Y position of each cell on the slide. Rescanning each cell from any selected dot plot region creates laser light-scatter images and red fluorescent images, which can be stored separately and later merged. This allows us to produce galleries of cells images (*see* **Note 2**).

The specific channel-splitting facility of this program allows each image to be split into three 8-bit images comprising red, green, and blue color components. The coassembly of the laser scan and red fluorescent images of the cells allows false color to be produced.

3.8. Technical Observations

These studies demonstrate the utility of LSC in correlating quantitative data on complex cell populations equivalent to that generated by FC, with highly informative qualitative imagery of the cells of interest.

Our work has been conducted on a first-generation commercial instrument (circa 1996) and early versions of the proprietary Wincyte software. Our system has been adapted for advanced image analysis using a simple link to the Kontron KS100 image analysis system. Our system has been robust, reliable, and straightforward to use. We understand that more recent versions of the instrument offer considerable advances in capability.

We consider that LSC offers a considerable advance in many respects over conventional FC:

- It allows for the visualization of cells of interest and the quantitative fluorometric analysis of cell, subcellular, and particulate fluorescence to the limits of resolution of the instrument. This allows validation of rare events and the study of spatially related and temporally discrete events.
- It allows the study of viable and whole-cell preparations.
- It allows the retention, reexamination, reinterrogation, and reexperimentation of samples and cells of interest.
- It allows for novel forms of presentation of samples to the instrument, as, for example, in fluidic chamber slides.
- It can be made interoperable with other compatible instruments, such as confocal microscopes made by Olympus Optical, whereby the coordinates of cells of interest can be transferred to the stage of the confocal microscope.
- It makes possible broader forms of experimentation that are not possible on flow cytometers.

Specifically, in these studies, we have demonstrated, among others, the following:

- The normal operational mode of the LSC, using single-laser excitation, forward scatter, and two-color analyses
- Techniques of sample presentation on the microscope slide
- The use of the instrument to create galleries of images

3.9. Specific Biological Observations

Our ongoing series of studies have revealed the following:

- The kinetics of uptake, distribution, and elimination of free doxorubicin and encapsulated doxorubicin in wild-type (drug-sensitive) and resistant breast cancer cell lines. Such studies can clearly be extended to any experimental cell type or any fluorochromatic drug or reagent handled by cells (*see* **Fig. 4**; Color Plate 8, following p. 274).
- The facility to discriminate between single cells and aggregates of cells, and subcellular particles—in this instance, drug-loaded albumin microcapsules.
- The use of fluorochromatic cytotoxic drugs as their own biomarkers of binding, uptake, distribution, exclusion, and cell damage in cancer cell lines in vitro.
- The workings of drug-resistance mechanisms in cancer cells.

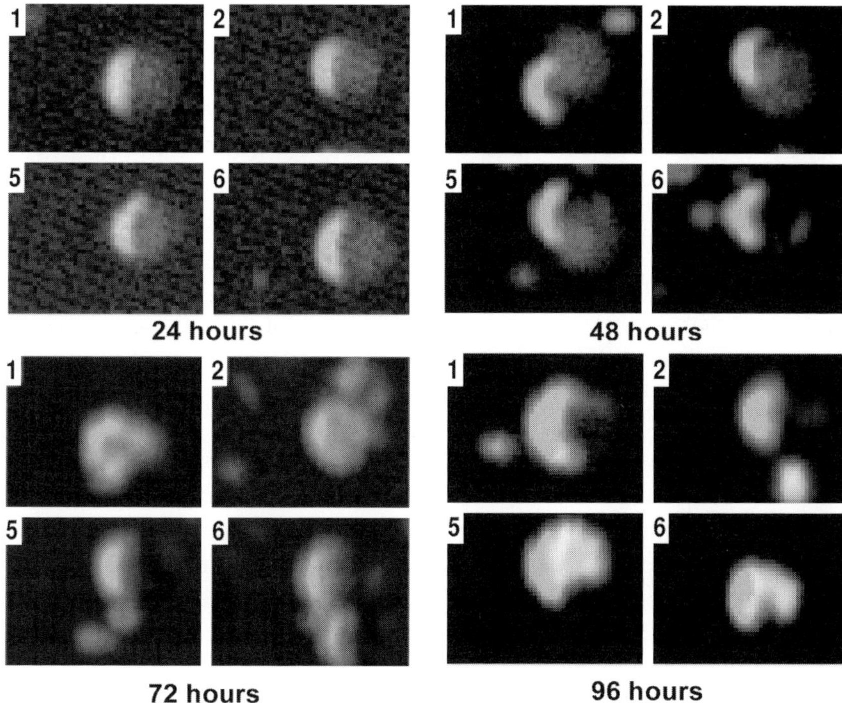

Fig. 4. Time-course experiments of doxorubicin-encapsulated microspheres. Microspheres were incubated with the target cells for 1 h under standard conditions. Cells were analyzed at various time intervals over 96 h. The images were captured on the LSC and then manipulated to produce the image galleries. The illustrations show the binding, uptake, and intracellular distribution of the bright red cytocaps and the gradual release of free doxorubicin into the cells. (*See* Color Plate 8, following p. 274.)

- The induction of cytostasis and apoptosis in target cells.
- The biological complexity of homogenous cell lines with respect to drug uptake and metabolism.

3.10. Quantitative Analysis of Bronchial and Peripheral Blood Eosinophils and Respiratory Epithelial Cells Using LSC

The objective, quantitative measurement of specific cell types in complex samples of clinical material has important diagnostic and therapeutic applications. Microscope-based studies with trained technical observers, such as used in the cervical smear cytology program or in genotoxicity testing, are laborious, inefficient, subjective, and expensive. Automation would offer considerable advantages. The quantitative measurement of eosinophils in the sputum of asthmatic patients has valuable research and diagnostic functions. Cellular and fluid markers of inflammation in sputum have been assessed by a number of investigators, and their validity, repeatability, and responsiveness have been demonstrated in small-scale studies. Sputum eosinophil counts as a marker of asthmatic airway inflammation have the potential for aiding diagnosis and monitoring treatment response to steroid medication in clinical asthma. Sputum cytology also aids the diagnosis of other diseases, including lung cancer. Sputum is a viscid and heterogeneous substance that is unsuitable for FC analysis. Cell preservation is poor and relatively few cells are found in a unit volume of sputum. Manual analysis is very tedious. We thus devised a procedure for objective measurement of sputum eosinophils and bronchial epithelial cells using the planar analysis system of the LSC.

3.10.1. Human Sputum Samples

Samples of human sputum suitable for analysis are obtained in specialist respiratory practice by sequential saline aerosol inhalation at concentrations of 3%, 4%, and 5% for 5 min each. Subjects expectorated for approx 2 min and sputum samples were collected in a sterile 30-mL universal container on ice ***(18)***.

Sputum was processed within 2 h of expectoration. Macroscopically visible sputum plugs were manually selected from saliva using blunt forceps and a Petri dish to minimize oral squamous cell contamination. The selected sample was weighed and processed on ice in 4 vol of 0.1% dithiothreitol (DTT) using gentle aspiration through a Pasteur pipet and subsequent vortexing for 15 s. Following this, the sample was rocked on a bench rocker on ice for 5 min and mixed thoroughly with equal volumes (to DTT) of Dulbecco's phosphate-buffered saline (D-PBS). The cell suspension was filtered through 45-µm nylon mesh and centrifuged for 10 min at 800*g* at 4°C. The cell pellet was resuspended in D-PBS to determine cell viability, oral squamous cell contamination, and absolute cell numbers. The volume of the cell suspension was adjusted for optimal cytospin dispersion with D-PBS to 0.25×10^6/mL. Cell samples can be extracted from sputum by processing on ice in 4 vol of 0.1% DTT, followed by 4 vol of D-PBS, filtering through a 45-µm nylon gauze, and centrifuging for 10 min at 800*g* at 4°C. Suspensions should be adjusted for optimal cytospin dispersion with D-PBS to 0.25×10^6/mL.

3.10.2. Use of the Octospot Cytospin System

To accomplish time-efficient analysis by LSC without the need for frequent slide replacement, we used the novel Octospot® cytospin system (*see* **Fig. 5**). This system is compatible with the Shandon cytocentrifuge and allows transfer of cell suspensions (40–80 µL per well) from eight-well microtiter plate strips onto microscope slides, thus generating eight separate cytospins on a single slide. Octospot slides are covered with a solvent-resistant hydrophobic coating surrounding each well, allowing differential immunostaining of adjacent

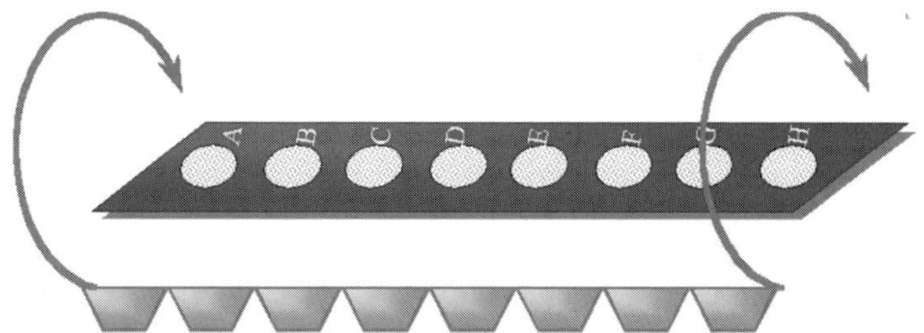

Fig. 5. The octospot® slide system allows the generation of eight individual cytospins on a single microscope slide using the Shandon microcentrifuge and standard eight-well microtiter plate strips.

cytospins. A 40-μL sputum cell suspension (0.25×10^6/mL) was added to each well and spun onto slides at 450 rpm for 6 min in a Shandon cytocentrifuge. Cytospins were air-dried and frozen at –20° until further analysis. After thawing, slides were fixed in high-grade acetone/methanol (50/50). Individual cytospins on nonproprietary standard microscope slides were circled with a hydrophobic wax marker to allow immunostaining of adjoining cytospins with 50 mL of antibody solutions (*see* **Note 3**).

Slides were incubated at room temperature for 1 h with mouse a human-MBP monoclonal antibody (Ab), mouse and human-cytokeratin Ab, or isotype-matched control Ab in PBS + 0.1% BSA. Each cytospin was washed with 20 vol (1 mL) of PBS using simultaneous suction and pipetting to avoid overflow and cross-contamination between adjacent cytospins. To reduce nonspecific binding of fluorochrome-conjugated second Ab to highly charged eosinophil proteins, cytospins were blocked with Chromotrope2R for 15 min, followed by a further washing step and incubation with goat and mouse–Oregon Green®-conjugated second Ab (20 mg/mL) and 0.2 mg/mL propidium iodide for 1 h at room temperature (*see* **Fig. 6**) (*see* **Notes 4–6**).

3.10.3. Analysis Using the LSC

Slides are scanned on the LSC stage using the 20× objective and argon laser light stimulation at 5 mW. Wincyte software is used to draw contours around areas of specified fluorescence intensity, relating to specific staining protocols. For all of the experiments, the red fluorescence channel was used for segmentation around propidium iodide (PI)-stained cell nuclei. Only the red and green sensors were used in channel 1 and 4, respectively. The scan data display in conjunction with scattergram windows for peak fluorescence (red and green max pixel) plotted against area was generally used for optimizing detector voltage and gain as well as background values (all set in *Setup → LSC settings*). Offset

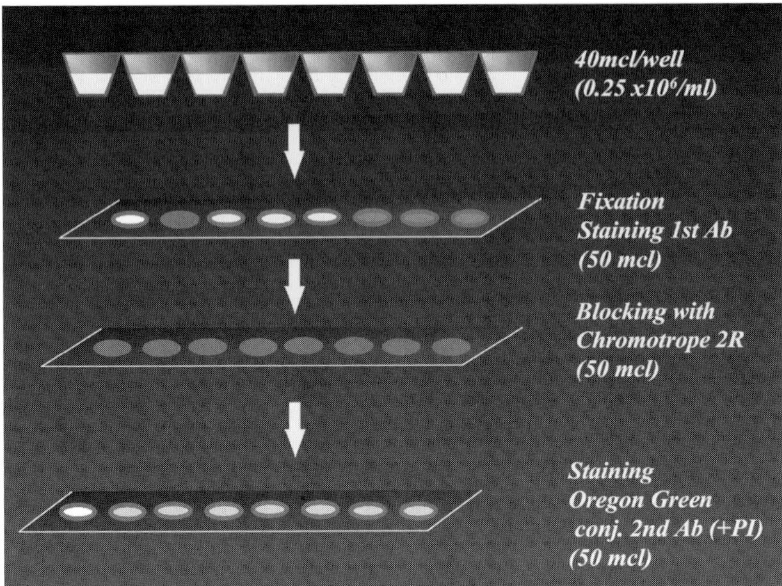

Fig. 6. Overview of indirect immunostaining protocol of sputum cytospins using the Octospot® slide system. Immunostaining with primary antibody was followed by a block step with Chromotrope 2R to reduce unspecific binding of conjugated secondary antibody to highly charged eosinophil-derived proteins. Finally, secondary antibody solution (Oregon Green® conjugated) is applied. Note that negative controls and differential antibody staining is performed on the same slide, thereby reducing between-slide variability.

values of 2000 for red and 2035 for green fluorescence were used. The PMT power setting ("gain") was varied depending on fluorescent-staining intensity between 15 and 60 for red and green fluorescence aiming to utilize the dynamic range of the instrument optimally for individual experiments. For sputum cytospin analysis, red gain was adjusted to achieve mean area values for single cells of 100–200 (25–50 mm^2 for scanning at ×20 magnification).

To allow accurate immunophenotyping of sputum cytospins, detection of a maximal number of single-cell events within the initial cell suspension was critically important. To minimize software contouring around cell clumps and overlapping cells, great care was taken during cytospin preparation to achieve adequate cell dispersion on the glass slide. Preliminary experiments established that cell dispersion for sputum cytospins was optimal at initial cell concentrations of 0.5 × 106/mL. Acytospin index (CI),

$$CI = \frac{SCC}{TCC}$$

where SCC is the total number of single-cell events contoured and TCC is the total number of events contoured, was determined for all cytospins to monitor cell dispersion. Single-cell events were determined by combining the software algorithm for single-cell detection with a gating step based on nuclear size, as PI stained nuclei were used as contouring parameter. Cell dispersion within single-cytospin areas was variable but could be optimized by adding a second cardboard filter to the clip mechanism during cytocentrifugation. In the case of the Octospot slide, eight regions can be set around the eight octospots, thus allowing for automated analysis and comparison of the entire content of each cytospin on the slide. Thus, for example, controls can be analyzed alongside the test samples to minimize experimental variation and error (*see* **Note 7**).

3.10.4. Results

In order to calculate the proportion of a particular cell subset as a percentage of the total blood and sputum samples were labeled with PI to utilize its bright red nuclear staining pattern for software contouring and, hence, capture of all nucleated cells or nucleated cell remnants. The WinCyte software that controls the LSC draws contours around areas of specified fluorescence intensity, if sufficiently contrasted from the background, and registers these contours as objects. Within the area of an object and within a customizable distance surrounding it, other cell parameters can be sampled. The software program distinguishes and optionally excludes cell aggregates based on bit-pattern data. By using this exclusion algorithm and, in addition, setting a nuclear size gate based on the measured area or total integrated red fluorescence of captured events, it was possible to focus on a relatively pure single-cell population within each cytospin (*see* **Figs. 7B**, **8A**, and **9A**). Green fluorescence was sampled within this single cell gate.

For sputum eosinophils, the peak fluorescence signal within each data contour was plotted against contour (nuclear) size (*see* **Figs. 7** and **8**), whereas the more homogenous and generally brighter staining of sputum epithelial cells was plotted as integrated green fluorescence within each data contour against peak green fluorescence signal (*see* **Fig. 9B**; Color Plate 9 following p. 274). To exclude oral squamous epithelial cells, events within the high green peak/high green integrated fluorescence, gate were plotted in addition for their red peak and integrated fluorescence, as bronchial epithelial cells feature more uniform and generally brighter nuclear staining than contaminating oral squamous epithelial cells (*see* **Fig. 9B**). For all sputum samples, at least three cytospins per slide and antibody were analyzed and specific mean percentages were calculated after subtraction of negative controls.

More detailed results and a method comparison using this method and traditional manual differential cell counting are presented elsewhere ***(18)***.

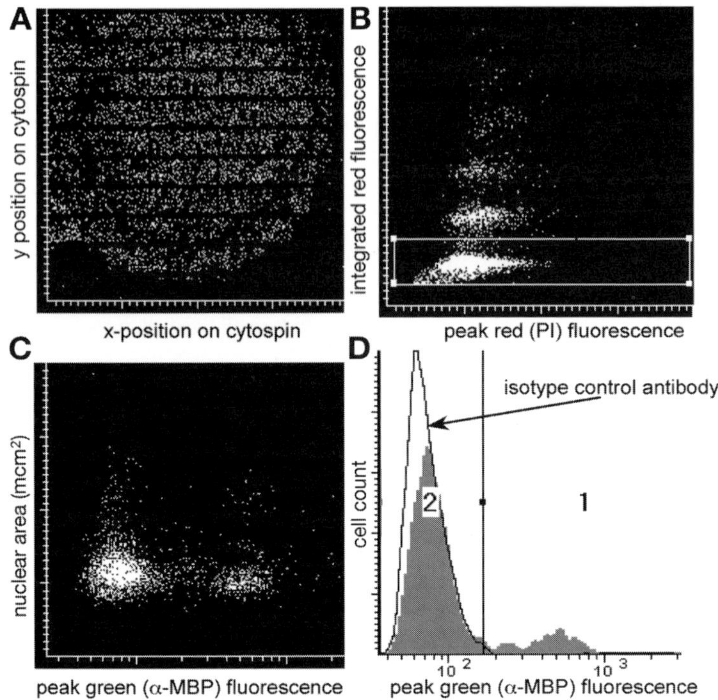

Fig. 7. Determination of blood eosinophil count by LSC. **(A)** *x*/*y*-position of propidium iodide labelled cells detected in a single cytospin. **(B)** Multiple cell objects are excluded on the basis of their higher integrated propidium iodide fluorescence. **(C)** Single cells are scanned for specific green fluorescence after labelling with α-MBP antibody and FITC-conjugated secondary antibody. **(D)** Histogram of plot in C with isotype matched control antibody overlaid.

3.10.5. Technical Observations: LSC in Sputum Analysis

- Octospot cytospin slides are a valuable tool for the presentation and analysis of samples to the LSC.
- Selective gating on peak and integral fluorescence and nuclear and whole-cell fluorescence allows discrimination of cell populations of interest, for both green and red fluorochromes.
- Gating strategies can be validated by direct visualization of the selected cells.
- Perseverance and attention to detail can allow quite challenging sample presentation problems to be overcome by LSC.
- Sample dispersion and cytospin density are critical to the optimal analysis.
- Propidium iodide staining for DNA is a valuable technique for gating on dispersed cells using their nuclear fluorescence.
- Repeated scans are simple to perform and overcome the problem of interobserver error, which complicates manual counts.

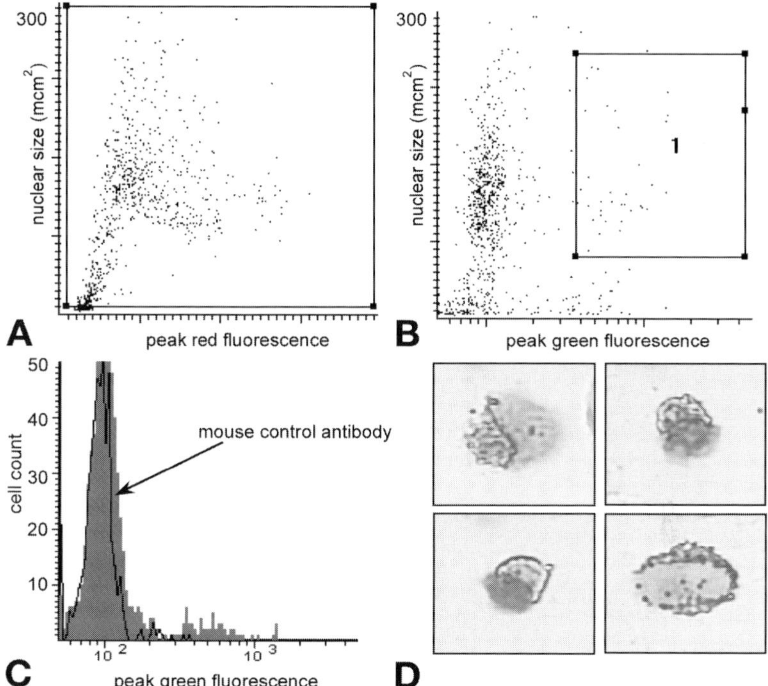

Fig. 8. Gate settings for the analysis of sputum eosinophil cells by LSC. **(A)** Single cells gated by detection of niuclear staining with propidium iodide. **(B)** Eosinophil gate (major basic protein positive cells in gate 1). **(C)** Histogram of plot B with isotype matched control antibody overlaid. **(D)** Brightfield microscopy video captures of relocated cells within eosinophil gate restained with Romanowski stain. More than 90% of cells within gate displayed eosinophil morphology. Outside this gate cells of this morphology were rare.

3.10.6. Biological Observations: LSC in Sputum Analysis

- Eosinophils and bronchial epithelial cells can be successfully identified and quantified on the LSC.
- Compared to manual counting—the current gold standard—estimation of the sputum eosinophil count by LSC was equivalent.
- The assay has demonstrated higher counts of eosinophils in the sputum of asthmatic than in normal subjects.
- Immunophenotyping using techniques demonstrated in this study could be extended to other cell types in complex clinical samples.
- The addition of a third photomultiplier tube to the instrument would allow double staining of other intracellular and extracellular antigens, such as is now possible.

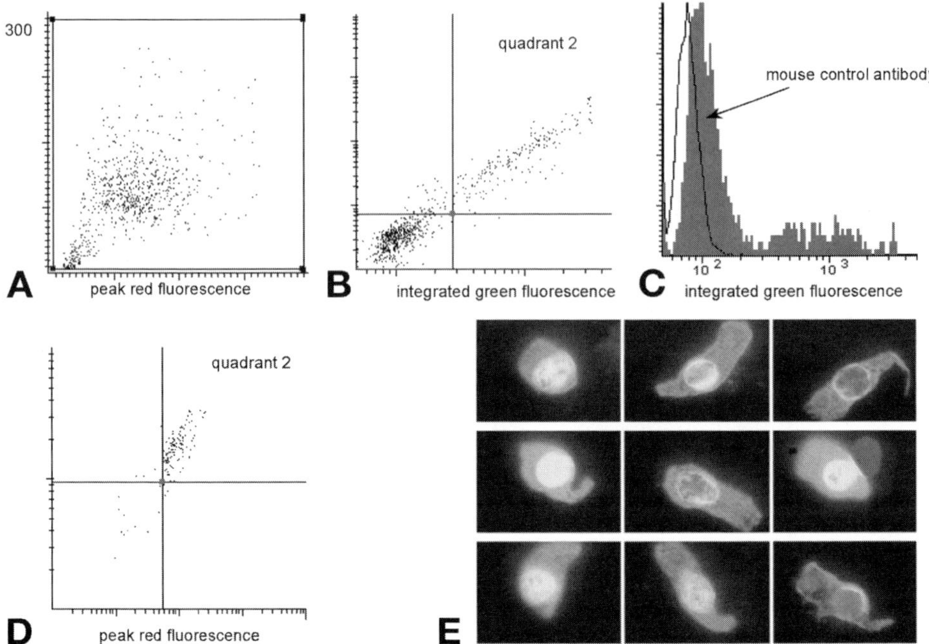

Fig. 9. Gate settings for the analysis of sputum bronchial epithelial cells. **(A)** Single cells gated by detection of nuclear staining with PI. **(B)** Peak green fluorescence plotted against integrated green fluorescence for single cells stained with α-cytokeratin antibody (labeled with Oregon Green-conjugated secondary antibody. **(C)** Histogram of B with isotype-matched control antibody overlaid. **(D)** Cells within quadrant 2 of plot B are plotted for red fluorescence to exclude oral sqamous epithelial cells; **(E)** Epifluorescent video captures of relocated cells within quadrant 2 of plot D (typical of bronchial epithelial cells). In excess of 95% of cells within this gate had this morphology. Outside of this gate, cells with this morphology were very rare (*See* Color Plate 9, following p. 274.)

4. Notes

1. Chamber slides can be constructed from normal glass microscope slides. Multilayered adhesive tape is aligned across the two faces of the slide with a shorter piece of tape joining the two at the end. A glass cover slip is positioned carefully on the tape and the edges are sealed. A 20-μL aliquot of sample is pipetted onto the chamber slide at the end of the cover slip that is not sealed. The cells are allowed to settle for 5 min. The slide should be then sealed with transparent nail varnish before analysis.
2. To improve graphical presentation on older LSC instruments, digital color images of the cells can be captured in WinCyte and copied to a graphic package such as Paint Shop Pro 5™ for Windows as bitmap (bmp) images for display. There, the images of red fluorescence can be superimposed onto the laser light-scatter

images. False colors can be used to identify the cells (green or blue), doxorubicin (red), and doxorubicin microcapsules (red).

3. Samples can be stored for later analysis by freezing at –20°C. Cytospins can be circled with a hydrophobic wax marker to allow immunostaining with 50 µL of antibody solutions.
4. Oregon Green gives brighter staining and less photobleaching than FITC conjugates.
5. Each cytospin can be washed with 1.0 mL PBS with simultaneous microsuction to avoid overflow and cross-contamination between adjacent cytospins. Slides should be cover-slipped with glycerol 25% in PBS.
6. To reduce nonspecific binding of fluorochrome-conjugated secondary antibody to highly charged eosinophilic proteins, cytospins can be blocked with Chromotrope R (Sigma) for 15 min.
7. There must be sufficient contrast between the cells of interest and the background to achieve satisfactory contouring. By focusing on the cell nucleus rather than the entire cell, better contrasts can be achieved.

Acknowledgments

We thank Dr. Alison Goodall for her help with the doxorubicin microcapsule studies.

Elements of our work have been supported in turn by the UK's Cancer Research Campaign, Wessex Cancer Trust, and the NHS Research and Development Executive.

References

1. Kamentsky, L. A. and Kamentsky, L. D. (1991) Microscope based multiparameter laser scanning cytometer which yields data comparable to flow cytometry data. *Cytometry* **12**, 381–387.
2. Kamentsky, L. A., Kamentsky, L. D., Fletcher, J. A., Kurose, A., and Sasaki, K. (1997) Methods for automatic multiparameter analysis of fluorescence in situ hybridised specimens with a laser scanning cytometer. *Cytometry* **27**, 117–125.
3. Styles, J. A. and Rew, D. A. (2001) Automation of mouse micronucleus genotoxicity assays by laser scanning cytometry. *Cytometry* **44**, 153–155.
4. Clatch, R. J., Foreman, J. R., and Walloch, J. L. (1998a) Simplified immunophenotypic analysis by laser scanning cytometry. *Cytometry* **34**, 1, 3–16.
5. Clatch, R. J. and Foreman, J. R. (1998b) Five color immunophenotyping plus DNA content by laser scanning cytometry. *Cytometry* **34**, 1, 36–38.
6. Sasaki, K., Kurose, A., Miura, Y., Sato, T., and Ikeda, E. (1996) DNA ploidy analysis by laser scanning cytometry in colorectal cancers, and comparison with flow cytometry. *Cytometry* **23**, 106–109.
7. Reeve, L. and Rew, D.A. (1997) New technology in the analytical cell sciences: the laser scanning cytometer. *Eur. J. Surg. Oncol.* **23**, 445–450.
8. Woltmann, G., Wardlaw, A. J., and Rew, D. A. (1998) Image analysis enhancement of the laser scanning cytometer. *Cytometry* **33**, 362–365.

9. Rew, D. A., Reeve, L., and Wilson, G. D. (1998) A comparison of flow and laser scanning cytometry for the measurement of cell proliferation in human solid tumors. *Cytometry* **33**, 355–361.
10. Rew, D. A. (2000) In (Darzynkiewicz. Z., Robinson, P. J., and Crissman, H. A., eds.), *Cytometry*; 3rd ed. Wiley, New York, Chap. 68.
11. Reeve, L. (2000) The development of tumor specific assays for cellular response to anthracycline drugs using laser cytometry. PhD thesis, University of Leicester, UK.
12. Kaur, D. (2002) Investigation of cellular and molecular mechanisms involved in targeted drug delivery systems for human cancers. Doctoral thesis, University of Leicester, UK.
13. Krishan, A. and Sauerteig, A. (1992) Flow cytometric monitoring of cellular resistance to cancer chemotherapy, in *Flow Cytometry: Principles and Clinical Application* (Bauer, K.D., Duque, R.E., and Shankey, T.V., eds.), Williams and Wilkins, New York, pp. 459–467.
14. Landon, T. M. (1997) Multidrug resistance assays, in *Bioprobes*, (Landon, T. M., ed.), Molecular Probes, Eugene, OR, pp. 21, 22, 25.
15. Muller, I., Jenner, A., Bruchelt, G., Niethammer, D., and Halliwell, B. (1997) Effect of concentration on the cytotoxic mechanism of doxorubicin—apoptosis and oxidative DNA damage. *Biochem. Biophys. Res. Commun.* **230**, 254–257.
16. Soule, H. D., Vazguez, J., Long, A., Albert, S., and Brennan, M. (1973) A human cell line from a pleural effusion derived from a breast carcinoma. *J. Natl. Cancer Inst.* **51**, 1409–1416.
17. Batist, G., Tulpule, A., Sinha, B. K., Katki, A. G., Myers, C. E., and Cowan, K. H. (1986) Overexpression of a novel anionic glutathione transferase in multidrug-resistant human breast cancer cells. *J. Biol. Chem.* **261**, 15544–15549.
18. Woltmann, G., Ward, R. J., Symon, F. A., Rew, D. A., Pavord, I., and Wardlaw, A. J. (1999) Objective quantitative analysis of eosinophils and bronchial epithelial cells in induced sputum by laser scanning cytometry. *Thorax* **54**, 124–130.

10

Laser Capture Microdissection

Virginia Espina, John Milia, Glendon Wu, Stacy Cowherd, and Lance A. Liotta

Summary

Laser capture microdissection (LCM) is a technique for isolating pure cell populations from a heterogeneous tissue section or cytological preparation via direct visualization of the cells. This technique is applicable to molecular profiling of diseased and disease-free tissue, permitting correlation of cellular molecular signatures with specific cell populations. DNA, RNA, or protein analysis can be performed with the microdissected tissue by any method with adequate sensitivity. The principle components of LCM technology are (1) visualization of the cells of interest via microscopy, (2) transfer of laser energy to a thermolabile polymer with formation of a polymer–cell composite, and (3) removal of the cells of interest from the heterogeneous tissue section. LCM is compatible with a variety of tissue types, cellular staining methods, and tissue-preservation protocols that allow microdissection of fresh or archival specimens. LCM platforms are available as a manual system (PixCell; Arcturus Bioscience) or as an automated system (AutoPix™).

Key Words: Cancer; DNA; laser capture microdissection; molecular profiling; proteomics; protein; RNA; tissue heterogeneity.

1. Introduction

Laser capture microdissection (LCM) is a technique for isolating pure cell populations from a heterogeneous tissue section or cytological preparation via direct visualization of the cells. A common problem encountered by genomic and proteomic researchers in the analysis of tissue arises from the heterogeneous nature of the tissue. Molecular profiling of a pure cell population, which is reflective of the cell population's in vivo genomic and proteomic state, is essential for correlating molecular signatures in diseased and disease-free cells *(1–4)*. Direct microscopic visualization of the cells permits the selection of normal, premalignant, and malignant cells or disease and disease-free cells as

From: *Methods in Molecular Biology, vol. 319: Cell Imaging Techniques: Methods and Protocols*
Edited by: D. J. Taatjes and B. T. Mossman © Humana Press Inc., Totowa, NJ

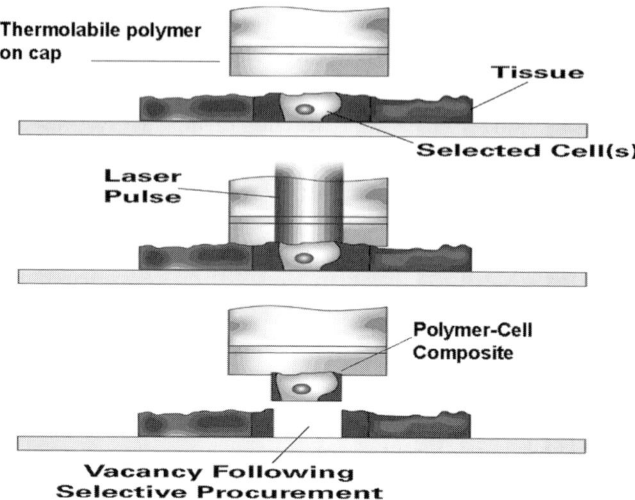

Fig. 1. Principles of LCM. A thermolabile polymer supported on an optical-quality plastic cap is positioned directly above the surface of a tissue section that is mounted on a glass microscope slide. A stationary, near-infrared laser is pulsed in the vicinity of the cells of interest. The polymer melts, or wets, in the vicinity of the laser pulse, forming a polymer–cell composite. Removal of the cap, away from the tissue surface, results in microdissection of the desired cells.

distinct cell populations from the heterogeneous tissue. Heterogeneous tissue might confuse molecular analysis because it is currently impossible to discern which cells contribute which cellular constituents to a given tissue lysate. LCM enables researchers to isolate specific cells of interest, without contamination from surrounding cells ***(5–8)***.

Laser capture microdissection instruments as developed at the National Institutes of Health exist in manual and automated (robotic) platforms (Arcturus Bioscience, Inc., Mountain View, CA) ***(9,10)***. The manual system, PixCell, and the automated system, AutoPix™ utilize identical principles for microdissection (*see* **Fig. 1**). The primary components of LCM technology are (1) visualization of the cells of interest via microscopy, (2) transfer of laser energy to a thermolabile polymer with formation of a polymer–cell composite, and (3) removal of the cells of interest from the heterogeneous tissue section.

A stationary near-infrared laser mounted in the optical axis of the microscope stage is used for melting, or wetting, a thermolabile polymer film (*see* **Note 1**). The polymer film is manufactured on the bottom surface of an optical-quality plastic support or cap. The cap acts as an optic for focusing the laser in the same plane as the tissue section. The polymer melts only in the vicinity of the laser pulse, forming a polymer–cell composite. A dye incorporated into the polymer serves two purposes: (1) It absorbs laser energy, preventing damage to the cellular constituents, and (2) It aids in visualizing areas of melted polymer. Removal of the polymer from the tissue surface shears the embedded cells of interest away from the heterogeneous tissue section. Extraction buffers applied to the polymer film solubilize the cells, liberating the molecules of interest. The microdissected cells can be analyzed for DNA, RNA, or protein by any method with appropriate sensitivity ***(6–9,11–15)***. Protein analysis of microdissected frozen-tissue sections will be used to illustrate LCM protocols.

2. Materials

2.1. Preparation of Tissue Sections or Cytospin Preps

1. Uncoated, precleaned glass microscope slides, 25 × 75 mm (A. Daigger & Co., Wheeling, IL).
2. Specimen for protein analysis of microdissected tissue: cytospin preps or frozen–tissue sections cut at 2–15 mm (*see* **Note 2**).
3. Specimen for DNA or RNA analysis of microdissected tissue: cytospin preps, frozen-tissue sections, ethanol- or formalin-fixed paraffin-embedded tissue sections cut at 2–15 mm (5–8 mm is optimal) (*see* **Note 2**).
4. Cryopreservation solution (OCT) (Sakura Finetek Corp., Torrance, CA).

2.2. H&E Staining of Tissue Sections

1. Mayer's Hematoxylin Solution (Sigma Diagnostics, St. Louis, MO). Hematoxylin is an inhalation and contact hazard. Wear gloves when handling.
2. Eosin Y Solution, alcoholic (Sigma Diagnostics). Eosin Y is flammable. Store away from heat, sparks, and open flames. Contact hazard; wear gloves when handling.
3. Scott's Tap Water Substitute Blueing Solution (Fisher Scientific, Pittsburgh, PA).
4. Ethanol gradient: 70% (v/v in purified H_2O), 95% and 100% ethanol. Prepare fresh ethanol solutions weekly, or sooner if staining more than 20 slides/wk or if the ambient humidity is greater than 40%.
5. Ethyl alcohol, absolute, 200 proof for molecular biology (Sigma-Aldrich, Milwaukee, WI). Ethanol is flammable. Store away from heat, sparks, and open flames. Do not ingest. Contact hazard; wear gloves when handling.
6. Purified water (Type I reagent-grade water).
7. Xylene (Mallinckrodt Baker, Inc., Phillipsburg, NJ). Xylene vapor is harmful or fatal; use with appropriate ventilation and discard in appropriate hazardous waste container. Xylene is flammable; store and use away from heat, sparks, and open flame. Contact hazard; wear gloves when handling.
8. Protease inhibitors (Complete Protease Inhibitor Cocktail Tablets; Roche, Mannheim, Germany).

2.3. Laser Capture Microdissection

1. PixCell II, PixCell IIe, or AutoPix laser capture microdissection system (Arcturus Engineering, Mountain View, CA).
2. CapSure™ Macro LCM Caps (Arcturus Bioscience) (*see* **Note 3**).
3. PrepStrip™ Tissue Preparation Strip (Arcturus Bioscience).
4. CapSure™ Cleanup pad (Arcturus Bioscience).
5. 500-µL Microcentrifuge tubes: Safe-Lock Eppendorf tubes (cat. no. 22 36 361-1; Brinkmann Instruments Inc., Westbury, NY) or MicroAmp™ 500 µL Thin-walled PCR Reaction Tubes (cat. no. 9N801-0611; Applied Biosystems, Foster City, CA).
6. Extraction buffer for cellular constituent of interest.

3. Methods

The following protocols illustrate (1) frozen-section sample preparation, (2) hematoxylin and eosin (H&E) tissue staining, (3) manual LCM for protein analysis, and (4) automated LCM. Alternative tissue preparation methods, such as ethanol or formalin fixation with paraffin embedding, are acceptable for DNA analysis of microdissected tissue ***(16)***.

3.1. Frozen-Tissue Sectioning

Frozen surgical biopsy material should be embedded directly in a cryopreservative solution or liquid nitrogen as soon as the specimen is procured. Prompt preservation of the sample limits protein degradation as a result of protease activity. The samples should be cut to 2–15 mm thickness (5–8 mm is optimal) on plain, uncharged, precleaned glass microscope slides. Position the tissue section near the center of the slide, avoiding the top and bottom third of the slide (*see* **Notes 2**, **4**, and **5**). Do not allow the tissue section to dry on the slide. Place the slide directly on dry ice or keep the slide in the cryostat at –20ºC or colder until the slides can be stored at –80ºC (*see* **Note 6**).

3.2. H&E Staining

Classic tissue-staining protocols allow visualization of the tissue or cells of interest with a standard inverted light microscope. Fixation of the tissue in 70% ethanol is followed by staining of the cellular constituents, with final dehydration in an ethanol gradient. Incorporation of protease inhibitors in the staining reagents, along with a microdissection session limited to 1 h, minimizes protein degradation during the staining process ***(17)*** (*see* **Note 7**). Most cellular staining protocols are compatible with LCM (*see* **Note 8**). Complete dehydration of the tissue is necessary for minimizing the upward adhesive forces between the tissue section and the slide (*see* **Note 9**). Fluorescent stains are compatible with fluorescence-equipped PixCell systems and all AutoPix systems ***(18)*** (*see* **Note 10**).

3.2.1. H&E Staining for Frozen Tissue Sections

1. Remove slide from freezer and place on dry ice or directly into the 70% ethanol fixative bath.
2. Dip, or gently shake, the slide in each of the following solutions, for the time indicated. Blot the slide on absorbent paper in between each solution, preventing carryover from the previous solution.
3. 70% Ethanol fixative: 3–10 s.
4. Distilled (d)H_2O: 10 s.
5. Mayer's hematoxylin: 15 s.
6. dH_2O: 10 s.
7. Scott's tap water substitute: 10 s.
8. 70% Ethanol: 10 s.
9. Eosin Y (optional): 3–10 s.
10. 95% Ethanol: 10 s.
11. 95% Ethanol: 10 s.
12. 100% Ethanol: 30 s to 1 min.
13. 100% Ethanol: 30 s to 1 min.
14. Xylene: 30 s to 1 min.
15. Xylene: 30 s to 1 min.

3.2.2. H&E Staining for Formalin-Fixed Paraffin-Embedded Tissue Sections

Paraffin-embedded tissue sections must be deparaffinized prior to staining. Xylene acts as a solvent, removing the paraffin. Rehydration of the slide allows staining of the tissue elements.

1. Xylene: 5 min.
2. Xylene: 5 min.
3. 100% Ethanol: 30 s.
4. 95% Ethanol: 30 s.
5. 70% Ethanol: 30 s.
6. dH_2O: 10 s.
7. Mayer's hematoxylin: 15 s.
8. dH_2O: 10 s.
9. Scott's tap water substitute: 10 s.
10. 70% Ethanol: 10 s.
11. Eosin Y (optional): 3–10 s.
12. 95% Ethanol: 10 s.
13. 95% Ethanol: 10 s.
14. 100% Ethanol: 30 s to 1 min.
15. 100% Ethanol: 30 s to 1 min.

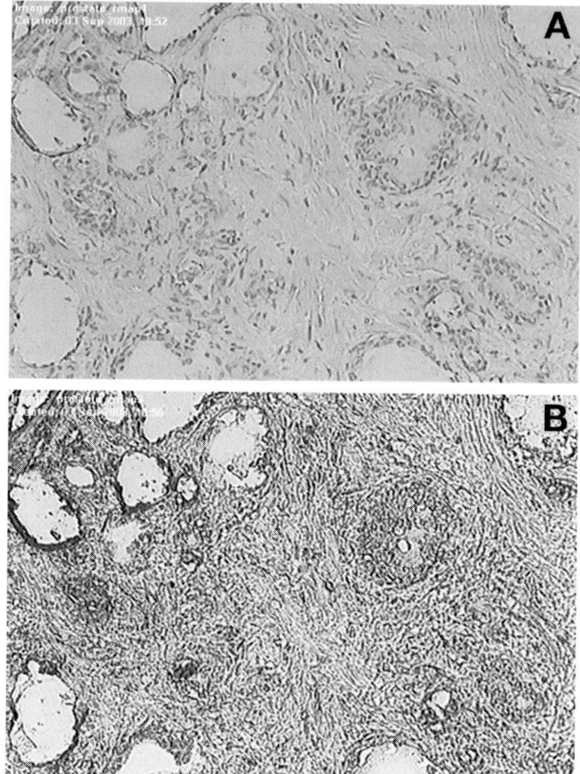

Fig. 2. Comparison of index-matched and non-index-matched tissue images. **(A)** Fluid, such as xylene, on the surface of the tissue acts as a refraction medium, providing an index-matched image of the tissue. As the slide dries **(B)**, the refractive index of the fluid is no longer present. The dry-slide image appears as shades of gray. This issue is overcome on the PixCell system by digitally saving the index-matched images of the tissue, or cells of interest, and using these images as a guide or map for microdissection.

16. Xylene: 30 s to 1 min.
17. Xylene: 30 s to 1 min.

3.3. Manual Laser Capture Microdissection (PixCell System)

There is no warm-up period required for the PixCell II/IIe. The instrument is ready for operation after turning on the power. Visualization of cellular morphological features and the tissue, in general, are achievable with image magnifications up to ×40 via optical and color digital imaging. Cover slips and mounting media are not compatible with microdissection. In addition, a lack of immersion fluids on any of the optics prevents refraction of light from the tissue image. Thus, the color and detail of a given tissue stain is lost as the stained slide dries (*see* **Fig. 2**). Manual LCM methods capitalize on the index refraction of a wet tissue slide for visualizing and reviewing an index-matched image of the tissue ***(19,20)***. An index-matched image or images can be digitally saved and used as a guide, or scout, to locate the cells of interest after the stained slide dries.

3.3.1. Slide Preparation

Allow the stained slide to air-dry. Dust or debris on the slide can be removed by blotting the dried slide with a PrepStrip sample preparation strip.

3.3.2. PixCell Instrument Procedure for Microdissection

1. Load the CapSure cassette module with a CapSure cartridge.
   a. Remove the CapSure cassette module from the platform.
   b. Press in the locking pins on each end to hold the cassette in the load position.
   c. Slide a CapSure cartridge into the cassette until it stops. Two cartridges can be loaded onto the cassette module.
   d. After the cartridges are loaded, pull the locking pins out to lock the cartridges in place. Load the cassette module onto the PixCell II/IIe.
2. Access the LCM software program by double-clicking on the Arcturus software icon.
3. Enter your user name or select a name from the list. Click on "Acquire data."
4. Enter a study name or select a study name from the list. Click on "Select."
5. Enter the Slide # and Cap Lot #. If desired, notes concerning the slide or study can be entered as "Notes."
6. Click the checkbox for "Stamp images with name, date & time" if this information is to be imprinted on the images created during LCM. Click "Continue."
7. Move the joystick into the vertical position to ensure proper positioning of the cap in relation to the capture zone.
8. Place the microscope slide containing the prepared and stained specimen for microdissection on the stage.
9. Locate the cells of interest using either the oculars or the monitor. After the target area for dissection is in the viewing area, with the joystick still in the vertical position, press the "Vacuum" switch on the front of the Controller to activate the vacuum and hold the slide in place during microdissection.
10. The Live Video and PC screen displays the current image on the microscope. The images can be saved as you work by selecting the appropriate icon from the toolbar (*see* **Note 11**).

    Map image: low/high-power objective image of the general area or specific cells to be microdissected (*see* **Fig. 2**);
    Before image: intact tissue prior to microdissection;
    After image: tissue after microdissection;
    Cap image: microdissected tissue only.
11. Slide the CapSure cassette backward or forward so that a cap is sitting at the "Load" position. Swing the placement arm over the cap. While placing one hand over the counterbalance to prevent jarring of the cap and improper seating, lift the placement arm and place the cap onto the slide. This is the transfer position.

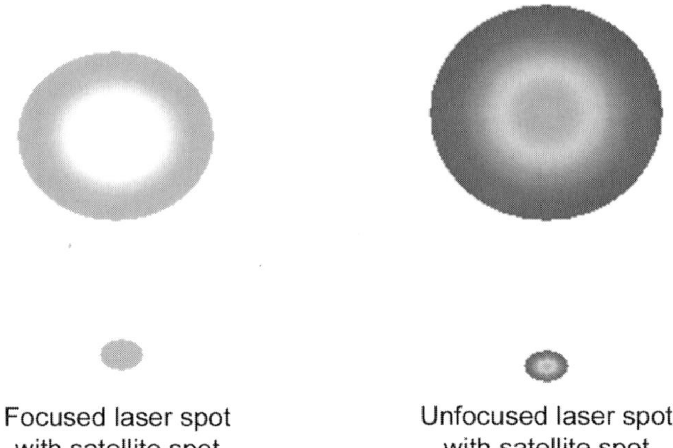

Fig. 3. Laser spot focus. A focused laser spot is essential for efficient microdissection. A focused laser appears as a bright, well-defined circle without a halo or corona. Unfocused laser spots appear as blurry spots with halos or coronas around a center bright spot. AutoPix instruments exhibit a satellite spot near the principle laser spot. The satellite spot is an optical phenomenon because of second-surface reflection of the laser beam as it propagates through the optical system. The satellite spot is a byproduct of the optical system but is helpful for determining the proper focus of the primary laser spot.

12. Enable the laser by turning the key switch located on the front of the controller and then press the "LASER ENABLE" button. Pressing this button will activate the target beam when the placement arm is in the transfer position.
13. Verify that the laser is in focus.
    a. Select the small spot size (7.5 μm), using the "Spot Size Adjust" lever found on the left side of the microscope.
    b. Rotate the objectives of the microscope until the 10x objective is in use.
    c. Using the joystick, move the slide under the laser so that target beam is located in an area without tissue.
    d. Reduce the intensity of the light through the optics until the field viewed on the monitor is almost dark and the target beam is easily viewed.
    e. Using the "Laser Focus Adjust" located just below the size adjustment lever, adjust the target beam until the beam reaches the point of sharpest intensity and most concentrated light with little or no "haloing" or coronas (*see* **Fig. 3**). The laser should now be focused for any of the three laser sizes and objectives. Select the laser spot size suitable for the microdissection and cell size. (*See* **Notes 12** and **13.**)
14. Press the red pendant button to fire a test laser pulse. Observe the wetted polymer after the laser is fired. Firing the laser pulse causes the polymer to melt in the vicinity of the laser pulse. There should be a distinct clear circle surrounded by a dark ring (*see* **Fig. 4**) (*see* **Note 14**).

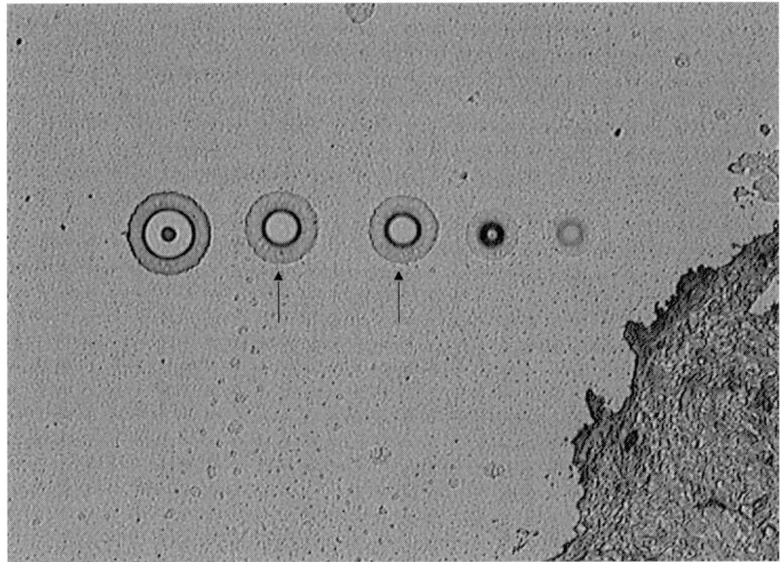

Fig. 4. Wetted polymer for effective microdissection. Adequate power, duration, proper tissue thickness, and cap placement on the tissue are parameters affecting the melting of the polymer. An ideal wetted polymer appears as a distinct dark ring with a clear center (arrows). The presence of a dark area in the center of the spot indicates that the power and/or duration settings are too high (leftmost spot). Inadequate power and duration result in failure of the laser to melt the polymer, creating a spot that appears as a gray, fuzzy circle (rightmost spot). Adequate power with inadequate duration might lead to spots with minute diameters, as shown by the spot second from the right.

15. Adjust the "Power" and "Duration" of the laser pulse with the up and down arrows on the front of the controller to obtain a melted polymer spot with a diameter similar in size to the selected laser spot size. Use the suggested ranges as a reference point. These settings can be adjusted to customize the melted polymer spot to the type and thickness of the tissue to be dissected. The suggested settings for the PixCell II system with a Macro cap are as follows:

Spot size	Power	Duration
Small, 7.5 µm	45 mW	750 µs
Medium, 15 µm	35 mW	1.5 ms
Large, 30 µm	25 mW	5.0 ms

16. Single-cell microdissection is possible by adjusting the power and duration settings such that a very narrow area of the polymer is melted with each laser pulse. Suggested settings for single-cell microdissection are approx 45 mW power and 650 µsec duration for PixCell II and Macro caps.

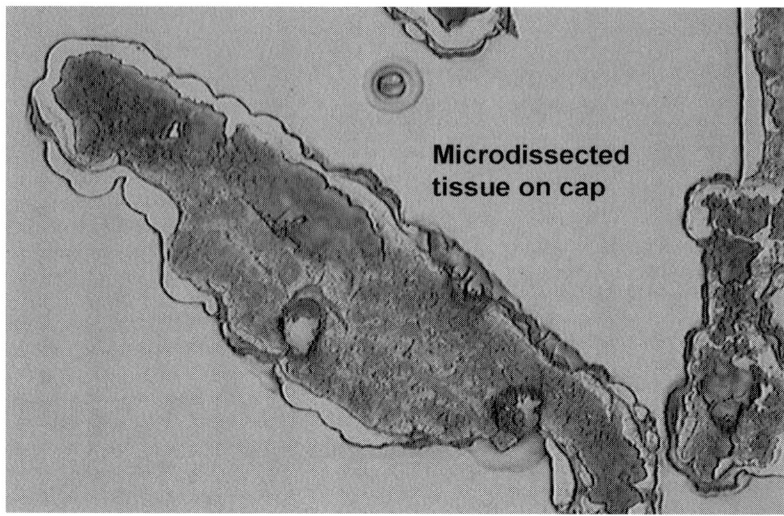

Fig. 5. Microdissected tissue embedded in the polymer cap. After microdissection, the polymer–cell composite can be visualized by placing the cap on the glass microscope slide in an area lacking tissue. The presence of stained cellular material inside the wetted margins of the polymer indicates effective microdissection.

17. To perform the microdissection, visually locate the cells of interest. Align the target beam directly over the cells of interest. Using the target beam to guide the dissection, press the pendant switch for single shots. For a rapid fire of pulses, press and hold the pendent switch. The laser pulse frequency interval can be adjusted on the controller by selecting "REPEAT" and then selecting the desired time between laser pulses.
18. After the desired number of cells has been collected on a cap, remove the cap by lifting it off the slide using the placement arm (*see* **Note 15**). Swing the placement arm away from the slide.
19. If desired, the microdissected material can be viewed by placing the cap on an area of the slide without tissue. Turn the vacuum off by depressing the "vacuum" button on the controller. Position the slide so the tissue is not in view on the monitor or through the oculars. Swing the placement arm, containing the cap, back onto the slide. Use the joystick to manipulate the cap above the objective. Observe the cap for microdissection of the desired cells and for debris and/or adhesion of nonspecific tissue to the polymer surface (*see* **Fig. 5**).
20. If viewing the cap reveals debris or nonspecific tissue adhesion, the said material can be removed by blotting the polymer surface with the CapSure Cleanup pad or an adhesive note.
21. Lift and rotate the cap arm until the cap is over the "cap removal site." Lower the arm and then rotate the arm back toward the slide. The cap will remain at the cap removal site. Blot the cap as described in **step 20** if necessary.

22. Insert the polymer end of the cap into the top of a 500-μL microcentrifuge tube. The sample is now ready for extraction of the desired components or the cap–tube assembly can be stored for extraction at a later date.
23. After all dissections are completed, the PixCell II/IIe should be put in shutdown by first pressing the "Laser Enable" button to disable the laser. Turn off the power to the PixCell II/IIe, the controller, and the video monitor.

3.3.3. Saving Images

Images saved during the microdissection can be saved on the computer hard drive.

1. Click the "Done" button on the image toolbar. Another slide can be microdissected, another study initiated, or the program terminated.
2. Click "Save Images" to save the images on the C drive of the PC. The images can be copied to a PC-formatted, 100-Mb zip disk as a .JPEG or .TIFF format after saving the image on the C drive. The AutoPix system is equipped with a CD-RW drive for file transfer and sharing.
3. Click on "Save Data."
4. Archived LCM images on the PixCell system are stored under the following file directory: C:\\LCMdata\user name\study name\date.

3.3.4. Storing Samples for Downstream Analysis

Microdissected cells for protein analysis can be stored at –80°C prior to extraction. Microdissected cells for DNA analysis can be stored desiccated at room temperature up to 1 wk prior to extraction. Samples for RNA analysis should not be stored prior to extraction. Condensation in the microcentrifuge tube during storage can be a potential source of RNAse contamination.

3.4. Automated Laser Capture Microdissection (AutoPix System)

The AutoPix combines robotics and optical scanning software for automated microdissection of selected cells (*see* **Note 16**). The AutoPix incorporates imaging software for creating index-matched, stitched images of the tissue, permitting more accurate identification of cellular morphology during cell selection. The resolution of the stitched images is constant, but the area of the images changes with magnification, permitting precise areas of tissue to be annotated for microdissection.

Annotation software coupled with the index-matched image permits single-point dissection, line dissection or, polygon dissection (*see* **Fig. 6**; Color Plate 10, following p. 274). Algorithm-based, cell image recognition software on the AutoPix platform enhances the LCM technology. The algorithm is based on texture, morphology, size, color, and contrast of the tissue, permitting automated cell selection in addition to automated microdissection ***(21)***.

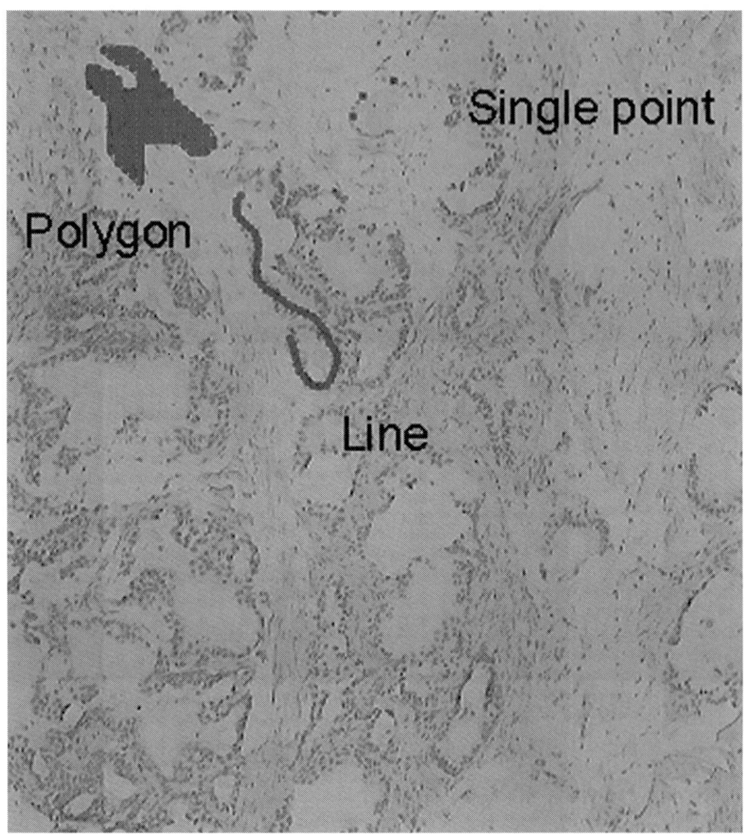

Fig. 6. Annotation of stitched images on AutoPix system. Annotation software permits single-point, line, or polygon area selection for microdissection on the index-matched stitched image. The ability to definitively select the cells of interest enhances the accuracy of microdissection. (*See* Color Plate 10, following p. 274.)

The AutoPix visualization system does not include oculars because of the enclosed system configuration. Instead, a PAL-format color camera permits visualization of the slide as a "roadmap image" for determining the target area of microdissection ***(21)***. The enhancements of the automated system include multiple-slide capacity (three slides), area quantitation of microdissected tissue, wetted polymer spot measurement, and cell recognition software.

4. Notes

1. The short laser pulse widths utilized, the low laser power levels required, the absorption of the laser pulse by the polymer and dye, and the long elapsed time between laser pulses combine to prevent the experimenter from depositing any

significant amount of heat at the tissue surface that might affect later laboratory analysis. The near-infrared laser diode has a maximum laser output of 100 mW.

2. Optimal tissue thickness for microdissection is 5–8 μm. Tissue sections less than 5 μm might not provide a full cell thickness, necessitating microdissection of more cells for a given assay. Tissue sections thicker than 8 μm might not microdissect completely, leaving integral cellular components adhering to the slide.
3. CapSure HS caps (Arcturus Bioscience) are often utilized in microdissection of tissue for RNA analysis. A 12-μm rail on the surface of the polymer prevents the polymer from touching the tissue except in the vicinity of the laser pulse. The HS caps are designed with an extraction device, allowing extraction buffer to contact the polymer within a centrally designated area. These features limit any potential RNA contamination from surrounding cells. CapSure HS caps can be used successfully for DNA or protein extraction. In contrast, CapSure Macro caps are placed in direct contact with the tissue and are not equipped with an extraction device. Any cellular material on the surface of the polymer of a Macro cap will be available for extraction.
4. The inverted light microscope in the PixCell platform utilizes a vacuum to immobilize the slide on the microscope stage. The size of the objective opening limits the microdissectable area on the microscope slide to the middle third of the slide. In contrast, the AutoPix stage has a capacity for one to three slides and the microdissectable area of the slide is approximately 19 × 45 mm.
5. Lung tissue or other tissue with a thin, open architecture can be cut on charged or silanized slides to prevent the tissue from nonspecifically adhering to the polymer during microdissection. In general, coated slides are not used for microdissection because of the increased adhesive forces between the tissue and the slide. Effective microdissection is a balance between three adhesive forces: (1) maximizing downward adhesive forces between the polymer and the tissue, (2) minimizing lateral adhesive forces between the cells, and (3) minimizing upward adhesive forces between the slide and the tissue.
6. Avoid repeated temperature fluctuations of the cut frozen sections. Store the frozen-section slides at –80°C until the time of microdissection. Repeated fluctuations in temperature might cause the tissue to adhere more tightly to the slide, limiting the effectiveness of microdissection.
7. Protease and or phosphatase inhibitors can be added to compatible staining solutions ***(17)***. Complete Protease Inhibitor Cocktail tablets are water soluble. For protease inhibitor addition to the 70% ethanol solution, dissolve the tablet in 15 mL of dH_2O, and then add 35 mL ethanol. Limiting the time from staining to completion of microdissection also ensures preservation of cellular constituents. Frozen-section samples for RNA and protein analysis should be stained and microdissected within 1 hr.
8. Examples of LCM compatible stains are H&E, methylene blue, Wright-Giemsa or toluidine blue. Eosin staining of the cytoplasm is not necessary for visualization of cells during microdissection. Minimal staining times, in whichever staining protocol is utilized, limit potential protein alterations as result of contact with

the staining reagents. Selection of tissue-staining protocols should be based on compatibility with the downstream analysis to be performed with the microdissected tissue.

9. Prolonging the 100% ethanol dehydration steps of the staining protocol to a maximum of 5 min could enhance tissue dehydration, maximizing microdissection efficiency. Skin tissue, cartilage, and samples prepared on charged slides might be difficult to microdissect. Additional slide or tissue treatments such as incorporation of glycerol slide coatings or modified staining protocols with glycerol might be required ***(22)***. The following is one such approach for frozen-tissue sections as adapted from **ref. *22***. Stain the slide in Mayer's hematoxylin for 30 s. Rinse in dH_2O for 15 s. Fix the slide in 70% and 95% ethanol solutions for 10 s each. Rinse in dH_2O for 10 s. Place the slide in Scott's Tap Water (Blueing) solution for 15 s. Dehydrate the slide in 70% ethanol for 2 min. Soak the slide in 3% glycerol in phosphate-buffered saline for 5–10 min. Dehydrate in two solutions of 100% ethanol: first solution for 10 s and second solution for 1 min. Clear the slide in two changes of xylene or xylene substitute for 1 min each. Allow the slide to air-dry before proceeding to microdissection.
10. PixCell instruments equipped with fluorescent modules incorporate mercury vapor lamps with blue, green, and red filter cubes. Additional filter cube positions are available for end-user modifications. Blue filter cubes use 455- to 495-nm excitation wavelengths, with emission greater than 510 nm. Green filter cubes use 503- to 547-nm excitation wavelengths, with emission greater than 565 nm. The red filter cube excitation wavelengths are 590–650 nm, with emission greater than 667 nm. These filter cubes can be used with immuno-LCM protocols ***(18)***.
11. A drawback of the PixCell system is the inability to microdissect directly from an index-matched image of the tissue (*see* **Fig. 2**). Map images can be saved while the image is wet, providing a guide for microdissection ***(19,20)***. This is not an issue with the AutoPix platform. Cells for microdissection are selected via annotation software directly from an index-matched, stitched image (*see* **Fig. 6**). Index-matched images can be obtained with either system by rewetting the tissue with a drop of xylene prior to microdissection. It is imperative that the slide be completely dry prior to cap placement for microdissection because xylene dissolves the polymer.
12. The laser should not be refocused when changing objectives or spot sizes. It is only necessary to focus the laser, with the small 7.5-μm spot size setting and the 10× objective, for each initial cap placement and any time the cap is repositioned on the tissue. Microdissection can be performed with any suitable laser spot size and a 4×, 10× or 20× objective.
13. Satellite laser spots are a phenomenon noted on the AutoPix platform (*see* **Fig. 3**). The satellite laser spot is a second-surface reflection of the laser as the laser beam propagates through the optics. The laser is reflected from the back (second) surface of a coated optic. The coating is required to allow the laser to change direction in the optical path. Imperfections in the coating allow a portion of the laser beam to pass through the coating and reflect back to the viewer from the second

optical surface. The power of the satellite spot is too low to melt the polymer and does not interfere with efficient microdissection. The satellite spot is a byproduct of the light amplification system and was not designed as a system component, but the satellite spot is helpful for accurately focusing the primary laser spot. If the satellite spot is in focus, the primary laser spot will be in focus.

14. The dark ring produced by pulsing the laser is a combination of migration of the dye and changes in the thickness of the polymer wall at the site of the laser pulse (*see* **Fig. 4**). These changes in the polymer permit visualization of the melted polymer. The black ring should be sharp in appearance with a clear center. This pattern indicates: (1) proper laser focusing, (2) adequate laser operation, and (3) acceptable performance of the CapSure polymer. A "fuzzy" ring could indicate improper focusing of the laser, uneven placement of the CapSure cap on the tissue, or inadequate power and/or duration of the laser pulse. The first step in troubleshooting a poorly wetted polymer spot is repositioning the cap on the tissue. Often, the cap is crooked or uneven in relation to the tissue. The second step in resolving poorly wetted polymer spots is refocusing the laser. The third action to correct poor polymer wetting is adjustment of the power and duration. Increase the laser power by approx 10 mW and the duration by 2.0 ms and fire another laser test pulse. Observe the wetted polymer for the appropriate appearance. If the above steps fail to resolve the problem, discard the cap and repeat the process with a fresh cap.
15. Assuming an average epithelial cell diameter of 7 µm and a 30-µm laser spot size, the operator can expect to collect, on average, five to six cells per laser pulse. Using this information, it is possible to estimate the number of cells captured based on the number of laser pulses counted during microdissection. The number of laser pulses is automatically counted on the toolbar on both the PixCell and AutoPix systems.
    a. 30-µm laser spot size: Number of pulses × 5 = total cells captured.
    b. 15-µm laser spot size: Number of pulses × 3 = total cells captured.
    c. 7.5-µm laser spot size: Number of pulses × 1 = total cells captured.
16. The AutoPix requires a warm-up period of 1 h prior to use if the instrument has been turned off. The instrument power can remain on for daily operation eliminating the need for a warm-up period.
17. The AutoPix is outfitted with a xenon lamp for fluorescence visualization, with similar filter cube configurations as the PixCell LCM system. The AutoPix is equipped with a high-sensitivity black-and-white camera in addition to the color camera.

Acknowledgments

The authors thank Don Armstrong, Ph.D., Arcturus Bioscience Inc., for critical review of the manuscript. They also thank Geoff Goodrich, Ashi Malekafzali, and Theresa Taylor for consistent, helpful advice.

References

1. Liotta, L. A. and Kohn, E. C. (2001) The microenvironment of the tumour–host interface. *Nature* **411,** 375–379.
2. Liotta, L. A. and Kohn, E. C. (2002) Stromal therapy: the next step in ovarian cancer treatment. *J. Natl. Cancer Inst.* **94,** 1113–1114.
3. Michener, C. M., Ardekani, A. M., Petricoin, E. F., 3rd, Liotta, L. A., and Kohn, E. C. (2002) Genomics and proteomics: application of novel technology to early detection and prevention of cancer. *Cancer Detect. Prev.* **26,** 249–255.
4. Wulfkuhle, J. D., Sgroi, D. C., Krutzsch, H., et al. (2002) Proteomics of human breast ductal carcinoma in situ. *Cancer Res.* **62,** 6740–6749.
5. Emmert-Buck, M. R., Strausberg, R. L., Krizman, D. B., et al. (2000) Molecular profiling of clinical tissue specimens: feasibility and applications. *Am. J. Pathol.* **156,** 1109–1115.
6. Gillespie, J. W., Ahram, M., Best, C. J., et al. (2001) The role of tissue microdissection in cancer research. *Cancer J.* **7,** 32–39.
7. Paweletz, C. P., Liotta, L. A., and Petricoin, E. F., 3rd. (2001) New technologies for biomarker analysis of prostate cancer progression: laser capture microdissection and tissue proteomics. *Urology* **57,** 160–163.
8. Simone, N. L., Paweletz, C. P., Charboneau, L., Petricoin, E. F., 3rd, and Liotta, L. A. (2000) Laser capture microdissection: beyond functional genomics to proteomics. *Mol. Diagn.* **5,** 301–307.
9. Bonner, R. F., Emmert-Buck, M., Cole, K., et al. (1997) Laser capture microdissection: molecular analysis of tissue. *Science* **278,** 1481–1483.
10. Emmert-Buck, M. R., Bonner, R. F., Smith, P. D., et al. (1996) Laser capture microdissection. *Science* **274,** 998–1001.
11. Chuaqui, R., Vargas, M. P., Castiglioni, T., et al. (1996) Detection of heterozygosity loss in microdissected fine needle aspiration specimens of breast carcinoma. *Acta. Cytol.* **40,** 642–648.
12. Bichsel, V. E., Petricion, E. F., 3rd, and Liotta, L. A. (2000) The state of the art microdissection and its downstream applications. *J. Mol. Med.* **78,** B20.
13. Miura, K., Bowman, E. D., Simon, R. (2002) Laser capture microdissection and microarray expression analysis of lung adenocarcinoma reveals tobacco smoking- and prognosis-related molecular profiles. *Cancer Res.* **62,** 3244–3250.
14. Ornstein, D. K., Englert, C., Gillespie, J. W., et al. (2000) Characterization of intracellular prostate-specific antigen from laser capture microdissected benign and malignant prostatic epithelium. *Clin. Cancer Res.* **6,** 353–356.
15. DiFrancesco, L. M., Murthy, S. K., Luider, J., and Demetrick, D. J. (2000) Laser capture microdissection-guided fluorescence in situ hybridization and flow cytometric cell cycle analysis of purified nuclei from paraffin sections. *Mod. Pathol.* **13,** 705–711.
16. Gillespie, J. W., Best, C. J., Bichsel, V. E., et al. (2002) Evaluation of non-formalin tissue fixation for molecular profiling studies. *Am. J. Pathol.* **160,** 449–457.
17. Simone, N. L., Remaley, A. T., Charboneau, L., et al. (2000) Sensitive immunoassay of tissue cell proteins procured by laser capture microdissection. *Am. J. Pathol.* **156,** 445–452.

18. Fend, F., Emmert-Buck, M. R., Chuaqui, R., et al. (1999) Immuno-LCM: laser capture microdissection of immunostained frozen sections for mRNA analysis. *Am. J. Pathol.* **154,** 61–66.
19. Wong, M. H., Saam, J. R., Stappenbeck, T. S., Rexer, C. H., and Gordon, J. I. (2000) Genetic mosaic analysis based on Cre recombinase and navigated laser capture microdissection. *Proc. Natl. Acad. Sci. USA* **97,** 12,601–12,606.
20. Mouledous, L., Hunt, S., Harcourt, R., Harry, J., Williams, K. L., and Gutstein, H. B. (2003) Navigated laser capture microdissection as an alternative to direct histological staining for proteomic analysis of brain samples. *Proteomics* **3,** 610–615.
21. Arcturus Engineering (2002) *User Guide AutoPix Automated Laser Capture Microdissection System,* Arcturus Engineering, Mountain View, CA, p. 5.
22. Agar, N. S., Halliday, G. M., Barnetson, R. S., and Jones, A. M. (2003) A novel technique for the examination of skin biopsies by laser capture microdissection. *J. Cutan. Pathol.* **30,** 265–270.

11

Analysis of Asbestos-Induced Gene Expression Changes in Bronchiolar Epithelial Cells Using Laser Capture Microdissection and Quantitative Reverse Transcriptase–Polymerase Chain Reaction

Christopher B. Manning, Brooke T. Mossman, and Douglas J. Taatjes

Summary

Laser capture microdissection (LCM) enables the removal of discrete microstructures or cell types from properly prepared histological sections. Extraction of RNA from microdissected tissue followed by quantitative reverse transcriptase–polymerase chain (QRT-PCR) reaction permits the analysis of cell-type or microstructure-specific gene expression changes that occur in response to various stimuli in the environment. In our lab, the combination of LCM and QRT-PCR has proven very useful in the determination of the in vivo gene expression changes that occur in bronchiolar epithelium in response to inhalation of crocidolite asbestos. A detailed description of the preparation of cDNA from bronchiolar epithelial cells obtained by LCM is described in this work.

Key Words: Asbestos; bronchiolar epithelium; environmental pathology; laser capture microdissection; lung disease; microgenomics; quantitative reverse transcriptase–polymerase chain reaction.

1. Introduction

Laser capture microdissection (LCM) is a relatively new and unique technique of increasing popularity that facilitates the identification of cell-type or microstructure-specific changes in gene expression by providing a means to selectively remove individual cells or small groups of cells from properly prepared tissue sections *(1,2)*. The microdissection instrument that has been most widely used, the Arcturus PixCell II, relies on a polymer film that is placed over the tissue of interest. A short-duration pulse from an infrared laser is used to select the desired cells by melting the area of the film with which they are in

From: *Methods in Molecular Biology, vol. 319: Cell Imaging Techniques: Methods and Protocols*
Edited by: D. J. Taatjes and B. T. Mossman © Humana Press Inc., Totowa, NJ

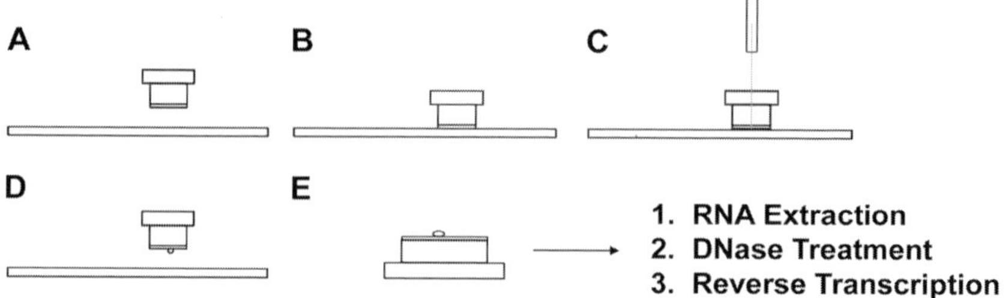

Fig. 1. Laser capture microdissection and RT–PCR. **(A)** Dehydrated tissue section and cap with polymer film on surface facing slide. **(B)** The cap is placed over the tissue to be microdissected. **(C)** An infrared laser pulse is directed at the film in contact with the cells of interest. **(D)** Removal of the cap after the film is melted also removes the microdissected cells. **(E)** The microdissected cells can be used for analysis of gene expression using RT-PCR.

contact. RNA can then be extracted from these cells for quantitation via quantitative reverse transcriptase–polymerase chain reaction (QRT-PCR) or qualitative assessment for the presence or absence of expression of a gene by gel-based RT-PCR.

One of the challenges facing researchers interested in understanding the in vivo signaling mechanisms and changes in gene expression involved in the responses of the lung to environmental insults, such as asbestos, is its complex architecture. The lung is a highly heterogeneous organ with over 40 cell types. An important consequence of this heterogeneity is that the responses of individual cell types to pollutants are diluted in measurements of gene expression in relatively crude preparations such as whole-lung extracts. The combination of LCM and QRT-PCR (*see* **Fig. 1**) is particularly useful for measuring changes in expression of transcription factors and markers of proliferation and injury in lung epithelial cells following exposure to asbestos *(3)*. In this chapter, the methodology for the isolation of DNA-free RNA from bronchiolar epithelial cells (*see* **Fig. 2**) and subsequent preparation of cDNA are described in detail *(4–6)*.

2. Materials

2.1. Kits and Reagents

1. Arcturus PicoPure RNA isolation kit: Components include conditioning buffer (CB), extraction buffer (XB), 70% ethanol (EtOH), wash buffer 1 (W1), wash buffer 2 (W2), elution buffer (EB), and RNA purification columns.
2. Ethanol (ETOH) series dilutions (70% and 95% ETOH): Add 30 or 5 mL of autoclaved distilled-deionized water ($ddH_2 0$) to prepare. 70 or 95 mL of 100% ETOH to 100 mL.
3. Arcturus CapSure LCM caps, CapSure pads, and prep strips (included with caps).

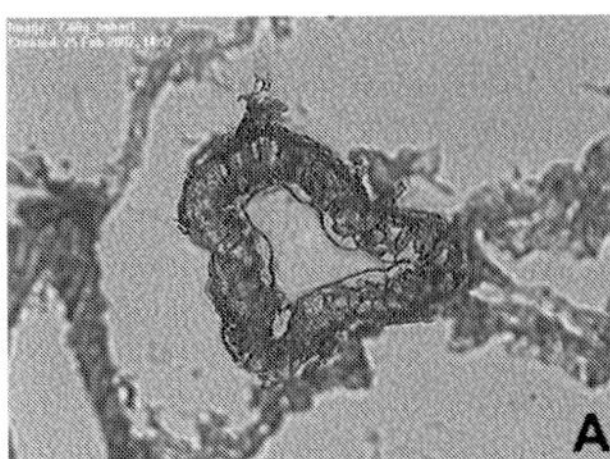

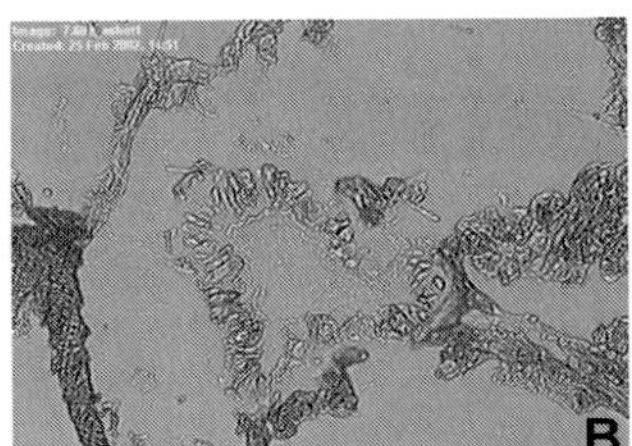

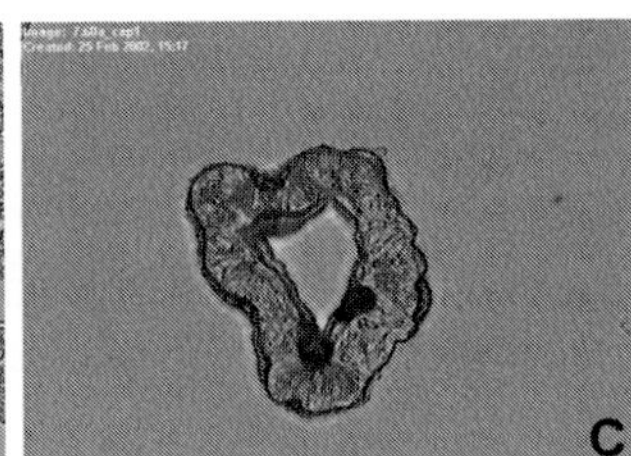

Fig. 2. Removal of a bronchiole from a lung section using laser capture microdissection: before microdissection **(A)**, after microdissection **(B)**, and the isolated bronchiole on the surface of the cap **(C)**.

4. Promega RQ1 RNase-free DNase.
5. Promega Reverse Transcription system.

2.2. Equipment

1. A cryostat suitable for cutting 10-μm thick sections from tissue embedded in optical cutting temperature (OCT) at –20°C is required for generating sections that can be properly microdissected.
2. An Arcturus PixCell II laser capture microdissector is used to remove bronchiolar epithelium from prepared sections.
3. A dessicator, preferably with a vacuum pump, is needed to keep the prepared tissue sections dehydrated until use.

3. Methods

3.1. Specimen Preparation

Following euthanasia by a combination of 5 mg sodium pentobarbital and pneumothorax, the lungs are inflated with calcium/magnesium-free phosphate-buffered saline (PBS) to a pressure of 30 cm H_2O. After occluding the primary bronchi by tying off with surgical suture, one lung is immersed in a cryomold containing OCT embedding compound. The cryomold is then dipped into liquid-nitrogen-cooled isopentane until the appearance of the OCT changes from clear to opaque. Immediately remove the new specimen block and transfer to a –80°C freezer.

3.2. Slide Preparation

3.2.1. Sectioning

Specimen OCT blocks should be equilibrated at –20°C for 20 min prior to sectioning. Sections should be cut at 10 μm and transferred within min to a –80°C freezer for storage or processed immediately for microdissection. The sections must be kept cold to minimize RNA degradation in the tissue of interest (*see* **Notes 1** and **2**).

3.2.2. Dehydration

It is essential that tissue be completely dehydrated for efficient transfer of cells from the tissue section to the polymer film on the CapSure caps. The following procedure should result in tissue ready for microdissection (typically done in groups of 5–10 slides with Coplin jars):

1. Immerse in cold 70% ETOH (chilled to –20°C) for 60 s.
2. Immerse in ddH_20 for 30 s.
3. Immerse in 70% ETOH for 60 s.
4. Immerse in 95% ETOH for 60 s.
5. Immerse in 100% ETOH for 2 min.
6. Immerse in xylene twice for 10 min each.
7. After removing sections from xylene, allow them to dry for 5 min in a fume hood.
8. Keep slides in a vacuum dessicator until use (for no more than 12 h).

3.3. Microdissection

1. Prior to microdissection, weakly adhering tissue should be removed from the slide using the prep strips included with the CapSure caps. This is accomplished by peeling off the backing of the prep strip and pressing it onto the section. Next, apply pressure on the strip with two fingers and move along the strip in one direction, taking care not to apply so much force as to damage the tissue on the slide.
2. For bronchiolar epithelium, a 7.5-µm laser spot size is best to achieve capture. The power and duration of laser pulses should be adjusted as necessary to get good capture of the desired cells. To test if the settings are correct, position the laser over the lumen of a bronchiole and pulse the laser. If the settings are correct, the laser will leave a dark ring with a clear center in the film.
3. After collecting the cells of interest onto a cap, gently roll the film surface of the cap over the exposed surface of a CapSure pad. This will remove tissue that was not selected but adhered to the cap nonspecifically.
4. Immediately snap the cap into the opening of a 0.5-mL RNase-free Eppendorf tube that contains 50 µL of extraction buffer. Invert the tube and, in that position, incubate at 42°C for 30 min.
5. Centrifuge at 800*g* for 2 min to pull the extraction buffer to the bottom of the tube. Pull off the cap and close the lid to the tube. Immediately snap-freeze the extract using dry ice and place in a freezer at –80°C for storage. If desired, the rest of the isolation can be performed immediately.
6. Apply 250 µL of conditioning buffer to the surface of the filter in the purification columns. Allow the column to equilibrate for 5 min at room temperature. After the column is equilibrated, centrifuge at 16,000*g* for 1 min.
7. Thaw the samples and pipet 50 µL of 70% ethanol (from the kit) to each sample. Mix well using the pipet.
8. Add the mixture of ethanol and extract to the purification column and centrifuge for 2 min at 100*g*, followed by 30 s at 16,000*g*.

9. Add 100 µL of W1 to each column and centrifuge at 8000*g* for 1 min.
10. Add 100 µL of W2 to each column and centrifuge at 8000*g* for 1 min.
11. Add another 100 µL of W2 to each column and centrifuge at 16,000*g* for 1 min.
12. Remove all filtrate from the tube and replace the purification filter. Centrifuge at 16,000*g* for 1 min to ensure that the filter is dry.
13. Transfer the purification filter to a new tube (from the kit) and elute by applying 24 µL of elution buffer to the filter, allowing equilibration for 1 min, and centrifuging for 1 min at 16,000*g*.

3.4. DNase Treatment

1. Add 3 µL of Promega RQ1 DNase 10X reaction buffer to each sample.
2. Add 3 µL of RQ1 RNase-Free DNase to each sample.
3. Incubate at 37°C for 30 min.
4. Add 3 µL of RQ1 DNase stop solution to each sample.
5. Incubate at 65°C for 10 min.

3.5. Reverse Transcription

To each sample add the following reagents from the Promega Reverse Transcription kit and follow **steps 6–13**:

1. 12 µL of 25 m*M* $MgCl_2$.
2. 6 µL of 10X reverse transcription buffer.
3. 6 µL of 10 m*M* dNTP mixture.
4. 1.5 µL of Rnasin.
5. 3 µL of random hexamers.
6. Mix by pipetting.
7. Remove 10 µL from each sample to serve as that sample's "no reverse transcription" (NRT) control.
8. Add 45 U AMV reverse transcriptase to each sample and 1 µL of nuclease-free water to each NRT control.
9. Incubate all tubes at room temperature for 10 min.
10. Incubate all tubes at 42°C for 1 h.
11. Incubate all tubes at 95°C for 5 min.
12. Cool tubes on ice for 5 min.
13. Freeze tubes at –20°C. The cDNA is now ready for analysis using either Q-PCR or gel-based PCR.

4. Notes

1. Sectioning should be performed as quickly as possible to preserve the integrity of the RNA within specimen blocks.
2. As with any technique involving RNA, special care must be employed to prevent RNase contamination. Gloves should be changed regularly. Benchtops should be cleaned with an RNase-destroyed agent such as RnaseZap from Ambion. The surfaces of the cryostat should also be cleaned prior to sectioning. RNase-free filter tips for pipets and RNase-free tubes are also essential.

References

1. Emmert-Buck, M. R., Bonner, R. F., Smith, P. D., et al. (1996) Laser capture microdissection. *Science* **274,** 998–1001.
2. Taatjes, D. J., Palmer, C. J., Pantano, C., Hoffmann, S.B., Cummins, A., and Mossman, B.T. (2001) Laser-based microscopic approaches: application to cell signaling in environmental lung disease. *Biotechniques* **31,** 880–894.
3. Manning, C. B., Cummins, A. B., Jung, M. W., et al. (2002) A mutant epidermal growth factor receptor targeted to lung epithelium inhibits asbestos-induced proliferation and proto-oncogene expression. *Cancer Res.* **62,** 4169–4175.
4. *PicoPure RNA Isolation Kit User Guide Version B*, Arcturus Engineering, Mountain View, CA.
5. *RQ1 RNase-Free DNase Technical Bulletin TB518*, Promega Corporation.
6. *Reverse Transcription System Technical Bulletin TB099*, Promega Corporation.

12

New Approaches to Fluorescence *In Situ* Hybridization

Sabita K. Murthy and Douglas J. Demetrick

Summary

Fluorescence *in situ* hybridization (FISH) is a nonisotopic labeling and detection method that provides a direct way to determine the relative location or copy number of specific DNA sequences in nuclei or chromosomes. With recent advancements, this technique has found increased application in a number of research areas, including cytogenetics, prenatal diagnosis, cancer research and diagnosis, nuclear organization, gene loss and/or amplification, and gene mapping. The availability of different types of probe and the increasing number of FISH techniques has made it a widespread and diversely applied technology. Multicolor karyotyping by multicolor FISH and spectral karyotyping interphase FISH and comparative genomic hybridization allow genetic analysis of previously intractable targets. We present a brief overview of FISH technology and describe in detail methods of probe labeling and detection for different types of tissue sample, including microdissected nuclei from formalin-fixed paraffin-embedded tissue sections.

Key Words: Fluorescence *in situ* hybridization (FISH); interphase FISH; laser capture microdissection (LCM); LCM-FISH.

1. Introduction

Fluorescence *in situ* hybridization (FISH) is a powerful molecular–cytogenetic detection technique that utilizes a fluorescent-labeled DNA probe, which is hybridized to a genomic target in the nuclei of fixed cells, to ascertain the presence or absence of a particular DNA sequence. Hybridization to the target loci is visualized by the detection of fluorescent signals on metaphase chromosomes or interphase nuclei (*see* **Fig. 1**; Color Plate 11, following p. 274). Using appropriate probes, aberrations such as chromosomal aneuploidy, deletions, duplications, and translocations can be detected on a cell-to-cell basis in metaphase chromosome preparations or in nondividing interphase nuclei. Since the first demonstration of this fluorescence labeling technique *(1–3)*, FISH has developed very rapidly to be a valuable tool in basic cytogenetic research and diagnosis. Advances in probe generation, hybridization technology, fluorescence

From: *Methods in Molecular Biology, vol. 319: Cell Imaging Techniques: Methods and Protocols*
Edited by: D. J. Taatjes and B. T. Mossman © Humana Press Inc., Totowa, NJ

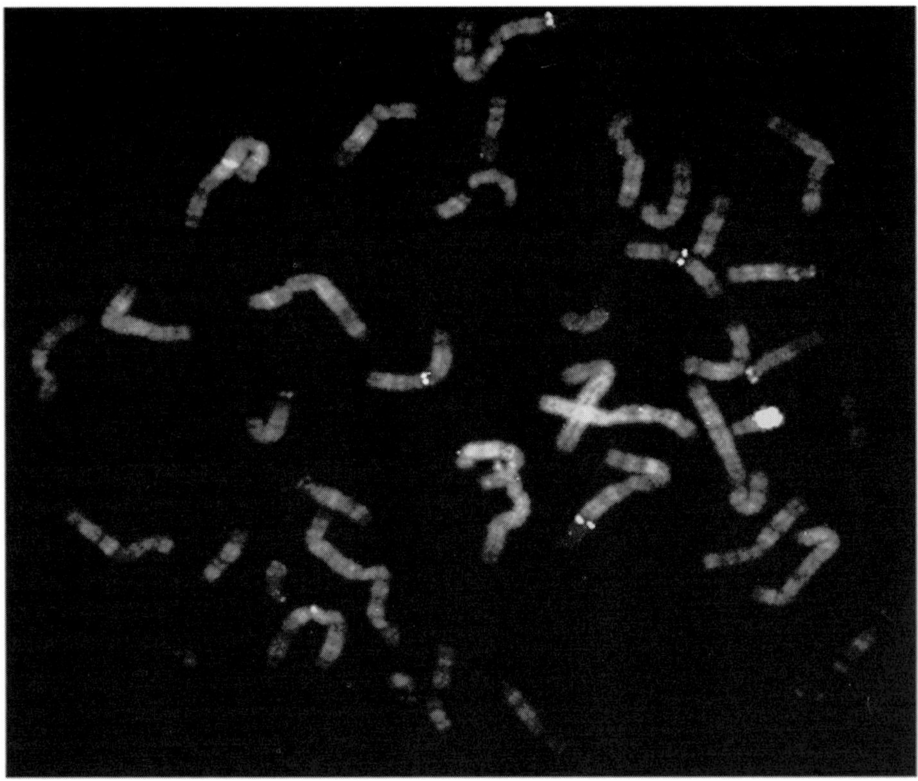

Fig. 1. Metaphase analysis with multiple single-gene probes. Multicolor FISH showing simultaneous localization of three human genomic BAC probes, CDKN2c (green), RB1 (red), and CCND1 (yellow), on a normal human metaphase spread. (*See* Color Plate 11, following p. 274.)

microscopy, and digital and spectral imaging have led to the rapid and improved development of molecular cytogenetic approaches that allow identification of chromosomal aberrations with unprecedented accuracy. Several reviews are available that emphasize the role of FISH in biological research as well as in genetic diagnostic applications ***(4–9)***.

1.1. FISH Probes

1.1.1. Centromeric Probes

Specific or restricted specificity alpha satellite probes are used to identify repeat sequence targets in the centromeric region of each chromosome. Chromosome-specific centromeric probes are now readily available commercially, enabling us to identify individual chromosomes and to analyze numerical chromosomal abnormalities in interphase and metaphase preparations.

1.1.2. Gene-Specific Probes

Locus-specific or unique probes are cosmid (with insert of about 40 kb) and BAC, PAC, or P1 (with inserts of about 125 kb) genomic clones and are used for gene mapping, deletion, and amplification analysis, as well as analysis of chromosome rearrangements such as duplication and translocation. These probes have wide application in genetic diagnosis such as identification of microdeletions, deletion of tumor suppressor genes ***(10)***, amplification of oncogenes ***(11,12)***, and the demonstration of fusion genes involved in the various translocations in cancer ***(13–15)***.

1.1.3. Whole-Chromosome Paints

Whole-chromosome painting probes are polymerase chain reaction (PCR)-generated from flow-sorted genomic DNA libraries that are used to identify individual chromosomes or chromosome segments ***(16,17)***. Chromosome painting probes can also be generated from PCR-amplified DNA of microdissected chromosome segments or markers of unknown origin ***(18,19)***. These chromosome-specific probes are either labeled with a single fluorochrome or with specific combinations of two or more fluorochromes to generate increasing numbers of discernable targets beyond the number of fluorochromes that are available. Using combinatorial labeling with 5 different fluorochromes, up to 31 different targets can be distinguished ***(20,21)***.

1.1.4. Telomere Probes

Cytogenetic analysis of chromosome ends is often challenging, as most of the chromosome ends are G-band negative and, thus, lightly stained. Cryptic rearrangements involving these regions are often difficult to detect by routine cytogenetic banding methods. Submicroscopic deletions or rearrangements of the telomeres have deleterious consequences leading to genomic instability and cancer. FISH probes specific to unique telomere sequences are now available commercially, which enable the identification of individual telomeres and their rearrangements. Also, peptide nucleic acid probes can be artificially generated and used instead of natural nucleic acid probes ***(22)***.

1.2. Applications of FISH

1.2.1. Multicolor FISH of Metaphase Preparations (M-FISH, SKY, COBRA-FISH, Rx-FISH, mBAND-FISH)

Conventional cytogenetic analysis provides essential information of diagnostic and prognostic importance in patients with genetic abnormalities. However, it has major limitations with respect to the detection of subtle or cryptic chromosomal aberrations and the analysis of highly rearranged chromosomes, particularly in poorly spread and/or contracted metaphase preparations. Complex chromosomal rearrangements (CCRs) involving whole chromosomes or parts of chromosomes are common in cancer cells, where it is often impossible to identify the chromosomes or chromosome bands that are involved in the rearrangement. New molecular cytogenetic approaches such as multicolor FISH (M-FISH), spectral karyotyping (SKY), combined binary ratio labeling FISH (COBRA-FISH), and multicolor banding (m-BAND FISH) ***(21,23–27)*** enable improved characterization of these CCRs.

1.2.1.1. Spectral Karyotyping

Spectral karyotyping is a multicolor FISH technique that allows one to simultaneously visualize all human chromosomes represented in different colors by the computer. This facilitates karyotype analysis, particularly for the identification of ambiguous and hidden chromosomal abnormalities. Applications of SKY to screening genomes for chromosomal aberrations in human disease and cancer are numerous. It is useful in mapping chromosomal break points, detecting subtle translocations, identifying marker chromosomes, homogeneously stained regions, and double minutes, and characterizing CCRs ***(21,24)***. Although SKY enables the identification of individual chromosomes and small chromosome segments, it is usually limited to the analysis of chromosome translocations without further differentiation of chromosomal subregion. Subregional human genomic probes "painting" several colored chromosome bands have been established by microdissection ***(19)*** and "reverse painting" ***(18)***. However, these probes cannot identify intrachromosomal rearrangements such as inversions. Subchromosomal paints for human chromosomes derived by reverse painting of DNA from primates such as gibbons, which have very close homology, are found to be useful in delineating intrachromosomal rearrangements ***(28)***.

1.2.1.2. Comparative Genomic Hybridization

Comparative genomic hybridization (CGH), yet another specialized method of FISH application, is widely used in cancer research and diagnosis ***(12,29–31)***. CGH utilizes the hybridization of differentially labeled tumor and reference DNA to normal metaphase chromosomes to generate a map of DNA copy number changes in tumor genomes in a single hybridization experiment. It has been used to complement immunohistochemical, DNA content measurement, and histomorphology to establish a phenotype/genotype correlation in solid tumor progression ***(32–37)***. CGH has now become a routine approach and is widely accepted in cancer research. It is the method of choice for obtaining an overview of genetic alterations in cancer cells, especially in solid tumors, where highly abnormal chromosomes and complex karyotypes are common, but good metaphase preparations are difficult to obtain ***(37,38)***. CGH can also be applied to archival paraffin material for retrospective studies ***(39–41)***, and because of the ability to amplify and subsequently label small amounts of DNA by degenerate oligonucleotide primer PCR (DOP-PCR) ***(42)***, microdissected or laser capture microdissected material can also be used ***(43–46)***.

1.2.1.3. Telomere and Telomere Length Measurement (tel-FISH, Q-FISH, Flow-FISH)

Cryptic chromosome rearrangements that are otherwise sometimes missed by routine cytogenetic analysis can be easily identified by chromosome-specific telomeric probes (tel-FISH), which are available commercially. Telomeres containing noncoding DNA repeats at the end of the chromosomes are essential for chromosomal stability and are implicated in regulating the replication and senescence of cells. The gradual loss of telomere repeats in cells has been linked to aging and tumor development. The most important indicator of correct telomere function is telomere length maintenance within the range typical for each species. Quantitative fluorescence *in situ* hybridization (Q-FISH) provides an estimate of telomere length in each individual chromosome with a resolution of 200 basepairs ***(47)***. A variation of Q-FISH is the flow-FISH technique, in which a fluorescein isothiocyanate (FITC)-labeled telomere-specific peptide nucleic acid probe is hybridized in a quantitative way to telomere repeats, followed by telomere fluorescence measurements on individual cells by flow cytometry ***(48,49)***. It offers the advantage of looking at telomere fluorescence in different subpopulations of cells.

1.2.2. Interphase FISH

One of the theoretical advantages of FISH over conventional cytogenetics is the ability to identify small genetic alterations in both dividing and nondividing cells. Interphase FISH has opened up a wide range of applications for clinical diagnosis. It can be used to define chromosomal numerical abnormalities, chromosome rearrangements involving amplification, deletion or translocation of specific chromosome regions, identification of marker chromosomes ***(50)***, and copy number of individual genes.

Interphase FISH is of special interest in cancer diagnosis. Cancer tissues, particularly solid tumors, frequently fail to grow and divide in tissue culture and often have a very poor mitotic index and poor-quality metaphase chromosomes for cytogenetic analysis. Interphase FISH allows cytogenetic diagnosis from these specimens that cannot be used in conventional dye-banding analyses. Furthermore, translocation break-point analysis ***(51)*** and gene copy number analysis of cancer specimens ***(52)***, using single-gene probes, have proven to be useful predictors of treatment response and this area of molecular diagnosis is rapidly growing.

Unlike conventional blood or other specimens submitted for conventional genetic analysis, cancer specimens are usually contaminated with large numbers of normal cells which could lead to false-negative results in a test evaluating the presence of a particular cancer abnormality.

1.2.2.1. FISH of Paraffin-Embedded Tissues

Fluorescence *in situ* hybridization of paraffin sections has been described ***(53)***, and commercial kits are currently available to perform such analyses. If the FISH target is highly amplified, one may be able to directly identify amplification within the cancer cells of paraffin sections. Background noise from autofluorescence in the tissue section, difficulty assessing an appropriate tissue plane in a high-power view of a thick section, or poor probe penetration, however, could seriously compromise interpretation of low-level amplification, translocation, or deletion ***(54,55)***. However, improved techniques for analysis of such fixed tissues by FISH have been reported ***(55)***.

High-throughput genetic profiling of tumor tissues by FISH on tissue microarray is a very promising improvement to conventional paraffin section FISH ***(56–58)***, where a large number of tissue sections can be labeled and analyzed from one slide (*see* **Fig. 2**; Color Plate 12 following p. 274). This is very economical, time saving and very useful for retrospective studies of prognostic or predictive markers. Wholesale purification of nuclei from paraffin sections could decrease the effects of tissue autofluorescence, probe penetration, or specimen thickness on FISH signal visualization ***(59)***; however, the *in situ* verification of the source of the cells is lost when the tissue is destroyed during nuclear isolation procedures. Laser capture microdissection (LCM), a new technique to purify individual or groups of cells from paraffin-embedded tissue sections ***(60)***, was used to prepare nuclei from breast carcinoma cells for both FISH and flow cytometric analysis ***(61)***. This technique, LCM-FISH, allows identification of normal interphase copy numbers, permitting detection of low-level amplification or potentially single-copy deletions, which would be very difficult in paraffin section FISH (*see* **Fig. 3**; Color Plate 13, following p. 274). LCM-FISH was also very useful in evaluating deletion as a mechanism for loss of heterozygosity (LOH) by allowing comparison between microsatellite analysis and interphase FISH in microdissected breast cancer cells ***(62)***. In our experience, purification of nuclei from either whole or microdissected tissue results in much less background staining, likely because most of the cellular RNA is removed from the specimen.

1.2.2.2. Interphase FISH and Prenatal Diagnosis

Interphase FISH also has considerable application to prenatal diagnosis. Prenatal assessments of common aneuploidy can be determined rapidly by interphase FISH, using uncultured amniocyte cells. Noninvasive prenatal diagnosis is also becoming possible with the analysis of fetal cells in maternal circulation.

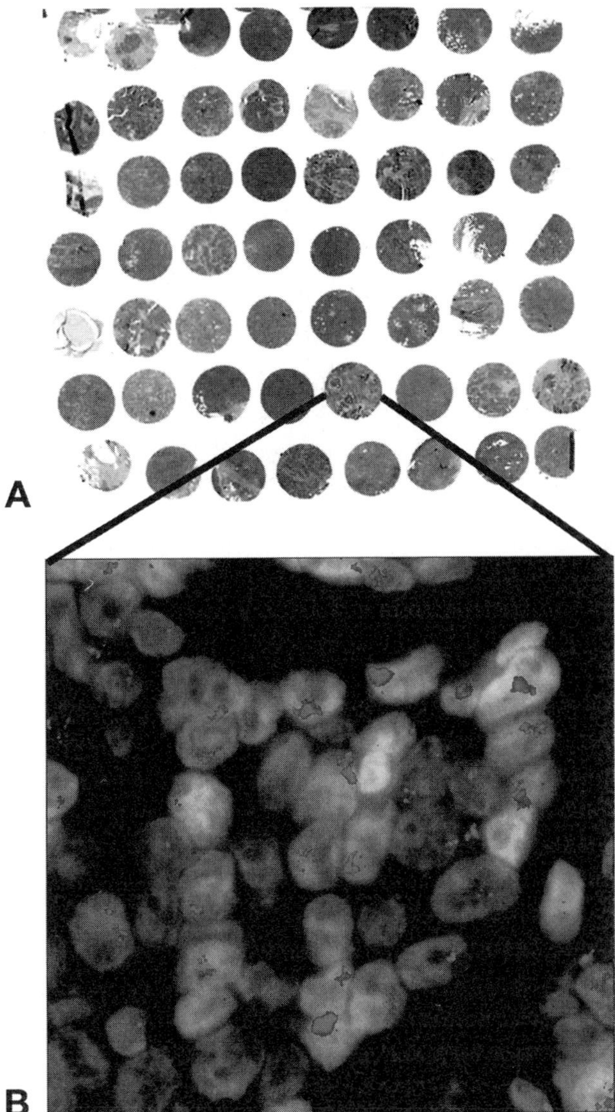

Fig. 2. FISH of tissue microarray breast carcinoma specimens. A tissue microarray block was prepared from 1.5-mm plugs of paraffin-embedded, formalin-fixed human breast cancer specimens. A 6-µm section was cut, deparaffinized, and subjected to FISH with a human erbB-2 genomic BAC probe (red). A hematoxylin/eosin figure **(A)** shows the location of the higher-magnification FISH image **(B)**, which illustrates amplification of the *erbB-2* gene. (*See* Color Plate 12, following p. 274.)

FISH allows rapid detection of specific chromosome abnormalities in uncultured amniotic fluid cells within 2 d of amniocentesis. The technique is typically used in pregnancies at high genetic risk for which a quick result is important in future

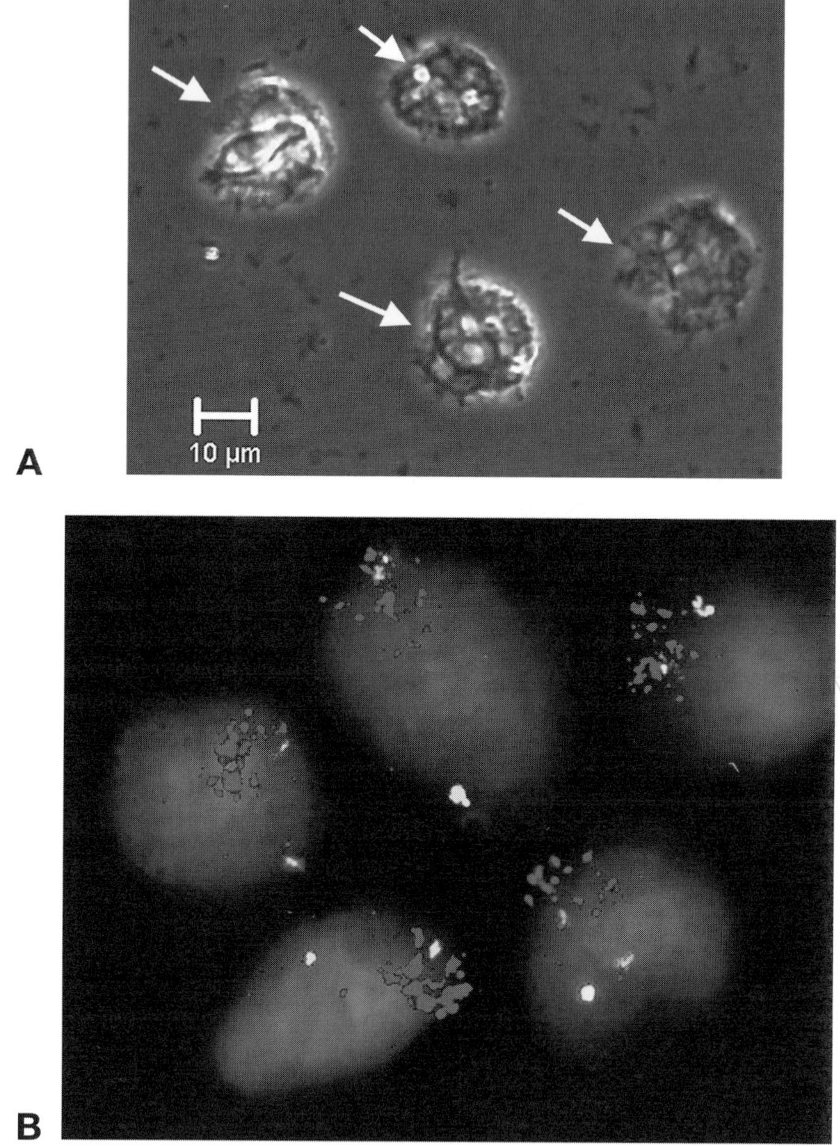

Fig. 3. FISH of LCM-prepared nuclei. Cells were harvested from 30-µm sections of hematoxylin-stained human breast carcinoma by LCM. The nuclei were purified using organic solvents, rehydration, and proteinase K digestion, as described previously ***(61)***, dropped onto clean glass slides, and fixed. The specimens were photographed under phase contrast with the "naked nuclei" indicated by arrows **(A)**. The scale bar indicates 10 µm. The cells were subjected to hybridization with a genomic BAC probe for human erbB-2 (red) and a human centromeric probe for chromosome 17 (green) and show amplification of erbB-2, as in Fig. 2. (*See* Color Plate 13, following p. 274.)

decision making ***(63–66)***. Interphase FISH is also a very useful follow-up tool on investigations of uncultured amniotic fluid (AF) cells after finding an uncertain chromosome aberration in a first chorionvilli (CV sample) or AF sample ***(67–69)***.

1.2.3. Fiber FISH

Direct visualization of fluorescent probes on extended DNA has been used in high-resolution gene mapping and identification of chromosome break points ***(70)***. Recent progress in enhancing the resolution and sensitivity of the FISH technique has further facilitated the physical map construction ***(71–73)*** of the human genome. For example, fiber-FISH has been used to prepare contigs of the human Y chromosome covering areas that were resistant to easy physical mapping ***(74)***. This technique is now only rarely used.

1.3. Microarray-Based CGH

Utilizing the recent advancement of expression microarray technology, microarray-based CGH provides a means to quantitatively measure DNA copy number aberrations and to map them directly onto genomic sequences where genomic DNA is hybridized onto arrayed genomic clones ***(75,76)***. Ligation-mediated PCR is an improved method of generating homogenous and highly reproducible amplified DNA from a single cell, with fragment size of 0.2–2 kb ***(77,78)***. It has been demonstrated by these authors that arrays generated from ligation-mediated PCR of BAC genomic DNAs provided precise measurements in cell lines and clinical material so that high-level amplification and single-copy alterations could be reliably detected in diploid, polyploid, and heterogeneous backgrounds.

An alternate technique, namely hybridization onto cDNA microarrays, offers significant advantages, most importantly the ability to identify amplified genes rather than amplified genomic regions and to perform expression studies on the same arrays using standard expression microarray approaches ***(79–81)***. Thousands of mapped cDNAs are readily available, which facilitate amplicon mapping and identification of new cancer genes (http://www.bcgsc.bc.ca/; http://www.nhgri.nih.gov/; http://www.ncbi.nlm.nih.gov/; http://www.cephb.fr/bio/ ceph-genethon-map.html). A powerful approach to defining putative amplification target genes might combine CGH with results from cDNA microarray analysis, followed by a quick survey of large numbers of uncultured tumors with tissue microarray technology to study the clinical significance of such newly discovered gene amplifications. Array-based CGH, in which fluorescence ratios at arrayed DNA elements provide a locus-by-locus measure of DNA copy number variation, represents another means of achieving increased mapping resolution. Whole genomes of organisms can be prepared using bacterial artificial chromosomes and arrayed onto glass for CGH with 1 Mb resolution, including human ***(82)*** and mouse ***(83)***. It is hypothesized that matrix-based

DNA or RNA hybridization techniques could potentially replace the need for chromosome preparation ***(58,75,79–81)***. Although CGH has the advantage of requiring only genomic DNA and has proven to be an important initial screening test for chromosomal gains and losses in tumor progression, array CGH cannot at present detect structural alterations such as isochromosome formation, double minutes, homogenously stained regions, clonal heterogeneity, and so forth.

2. Materials

A number of labeled FISH probes are available commercially (e.g., Vysis, Cytocell, etc.) that are more often used for clinical diagnostic purposes. They are used directly for hybridization as per the manufacturer's instructions. Here, detail labeling and detection protocols for FISH, much of which was previously described by Trask ***(84)*** and have been subsequently modified ***(85)***.

2.1. Reagents

1. Nick translation labeling kit (Gibco, cat. no.18160-010).
2. For indirect labeling: Digoxygenin-11-dUTP, Fluorescein-12-dUTP or Biotin-16-dUTP (Roche, cat. nos. 1-093-088, 1-427-857, and 1-093-070, respectively).
3. For direct labeling: Cy3-dCTP, Cy5-dCTP, or FluorX-dCTP (Amersham, cat. nos. PA53021, PA55021, and PA58021, respectively).
4. DNAse I (Gibco, cat. nos.18047-019).
5. Microspin G50 column (Pharmacia, cat. no. 27-5330).
6. Human DNA: human Cot-1, 250 μL (1 mg/mL; Gibco, cat. no. 15279-011) + 50 μL.
7. Human placental DNA (10 mg/mL; Sigma, cat. no. D4642); Mix and ethanol precipitate with 30 μL 3*M* sodium acetate, pH 5.2. Wash with 70% ethanol and resuspend the pellet in 50 μL TE (*see* **Subheading 2.2., item 9**).
8. Herring sperm DNA (10 mg/mL) (Gibco, cat. no. 15634-017).
9. Sheep anti-digoxygenin Fab fragments: 1/100 dilution (Roche, cat. no. 1214 667).
10. Cy3-conjugated donkey anti-sheep IgG (Jackson ImmunoResearch, cat. no. 713-165-147).
11. Antifade (Molecular Probes, cat. no. S7461).
12. Proteinase K buffer: 50 m*M* Tris-HCl, 10 m*M* NaCl, 10 m*M* EDTA, pH 8.0.
13. Proteinase K (Invitrogen, cat. no. 25530-031).
14. 4′-6-Diamidino-2-phenyindole dilactate (DAPI) stain 1 mg/mL (Sigma, cat. no. D9564).
15. Tissue culture media: modified Eagle's medium (MEM) and RPMI-1640 (Gibco, cat. nos. 11090-081 and 18875-093, respectively).
16. Fetal bovine serum (FBS) (Gibco, cat. no.10437-028).
17. Antibiotic/antimycotic 100X: 10,000 U penicillin/10,000 μg streptomycin/25 μg/mL fungizone (Gibco, cat. no. 15240-062).
18. HEPES buffer, 1 *M* (Gibco, cat. no. 15630-080).
19. L-Glutamine: 200 m*M* (100X) (Gibco, cat. no. 25030-081).

20. Phytohemagglutinin (PHA-M form): Reconstituted in 10 mL sterile water (Gibco, cat. no. 10576-015).
21. KaryoMax™ Colcemid solution: 10 µg/mL (Gibco, cat. no. 15210-016).
22. Actinomycin D: 250 µg/mL in PBS (Sigma, cat. no. A1410).
23. RNAse A (Sigma, cat. no. P8038): Prepare 20 mg/mL in TE and heat-inactivate in a boiling water bath for 20 min to remove DNAse activity.
24. Drierite (BDH, cat. no. B26998).

2.2. Solutions

1. Hybridization mix: 2 mL of 50% dextran sulfate, 1 mL of 20X SSC (pH 7.0), and 5 mL good quality formamide; pH with 1 *M* HCl to 7.00.
2. Denaturing solution: 70 mL formamide, 10 mL 20X SSC (*see* **item 8**), and 20 mL high-quality distilled water (pH to 7.0).
3. Formamide wash solution: 50 mL formamide, 10 mL 20X SSC, and 40 mL high-quality distilled water (pH solution to 7.0).
4. PBS (phosphate-buffered saline) 10X: 80 g NaCl, 2 g KCl, 11.5 g $Na_2HPO_4 \cdot 7H_2O$, 2 g KH_2PO_4. Make up to 1 L with distilled water and autoclave.
5. Stop buffer: 0.5 *M* EDTA (pH 8.0).
6. McIlvaine's buffer pH 7.0: 0.63 g citric acid (anhydrous) and 6.19 g sodium dibasic phosphate (anhydrous) in 500 mL water.
7. Counterstain: 2 µL DAPI stain (1 mg/mL stock) (Sigma, cat. no. D9564) in 50 mL McIlvaine's buffer.
8. 20X SSC: 175 g NaCl and 88 g Na-citrate (anhydrous) in 1 L water.
9. Tris-EDTA (TE): 10 m*M* Tris-HCl, 1 m*M* EDTA.
10. Antibody blocking solution: 0.1 *M* sodium phosphate buffer, pH 8.0, 0.1% NP-40 detergent, 0.02% sodium azide, 5% nonfat dried milk. Mix and let it stand several days at 4°C. Centrifuge for 15 min at 200*g* to remove insoluble milk proteins. Store at 4°C.
11. Antibody wash solution: 2X SSC, pH 7.00, and 0.005% 3[(cholamidipropyl) dimethyl ammonio]-propane sulfonate (CHAPS) detergent (Calbiochem, cat. no. 220 201).
12. Trypsin-EDTA: 0.25% trypsin and 1 m*M* EDTA (Gibco, cat. no. 25200-056).
13. Cytogenetics fixative: 1 : 3 glacial acetic acid and anhydrous methanol.
14. Hypotonic solution: 0.075 *M* KCl solution, pH 7.0.
15. Carlsberg solution: 0.1% Sigma protease XXIV, 0.1 *M* Tris-HCl, 0.07 *M* NaCl, pH 7.2.

3. Methods

In this section, we discuss the actual methodology for specimen preparation and the performance of FISH for the more common applications that can be performed in the average laboratory or institution. We will discuss metaphase preparations from synchronized fibroblasts and lymphocytes and interphase preparations from tissue specimens. We will then discuss basic probe labeling, hybridization of the labeled probe to denatured genomic DNA in the specimen, detection of the labeled probe, and visualization of the specimen showing *in situ* hybridization. Finally, there will be a discussion of specific applications for performing FISH on nuclei isolated from formalin-fixed, paraffin-embedded whole-tissue sections, performing FISH on formalin-fixed, paraffin-embedded tissue sections or tissue microarrays, and, finally, FISH on nuclei from LCM specimens.

3.1 Specimen Preparation

3.1.1. Metaphase Preparation From Fibroblast Culture

1. Two to three days prior to performing the chromosome spreads, split the fibroblast cells 1 : 3 into new flasks with 10 mL MEM culture media and HEPES buffer, 20% FBS, and 1% penicillin/streptomycin and allow the cells to grow to 50–60% confluency. Feed the cells with fresh media 12–14 h prior to harvesting.
2. Add Colcemid to a final concentration of 0.02 μg/mL for 3–4 h.
3. Transfer the media and add 0.5 mL of trypsin-EDTA to detach the cells (1–2 min). Collect the cells in a tube and centrifuge at 400*g* for 10 min. To the gently vortex-resuspended pellet, add 10 mL of warm (37°C) freshly made KCl solution (0.075 *M*) and incubate at 37°C for 30 min.
4. Add 4–5 drops of cold fresh Cytogenetics fixative to the cells in KCl (*see* **Note 1**) and centrifuge at 400*g* for 10 min. To the gently vortex-resuspended pellet, add 10 mL of cold fixative dropwise and with gentle vortexing (4°C) and centrifuge again for 10 min. Repeat this twice. The pellet must be fully resuspended prior to adding the fixative in order to minimize clumps and maximize yield.
5. Resuspend the final pellet in a smaller volume of fixative (0.5–1 mL depending on the size of the pellet). Drop 1–2 drops of this final suspension onto a clean slide and immediately place it on a warm plate (at 37°C) to dry. Warm the slides for 4 h and let them age at room temperature for 3–4 d (*see* **Note 2**). Better spreading might require dipping the slides into cold water and dropping the suspension onto the wet slides and/or from greater height (0.3–1.0 m). Where available, a thermotron (Cytogenetic drying chamber) can be used, which provides an optimum and controlled temperature and humidity environment for achieving ideal spreading results.
6. Seal the slide box in a Ziploc™ plastic bag with silica gel or Drierite and store them at –20°C until further use (*see* **Note 3**).

3.1.2. Metaphase Preparation From Lymphocyte Culture

1. For metaphase preparations from lymphocytes, add 0.5 mL of whole blood in 10 mL of RPMI 1640 tissue culture media with 20% fetal calf serum, 1% penicillin/streptomycin, 1% L-glutamine, 20 U of heparin, 20 U of HEPES buffer, and 4% PHA. Incubate at 37°C for 72 h.
2. Add Colcemid to 1 μg/mL final concentration at the 68th h of culture. Incubate at 37°C for 4–5 h.
3. Centrifuge at 400*g* for 10 min. To the gently vortex-resuspended pellet, add 10 mL of prewarmed 0.075 *M* KCl and incubate at 37°C for 30 min.
4. Continue as in **step 4** of **Subheading 3.1.1.**

3.1.3. Interphase Nuclei Preparation From Fresh or Frozen Tissue Samples

1. With a sterile scalpel, cut a small piece (0.1–0.2 cm) of fresh or frozen tissue and gently touch it onto four to five spots onto a clean slide (one might want to score a circle on the underside of the slide with a diamond pencil to mark the spot for hybridization and visualization purposes).
2. Semidry (*see* **Note 4**) and fix the cells in fresh Cytogenetics fixative for 10 min. Air-dry and then heat on a warm plate at 37°C for 4 h. Age the slides for 3 to 4 d at room temperature and store them until used further for FISH.

3.2. Probe Labeling, Hybridization, and Visualization

Probes can be labeled either by direct or indirect labeling methods. In direct labeling, the detectable molecule (Cy3-dCTP or Cy5-dCTP) is bound directly to the nucleic acid probe so that after hybridization and posthybridization washing, it can be visualized immediately. This type of labeling gives less background, but the signal is often weak. In indirect labeling, the probe is first labeled enzymatically with a hapten (digoxygenin or biotin, **Subheading 3.2.1., step 1**), which in the second step is accessible to the respective antibody (antidigoxygenin or avidin, **Subheading 3.2.2.3., step 3**) to which the fluorescent molecule is tagged. This type of labeling gives a brighter signal because of the attachment of a large number of antibody molecules to the hapten, which leads to an amplification of the fluorescent signal.

3.2.1. Probe Labeling

Perform the nick translation labeling method using a Gibco Nick Translation Kit (*see* **Note 5**). The following protocol is as previously published with some modifications *(85)*:

1. In a microfuge tube, add the following:

   500 ng probe DNA and dH_2O to a total volume of 18.5 µL;
   2.5 µL of dTTP minus dNTP mix for indirect labeling or dCTP minus dNTP mix for direct labeling;
   1 µL Digoxygenin-11-dUTP for indirect labeling or 2.5 µL Cy3-dCTP for direct labeling.
2. In a separate tube, add 1 µL DNAse I to 20 mL of dH_2O (*see* **Note 6**). Add 1 µL of this to the above mixture. Mix well and add 2.5 µL Pol I/DNAse I mix (0.5 U/µL DNA Polymerase I, 0.4 mU/µL DNAse I). Incubate at 16°C for 90 min.
3. Add 7 µL of FISH stop buffer (0.5 *M* EDTA, pH 8.0) and 25 µL TE. Desalt through a Microspin™ G50 column as per the manufacturer's instructions.
4. Dry the sample on a Speedvac™ and resuspend the sample in 20 µL TE.

   Labeled probes can be stored at –20°C in the dark for up to 1 yr.

3.2.2. Hybridization

3.2.2.1. Hybridization Mix and Probe Denaturation

3 µL of labeled probe (from the 20-µL labeled probe of **Subheading 3.2.1.**);
1 µL of human DNA (Human Cot + Placental DNA, final concentration per 20-µL reaction: 5 µg and 10 µg, respectively);
1 µL Herring sperm DNA (final concentration per 20-µL reaction: 10 µg);
15 µL Hybridization mix.

Mix well and denature at 80°C for 8 min (*see* **Note 7**). Incubate at 37°C for 15–30 min prior to application to the slide, to suppress repetitive DNA-associated background.

3.2.2.2. Denaturation of Target DNA

Incubate aged slides (at least 3–4 d old) at 80°C for 4 min in 70% formamide solution (pH 7.0). Immediately quench in ice-cold 70% ethanol and dehydrate through 80%, 90%, and 100% ethanol and dry the slides on a warm plate (37°C) for 5 min.

Add the hybridization mix with the probe from **Subheading 3.2.2.1.** onto the prewarmed, denatured slide and cover-slip it. Seal the cover slip with rubber cement. Incubate the slide in a humidified chamber at 37°C overnight (chamber can be made by wetting paper towels with water and putting the slides on a pipet-tip rack onto the towels within a sealed Tupperware™ or like container).

3.2.2.3. Visualization

3.2.2.3.1. Primary Antibody Reaction for Indirectly Labeled Slides

1. After overnight hybridization at 37°C, rinse in PBS to remove excess probe. Wash the slides with 50% formamide wash solution, three times for 5 min each, at 42°C and three times for 5 min each with 2X SSC at 42°C.
2. Add 100 µL antibody blocking solution, cover slip the slide, and incubate at 37°C for 30 min in a humidified chamber.
3. Make primary antibody solution by adding 1 µL sheep antidigoxygenin (1:100 dilution) to 99 µL antibody blocking solution. Centrifuge at 10,000*g* for 5 min. Add the supernatant to the slide after blocking for 30 min, cover slip, and incubate at room temperature for 1 h.
4. Rinse slide with PBS and wash three times with antibody wash solution for 5 min each. Again, block with blocking solution for 5 min. In a separate tube, add 1 µL Cy3-conjugated anti-sheep IgG (1:50 dilution) to 99 µL blocking solution and add this to the slide after centrifugation, as earlier, cover slip, and incubate for 45 min at room temperature.
5. Rinse in PBS and wash in antibody wash solution three times for 5 min each at room temperature.
6. Rinse in PBS and counterstain.

3.2.2.3.2. Posthybridization Washing for Direct Labeling

For directly labeled slides, after overnight hybridization, proceed as follows:

1. Rinse the slide in PBS and wash in 50% formamide solution at 42°C, three washes of 5 min each.
2. Rinse three times for 5 min each in 2X SSC.
3. Rinse in PBS and counterstain.

3.2.2.3.3. Counterstaining

After the washing steps for both of the labeling procedures, stain the metaphase chromosomes and/or interphase nuclei as follows. In 50 mL McIlvaine buffer, add 2 µL of DAPI stain (1 mg/mL stock). Stain the slides for 4 min for metaphase spreads and 2 min for interphase nuclei. The DAPI might need to be titrated to a much lower concentration (0.1–2 µL DAPI stock per 50 mL of buffer) to decrease very bright interphase nuclei staining.

After staining with DAPI, treat the metaphase preparations with Actinomycin D for 5 min for DA-DAPI banding (not necessary for interphase specimens). Rinse in PBS, air-dry, and mount in glycerol-based antifade. Slides can be stored at 4°C in the dark until visualization, if less than 12 h (*see* **Note 8**), or at –20°C for longer periods (up to several months). Examples of multicolor FISH on metaphase chromosomes are shown in **Fig. 1.**

3.2.3. FISH Analysis of Formalin-Fixed, Paraffin-Embedded Tumor Samples

Fluorescence *in situ* hybridization on a formalin-fixed paraffin-embedded tissue sample is difficult because of poor probe penetration as well as autofluorescence from the fixed tissue. The following protocol *(55)*, which results in the isolation of whole nuclei, has demonstrated good FISH results in such tissues. For treatment of paraffin sections in which the morphology is to be retained, *see* **Note 9**.

1. Deparaffinize the tissue in xylene for 30 min (at least two changes of fresh xylene).
2. Rehydrate through a 100%, 95%, 70%, and 50% ethanol series and finally wash in dH_2O.
3. Digest the tissue in Carlsberg solution for 1 h at room temperature. Rinse in water.
4. Treat with 100 µg/mL heat inactivated Rnase A (Sigma) in 10 m*M* Tris-HCl, pH 7.5, 1 m*M* EDTA, 0.3% NP-40 for 15 min at room temperature.
5. Remove nuclei (carefully draw supernatant through wide-bore pipet or filter through nylon mesh to remove tissue fragments).
6. Drop one to two drops of the above suspension onto a clean glass slide, air-dry, and fix in Cytogenetics fixative (can be stored at this time). Bake 4 h at 37°C and store as described previously.
7. Allow slides to come to room temperature and remove from storage box.
8. Incubate in 50% glycerol/0.1X SSC at 90°C for 3 min.
9. Cool to room temperature in 2X SSC.

10. Denature for 5 min at 80°C in 70% formamide in 2X SSC and quench in ice-cold 70% ethanol.
11. Dehydrate in ethanol series.
12. Incubate in Proteinase K (8 µg/mL) solution for 7.5 min at 37°C.
13. Dehydrate in ethanol series and air-dry.
14. Hybridize with probe and proceed as in routine FISH protocol for washing and visualizing.

3.2.4. FISH of LCM-Nuclei

For FISH of nuclei for which the cell of origin must be known, isolation of nuclei from whole tissue ***(55)*** is not acceptable. Cells can be microdissected using an Arcturus PixCell instrument; then the capture polymer is dissolved to release the tissue fragment, allowing enzymatic release of nuclei and subsequent FISH. The procedure is basically as described previously ***(86)***.

1. Cut 20- to 30-µm-thick paraffin sections from the blocks of interest. It is best to start with thicker sections to minimize sectioning of the large nuclei common to neoplasms (*see* **Note 9**).
2. Extensively deparaffinize (three changes of fresh xylene over 10–30 min) and stain with hematoxylin (*not* eosin, which leads to a broad-band fluorescent background).
3. Laser capture microdissect the dehydrated slide using the PixCell microdissection instrument (Arcturus Engineering Inc., Mountainview, CA) according to the manufacturer's protocols (www.arctur.com). You will need to increase the pulse energy and pulse time over the usual to allow capture of cells from the thick section (*see* **Note 10**).
4. Extract nuclei from the LCM caps as follows ***(86)***, using care to avoid flames or sparks and an appropriate fume safety cabinet (*see* **Note 11**). Add 100 µL of fresh chloroform into a 0.5-mL microfuge tube and cap with a CapShur LCM cap containing the microdissected specimen. Invert the tube for 10 s and centrifuge at 3000*g* for 30 s to release the tissue from the "capture" polymer of the LCM cap (*see* **Note 12**). Remove the LCM cap and add 200 µL of anhydrous ethyl ether and mix by inversion (necessary to lower density of the solvent, allowing tissue to pellet).
5. Centrifuge and remove the supernatant. Wash the pellet three times with 400 µL of fresh xylene to remove dissolved LCM cap polymers. Wash the pellet with 400 µL of 100%, 95%, and 70% ethanol, and two washes with TE, pH 8.0. Finally, wash the pellet with Proteinase K buffer (50 m*M* Tris-HCl, 10 m*M* NaCl, 10 m*M* EDTA, pH 8.0) and resuspend in the same solution to a final volume of 100 µL. Add Proteinase K (50 µL at 0.015% to a final concentration of 0.005%). Digest for 60–120 min at 37°C with gentle finger-vortexing every 10–20 min.
6. For FISH analysis, dilute the cells with 350 µL of TE and gently remove the liquid into another 0.5-mL microfuge tube with a large-bore pipet tip (0.5–1 mm) to leave the undigested tissue fragments in the tube. Centrifuge for 5 min at 10,000*g*, followed by careful removal of the supernatant. Resuspend the pellet in 50 µL of TE and drop 2–3 drops of this suspension onto a clean slide. Air-dry the slides and

then fix them in fresh Cytogenetics fix for 5 min and air-dry again. Warm at 37°C for 4 h and either store as usual or proceed to hybridization. (*See* **Notes 13** and **14**.)

4. Notes

1. Slow fixation of cells helps to retain a better morphology of the metaphase chromosomes.
2. Aging of slides is important, as it leads to slow and complete fixation of the cells. This helps in retaining good chromosome/nuclei morphology during the succeeding denaturation and staining procedures.
3. Slides stored at –20°C under desiccation can be used for more than 1 yr without any distortion of the metaphases plates or interphase nuclei, or loss of signal. Remember to allow the slides to come to room temperature before unsealing the storage bag, to minimize condensation.
4. Do not dry the tissue completely after making the imprints.
5. Although dogma suggests that nick translation labeling must be used to ensure optimal penetration of the specimen by the labeled probe, our personal experience has not identified significant differences in FISH signal intensity between random primer or nick translation hapten-labeled genomic probes.
6. Dilution of DNAse I should be optimized by digesting and aliquot of probe with various dilutions of DNAse I (1/1000 to 1/20,000) and electrophoresing a small aliquot of the sample in a 1% agarose gel. Optimal DNA digestion will yield a smear between 200 and 600 bp.
7. If aminoallyl-labeled nucleotides are used for labeling, it is important to denature at 65°C instead of 80°C. A higher denaturation temperature might cause loss of fluorescence.
8. Photobleaching seems to be lessened and the fluor signal seems better if the slides are left overnight prior to visualization.
9. If purification of nuclei is not desirable, bypassing **steps 3–7** will result in treatment of the tissue to improve probe penetration and decrease autofluorescence, without removing nuclei (*see* **Fig. 2**).
10. Some experimentation is required to optimize specimen thickness (for better visualization) vs nuclear yield. **Do not** use adhesive or other coated slides to pick up the paraffin sections from the microtome, or you will not be able to remove the microdissected cells from the glass slide.
11. Some experimentation is required to maximize capture, which will be dependent on section thickness and tissue composition. For the PixCell II, laser energy might require 70–80 mW or even higher for 30-μm-thick sections and pulse duration of 7–8 ms.
12. Be very careful using these solvents! **Diethyl ether is highly volatile, flammable, and potentially explosive if not fresh or stored properly**.
13. It is not uncommon for bits of the capture polymer to remain undissolved. Leave them in the tube for the following steps, but when resuspending the cells for harvest, add more TE up to 700 μL, or perform two pipet extractions (e.g., 350 μL each) to maximize the yield of nuclei, and then centrifuge the nuclei slightly longer (6 min instead of 5 min) to pellet them.

14. For LCM, familiarity with the specimen morphology is essential. For the inexperienced user, it might be helpful to have a picture of the tissue section marked up by a pathologist to guide the microdissection. The morphology of dried, 30-μm-thick tissue sections is often terrible. As well, the tissue thickness precludes dissections of single cells or tiny (four to eight) cell groups unless they are well separated from other cells, as the laser energy required to melt the polymer to penetrate a thick section usually expands the beam width beyond 30–40 μm. Make sure that collagen fibrils are not inadvertently captured, as they will cause lifting of adjacent tissue and contaminate the specimen.

    For poor yields, check that the section thickness is thick enough to avoid nuclear slicing, that the capture polymer is dissolving, that the released tissue sections are not floating and thus being removed with solvent supernatants, and that the centrifugation step is fast enough and long enough to ensure pelleting of the nuclei from your specimens. One can drop a sample of the nuclei on a slide and examine it after air-drying under phase-contrast optics ***(86)*** or stain it with methylene blue or hematoxylin to evaluate the quality of the preparation prior to using it for FISH.

References

1. Pinkel, D., Straume, T., and Gray, J. W. (1986) Cytogenetic analysis using quantitative, high-sensitivity, fluorescence hybridization. *Proc. Natl. Acad. Sci. USA* **38,** 2934–2938.
2. Pinkel, D., Gray, J. W., Trask, B., van den Engh, G., Fuscoe, J., and van Dekken, H. (1986) Cytogenetic analysis by in situ hybridization with fluorescently labeled nucleic acid probes. *Cold Spring Harbor Symp. Quant. Biol.* **51,** 151–157.
3. Gray, J. W., Lucas, J., Peters, D., et al. (1986) Flow karyotyping and sorting of human chromosomes. *Cold Spring Harbor Symp. Quant. Biol.* **51,** 141–149.
4. Carter, N. P. (1996) Fluorescence in situ hybridization-state of the art. *Bioimaging* **4,** 41–51.
5. Gray, J. W., Kallioniemi, A., Kallioniemi, O., Pallavicini, M., Waldman, F., and Pinkel, D. (1992) Molecular cytogenetics: diagnosis and prognostic assessment. *Curr. Opin. Biotechnol.* **3,** 623–631.
6. King, W., Proffitt, J., Morrison, L., Piper, J., Lane, D., and Seelig, S. (2000) The role of fluorescence in situ hybridization technologies in molecular diagnostics and disease management. *Mol. Diagn.* **5,** 309–319.
7. Lengauer, C., Kinzler, K. W., and Vogelstein, B. (1997) Genetic instability in colorectal cancers. *Nature* **386,** 623–627.
8. Speicher, M. R. and Ward, D. C. (1996) The coloring of cytogenetics. *Nature Med.* **2,** 1046–1048.
9. Kearney, L. (2001) Molecular cytogenetics. *Best Pract. Res. Clin. Haematol.* **14,** 645–669.
10. Thiagalingam, S., Laken, S., Willson, J. K., et al. (2001) Mechanisms underlying losses of heterozygosity in human colorectal cancers. *Proc. Natl. Acad. Sci. USA* **98,** 2698–2702.

11. Ross, J. S., Sheehan, C., Hayner-Buchan, A. M., et al. (1997) HER-2/neu gene amplification status in prostate cancer by fluorescence in situ hybridization. *Hum. Pathol.* **28,** 827–833.
12. Kallioniemi, O. P., Kallioniemi, A., Kurisu, W., et al. (1992) ERBB2 amplification in breast cancer analyzed by fluorescence in situ hybridization. *Proc. Natl. Acad. Sci. USA* **89,** 5321–5325.
13. Yehuda-Gafni, O., Cividalli, G., Abrahmov, A., et al. (2002) Fluorescence in situ hybridization analysis of the cryptic t(12;21) (p13;q22) in childhood B-lineage acute lymphoblastic leukemia. *Cancer Genet. Cytogenet.* **132,** 61–64.
14. Pelz, A. F., Kroning, H., Franke, A., Wieacker, P., and Stumm, M. (2002) High reliability and sensitivity of the BCR/ABL1 D-FISH test for the detection of BCR/ABL rearrangements. *Ann. Hematol.* **81,** 147–153.
15. Batanian, J. R., Bridge, J. A., Wickert, R., Vogler, C., Gadre, B., and Huang, Y. (2002) EWS/FLI-1 fusion signal inserted into chromosome 11 in one patient with morphologic features of Ewing sarcoma, but lacking t(11;22). *Cancer Genet. Cytogenet.* **133,** 72–75.
16. Pinkel, D., Landegent, J., Collins, C., et al. (1988) Fluorescence in situ hybridization with human chromosome-specific libraries: detection of trisomy 21 and translocations of chromosome 4. *Proc. Natl. Acad. Sci. USA* **85,** 9138–9142.
17. Collins, C., Kuo, W. L., Segraves, R., Fuscoe, J., Pinkel, D., and Gray, J. W. (1991) Construction and characterization of plasmid libraries enriched in sequences from single human chromosomes. *Genomics* **11,** 997–1006.
18. Carter, N. P., Ferguson-Smith, M. A., Perryman, M. T., et al. (1992) Reverse chromosome painting: a method for the rapid analysis of aberrant chromosomes in clinical cytogenetics. *J. Med. Genet.* **29,** 299–307.
19. Guan, X. Y., Trent, J. M., and Meltzer, P. S. (1993) Generation of band-specific painting probes from a single microdissected chromosome. *Hum. Mol. Genet.* **2,** 1117–1121.
20. Ried, T., Schrock, E., Ning, Y., and Wienberg, J. (1998) Chromosome painting: a useful art. *Hum. Mol. Genet.* **7,** 1619–1626.
21. Ried, T., Liyanage, M., du Manoir, S., et al. (1997) Tumor cytogenetics revisited: comparative genomic hybridization and spectral karyotyping. *J. Mol. Med.* **75,** 801–814.
22. Corey, D. R. (1997) Peptide nucleic acids: expanding the scope of nucleic acid recognition. *Trends Biotechnol.* **15,** 224–229.
23. Liyanage, M., Coleman, A., du Manoir, S., et al. (1996) Multicolour spectral karyotyping of mouse chromosomes. *Nature Genet.* **14,** 312–315.
24. Schrock, E., du Manoir, S., Veldman, T., et al. (1996) Multicolor spectral karyotyping of human chromosomes. *Science* **273,** 494–497.
25. Tanke, H. J., Wiegant, J., van Gijlswijk, R. P., et al. (1999) New strategy for multicolour fluorescence in situ hybridisation: COBRA: COmbined Binary RAtio labelling. *Eur. J. Hum. Genet.* **7,** 2–11.
26. Speicher, M. R., Gwyn Ballard, S., and Ward, D. C. (1996) Karyotyping human chromosomes by combinatorial multi-fluor FISH. *Nature Genet.* **12,** 368–375.

27. Johannes, C., Chudoba, I., and Obe, G. (1999) Analysis of X-ray-induced aberrations in human chromosome 5 using high-resolution multicolour banding FISH (mBAND). *Chromosome Res.* **7,** 625–633.
28. Muller, S., O'Brien, P. C., Ferguson-Smith, M. A., and Wienberg, J. (1998) Cross-species colour segmenting: a novel tool in human karyotype analysis. *Cytometry* **33,** 445–452.
29. Tirkkonen, M., Tanner, M., Karhu, R., Kallioniemi, A., Isola, J., and Kallioniemi, O. P. (1998) Molecular cytogenetics of primary breast cancer by CGH. *Genes Chromosomes Cancer* **21,** 177–184.
30. Kallioniemi, O. P., Kallioniemi, A., Sudar, D., et al. (1993) Comparative genomic hybridization: a rapid new method for detecting and mapping DNA amplification in tumors. *Semin. Cancer Biol.* **4,** 41–46.
31. Haas, O., Henn, T., Romanakis, K., du Manoir, S., and Lengauer, C. (1998) Comparative genomic hybridization as part of a new diagnostic strategy in childhood hyperdiploid acute lymphoblastic leukemia. *Leukemia* **12,** 474–481.
32. Kallioniemi, O. P., Kallioniemi, A., Piper, J., et al. (1994) Optimizing comparative genomic hybridization for analysis of DNA sequence copy number changes in solid tumors. *Genes Chromosomes Cancer* **10,** 231–243.
33. Heselmeyer, K., Schrock, E., du Manoir, S., et al. (1996) Gain of chromosome 3q defines the transition from severe dysplasia to invasive carcinoma of the uterine cervix. *Proc. Natl. Acad. Sci. USA* **93,** 479–484.
34. Ried, T., Knutzen, R., Steinbeck, R., et al. (1996) Comparative genomic hybridization reveals a specific pattern of chromosomal gains and losses during the genesis of colorectal tumors. *Genes Chromosomes Cancer* **15,** 234–245.
35. Schrock, E., Blume, C., Meffert, M. C., et al. (1996) Recurrent gain of chromosome arm 7q in low-grade astrocytic tumors studied by comparative genomic hybridization. *Genes Chromosomes Cancer* **15,** 199–205.
36. Jacobsen, A., Arnold, N., Weimer, J., and Kiechle, M. (2000) Comparison of comparative genomic hybridization and interphase fluorescence in situ hybridization in ovarian carcinomas: possibilities and limitations of both techniques. *Cancer Genet. Cytogenet.* **122,** 7–12.
37. Roylance, R., Gorman, P., Harris, W., et al. (1999) Comparative genomic hybridization of breast tumors stratified by histological grade reveals new insights into the biological progression of breast cancer. *Cancer Res.* **59,** 1433–1436.
38. Knuutila, S., Bjorkqvist, A. M., Autio, K., et al. (1998) DNA copy number amplifications in human neoplasms: review of comparative genomic hybridization studies. *Am. J. Pathol.* **152,** 1107–1123.
39. Isola, J. J., Kallioniemi, O. P., Chu, L. W., et al. (1995) Genetic aberrations detected by comparative genomic hybridization predict outcome in node-negative breast cancer. *Am. J. Pathol.* **147,** 905–911.
40. Speicher, M. R., Jauch, A., Walt, H., et al. (1995) Correlation of microscopic phenotype with genotype in a formalin-fixed, paraffin-embedded testicular germ cell tumor with universal DNA amplification, comparative genomic hybridization, and interphase cytogenetics. *Am. J. Pathol.* **146,** 1332–1340.

41. Ried, T., Just, K. E., Holtgreve-Grez, H., et al. (1995) Comparative genomic hybridization of formalin-fixed, paraffin-embedded breast tumors reveals different patterns of chromosomal gains and losses in fibroadenomas and diploid and aneuploid carcinomas. *Cancer Res.* **55,** 5415–5423.
42. Telenius, H., Carter, N. P., Bebb, C. E., Nordenskjold, M., Ponder, B. A., and Tunnacliffe, A. (1992) Degenerate oligonucleotide-primed PCR: general amplification of target DNA by a single degenerate primer. *Genomics* **13,** 718–725.
43. Becker, I., Becker, K. F., Rohrl, M. H., Minkus, G., Schutze, K., and Hofler, H. (1996) Single-cell mutation analysis of tumors from stained histologic slides. *Lab. Invest.* **75,** 801–807.
44. Bohm, M., Wieland, I., Schutze, K., and Rubben, H. (1997) Microbeam MOMeNT: non-contact laser microdissection of membrane-mounted native tissue. *Am. J. Pathol.* **151,** 63–67.
45. Kuukasjarvi, T., Tanner, M., Pennanen, S., Karhu, R., Visakorpi, T., and Isola, J. (1997) Optimizing DOP-PCR for universal amplification of small DNA samples in comparative genomic hybridization. *Genes Chromosomes Cancer* **18,** 94–101.
46. Liang, B. C., Meltzer, P. S., Guan, X. Y., and Trent, J. M. (1995) Gene amplification elucidated by combined chromosomal microdissection and comparative genomic hybridization. *Cancer Genet. Cytogenet.* **80,** 55–59.
47. Slijepcevic, P. (2001) Telomere length measurement by Q-FISH. *Methods Cell Sci.* **23,** 17–22.
48. Baerlocher, G. M., Mak, J., Tien, T., and Lansdorp, P. M. (2002) Telomere length measurement by fluorescence in situ hybridization and flow cytometry: tips and pitfalls. *Cytometry* **47,** 89–99.
49. Rufer, N., Brummendorf, T. H., Kolvraa, S., et al. (1999) Telomere fluorescence measurements in granulocytes and T lymphocyte subsets point to a high turnover of hematopoietic stem cells and memory T cells in early childhood. *J. Exp. Med.* **190,** 157–167.
50. Jiang, F. and Katz, R. L. (2002) Use of interphase fluorescence in situ hybridization as a powerful diagnostic tool in cytology. *Diagn. Mol. Pathol.* **11,** 47–57.
51. Gorre, M. E., Mohammed, M., Ellwood, K., et al. (2001) Clinical resistance to STI-571 cancer therapy caused by BCR-ABL gene mutation or amplification. *Science* **293,** 876–880.
52. Vogel, C. L., Cobleigh, M. A., Tripathy, D., et al. (2002) Efficacy and safety of trastuzumab as a single agent in first-line treatment of HER2-overexpressing metastatic breast cancer. *J. Clin. Oncol.* **20,** 719–726.
53. Lutz, S., Welter, C., Zang, K. D., Seitz, G., Blin, N., and Urbschat, K. (1992) A two-colour technique for chromosome in situ hybridization in tissue sections. *J. Pathol.* **167,** 279–282.
54. McKay, J. A., Murray, G. I., Keith, W. N., and McLeod, H. L. (1997) Amplification of fluorescent in situ hybridisation signals in formalin fixed paraffin wax embedded sections of colon tumour using biotinylated tyramide. *Mol. Pathol.* **50,** 322–325.
55. Hyytinen, E., Visakorpi, T., Kallioniemi, A., Kallioniemi, O. P., and Isola, J. J. (1994) Improved technique for analysis of formalin-fixed, paraffin-embedded tumors by fluorescence in situ hybridization. *Cytometry* **16,** 93–99.

56. Kononen, J., Bubendorf, L., Kallioniemi, A., et al. (1998) Tissue microarrays for high-throughput molecular profiling of tumor specimens. *Nature Med.* **4,** 844–847.
57. Barlund, M., Tirkkonen, M., Forozan, F., Tanner, M. M., Kallioniemi, O., and Kallioniemi, A. (1997) Increased copy number at 17q22-q24 by CGH in breast cancer is due to high-level amplification of two separate regions. *Genes Chromosomes Cancer* **20,** 372–376.
58. Barlund, M., Monni, O., Kononen, J., et al. (2000) Multiple genes at 17q23 undergo amplification and overexpression in breast cancer. *Cancer Res.* **60,** 5340–5344.
59. Schofield, D. E. and Fletcher, J. A. (1992) Trisomy 12 in pediatric granulosa-stromal cell tumors. Demonstration by a modified method of fluorescence in situ hybridization on paraffin-embedded material. *Am. J. Pathol.* **141,** 1265–1269.
60. Emmert-Buck, M. R., Bonner, R. F., Smith, P. D., et al. (1996) Laser capture microdissection. *Science* **274,** 998–1001.
61. DiFrancesco, L. M., Murthy, S. K., Luider, J., and Demetrick, D. J. (2000) Laser capture microdissection-guided fluorescence in situ hybridizatio and flow cytometric cell cycle analysis of purified nuclei from paraffin sections. *Mod. Pathol.* **13,** 705–711.
62. Murthy, S. K., DiFrancesco, L. M., Ogilvie, R. T., and Demetrick, D. J. (2002) Loss of heterozygosity associated with uniparental disomy in breast carcinoma. *Mod. Pathol.* **15,** 1241–1250.
63. Klinger, K., Landes, G., Shook, D., et al. (1992) Rapid detection of chromosome aneuploidies in uncultured amniocytes by using fluorescence in situ hybridization (FISH). *Am. J. Hum. Genet.* **51,** 55–65.
64. Van Opstal, D., Van Hemel, J. O., and Sachs, E. S. (1993) Fetal aneuploidy diagnosed by fluorescence in-situ hybridisation within 24 hours after amniocentesis. *Lancet* **342,** 802.
65. Ward, B. E., Gersen, S. L., Carelli, M. P., et al. (1993) Rapid prenatal diagnosis of chromosomal aneuploidies by fluorescence in situ hybridization: clinical experience with 4,500 specimens. *Am. J. Hum. Genet.* **52,** 854–865.
66. Philip, J., Bryndorf, T., and Christensen, B. (1994) Prenatal aneuploidy detection in interphase cells by fluorescence in situ hybridization (FISH). *Prenat. Diagn.* **14,** 1203–1215.
67. van den Berg, C., Van Opstal, D., et al. (2000) Accuracy of abnormal karyotypes after the analysis of both short- and long-term culture of chorionic villi. *Prenat. Diagn.* **20,** 956–969.
68. Los, F. J., van Den Berg, C., Wildschut, H. I., et al. (2001) The diagnostic performance of cytogenetic investigation in amniotic fluid cells and chorionic villi. *Prenat. Diagn.* **21,** 1150–1158.
69. Van Opstal, D., van den Berg, C., Galjaard, R. J., and Los, F. J. (2001) Follow-up investigations in uncultured amniotic fluid cells after uncertain cytogenetic results. *Prenat. Diagn.* **21,** 75–80.
70. Florijn, R. J., Bonden, L. A., Vrolijk, H., et al. (1995) High-resolution DNA fiber-FISH for genomic DNA mapping and colour bar-coding of large genes. *Hum. Mol. Genet.* **4,** 831–836.
71. Raap, A. K., Florijn, R. J., Blonden, L. A. J., et al. (1996) Fiber FISH as a DNA Mapping Tool. *Methods* **9,** 67–73.

72. Vaandrager, J. W., Schuuring, E., Zwikstra, E., et al. (1996) Direct visualization of dispersed 11q13 chromosomal translocations in mantle cell lymphoma by multicolor DNA fiber fluorescence in situ hybridization. *Blood* **88,** 1177–1182.
73. Vaandrager, J. W., Schuuring, E., Kluin-Nelemans, H. C., Dyer, M. J., Raap, A. K., and Kluin, P. M. (1998) DNA fiber fluorescence in situ hybridization analysis of immunoglobulin class switching in B-cell neoplasia: aberrant CH gene rearrangements in follicle center-cell lymphoma. *Blood* **92,** 2871–2878.
74. Rottger, S., Yen, P. H., and Schempp, W. (2002) A fiber-FISH contig spanning the non-recombining region of the human Y chromosome. *Chromosome Res.* **10,** 621–635.
75. Solinas-Toldo, S., Lampel, S., Stilgenbauer, S., et al. (1997) Matrix-based comparative genomic hybridization: biochips to screen for genomic imbalances. *Genes Chromosomes Cancer* **20,** 399–407.
76. Pinkel, D., Segraves, R., Sudar, D., et al. (1998) High resolution analysis of DNA copy number variation using comparative genomic hybridization to microarrays. *Nature Genet.* **20,** 207–211.
77. Klein, C. A., Schmidt-Kittler, O., Schardt, J. A., Pantel, K., Speicher, M. R., and Riethmuller, G. (1999) Comparative genomic hybridization, loss of heterozygosity, and DNA sequence analysis of single cells. *Proc. Natl. Acad. Sci. USA* **96,** 4494–4499.
78. Snijders, A. M., Nowak, N., Segraves, R., et al. (2001) Assembly of microarrays for genome-wide measurement of DNA copy number. *Nature Genet.* **29,** 263–264.
79. Heiskanen, M. A., Bittner, M. L., Chen, Y., et al. (2000) Detection of gene amplification by genomic hybridization to cDNA microarrays. *Cancer Res.* **60,** 799–802.
80. Forozan, F., Mahlamaki, E. H., Monni, O., et al. (2000) Comparative genomic hybridization analysis of 38 breast cancer cell lines: a basis for interpreting complementary DNA microarray data. *Cancer Res.* **60,** 4519–4525.
81. Pollack, J. R., Perou, C. M., Alizadeh, A. A., et al. (1999) Genome-wide analysis of DNA copy-number changes using cDNA microarrays. *Nature Genet.* **23,** 41–46.
82. Greshock, J., Naylor, T. L., Margolin, A., et al. (2004) 1-Mb Resolution array-based comparative genomic hybridization using a BAC clone set optimized for cancer gene analysis. *Genome Res.* **14,** 179–187.
83. Chung, Y. J., Jonkers, J., Kitson, H., et al. (2004) A whole-genome mouse BAC microarray with 1-Mb resolution for analysis of DNA copy number changes by array comparative genomic hybridization. *Genome Res.* **14,** 188–196.
84. Trask, B. J. (1991) DNA sequence localization in metaphase and interphase cells by fluorescence in situ hybridization. *Methods Cell Biol.* **35,** 3–35.
85. Demetrick, D. (1995) Fluorescence *in situ* hybridization and human cell cycle genes, in *The Cell Cycle: Materials and Methods* (Pagano, M., ed.), Springer-Verlag, New York.
86. Demetrick, D., Murthy, S., and DiFrancesco, L. (2002) Fluorescence in situ hybridization of LCM-isolated nuclei from paraffin sections. *Methods Enzymol.* **356,** 63–69.

13

Microarray Image Scanning

Latha Ramdas and Wei Zhang

Summary

Of the technologies available for measuring gene expression, microarrays using cDNA targets is one of the most common and well-developed high-throughput techniques. With this technique, the expression levels of thousands of genes are measured simultaneously. DNA probes are immobilized on solid surfaces, either membrane-based or chemically coated glass surfaces. On glass arrays, the probes are hybridized with fluorescent-labeled target samples. Fluorescence intensities, which reflect gene expression levels, are detected by imaging the array using a laser or white-light source and capturing the image using photomultiplier tube detection or a charge-coupled device camera. Different laser-based scanners are used in laboratories to scan microarray images. This chapter discusses the imaging process and the protocols being developed.

Key Words: Genomics; microarray; laser scanners; imaging; gene expression; fluorescence.

1. Introduction

From data derived from completion of the Human Genome Project, we know that there are approx 40,000 genes in human cells and that hundreds and thousands of these genes interact to perform diverse cellular activities. Microarray or DNAchip technology is a recently developed technique that allows biologists to obtain a bird's-eye view of the expression of thousands of genes in a single experiment. This technique has recently been used to address many biological issues, such as discovery of genes involved in certain diseases and molecular classification of cancers ***(1–5)***. The two platforms used with this technique are the membrane-based (porous surfaces like nylon) and chemically coated glass arrays. In both cases, thousands of DNA probes are immobilized robotically and hybridized with either ^{32}P- or ^{33}P-labeled targets in the case of membrane arrays or fluorescent-labeled cDNA targets in the case of glass arrays. With the glass microarrays, where fluorescent-labeled samples are hybridized, the signal

From: *Methods in Molecular Biology, vol. 319: Cell Imaging Techniques: Methods and Protocols*
Edited by: D. J. Taatjes and B. T. Mossman © Humana Press Inc., Totowa, NJ

intensities are measured with fluorescence laser scanners. Detection of signals in glass microarray images is accomplished by measuring fluorescence intensities with scanners that allow simultaneous determination of the relative expression levels of all the genes represented on the array. This technique is the focus in this chapter.

The imaging technique requires light of a specific wavelength to excite the fluorophores in the sample, and scanners can accomplish this by using one of two techniques. Most scanners use laser sources to excite each fluorescent probe, and emitted light is detected by a photomultiplier tube (PMT). This method uses a laser-based scanning system that involves the rapid movement of small points of laser light across the sample, from which the image is then reconstructed. The second imaging technique is based on a charge-coupled device (CCD). Scanners that use this method have a white-light arc lamp as their illumination source. Because white light encompasses all wavelengths in the visible spectrum, it is possible to select the wavelength of choice. Fluorescent light emitted from the sample is collected and imaged using CCD cameras.

In this chapter, the protocol for scanning will be elaborated for the laser scanner using PMT detection.

2. Materials

A total of 2304 known sequence-verified human cDNAs were prepared by polymerase chain reaction (PCR) from the cDNA clone library (Research Genetics, Huntsville, AL) using two primers on the vector. The DNA clones, in 384-well plates, were spotted onto poly-L-lysine-coated microscope slides using an arrayer (Genomic Solutions, Ann Arbor, MI). All clones except those of the controls were duplicated on the array. After printing, the slides were dried and crosslinked by ultraviolet radiation at 650 J/cm^2. The slides were then washed with water, dried, and stored for later use.

Cyanine-dye-labeled reverse-transcribed target cDNA sample in 130 μL of total volume of Express hyb solution (Clontech, Palo Alto, CA) with a mixture of blocking reagents was hybridized to the glass microarray slides. After hybridization, the slides were preprocessed for scanning. Laser scanners with appropriate laser sources for scanning (532-nm excitation for Cy3 and 695-nm excitation for Cy5) were used for scanning the microarray image. Quantification of the spots was carried out using an image analysis software product, ArrayVision (Imaging Research, Inc. Ontario, Canada).

3. Methods

The most frequently used method of labeling gene products or mRNA is to incorporate cyanine (Cy)-labeled nucleotides into cDNA during reverse transcription of the mRNA. The most common fluorophores are Cy3 (green-excitation light at 532 or 545 nm) and Cy5 (red-excitation light at 635 nm). Microarray scanners capture the Cy3 and Cy5 signals on each spot of the microarray, using either the laser light of the specific wavelength to excite the fluorophores or white light, for which the choice of wavelength is flexible. Superimposing the two images from Cy3 and Cy5 channels provides a composite image that allows straightforward visual estimation. If we color the Cy5 image red and the Cy3 image green, an intense green or red spot will signify a higher level of that particular gene in one of the two samples. A yellow spot will indicate similar expression of the gene in the two samples.

Several microarray scanners are available and appropriate for use in this technique, including a DNA microarray scanner (Agilent Technologies, Palo Alto, CA), ArrayWorx (Applied Precision, Issaquah, WA), ChipReader (Virtek Vision Corporation, Ontario, Canada), GenePix 4000B (Axon Instruments Inc., Foster City, CA), GeneTac LSIV and UC-4 (Genomic Solutions Inc., Ann Arbor, MI), and ScanArray (Packard Bioscience, Billerica, MA). The specifications for these scanners are listed in **Table 1**. Each scanner has unique features, and several factors should be considered when choosing one ***(6)***.

3.1. Signal Detection

3.1.1. Light Source

Laser light is a common light source in which a small point of light moves through the entire microarray slide and reconstructs the image from this point measurement. The laser source is preselected on the basis of excitation and the emission wavelength of the fluorescent probes. Another type of light source is broad-spectrum white light, which is used in scanners such as the ArrayWorx scanner. The advantage of this type of scanner is its flexibility: it offers selection from 330 to 700 nm for excitation and from 380 to 800 nm emission wavelengths via filter selection.

3.1.2. Photon Detection

Two main optical modes are used to detect signals from the microarray slides. Most laser-based scanners use a PMT detector and a scanning process that involves several steps:

$$\text{Dye} \xrightarrow[\text{Excitation}]{\text{Laser}} \text{Photons} \xrightarrow[\text{Amplification}]{\text{PMT}} \text{Electrons} \xrightarrow[\text{Filtration}]{\text{Converter}} \text{Signal}$$

The classic photomultiplier contains a light-sensitive photocathode that generates electrons when exposed to light. These electrons strike a charged electrode, called a dynode, and produce additional electrons. PMTs contain several

Table 1
Specifications for Six Microarray Scanners

	Agilent scanner	ArrayWorx	GenePix 4000B	Gene TAC LSIV	GeneTac UC-4	ScanArray Express	ChipReader
Light source	Laser (for Cy3 and Cy5)	White light	Laser (532 and 635 nm)	Laser (488, 494, 532, and 635 nm)	Laser (532 and 635 nm)	Laser	Laser
Detection	PMT	CCD	PMT and voltage adjustment	PMT	PMT	PMT and laser power adjustment	PMT
Imaging	Simultaneous	Stitch-by-position	Simultaneous	Dark-field, sequential	Dark-field, sequential	Confocal, sequential	Simultaneous
Field focus (μ*M*)	Dynamic autofocus		60 (thick specimen)	100	100	Autofocus	10 (small depth)
Pixel resolution	5, 10 μ*M*	3–26 μ*M*	5–100 μ*M*	1–100 μ*M*	1–100 μ*M*	5, 10, 20, 30 and 50 μ*M*	5, 10, 20, 30 μ*M*
Scan speed	~8 min/slide	>10 min/slide	~5 min/slide	~5 min/slide	~5 min/slide	~5 min/slide	~5 min/slide
Dimensions $l \times d \times h$ (in.)	24 × 36 × 24	31 × 23 × 23	13.5 × 8 × 17.5	36 × 24 × 22	24 × 16 × 12	30 × 16.5 × 14	11 × 9 × 12
Type of autoloader	48-Slide carousell	40-Slide carrier	Single slide	24-Slide carrier	4-Slide carrier	Single slide	Single slide

dynodes, so the electrons emitted are amplified at each dynode, thus amplifying the signal. The electrons are collected at the anode and output as a current that is proportional to the intensity of light that strikes the photocathode. Thus, the PMT amplifies and measures low levels of light. Some scanners, such as the GeneTac LSIV and UC-4 scanners, increase the efficiency of the PMT detector by using a side-window PMT detector to minimize photobleaching and dark-current noise.

Some laser/PMT-based systems, such as the ScanArray scanners, use confocal optics to sharpen the focus of the laser beam. This type of scanner focuses light at both the excitation end on the substrate and at the emission end on the PMT detector. The laser light, focused through a pinhole to restrict the focal length and reduce imaging artifacts, induces fluorescence. The emitted light is then collected by an objective lens and converged through another pinhole to the detector. However, if the substrate surface has large variations in thickness, the intensity measurements can differ depending on the focal plane.

The second method for collecting fluorescent light from a sample is by using a high-performance CCD camera. This is the technology used in the ArrayWorx scanner. A white-light beam is passed through an excitation filter, yielding monochromatic light of the appropriate wavelength to excite the molecules on the array slide. The emitted light is focused onto a CCD camera via emission filters. In the ArrayWorx scanner, the excitation beam is passed through a specifically configured fiberoptic bundle delivering highly stable light onto the sample. The CCD camera collects light from small panels on the slide, and these panels are tiled to reconstruct the image. This method is often referred to as "stitch-by-position." The disadvantage of this system is that fewer emission photons are generated than with the PMT, in which the signal is amplified. To overcome this weakness, CCD systems integrate the signals over a period of time, thus prolonging the scan time. This long exposure time increases the dark current, which is the generation of electrons from heat at a constant rate. Because dark-current electrons are indistinguishable from electrons generated by photons, more noise is introduced. To minimize this noise, the CCD system must be maintained at a low temperature. The above-discussed optics are shown in **Fig. 1**.

3.2. Scanning

3.2.1. Simultaneous and Sequential Scanning

Among the laser-based scanning systems available, some scan two dyes simultaneously, whereas others (sequential scanners) scan one dye at a time. Simultaneous scanning is fast and requires no image alignment after the scanning is completed, whereas sequential scanning acquires one image at a time and builds a composite image. Crosstalk between the two optical channels is

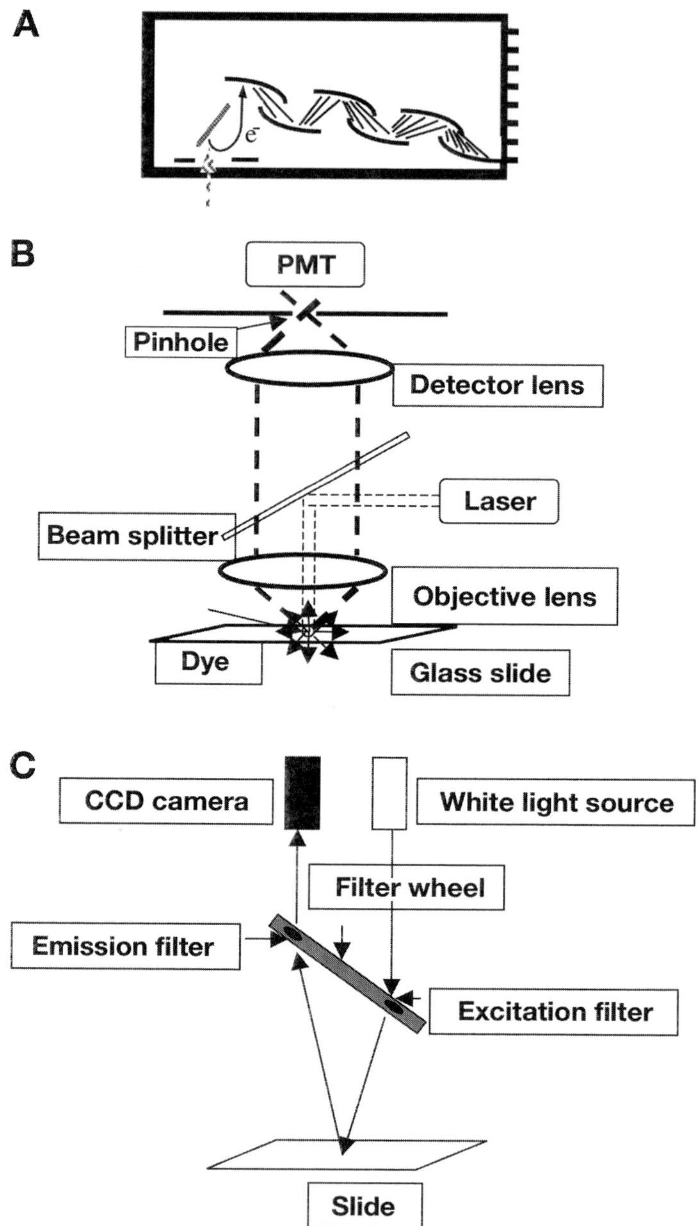

Fig.1. **(A)** The PMT amplifies electrons through dynodes, using the side window to minimize photobleaching, as in the Genomic Solutions GeneTAC UC-4 system. **(B)** The optical path in a confocal imaging technique showing the objective lens and the detector lens, and the convergence of the incident and the emission light through pinholes. **(C)** In the optical path of a CCD system, a white-light beam is passed through an excitation filter, and the chosen wavelength of light illuminates the sample on a glass slide. The light emitted through the emission filter is then focused onto a CCD camera.

not a concern for sequential scanning but could cause problems with simultaneous scanning. To overcome this problem, simultaneous-scanning devices use a laser wavelength of 532 nm instead of 545 nm to excite Cy3 and, thus, allow better spectral separation between Cy3 and Cy5, whose excitation wavelength is 635 nm.

3.2.2. Features of Scanning Devices

Two other scanning features are the stitch-by-position technology in CCD camera detection and the confocal scanning in PMT detection. Dark-field imaging and laser-beam scanning with PMT detection are combined in the GeneTAC LSIV and UC-4 scanners. In the dark-field approach, the laser beam is angled so that the backscattered light cannot reach the PMT, which improves the contrast and the signal sensitivity. As a drawback, these technologies yield images that are dimmer than those projected from other scanners, so the slides require longer exposure to the laser. However, the improved side-window PMT system compensates by preventing photobleaching. In addition, the system has an increased focal depth of approx 100 μm. Thus, there is no variability resulting from uneven slide surfaces.

3.3. Signal-to-Noise Ratio

The goal of scanning is to produce a high-quality image from which gene expression information can be acquired. The quality of an image is judged by its signal-to-noise ratio. Common sources of noise are thermal emissions or dark current, both of which occur in the PMT and CCD camera. The GeneTAC LSIV and UC-4 scanners use side-window PMT detection to reduce this noise, the GenePix 4000B scanner reduces the time the laser spends on each pixel, the ScanArray scanner uses a small depth of focus, and the ArrayWorx scanner uses a thermoelectrically cooled camera to reduce the dark current. Another kind of noise, called shot noise, is caused by the particle nature of light. This is the most dominant type of noise affecting the signal-to-noise ratio. The absolute magnitude of this noise increases with increasing signal intensity; however, the noise only increases by the square root of the signal intensity; thus, the signal-to-noise ratio decreases as the signal intensity increases.

3.4. Comparison of Scanners

The whole process of microarray technology involves many different steps, which is why many of the current protocols are standardized to minimize experimental variations. Standardized procedures are essential for compatible data production, quality control, and analysis. However, reliability of the data from experiment to experiment and comparison of data from laboratory to laboratory or instrument to instrument has not been extensively evaluated. A commonly accepted standard with which to compare the performances of various scanners is presently lacking.

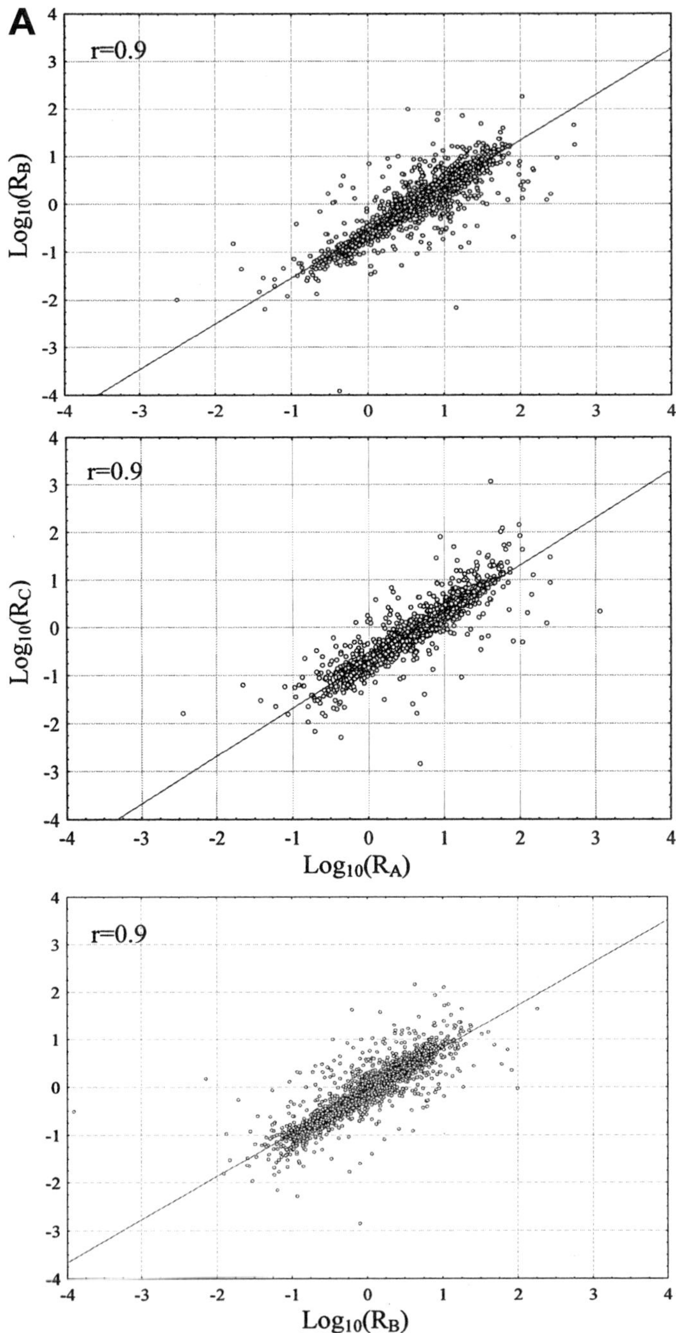

Fig. 2. Correlation analysis among three scanners at **(A)** 20 µm and **(B)** 10 µm resolution. The correlation coefficients *(r)* are labeled in each of the comparisons.

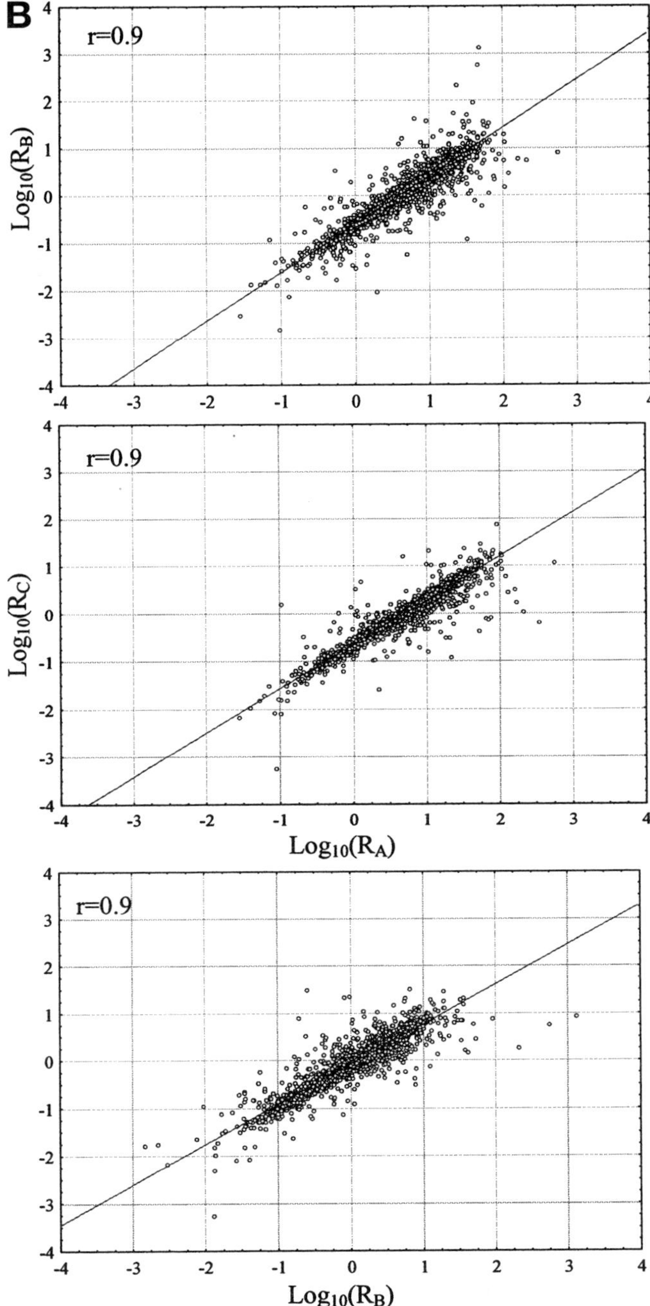

Fig. 2. *(Continued)* R_A, R_B, and R_C represent data obtained from scanners A, B, and C, respectively. $Log_{10}(R)$ is the logarithm-transformed intensity ratio of the two channels. A linear regression line is shown in each plot.

To evaluate the performance of different scanners and compare the results, we scanned a single microarray slide on three different scanners. Our analysis involved the GeneTAC LSIV scanner (A), which combines a laser-beam source, dark-field imaging, sequential scanning, and PMT detection, the GenePix 4000B scanner (B), which uses a laser source, simultaneous scanning, and PMT detection, and the ScanArray scanner (C), a laser scanner with confocal-imaging technology and PMT detection *(7)*. We scanned the slide on the three devices using comparable resolution, retrieved the tagged image file format (TIFF) images, and quantitated the thousands of spots on the array using the spot-finding software, ArrayVision (Imaging Research Inc., Ontario, Canada). The fluorescence intensities from each spot on the array were measured as sVol, which was the background-corrected volume, where volume is the density of each spot multiplied by its area and density is the average of all pixel intensities in the spot. The ratios of sVols from the two channels (Cy3 and Cy5) were used as measures of the relative expression levels, and the data were transformed into logarithmic values. The log-transformed intensity ratios from the three different instruments were then compared. When all three scanners were used, the correlation coefficient was between 0.90 and 0.96 (*see* **Fig. 2**); when images and data were obtained using the same scanner at different times, the correlation coefficient was approx 0.93. Thus, the instruments used did not cause any variability in results. We also compared the genes that were differentially expressed between the two cohybridized samples acquired from the three scanners. Among the 160 most differentially expressed genes, 95% of the genes were identified from all three analyses. Thus, all three instruments identified nearly all of the same differentially expressed genes.

3.5. Setting Scanning Parameters

This subsection tests the steps that need to be considered when using a sequential scanner, which combines dark-field imaging and laser-beam scanning with PMT detection, to obtain raw signal intensity data.

1. The computer and the scanner are turned on in that order to prewarm for scanning.
2. The scanning program controls the action on the laser scanner. Thus, using the program, the machine is set ready to scan. The microarray slides are loaded on the slide holder and, when ready, the shutter is closed to latch (*see* **Note 1**).
3. *Scan dimensions:* Scan dimensions, which define the area on the slide to be scanned, are set and the image captured. For a standard slide (25 mm × 76 mm), the dimensions depend on the area on the slide that the actual array occupies (approx 20 mm × 70 mm) and the values 1248 × 2348 are set on the program for 20 μm resolution.

4. *Scan resolution:* The resolution of the image is set from 5 to 50 μm, which also determines the scan dimension. For a preview of the image, a low resolution is set, but for actual measurement, the spot size and the number of pixels in the spot determine the scan resolution. To obtain valid data, the pixel resolutions should be such that the spot size is greater than 5 times the pixel size and preferably 10 times. Thus, for a spot size of 100 μm, the pixel size should be 10 μm.
5. *Pan and zoom:* Pan and zoom is another option that is available to increase the resolution of a small area on the array.
6. *Gains:* With the sequential scanner discussed here, there are no adjustments to the power of the laser. However, an adjustment to the gains can be made, which dictate the amplification of the signal as it passes through the dynodes and thus influence the signal (*see* **Note 2**).
7. *Palettes:* Another option on this arrayer is the viewing palette, which can assist in determining the optimum gain to be set for scanning. Generally, setting a red or green palette for Cy5 and Cy3, respectively, will produce individual images with red and green spots and a composite image that is mostly yellow if most of the expression is the same in both the samples. However, detecting the saturation of the spots is not possible with this color setup because the human eye cannot differentiate shades of green and red very well. For this reason, there is a feature that allows one to set a color to indicate saturation as white (*see* **Note 3**).
8. *Offset:* This is another option to modify the background to better visualize the spots. By changing the zero value to a finite offset value, the contrast between the spots and the background can be improved. However, this causes a filtering effect, and in order to obtain all the raw data from the array, it is best that this number remain at zero during scanning.

Several other scanners provide the option of varying the PMT voltage and the laser power to improve the signal-to-noise ratio of the spots. For example, in the GenePix 4000 scanner, the lasers operate at full power; hence, the voltage of the PMT controls the gains of the PMT.

4. Notes

1. The glass arrays need to be handled very carefully. While scanning, the slides are always handled with gloves. Care has to be taken not to touch the inner surface of the slide, as this can destroy the poly-L-lysine coat on the surface.
2. Caution must to be taken in increasing the gain, because this affects both signal and noise without discrimination. Also there is serious loss of information from the spots on the microarray as the spots become saturated in intensity. For example, we scanned a slide at a particular gain and several five-increment intervals of gain and analyzed the data. As the gain increases, more and more spots fall in the saturated range and expression variation for all the spots goes to a minimum (*see* **Fig. 3**; Color Plate 14, following p. 274). For best results, the signal intensities from the two channels should fall in the linear range of detection ***(8)***.

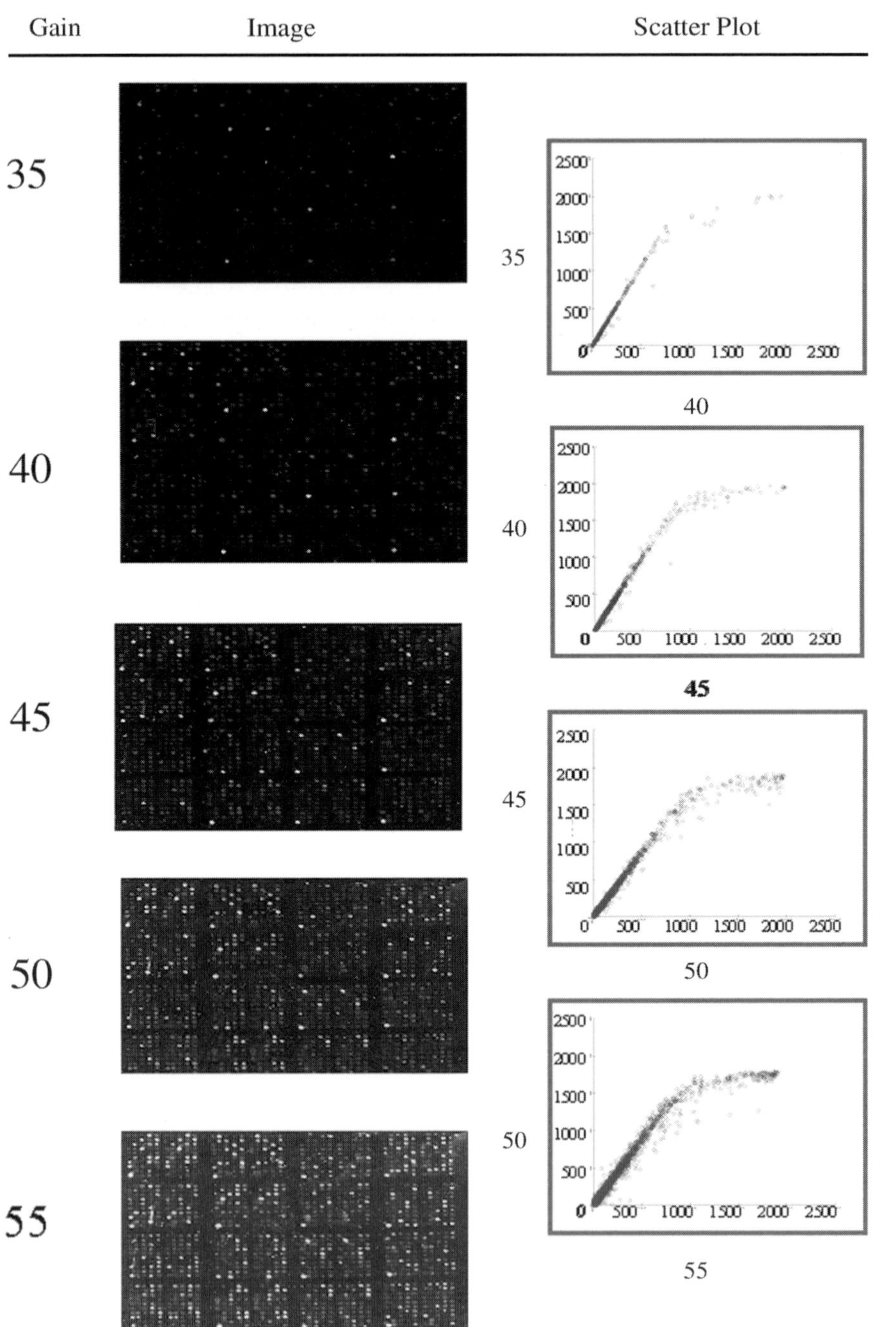

Fig. 3. The images for different gains used for scanning and the corresponding scatterplots are shown. The saturated spots in the images are shown as white spots. The scatterplot is the comparison of the signal intensities between two gains, and as the gain increases, the number of spots that are saturated in intensity increases. (*See* Color Plate 14, following p. 274.)

3. The palette option of red or green to white, where white indicates saturated spots, is a good choice while scanning. Reasonably high gains need to be used while scanning in order to see the spots of low fluorescence intensities. The best choice is the spectral palette with spectral colors that cover the whole range in the 16-bit image.

References

1. Bittner, M., Meltzer, P., Chen, Y., et al. (2000) Molecular classification of cutaneous malignant melanoma by gene expression profiling. *Nature* **406**, 536–540.
2. Eisen, M. B. and Brown, P. O. (1999) DNA arrays for analysis of gene expression. *Methods Enzymol.* **303**, 179–205.
3. Fuller, G. N., Rhee, C. H., Hess, K. R., et al. (1999) Reactivation of insulin-like growth factor binding protein 2 expression in glioblastoma multiforme: a revelation by parallel gene expression profiling. *Cancer Res.* **59**, 4228–4332.
4. Hegde, P., Qi, R., Abernathy, K., et al. (2000) A concise guide to cDNA micorarray analysis. *Biotechniques* **29**, 548–562.
5. Golub, T. R., Slonim, D. K., Tamayo, P., et al. (1999) Molecular classification of cancer: Class discovery and class prediction by gene expression monitoring. *Science* **286**, 531–537.
6. Ramdas, L. and Zhang W. (2002) What's happening inside your microarray scanner? *Biophotonics Int.* **9**, 42–47.
7. Ramdas, L., Wang, J., Hu, L., Cogdell, D., Taylor, E., and Zhang, W. (2001) Comparative evaluation of laser-based microarray scanners. *Biotechniques* **31**, 546–552.
8. Ramdas, L., Coombes, K. R., Baggerly, K., et al. (2001) Sources of nonlinearity in cDNA microarray expression measurements. *Genome Biol.* **2,** 47.

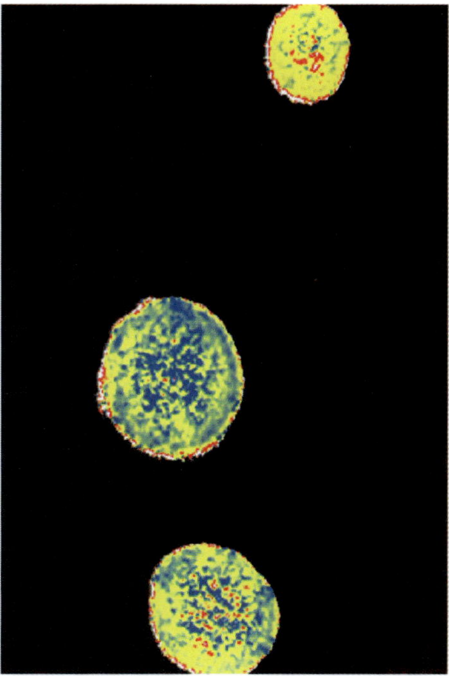

Color Plate 1, Fig. 1. Pseudocolored image of mouse islets loaded with Fura-2. (*See* full caption in Ch. 3 on p. 52 and discussion on p. 51.)

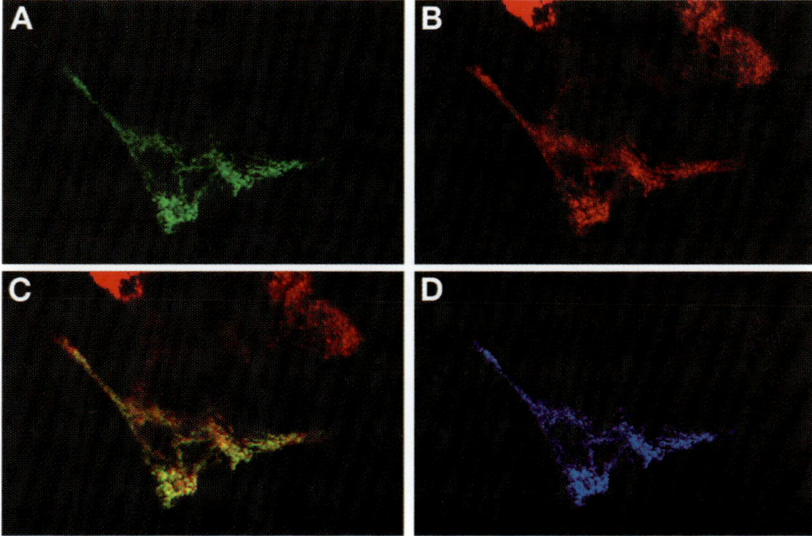

Color Plate 2, Fig. 3. Expression of mitochondrially-targeted ratiometric pericam (RPC-mt) in neuroendocrine cells: laser scanning confocal images of AtT20 cells. (*See* full caption and discussion in Ch. 3 on p. 58.)

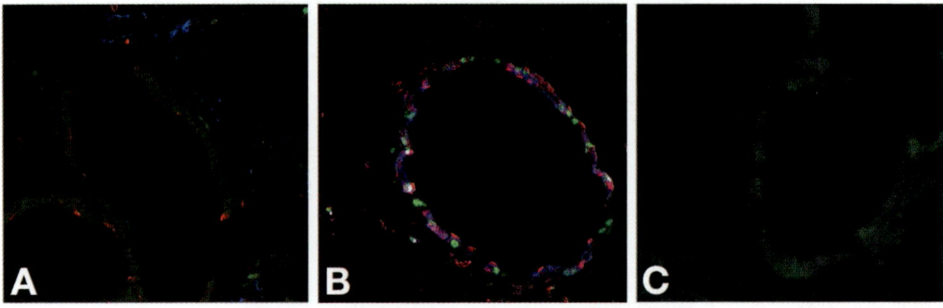

Color Plate 3, Fig. 1. Triple-labeling using mouse monoclonal PKC (blue), rabbit polyclonal p-ERK (red), and rat monoclonal Ki-67 (green) anti-bodies on sham animals **(A)** and animals exposed to crocidolite asbestos for 4 days **(B)**. A negative control omitting the primary antibody is shown in **(C)**. (*See* full caption in Ch. 4 on p. 74 and discussion on p. 73.)

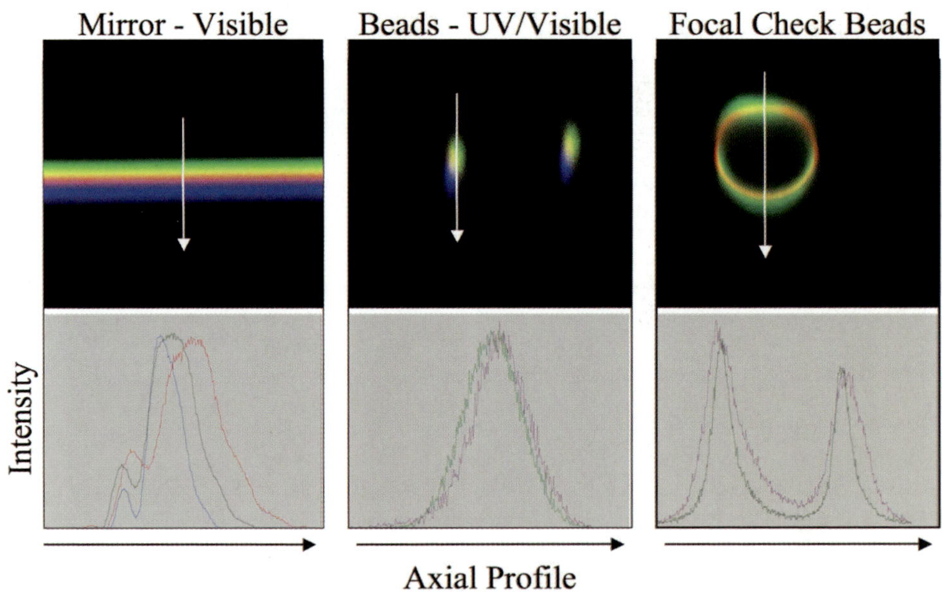

Color Plate 4, Fig. 6. Methods used to check spectral registration of different laser lines. (*See* full caption in Ch. 5 on p. 93 and discussion on p. 92.)

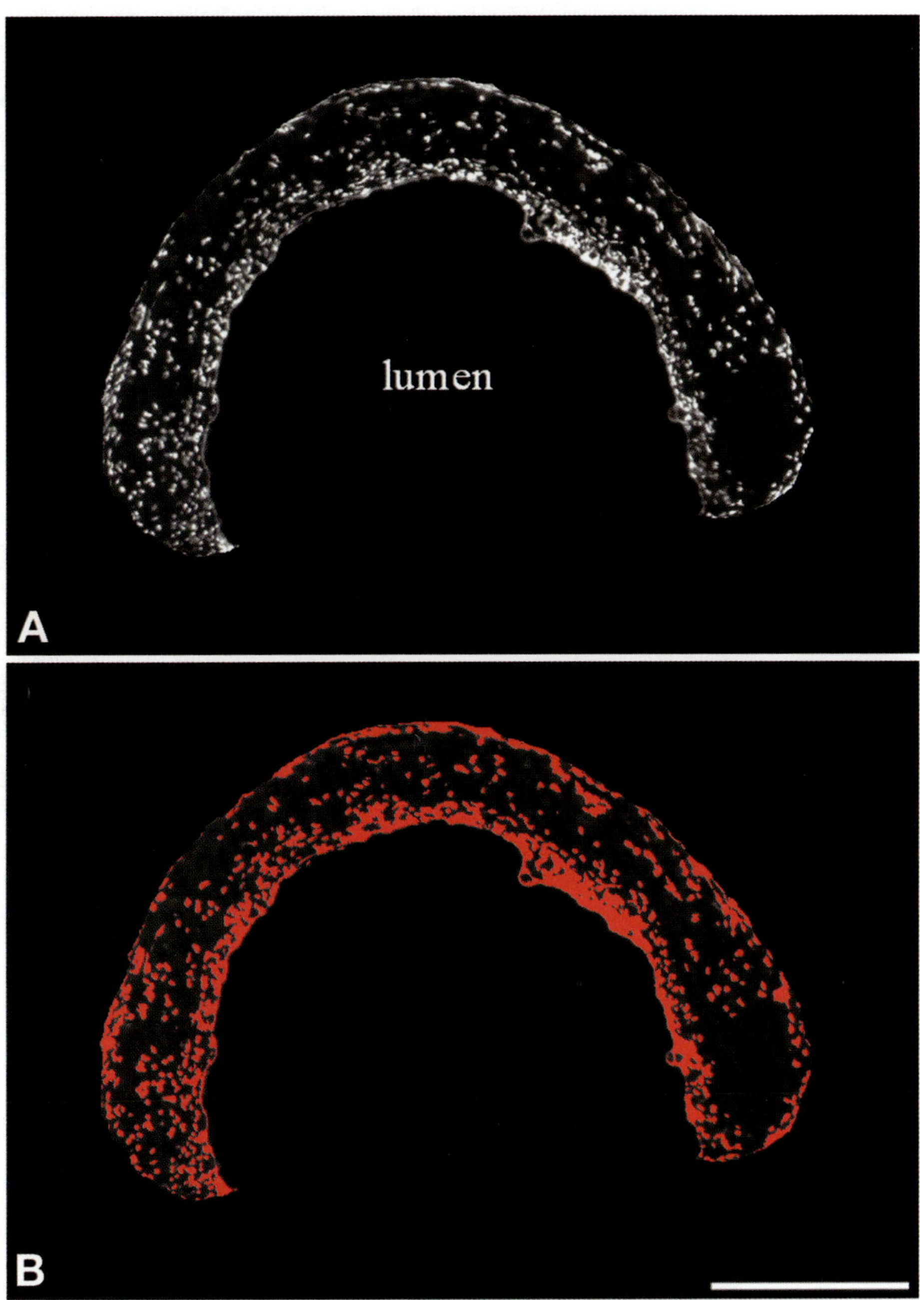

Color Plate 5, Fig. 2. Gray-scale fluorescent capture of DAPI-stained nuclei in mouse atherosclerotic lesion **(A)**, and with red pseudocolor overlay **(B)**. (*See* full caption in Ch. 6 on p. 144 and discussion on p. 142.)

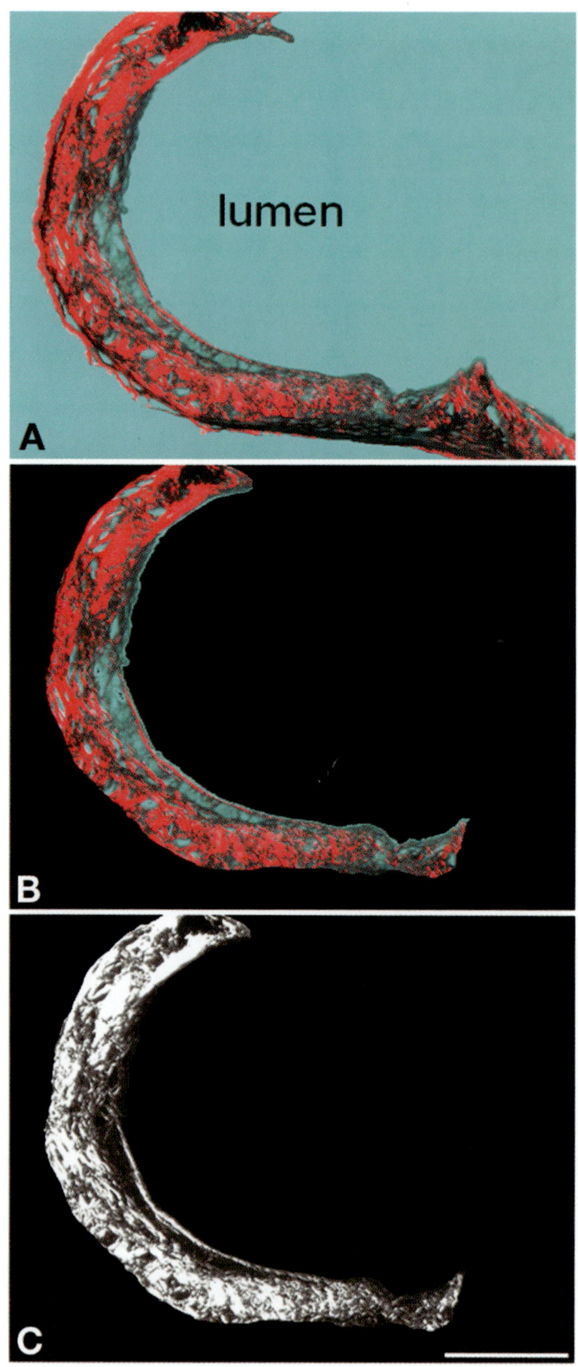

Color Plate 6, Fig. 3. Polarized light microscopy assists in more accurately discriminating atherosclerotic lesion borders. (*See* full caption in Ch. 6 on p. 149 and discussion on p. 146.)

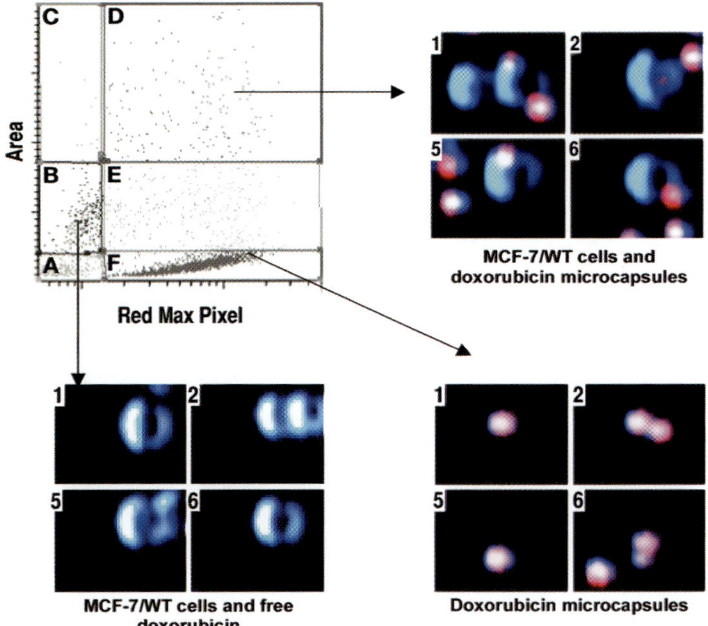

Color Plate 7, Fig. 3. Assay of cells incubated with either free drug or drug-encapsulated microspheres by laser scanning cytometry. (*See* full caption in Ch. 9 on p. 201 and discussion on p. 200.)

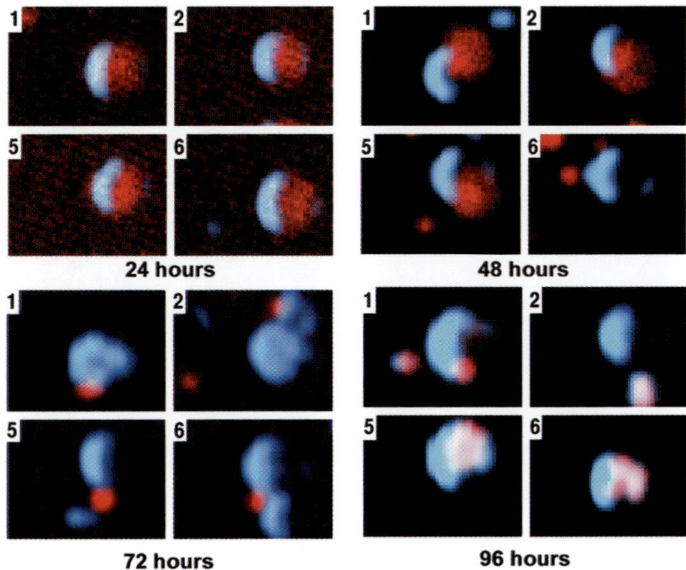

Color Plate 8, Fig. 4. Time-course experiments of doxorubicin encapsulated microspheres. (*See* full caption in Ch. 9 on p. 203 and discussion on p. 202.)

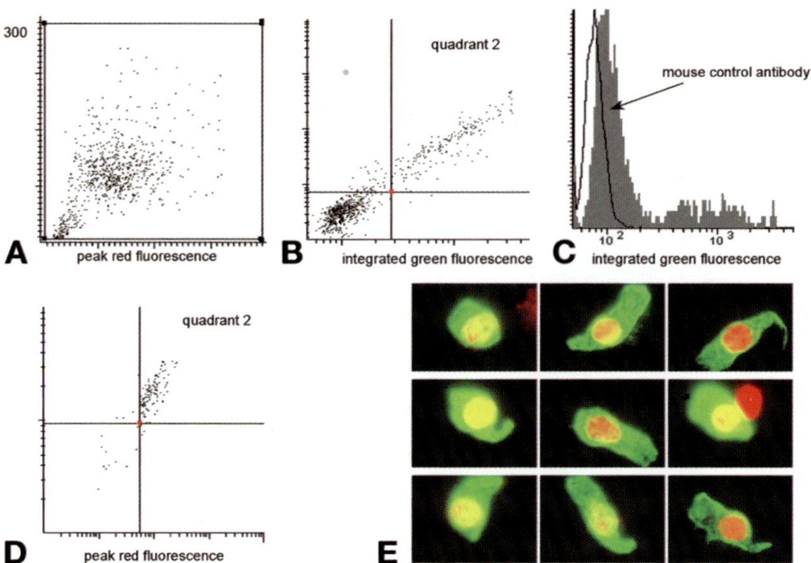

Color Plate 9, Fig. 9. Gate settings for the analysis of sputum bronchial epithelial cells by laser scanning cytometry. (*See* full caption in Ch. 9 on p. 210 and discussion on p. 207.)

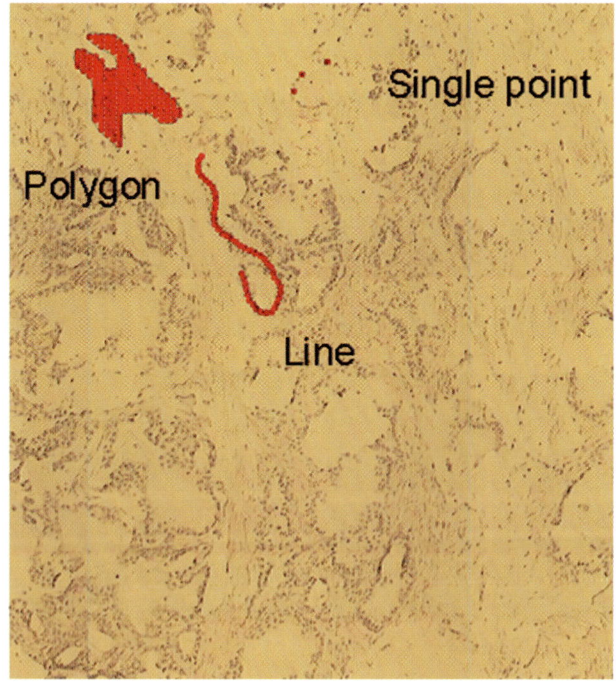

Color Plate 10, Fig. 6. Annotation of stitched images on AutoPix system. (*See* full caption in Ch. 10 on p. 224 and discussion on p. 223.)

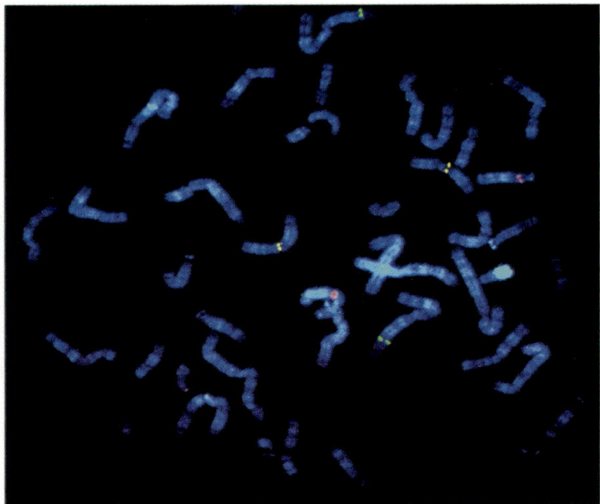

Color Plate 11, Fig. 1. Metaphase analysis with multiple single-gene probes. Multicolor FISH showing simultaneous localization of three human genomic BAC probes. (*See* full caption in Ch. 12 on p. 238 and discussion on p. 237.)

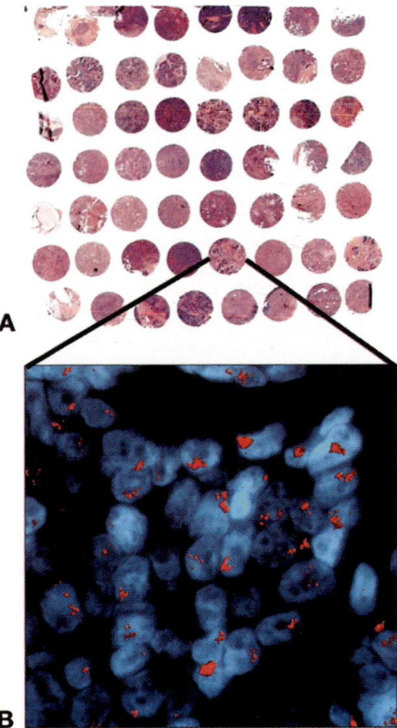

Color Plate 12, Fig. 2. FISH of tissue microarray breast carcinoma specimens. (*See* full caption in Ch. 12 on p. 243 and discussion on p. 242.)

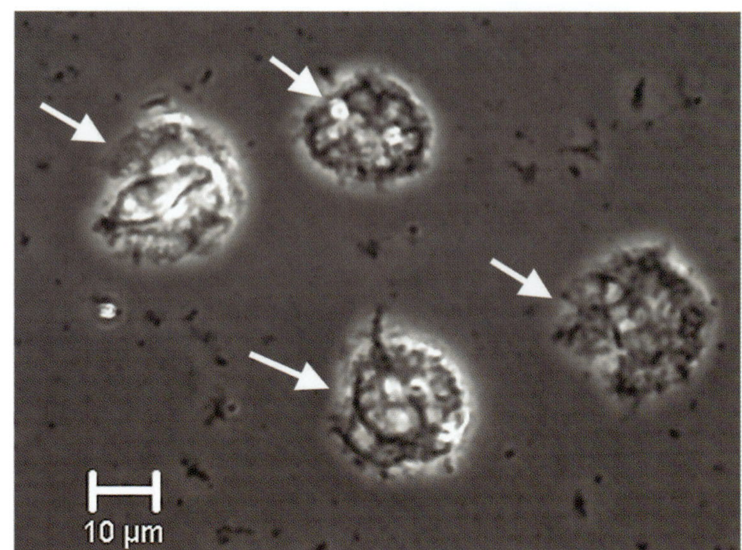

A

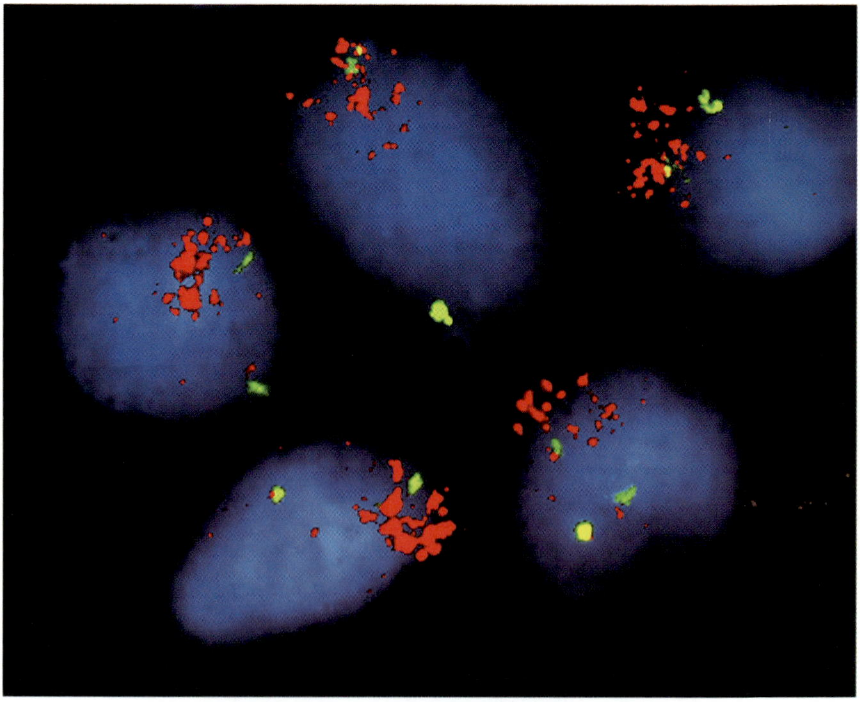

B

Color Plate 13, Fig. 3. FISH of LCM-prepared nuclei. Naked nuclei as photographed by phase contrast microscopy are shown in **(A)**. (*See* full caption in Ch. 12 on p. 244 and discussion on p. 242.)

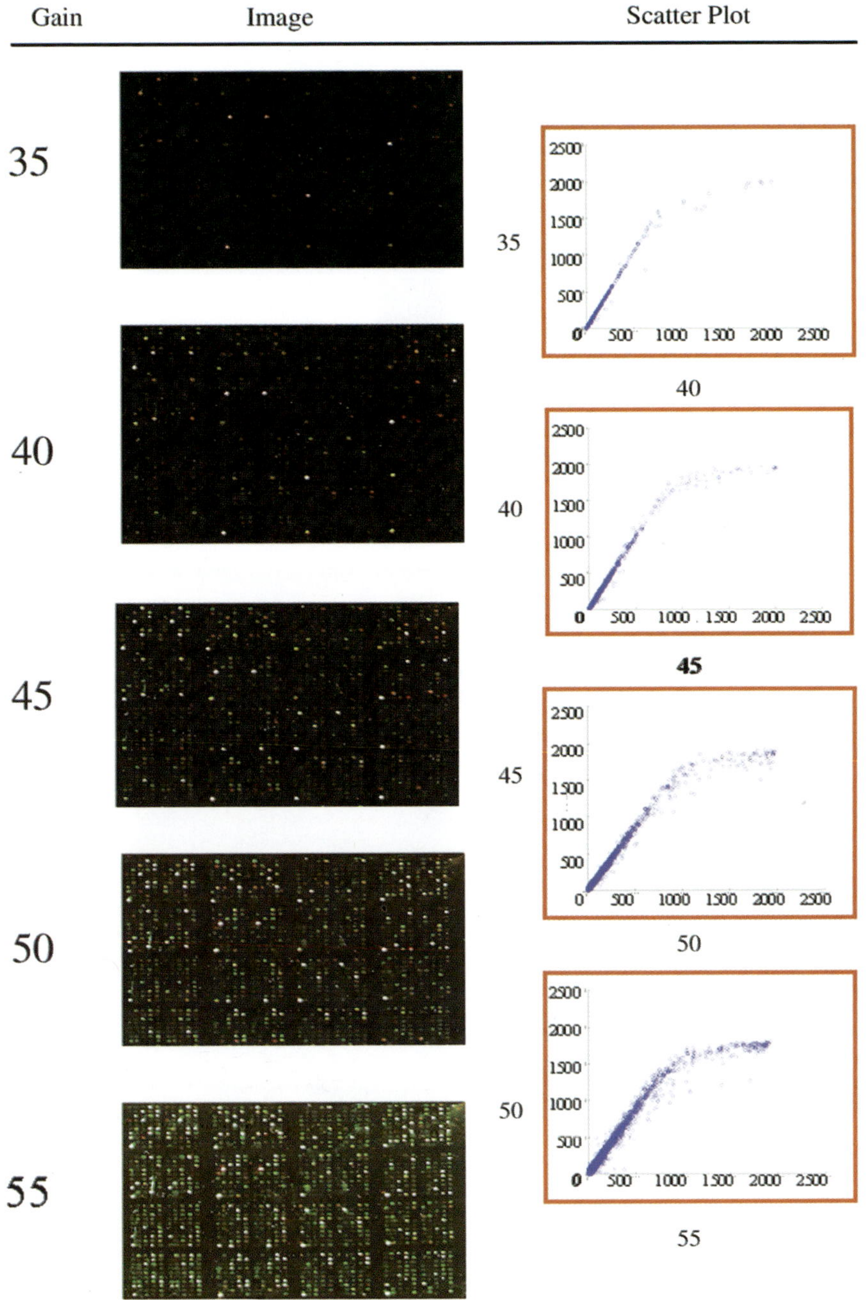

Color Plate 14, Fig. 3. Setting parameters for microarray image scanning: images for different gains and corresponding scatterplots. (*See* full caption in Ch. 13 on p. 272 and discussion on p. 271.)

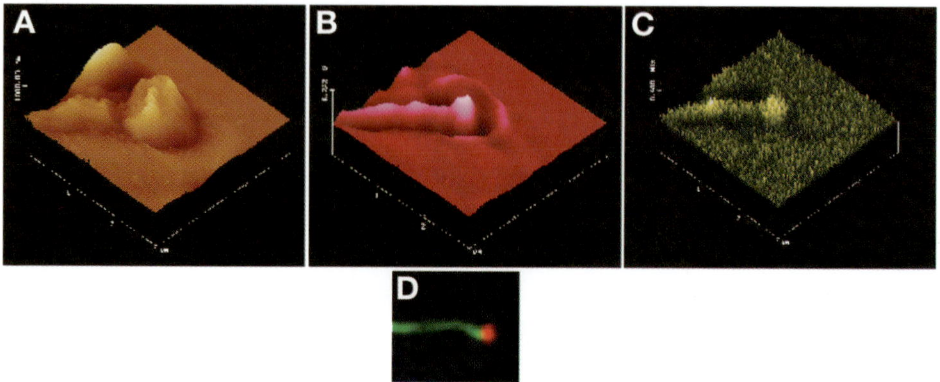

Color Plate 15, Fig. 6. NSOM images of the telomeric region of a human meiotic chromosome core after immunostaining of TRF2 by Cy3-labeled antibodies. (*See* full caption in Ch. 14 on p. 288 and discussion on p. 287.)

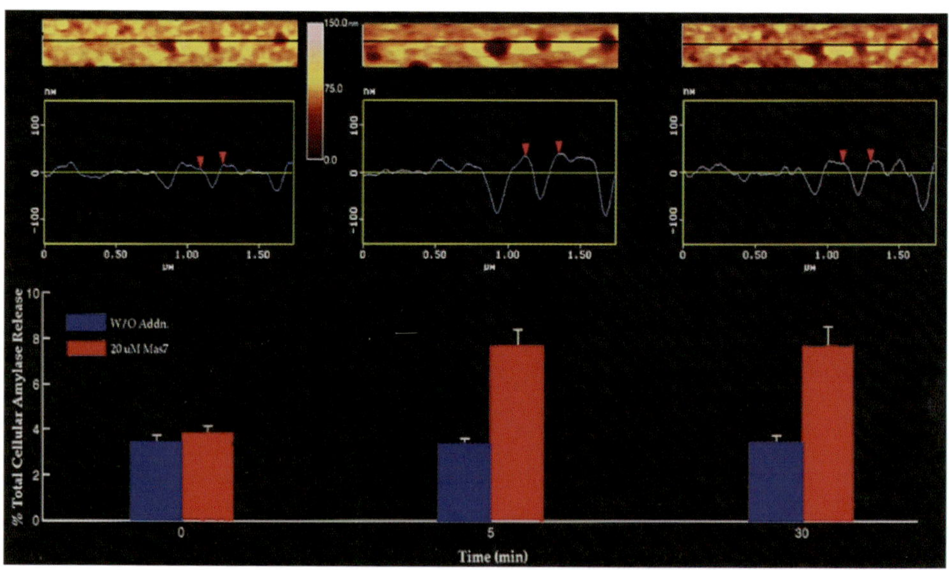

Color Plate 16, Fig. 3. Porosomes: dynamics of depressions following stimulation of secretion. (*See* full caption in Ch. 15 on p. 304 and discussion on p. 302.)

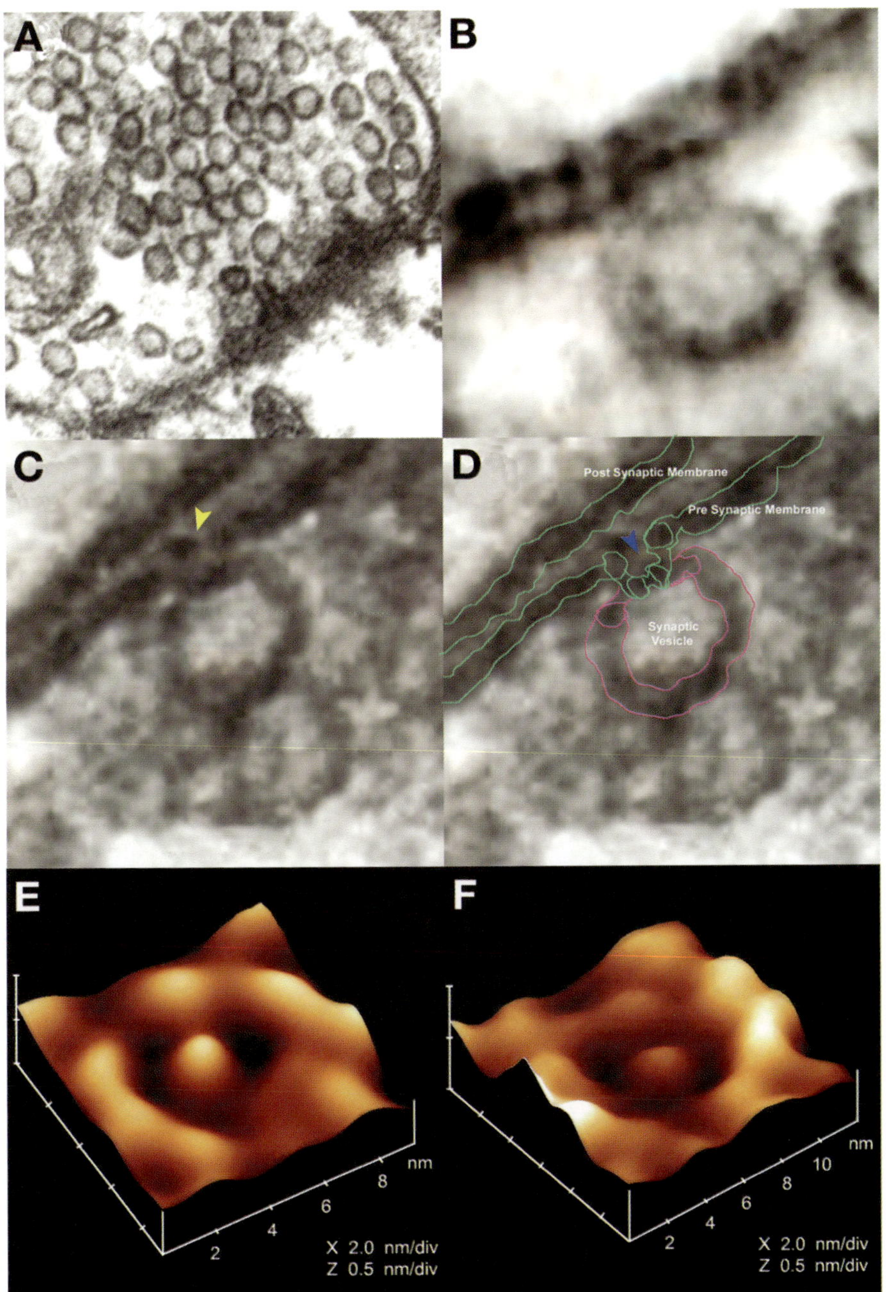

Color Plate 17, Fig. 5. Electron micrograph of porosomes in neurons. **(A–D)** AFM images of fusion pore in nerve terminal **(E)** and isolated neuronal porosome reconstituted into lipid membrane **(F)**. (*See* full caption in Ch. 15 on p. 306 and discussion on p. 304.)

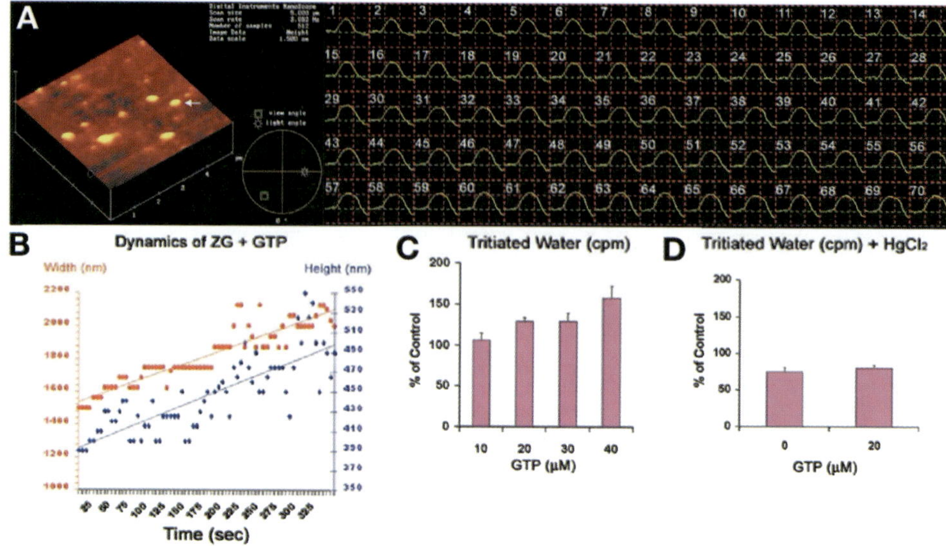

Color Plate 18, Fig. 6. Monitoring height and width of zymogen granule (arrow) after exposure to GTP. (*See* full caption and discussion in Ch. 16 on p. 325.)

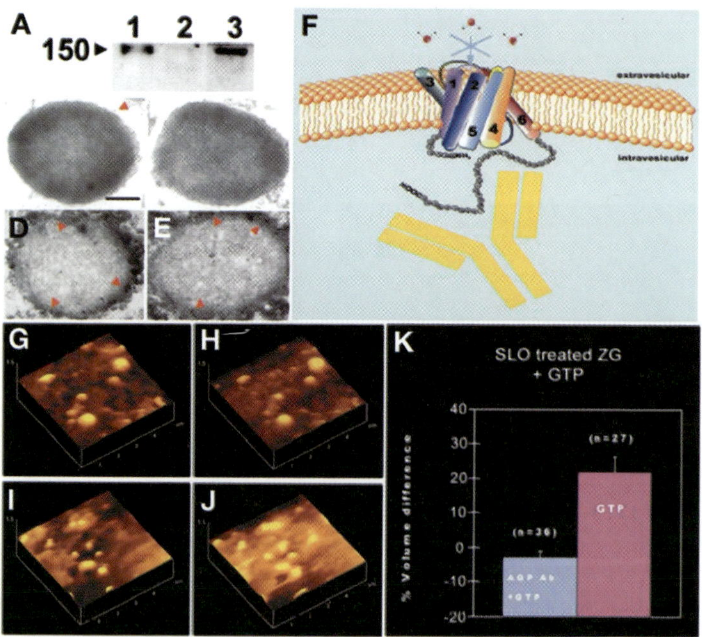

Color Plate 19, Fig. 7. Interaction of AQP1 antibody with zymogen granules. (*See* full caption in Ch. 16 on p. 327 and discussion on p. 326.)

14

Near-Field Scanning Optical Microscopy in Cell Biology and Cytogenetics

Michael Hausmann, Birgit Perner, Alexander Rapp, Leo Wollweber, Harry Scherthan, and Karl-Otto Greulich

Summary

Light microscopy has proven to be one of the most versatile analytical tools in cell biology and cytogenetics. The growing spectrum of scientific knowledge demands a continuous improvement of the optical resolution of the instruments. In far-field light microscopy, the attainable resolution is dictated by the limit of diffraction, which, in practice, is about 250 nm for high-numerical-aperture objective lenses. Near-field scanning optical microscopy (NSOM) was the first technique that has overcome this limit up to about one order of magnitude. Typically, the resolution range below 100 nm is accessed for biological applications. Using appropriately designed scanning probes allows for obtaining an extremely small near-field light excitation volume (some tens of nanometers in diameter). Because of the reduction of background illumination, high contrast imaging becomes feasible for light transmission and fluorescence microscopy. The height of the scanning probe is controlled by atomic force interactions between the specimen surface and the probe tip. The control signal can be used for the production of a topographic (nonoptical) image that can be acquired simultaneously. In this chapter, the principle of NSOM is described with respect to biological applications. A brief overview of some requirements in biology and applications described in the literature are given. Practical advice is focused on instruments with aperture-type illumination probes. Preparation protocols focussing on NSOM of cell surfaces and chromosomes are presented.

Key Words: Near-field scanning optical microscopy; NSOM; applications in biology; cell surfaces; metaphase chromosomes; meiotic chromosomes

1. Introduction

1.1. Diffraction Limit in Far-Field Light Microscopy

In biological settings, the usually applied microscope techniques (bright field, phase contrast, epifluorescence, confocal laser scanning, etc.) make use of the so-called far-field light microscopy. This means that the imaging process is

From: *Methods in Molecular Biology, vol. 319: Cell Imaging Techniques: Methods and Protocols*
Edited by: D. J. Taatjes and B. T. Mossman © Humana Press Inc., Totowa, NJ

determined by the physical behavior of *propagating* electromagnetic waves for specimen illumination and detection. Compared to the wavelength of light (typically in the range 400–800 nm), the distances between illumination source or detector system, respectively, and the specimen are hundreds of micrometers up to millimeters, that is they are large ("far field").

Because of the finite aperture of the microscope optics, light waves are diffracted, resulting in the well-known effect that an infinitesimal small pointlike object appears as a spread image of minimum diameter. The image intensity distribution of a pointlike object is given by the point spread function (PSF), from which a measure of spatial resolution can be defined for far-field optical microscopes ***(1)***. The classical diffraction limited optical resolution is given by

$$D = 0.61\left(\frac{\lambda}{\mathrm{NA}}\right)$$

(2,3) where λ is the wavelength of the light, NA is the numerical aperture of the objective lens, and *D* is the smallest distance between two pointlike objects that are imaged separately. A closely corresponding value also used as a measure of resolution is the full width at half-maximum of the PSF ***(4)***.

In addition to the physical constraints of the microscope optics, the real optical conditions of the specimen have also considerable influence on the attainable resolution in practice. Thus, in routine biological applications, the spatial resolution of far-field light microscopy is limited at about 250 nm ***(5,6)***. During the last two decades, several approaches have been described to overcome these imaging restrictions of far-field light microscopy ***(7)***. The first technique that successfully surpassed the diffraction limit has been near-field scanning optical microscopy (NSOM) (for a review, *see* **refs.** ***8*** and ***9***). Meanwhile, NSOM techniques have become a special discipline in optics, in which many laboratories develop highly sophisticated instruments and investigate their physical behavior (*see,* e.g., the 71 articles dicussed in Journal of Microscopy vol. 202, ***10***).

1.2. Principle of Near-Field Scanning Optical Microscopy

The principle of NSOM has been discovered and rediscovered several times. More than 70 yr ago, the theory of an instrument very similar in construction to modern NSOM systems was discussed ***(11)*** as a possibility for surpassing the diffraction limit of resolution ***(12)***. After the first experimental achievement of microscopic imaging with nearly atomic resolution by nonoptical scanning techniques (scanning tunneling microscopy, atomic force microscopy) ***(13,14)*** NSOM was realized ***(15,16)***.

Similar to atomic force microscopy, a sharp probe physically scans the sample surface in NSOM. The height movement of the scanning probe is controlled by the atomic force interaction between sample and probe tip, which generates

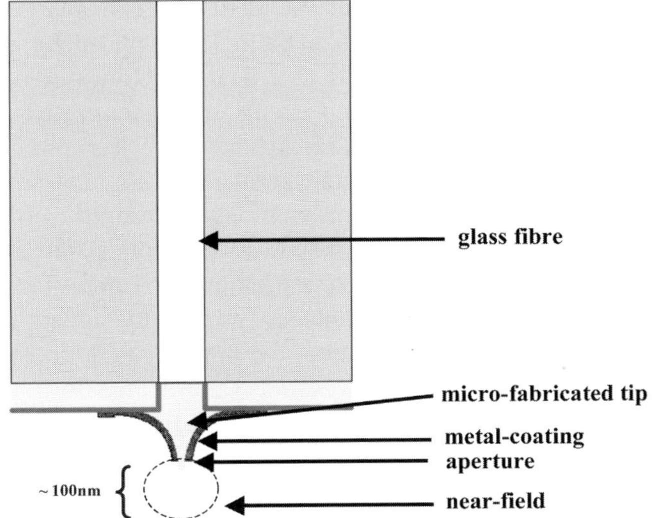

Fig. 1. Schematic representation of a NSOM probe.

a nonoptical image of the specimen topography. For biological applications, the most frequently applied NSOM probe consists of a metal (aluminum)-coated tapered optical fiber with a small aperture typically 20–130 nm in diameter at the end (*see* **Note 1**). This probe is used as a light source in optics, similar to the way the headset of a Walkman is used as a sound source in acoustics. In the latter case, the sound source has dimensions in centimeters, which is much smaller than the wavelengths of sound. Thus, the real acoustic figures and timber can only be heard if the ear is very close to the sound source (i.e., in a range much smaller than the wavelength).

In a NSOM probe tip, the incident light wave coming from a laser light source via a glass fiber is propagating into a funnel of dimensions below the diffraction limit (*see* **Fig. 1**). The light emitted by the aperture is, therefore, predominantly composed of *evanescent* waves rather than propagating waves. The intensity of evanescent waves decays exponentially with the distance from the NSOM probe tip. The region around the tip in which a significant intensity level of the evanescent wave can be detected is called the *near field* (*see* **Fig. 1**). The near-field region typically has dimensions less than 100 nm. This limits the tip-to-sample distance to considerably less than 100 nm in order to obtain high-intensity evanescent waves.

In principle, the NSOM probe tip illuminating the sample can also be used to detect the light emitted from the sample. This means that the probe tip works in the same way a doctor's stethoscope is used to detect sound generated by a lung or a heart. In this case, sound waves with a wavelength of meters are localized with an accuracy of centimeters. Therefore, the NSOM has occasionally been termed an "optical stethoscope" ***(15)***. Nevertheless, in biological applications, near-field illumination systems are more frequent than near-field detection systems.

In the case of near-field illumination systems, the light emitted by the NSOM tip is absorbed by the sample or by fluorochromes of the sample. The light subsequently transmitted through the sample (transmitted light) or emitted by the fluorochromes (fluorescent light) is predominantly composed of propagating waves and can be collected by conventional far-field optics and sensitive detectors. However, the NSOM probe is scanning the object point by point, so that the acquired image is the result of a sequence of intensity signals obtained from each precisely localized point. Therefore, the optical resolution is determined by the size of the illumination point. Because the near-field intensity decays rapidly with the probe distance, the aperture size is an appropriate parameter for estimating the optical resolution ***(17)*** (*see* **Note 2**). According to our experience, in practice, sub-hundred-nanometer resolution is feasible for many biological applications if the aperture has a diameter of 80 nm or less.

1.3. Requirements and Applications in Biology

A NSOM image usually records a small section of a sample (typically less than 10 μm × 10 μm) with high topographic and high optical resolution (better than 100 nm). To obtain an overview of the sample and preparation conditions, it is often helpful to image the same sample with lower magnification and resolution. This requires the addition of a far-field microscope. Because image detection of NSOM is based on far-field optics, the scanning unit can be adapted to a far-field microscope, so that both imaging modes can be run with the same instrument under identical specimen conditions (*see* **Note 3**).

Near-field scanning optical microscopic imaging is a combination of topographic (atomic force) imaging and high-resolution optical (light transmission, fluorescence) imaging. Both nonoptical and optical images are acquired simultaneously. Similar to atomic force microscopy, it is essential that the probe-to-sample distance (typically in the range of 10 nm) is accurately controlled by using a force feedback loop. The optical image, however, is very sensitive to small axial movements of the tip, because the near-field illumination intensity decays exponentially with the tip-to-sample distance. Therefore, the most commonly applied control mechanism in NSOM is, in contrast to atomic force microscopy, the so-called shear-force feedback control. The NSOM probe is mounted into a piezoelectric tube so that it oscillates at its resonance frequency in a lateral vibrational mode with an amplitude typically less than 1 nm. In close proximity to the sample, the shear forces dampen this motion and induce a measurable change in the oscillation amplitude and phase, which can be used as a control signal in an electronic feedback system and for the generation of the topographic image.

The physical constraints of the shear-force feedback control require dry specimens, which seems to preclude biological applications in which a fluid environment is of central importance for structure conservation. However, structure-conserving drying procedures are known from electron microscopy *(18)* and can be modified in an appropriate way (*see also* **Subheading 3.2.**) (*see* **Note 4**). The requirement for specimen drying in NSOM fluorescence imaging imposes restrictions on the choice of fluorochromes that can be used for labeling. Furthermore, in dry samples, photobleaching might increase as a result of the direct contact with oxygen from air.

Despite these so far existing limitations, NSOM has been utilized in many applications in different fields of biology and has proven a powerful technique when surface structures or structures very near to the surface need to be visualized by means of light microscopy at high spatial resolution and contrast. So far, NSOM has, for instance, been applied for the analysis of the following:

1. Cytoskeletal actin *(19)*.
2. Green fluorescent proteins in bacteria *(20)*.
3. Cortical neurons *(21)*.
4. Membranes of erythrocytes *(22)* and skin fibroblasts *(23)*.
5. Cell surfaces of mouse fibroblasts *(24)* and lymphocytes *(25)*.
6. Cell surfaces of human breast tumor cells *(26,27)* and cardiomyocytes *(28)*.
7. Metaphase chromosomes after G-banding *(29,30)* and fluorescence *in situ* hybridization (FISH) labeling *(31–34)*.
8. Meiotic chromosomes *(33)*.
9. DNA conformation *(35,36)*.

Especially in single-molecule fluorescence studies on cells and subcellular components (e.g., single green fluorescent protein photophysics) *(19,32, 36–39)*, NSOM has a superior advantage over far-field light microscopy because the topographic (nonoptical) image simultaneously acquired with the fluorescence image allows a precise localization of the fluorescence signal on the specimen.

2. Materials

2.1. Instrumentation

1. Inverted far-field microscope with Hg-illumination system and appropriate filters for fluorescence microscopy (*see* **Notes 5** and **6**).
2. Charge-coupled device (CCD) camera and a frame grabber board (if necessary) for far-field image detection.
3. NSOM unit (*see* **Fig. 2**) implemented in the far-field microscope.
4. Motor-driven stage and sensitive detectors for light detection in near-filed imaging (photomultiplier, avalanche diode).

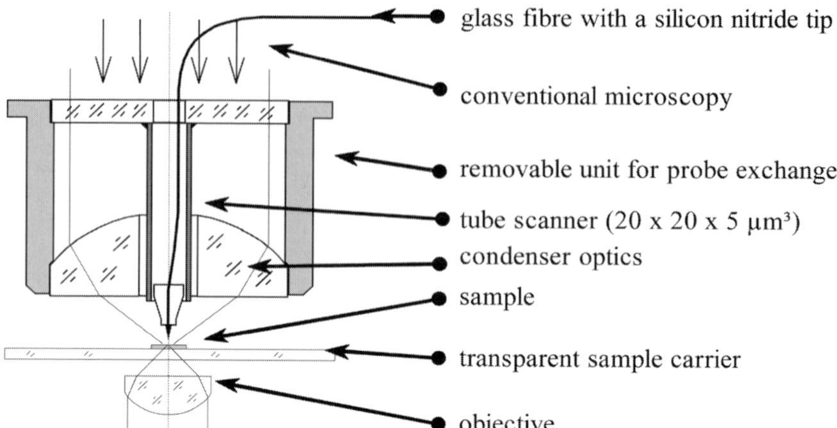

Fig. 2. Schematic representation of a NSOM unit implemented in the microscope condenser as used in a ß-type SNOM 210 from Carl Zeiss Jena (*see also* **ref.** ***33***).

5. Laser module (He–Ne laser, Ar laser, etc.) with an acousto-optical tunable filter (AOTF) to select the appropriate laser wavelength and to tune the intensity.
6. Laser fiber unit to couple the light into the NSOM illumination fiber (*see* **Fig. 1**).
7. NSOM control unit.
8. Computer with software to drive the NSOM unit.
9. Image detection and evaluation software for far-field images obtained by the CCD camera.
10. Image detection and evaluation software for near-field images obtained by point detectors (optical near-field signal) and the shear-force feedback control unit (atomic force signal).

Because the aforementioned components can be obtained from various companies, we do not indicate a special brand in this section.

2.2. Solutions

1. PBS (phosphate-buffered saline).
2. 3.7% and 4% Formaldehyde prepared from fresh paraformaldehyde.
3. Ethanol (70%, 80%, 85%, 90%, 95%, 100%).
4. HMDS (hexamethyldisilazane).
5. Formamide (70%).
6. 2X SSC, 0.5X SSC (standard sodium citrate).
7. 4X SSC/0.2% Tween-20.
8. 1X PBD (phosphate-buffered detergent).
9. 10 m*M* HCl.
10. MEM medium (Life Technologies) / 0.5% mammalian protease inhibitor (Sigma).
11. 1% Lipsol.
12. Fixative I: 3.7% acid-free formaldehyde, 0.1 *M* sucrose, pH 7.4.

13. PBS/0.5% Triton X-100.
14. PBTG: PBS, 0.1% Tween-20, 0.2% bovine serum albumin (BSA), 0.1% fish gelatin.

3. Methods

3.1. Image Acquisition

The operations of the different instruments differ in many details and have to be adapted in detail to a particular setting. Therefore, the following protocol only describes essential steps. Further details have to be taken from the manufacturer's operating instructions for the given instrument.

1. Mount the NSOM probe into the NSOM unit of the microscope (*see* **Note 7**).
2. Connect the glass fiber to the laser unit and adjust the fiber so that enough light intensity is coupled into the fiber.
3. Measure the resonance frequency of the NSOM probe (*see* **Note 8**).
4. Adjust the shear-force feedback control to this frequency.
5. Move the specimen into the center of the far-field image.
6. Approach the NSOM probe to the specimen surface (*see* **Note 9**).
7. Select the field of detection.
8. Make a fast scan in order to adjust the detector gain.
9. Scan the specimen (*see* **Note 10**).

The following typical examples for the application of NSOM were performed on a SNOM 210 in the β-type version (Carl Zeiss Jena GmbH, Digital Instruments Veeco GmbH). The piezo scanning unit was integrated into the microscope condenser of an Axiovert 135 microscope. Microfabricated probes (Institut für Mikrotechnik Mainz) with silicon nitride tips coated with aluminum were mounted in a shear-force sensor support. The probes typically had an aperture of 80–100 nm. The instrument was equipped with an argon-ion laser (λ = 458 nm, 488 nm) and two He–Ne lasers (λ = 543 nm, 633 nm) for near-field illumination. The illumination intensity was tuned by an AOTF for each laser wavelength independently. Fluorescence or transmission light was detected in air by an Achroplan long-distance objective 40×/NA 0.6 and transferred to a photomultiplier or an avalanche photodiode, respectively, using appropriate filter settings. The instrument was controlled by the NanoScope IIIa controller. Further details of the instrument are described in **ref.** ***33***.

3.2. NSOM of Cell Surfaces (25,27)

For the examples shown here (*see* **Figs. 3,4**), two types of cell systems were used:

1. Peritoneal cells (mouse lymphocytes and macrophages) dropped on glass slides.
2. Breast cancer cells of the cell line T-47D (ATCC HTB 133) (*see* **Note 11**) grown on chambered glass slides.

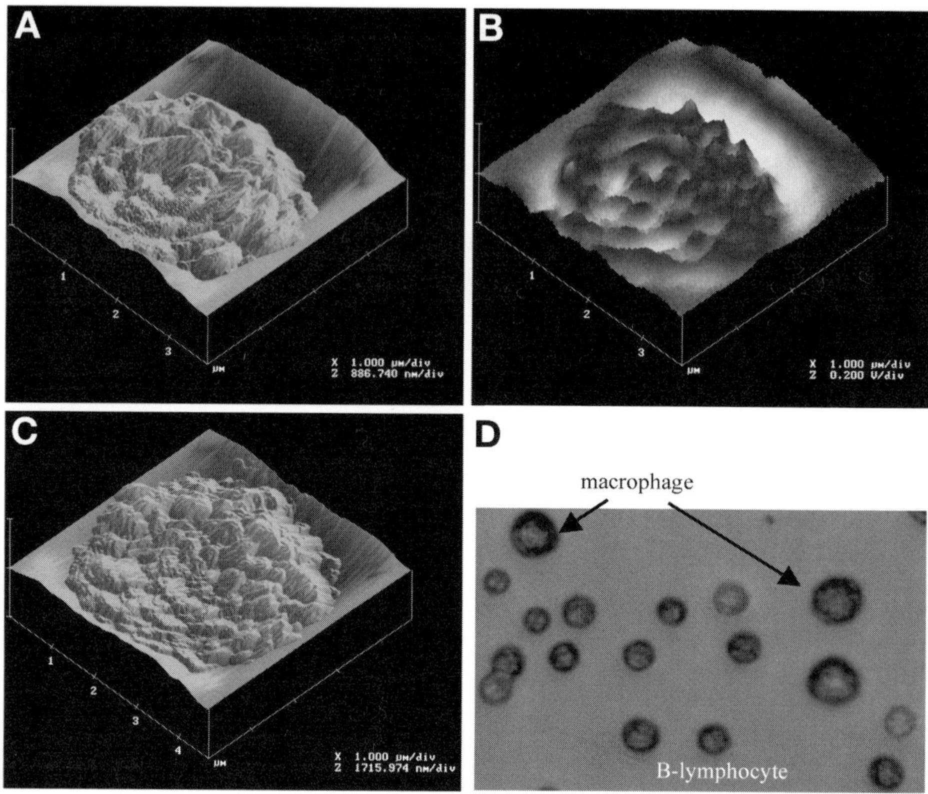

Fig. 3. NSOM images of an antibody-labeled mouse B-lymphocyte **(A,B)** and a mouse macrophage **(C)** of the same preparation. In the topographic images **(A,C)**, the different structures of the cell surfaces are visible; in the optical image **(B)**, regions of high Cy3 fluorescence intensity are shown as dark gray regions. In **(D)**, a far-field image of the same specimen is shown. The macrophages are larger than the B-lymphocytes. (Part of this figure was taken from **ref.** ***25*** with permission of SPIE.)

Prior to NSOM, the following preparation steps have to be done:

1. Seed and grow cells on ethanol-cleaned standard slides or cover glasses.
2. Remove cell culture medium.
3. Wash twice in PBS.
4. Fix cells with 4% formaldehyde in PBS for 15 min at 4°C.
5. Wash in PBS for 15 min at room temperature.
6. Optional: Label cell membrane with fluorescent antibodies (*see* **Note 12**).
7. Wash 15 min in PBS.
8. Dehydrate by an ethanol series (70%, 80%, 90%, 100%) at room temperature for 5 min each (*see* **Note 13**).

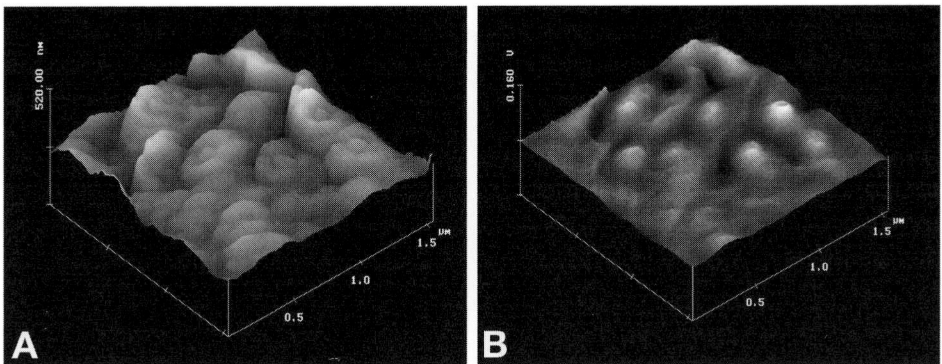

Fig. 4. Surface sections (1.5 µm x 1.5 µm) of a breast cancer cell of the line T-47D are shown. In the topographic image **(A)**, typical rosettelike cell surface structures are visible; in the optical image **(B)**, the light transmission image obtained for the He–Ne laser at 543 nm is shown (dark = high absorbance, light = high transmission). (From **ref.** ***25*** with permission of SPIE.)

9. Expose to HMDS (*see* **Note 13**). HMDS is known to reduce surface tension and to induce crosslinking in proteins.
10. Dry specimen at room temperature.

As an example, surfaces of peritoneal cells (mouse) (*see* **Fig. 3**) and breast tumor cells (human) (*see* **Fig. 4**) were visualized by NSOM 210. The scans were performed with a velocity of the NSOM probe of less than 1 µm/s. The images were processed (flattened, low-pass filtered) and visualized in three-dimensional topographic false color plots using the NanoScope IIIa software (version 4.42r1) running under Windows on a PC (**Figs. 3A–C** and **4**). Because the instrument is based on a standard Zeiss Axiovert microscope, far-field imaging (*see* **Fig. 3D**) was also possible using a CCD camera and a frame grabber board (Scion-Imaging). Image recording was controlled by the Scion Imaging software running under Windows on a PC.

The images obtained (*see* **Figs. 3** and **4**) show typical rosettelike and cylinderlike structures on the cell surface in the sub-hundred-nanometer range. Such studies will allow one to investigate the variations in surface morphology on a single-cell level, (e.g., after chemical or pharmacological treatment) (e.g., ***27***).

3.3. NSOM of Metaphase Chromosomes After FISH (29,33,34)

Fluorescence *in situ* hybridization ***(40,41)*** has become a routine technique in biomedical research and clinical diagnostics. For NSOM, standard FISH techniques can be applied. However, after specific DNA labeling, the specimen has to be dried carefully, which precludes the use of certain fluorochromes in NSOM applications.

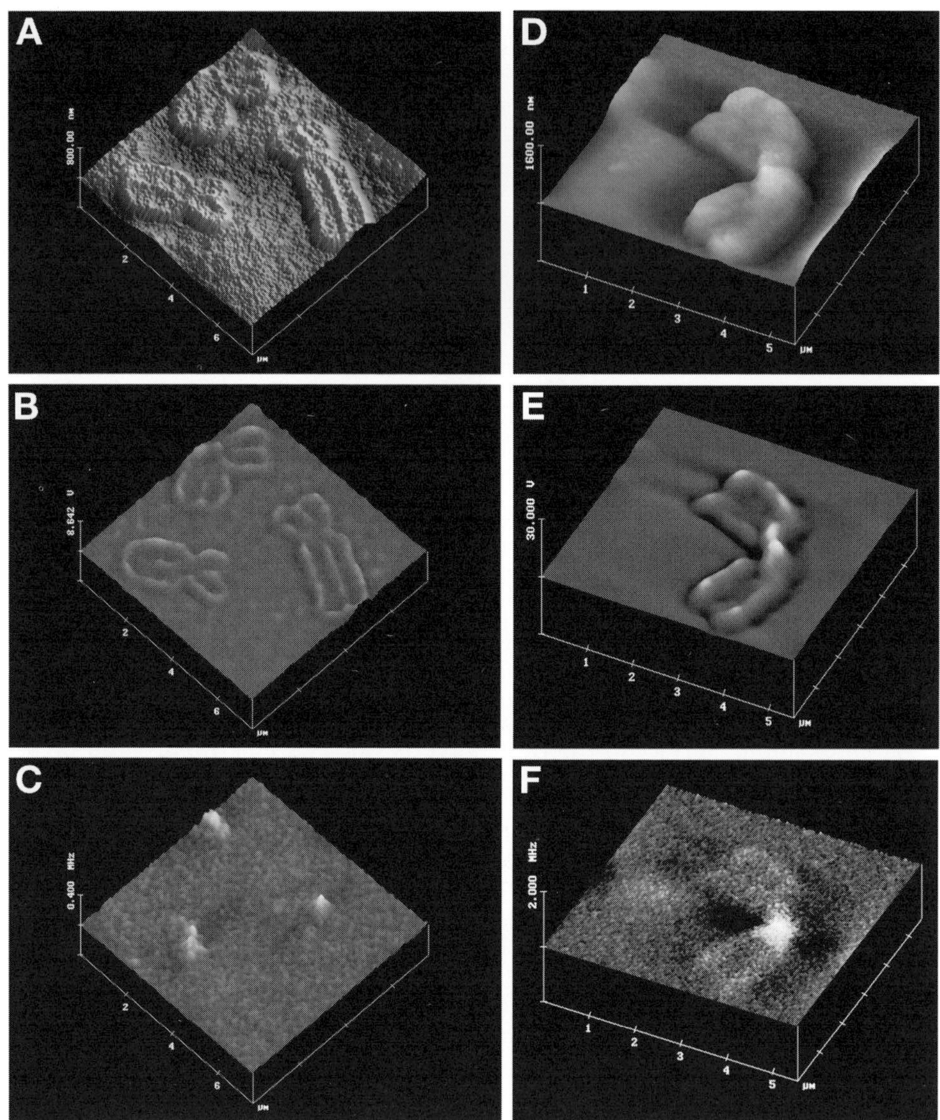

Fig. 5. NSOM images of human metaphase chromosomes after "standard" FISH **(A–C)** and low-temperature FISH **(D–F)** (*see* **ref. *34***) with a Cy3-labeled centromeric DNA probe. **(A)** The topographic near-field image reveals that chromosome arms display a collapsed structure, particularly of the central region of the chromatids, leaving higher rims at the chromosomal border. **(B)** NSOM transmitted light image at 543 nm showing a high intensity at the chromosomal border only, which verifies the topographic impression. **(C)** NSOM fluorescence image of Cy3 fluorescence in the centromeric region detected via a 590-nm long-pass filter. The high fluorescence intensity corresponds to the absorption at

The following protocol is based on a commercially available protocol:

1. Obtain chromosome spreads by standard fixation techniques (e.g., methanol/acetic acid fixation, formaldehyde fixation, etc.) (*see* **Note 14**).
2. Prepare and label respective DNA probe and dissolve in hybridization solution according to the protocol of the manufacturer of a labelling kit of your choice (*see* **Note 15**).
3. Denature preparation in 70% formamide/2X SSC at 70°C for 2 min, and DNA probe for 3 min at 95°C.
4. Dehydrate preparations through an ethanol series (70%, 80%, 95%) at 4°C for 2 min each.
5. Add DNA probe mixture to the slide.
6. Cover the hybridization mixture and preparations with a coverglass, seal with rubber cement, and hybridize in a humidified chamber at 37°C overnight.
7. Remove the seal and coverglass.
8. Wash in 0.5X SSC (pH 7.0) at 72°C for 5 min.
9. Transfer preparations to 1X PBD at room temperature.
10. Detect the hybrid molecules by incubation with fluorochrome-labeled secondary antibodies (*see* **Note 16**).
11. Counterstaining of the chromosomes is not necessary.
12. Rinse slides three times 2 min each in 1X PBD.
13. Air-dry preparations at room temperature.

This standard FISH technique involves denaturation of the probe and target DNA at temperatures ≥70°C, thereby creating single-stranded DNA molecules that form hybrids upon reassociation of the complementary DNA sequences. A typical NSOM image series of thermally denatured metaphase chromosome preparations and specific fluorescent labeling of all centromeres by FISH is shown in **Fig. 5**. The granularity of the chromosome morphology (see **Fig. 5A–C**) indicates that thermal denaturation processes of standard FISH techniques might have detrimental effects on chromatin organization. Therefore, an alternative to standard FISH protocols was introduced that omits denaturing chemical agents and thermal treatment ("low-temperature FISH") *(42)*. The following protocol is particularly suitable for detecting repetitive centromere probes:

1. Obtain chromosome spreads and DNA probe as described above (*see* **Note 14**).
2. Treat specimen with RNase in 2X SSC at 37°C for 1 h.

Fig. 5. *(Continued)* the centromere in (b). **(D–F)** Human metaphase chromosome 1 after visualization of the 1q12 pericentromeric region by "low-temperature" FISH using the DNA probe pUC1.77. **(D)** Topographic near-field image indicates that the FISH method applied maintains a filled appearance of the chromatids. **(E)** NSOM transmitted light image at 543 nm. High transmission is apparent along the chromatids. **(F)** NSOM fluorescence image (detected via a 590-nm long-pass filter) reveals strong signal at the hybridisation site.

3. Wash 3 min in 2X SSC at 37°C.
4. Incubate in 10 m*M* HCl for 2 min.
5. Add freshly prepared pepsin (final conc. 0.2 mg/mL) to prewarmed 0.01 *N* HCl in a Coplin jar at 37°C. Submerse preparations and incubate for 10 min.
6. Rinse the slides with distilled H_2O and wash twice in 2X SSC for 5 min.
7. Subject to a short fixation with freshly prepared 4% formaldehyde in PBS for 10 min at room temperature.
8. Wash twice again in 2X SSC.
9. Dehydrate through an ethanol series (70%, 85%, 95%) for 5 min each.
10. Air-dry preparations at room temperature.
11. Add 5 µL of the previously denatured probe mixture to the preparation and seal it with rubber cement under a plastic cover glass.
12. Hybridize for 15 h at 37°C.
13. Peel off rubber cement, float off cover glasses in 4X SSC/0.2% Tween-20 and wash preparations again for 5 min in the same solution at 37°C.
14. Incubate in 10% blocking solution in PBS or in 1.5% dry skim milk in PBS for 5 min.
15. Detect the hybrid molecules by an appropriate antibody–fluorochrome system (e.g., antidigoxigenin–Cy3 Fab fragments, or avidin–Cy3 in the case of biotin as the label of the DNA probe).
16. Wash in PBS for 10 min at room temperature.
17. Subject to another ethanol series (70%, 85%, 95%) for 5 min each.
18. Counterstaining of the chromosomes is not necessary for NSOM.
19. Air-dry the sample at room temperature.

A typical example of a metaphase chromosome after low-temperature FISH *(42)* is shown in **Fig. 5D–F**. In this case, chromosome 1 was labeled with a subcentromeric DNA probe (pUC1.77; *see* **Note 17**). The chromosome appears more voluminous with a "smooth" (i.e., less granular) surface as compared to thermal denaturation (compare **Fig. 5A,D**). After standard FISH, the chromosomes are generally very flat; signal and background intensities are at about the same levels. In contrast, an intensive transmitted light signal was restricted to the chromatids of the low-temperature FISH chromosomes, indicating an intensive near-field object interaction (*see* **Fig. 5E**). The fluorescence image shows a clearly visible labelling site at the centromeric region (*see* **Fig. 5F**).

3.4. NSOM of Meiotic Chromosomes (33)

Meiosis is the cell division type that contributes to sexual reproduction by reducing the diploid chromosome number to the haploid. Meiotic chromosomes display a unique organization in that they exhibit a proteinaceuos chromosome core along replicated sister chromatids, and the DNA emanates as loops from these cores. Cores of homologous chromosomes are connected during pachytene by a protein zipper called the synaptonemal complex (SC). The ends of the SCs are capped by the telomeres, which (in addition to other aspects) play an important role in the chromosome-pairing process at first meiotic prophase. So far, little is known about the relative distribution of telomere proteins within ends of the meiotic chromosome cores. NSOM might be a means to gain a better insight into the spatial organization of the meiotic telomere. The following (for details, *see* **ref. *43***) describes the preparation of meiotic chromosome spreads for NSOM:

1. Mince fresh or frozen testicular tissue in MEM medium/0.5% mammalian protease inhibitor (Sigma) at 4°C.
2. Remove tissue pieces and place a drop of the suspension on clean, aminosilane-coated glass slides (Super plus; Menzel Gläser).
3. Mix 50 µL of cell suspension and 150 µL of 1% Lipsol on a slide.
4. Briefly tilt the slide after each step to mix the solutions evenly within the resulting drop.
5. After 5 min, add fixative I and distribute the suspension evenly by streaking with the side of a pipet tip over the slide without touching the surface.
6. Fix specimen by adding 200 µL fixative I. Allow to air-dry in a fume hood.
7. Wash for 30 min with 0.5% Triton X-100/PBS.
8. Wash with PBS.
9. Incubate for 10 min in PBTG.
10. Incubate at 4°C overnight with 100 µL PBTG containing the primary antibody under a cover slip in a humid chamber (*see* **Note 18**).
11. Rinse three times 3 min with PBTG at 37°C.
12. Incubate for 30 min at 37°C with secondary, fluorochrome-conjugated antibody of choice (*see* **Note 19**).
13. Wash three times 3 min in PBS.
14. Dehydrate through an ethanol series (70%, 85%, 95% for 5 min each).
15. Air-dry the sample at room temperature.

For an example of successful labeling at a meiotic telomere, see **Fig. 6** and Color Plate 15, following p. 274. The meandering parts of the SC are visible in the topographic and light transmission image. From the fluorescence image, TRF2 signals can be localized in the telomere knob.

3.5. Discussion and Perspective

Many applications have shown that the intrinsic advantages of NSOM make this technique extremely useful for high-resolution imaging in life sciences. Nevertheless, it seems that NSOM is rarely applied in biology. Some features might be responsible for this situation: First, compared to far-field light microscopy, NSOM is a complex technique that requires detailed training of the operator before images of good quality can be obtained. Second, most near-field microscopes are devices in experimental physics laboratories and not widely accessible to the biologist. The commercially available instruments are primarily designed for measurements of solid-state physics and are not always

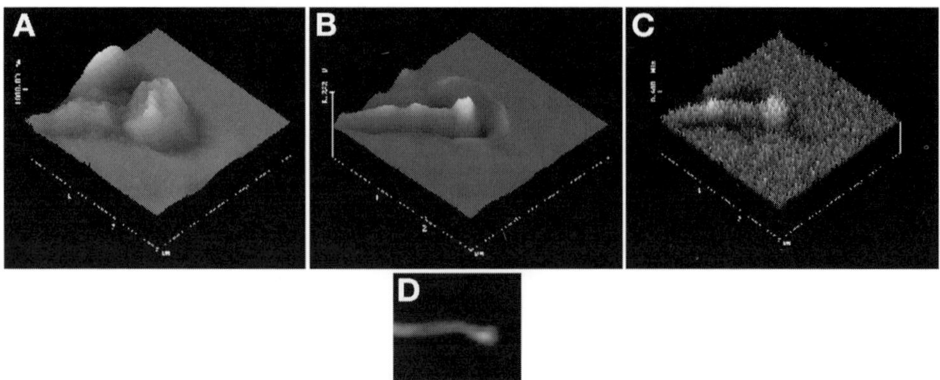

Fig. 6. NSOM images of the telomeric region of a human meiotic chromosome core after immunostaining of TRF2 by Cy3-labeled antibodies. **(A)** The topographic image shows a thickened telomere at the end of the core (called the attachment plate); **(B)** a telomeric knob as regions of higher intensity of transmitted light at 543 nm; **(C)** Cy3 fluorescence of the TRF2 antibody labeling site, which was recorded via a 590-nm low-pass filter; **(D)** far-field fluorescence image of the synaptonemal complex and the TRF2 label at the telomere (*see* Color Plate 15, following p. 274).

adapted to the requirements of biological experiments—for instance, the use of standard glass slides as specimen carrier. Third, the resolution and sensitivity of NSOM depends not only on the quality of the probe but also on the quality of the specimen. Especially, the drying process required because of the shear-force feedback control has to be done with great care in order to conserve the native supra molecular structures of the specimen. Despite great efforts of several groups ***(44–46)***, it remains a technical challenge to develop an instrument that works in a liquid (i.e., more physiological environment) and allows for near-field analysis of "soft objects."

4. Notes

1. The scanning probe is the most critical element of the complete technique. It is difficult to produce such probes in such a way that NSOM imaging can reproducibly be performed for the same specimen. Therefore, a reliable procedure to produce large quantities of high-quality probes is microfabrication of metal-coated silicon oxide or silicon nitride tips that are glued on glass fibers. However, examples of biological applications have been described using non-metal-coated aperture-type probes. Other types of probe referring to the so-called "apertureless" types are described in the literature, but, to our knowledge, they have not been applied for any routine application in biology.
2. As in far-field light microscopy, in practice the optical resolution of NSOM strongly depends on the experimental conditions applied (NSOM probe, light

intensity, optical conditions of the sample, sample topography, etc.). Structurally conserved surfaces of cells or organelles often show gross topological changes (typically several micrometers). In this case, the topography might modulate the optical signal because the axial tip-to-sample distance is not always the shortest tip-to-sample distance that controls the height movement of the NSOM probe. This effect can modulate the optical resolution within the image and could also cause artifacts in the image ***(47,48)***. However, theoretical approaches exist that allow for an approximate determination of the resolution of NSOM also from images of biological samples ***(49)***.

3. The combination of far-field and near-field microscopy in one instrument is an advantage for the user, especially for biological applications, if switching between both instrumentation modes can be done without changing the sample or sample carrier. Therefore, it is of practical importance in biological applications that the standard glass slides or cover glasses, which are routinely applied in far-field light microscopy, can also be used for specimen preparation under NSOM conditions.
4. The shear-force feedback control reacts also on transparent surface material. Hence, care has to be taken not to cover the specimen by some material that should not be imaged.
5. The microscope has to be protected against low-frequency vibrations of the surroundings by means of an active optical table.
6. If standard glass slides are used for far-field and near-field microscopy, long-distance objectives are useful to focus through the slide.
7. Several manufacturers offer NSOM probes. They are usually delivered ready to use for a given instrument. In all cases, special care has to be taken in handling the NSOM probe in order not to destroy the probe tip.
8. Each NSOM probe has its own resonance frequency that has to be determined after changing the NSOM probe. Sometimes more than one resonance can be detected. In this case, the frequency that displays the highest resonance amplitude should be utilized for the NSOM control.
9. By visual far-field microscopy, the NSOM probe can be manually adjusted into a focal plane above the specimen surface. The final approach of the NSOM tip should be done by the instrument under shear-force control.
10. Most biological samples are so-called soft samples (compared to solid-state surfaces) with large height modulations (typically in the micrometer range). Therefore, a very slow scan velocity is recommended (less than 1 µm/s).
11. Breast cancer cells of the cell line T-47D (ATCC HTB 133) were grown on chambered glass slides (Nalge Nunc International, Naperville, IL USA) and cultivated in RPMI 1640 cell culture medium with 10% fetal calf serum at 37°C with 5% CO_2 in a humidified atmosphere. Three to four days before reaching a confluent cell monolayer, the medium was changed to phenol red-free RPMI 1640, including 5% charcoal-stripped fetal calf serum, 1% antibiotic/antimycotic solution (GIbco BRL/Life Technologies), 2 m*M* glutamine, and 1% nonessential amino acids solution. After 6 d of further cultivation, aliquots of the cells were treated for

48 h with 17ß-estradiol (Sigma-Aldrich Chemie GmbH, Deisenhofen, Germany) at concentrations of 5×10^{-9} *M*, 5×10^{-7} *M*, or 5×10^{-5} *M*. The 17ß-estradiol stock solution was prepared in ethanol ***(27)***.

12. Mouse peritoneal cells: Immunoglobulin G on the mouse lymphocytes was visualized by biotinylated goat–anti-mouse IgG antibodies and Cy3-conjugated streptavidin. The incubation time was 15 min for each step. Cy3 is well suited for NSOM because of its high photostability (*see* **Subheading 1.3.**).
13. The duration of the ethanol and HMDS exposure is very critical on the conservation of cell surface structures. The appropriate exposure times have to be tested experimentally. For the examples described here, the optimum was as follows:

	Duration of each ethanol step	Duration of HMDS exposure
Fresh peritoneal cells	2.5 min	1.0 min
Cancer cells of T-47D	5.0 min	3.0 min

14. For the experiments presented here, lymphocytes were prepared from fresh peripheral blood and stimulated with phytohemagglutinin M to grow for 72 h. The cells were synchronized and arrested in mitosis by a colcemid block for the last 2 h of cultivation. After hypotonic treatment with prewarmed KCl (75 m*M*), the cells were fixed with cold methanol/acetic acid (3:1, v/v). Metaphase chromosomes and interphase cell nuclei were spread on precleaned slides. After evaporation of the fixative, the slides were stored in 100% ethanol at 4°C. Prior to FISH, they were rinsed with 100% ethanol and air-dried.
15. A commercially available DNA probe (Appligene Oncor) specific for all centromeres was used. This probe was labeled with digoxigenin. According to the manufacturer´s instruction, 1.5 µL of probe DNA was denatured in 30 µL Oncor Hybrisol VI (containing 50% formamide) at 72°C for 5 min.
16. For detection of the hybrid molecules, 60 µL of Cy3-labeled antidigoxigenin was applied and incubated under a plastic cover slip at 37°C for 45 min.
17. A DNA probe (pUC 1.77) specific for the region q12 on chromosome 1 was used for the experiments shown in **Fig. 5D–F**. The pUC 1.77 probe was labeled with biotin-11-dUTP. Five microliters of DNA probe (about 100 ng) were diluted with 3 µL 20X SSC, 3 µL of 10X HCl-Tris, and 19 µL of H_2O and denatured at 95°C for 4 min. The hybridization solution was cooled down to approx 60°C and kept at this temperature until use.
18. To our current knowledge, the NSOM data described here and in **ref.** ***33*** are the very first experiments in which NSOM imaging was performed to visualize telomere proteins of meiotic chromosomes. The labeling pattern shown was obtained with different primary antibodies, such as rabbit anti-human TRF1 or rabbit anti-human TRF2 ***(43)***.
19. For the detection of the primary rabbit antibodies consecutive incubations with a secondary biotinylated anti-rabbit antibody and avidin–Cy3 were performed.

The Cy3 fluorochrome proved effective in all NSOM experiments and was superior to FITC or Cy5.

Acknowledgments

Funding by the BMBF (Bundesminister für Bildung und Forschung) and the instrumental support of Carl Zeiss Jena GmbH are gratefully acknowledged. H.S. is grateful to C. Heyting, Wageningen, NL and T. de Lange, RU, New York, USA, for help with the antibodies and acknowledges support from the DFG (grant no. Sche350/8-3). The authors are indebted to J. Beuthan and C. Dressler, Berlin, for providing the breast cancer cells and for stimulating discussions. The authors thank H. Dittmar, G. Günther, B. Lanick, IMB, Jena, and M. Jerratsch, University of Kaiserslautern for technical assistance. The work of M.H. was partly supported by a NCI/CCR Intramural Research Award to S. Janz.

References

1. Abbe, E. (1873) Beiträge zur Theorie des Mikroskops und der mikroskopischen Wahrnehmung. *Arch. Mikrosk. Anat.* **9,** 413–468.
2. Lord Rayleigh, F. R. S. (1879) Investigation in optics, with special reference to the spectroscope. *Philos. Mag.* **8,** 261–274.
3. Born, M. and Wolf, E. (1970) *Principles in Optics*, 4th ed., Pergamon, Oxford, pp. 414–419.
4. Stelzer, E. H. K. (1998) Contrast, resolution, pixelation, dynamic range and signal-to-noise ratio: fundamental limits to resolution in fluorescence microscopy. *J. Microsc.* **189,** 15–24.
5. Kozubek, M. (2001) Theoretical versus experimental resolution in optical microscopy. *Microsc. Res. Tech.* **53,** 157–166.
6. Edelmann, P., Esa, A., Hausmann, M., and Cremer, C. (1999) Confocal laser-scanning fluorescence microscopy: in situ determination of the confocal point-spread function and the chromatic shifts in intact cell nuclei. *Optik* **110,** 194–198.
7. Sarikaya, M. (1992) Evolution of resolution in microscopy. *Ultramicroscopy* **47,** 1–14.
8. Zhang, P., Kopelman, R., and Tan, W. (2000) Subwavelength optical microscopy and spectroscopy using near-field optics. *Crit. Rev. Solid State Mater. Sci.* **25,** 87–162.
9. De Lange, F., Cambi, A., Huijbens, R., et al. (2001) Cell biology beyond the diffraction limit: near-field scanning optical microscopy. *J. Cell Sci.* **114,** 4153–4160.
10. Wilson, T., ed. Journal of microscopy, vol. **202,** 1–450.
11. Synge, E. H. (1928) A suggested method for extending microscopic resolution into the ultra-microscopic region. *Philos. Mag.* **6,** 356–362.
12. McCutchen, C. W. (1967) Superresolution in microscopy and the Abbe resolution limit. *J. Opt. Soc. Am.* **57,** 1190–1192.
13. Binnig, G., Rohrer, H., Gerber, C., and Weibel, E. (1982) Surface studies by scanning tunnelling microscopy. *Phys. Rev. Lett.* **49,** 57–61.
14. Binnig, G., Quate, C. F., and Gerber C. (1986) Atomic force microscope. *Phys. Rev. Lett.* **56,** 930–933.

15. Pohl, D. W., Denk, W., and Lanz, M. (1984) Optical stethoscopy: image recording with resolution λ/20. *Appl. Phys. Lett.* **44,** 651–653.
16. Betzig, E., Lewis, A., Harootuniam, A., Isaacson, M., and Kratschmer, E. (1986) Near-field scanning optical microscopy (NSOM): development and biophysical applications. *Biophys. J.* **49,** 269–279.
17. Dürig, U., Pohl, D. W., and Rohner, F. (1986) Near-field optical-scanning microscopy. *J. Appl. Phys.* **59,** 3318–3327.
18. Boyde, A. (1980) Review of basic preparation techniques for biological scanning electron microscopy, in *Electron Microscopy, Vol II* (Brederoo, P. and de Priester, W. eds.), Electron Microcopy Foundation, Leiden, pp. 768–777.
19. Betzig, E. and Chichester, R. J. (1993) Single molecules observed by near-field scanning optical microscopy. *Science* **262,** 1422–1425.
20. Subramaniam, V., Kirsch, A. K., and Jovin, T. M. (1998) Cell biological applications of scanning near-field optical microscopy (SNOM). *Cell Mol. Biol.* **44,** 689–700.
21. Talley, C. E., Cooksey, G. A., and Dunn, R. C. (1996) High-resolution fluorescence imaging with cantilevered near-field fiber optic probes. *Appl. Phys. Lett.* **69,** 3809–3811.
22. Enderle, T., Ha, T., Ogletree, D. F., Chemla, D. S., Magowan C., and Weiss, S. (1997) Membrane specific mapping and colocalization of malarial and host skeletal proteins in the *Plasmodium falciparum* infected erythrocyte by dual-color near-field scanning optical microscopy. *Proc. Natl. Acad. Sci. USA* **94,** 520–525.
23. Hwang, J., Gheber, L. A., Margolis, L., and Edidin, M. (1998) Domains in cell plasma membranes investigated by near-field scanning optical microscopy. *Biophys. J.* **74,** 2184–2190.
24. Kirsch, A. K., Subramaniam, V., Jenei, A., and Jovin, T. M. (1999) Fluorescence resonance energy transfer detected by scanning near-field optical microscopy. *J. Microsc.* **194,** 448–454.
25. Perner, B., Hausmann, M., Wollweber, L., Rapp, A., Monajembashi, S., and Greulich, K. O. (2000) Scanning near-field optical microscopy after structure conserving air-drying. *Proc. SPIE* **4164,** 10–17.
26. Nagy, P., Jenei, A., Kirsch, A. K., Szöllösi, J., Damjanovich, S., and Jovin, T. M. (1999) Activation-dependent clustering of the erbB2 receptor thyrosine kinase detected by scanning near-field optical microscopy. *J. Cell Sci.* **112,** 1733–1741.
27. Perner, B., Rapp, A., Dressler, C., et al. (2002) Variations in cell surfaces of estrogen treated breast cancer cells detected by a combined instrument for far-field and near-field microscopy. *Analyt. Cell. Pathol.* **24,** 89–100.
28. Micheletto, R., Denyer, M., Scholl, M., et al. (1999) Observation of the dynamics of live cardiomyocytes through a free-running scanning near-field optical microscopy setup. *Appl. Opt.* **38,** 6648–6652.
29. Wiegräbe, W., Monajembashi, S., Dittmar, H., et al. (1997) Scanning near-field optical microscope—a method for investigating chromosomes. *Surface Interface Anal.* **25,** 510–513.
30. Held, N., Hausmann, M., Perner, B., and Greulich, K. O. (2000) Optische Rasternahfeld-mikroskopie in der Zytogenetik. *CLB Chem. Labor Biotechn.* **51,** (9/2000), 324–327.

31. Moers, M. H. P., Kalle, W. H. J., Ruiter, A. G. T., et al. (1996) Fluorescence in situ hybridization of human metaphase chromosomes detected by near-field scanning optical microscopy. *J. Microsc.* **182,** 40–45.
32. Meixner, A. J. and Kneppe, H. (1998) Scanning near-field optical microscopy in cell biology and microbiology. *Cell. Mol. Biol.* **44,** 673–688.
33. Hausmann, M., Perner, B., Rapp, A., Scherthan, H., and Greulich, K. O. (2001) SNOM imaging of mitotic and meiotic chromosomes. *Eur. Microsc. Anal.* **71,** (5/2001), 5–7.
34. Winkler, R., Perner, B., Rapp, A., et al. (2002) Labelling quality and chromosome morphology after low temperature FISH analysed by scanning far-field and scanning near-field optical microscopy. *J. Microsc.* **209,** 23–33.
35. Ha, T., Enderle, T., Ogletree, D. F., Chemla, D. S., Selva, P. R., and Weiss, S. (1996) Probing the interaction between two single molecules: fluorescence resonance energy transfer between a single donor and a single acceptor. *Proc. Natl. Acad. Sci. USA* **93,** 6264–6268.
36. Garcia-Parajo, M. F., Veerman, J. A., Ruiter, A. G., and van Hulst, N. F. (1998) Near-field optical and shear-field microscopy of single fluorophores and DNA molecules. *Ultramicroscopy* **71,** 311–319.
37. Ambrose, W. P., Affleck, R. L., Goodwin, P. M., et al. (1995) Imaging of biological molecules with single molecule sensitivity using near-field scanning optical microscopy. *Exp. Tech. Phys.* **41,** 237–248.
38. Garcia-Parajo, M. F., Veerman, J. A., Segers-Nolten, G. M., de Grooth, B., Greve, J., and van Hulst, N. F. (1999) Visualising individual green fluorescent proteins with a near-field optical microscope. *Cytometry* **36,** 239–246.
39. Van Hulst, N. F., Veerman, J. A., Garcia-Parajo, M. F., and Kuipers, L. (2000) Analysis of individual (macro)molecules and proteins using near-field optics. *J. Chem. Phys.* **112,** 7799–7810.
40. Clark, M. (ed.) (1996) *In Situ Hybridization.* Chapman & Hall, Weinheim.
41. Van der Ploeg, M. (2000) Cytochemical nucleic acid research during the twentieth century. *Eur. J. Histochem.* **44,** 7–42.
42. Durm, M., Haar, F. -M., Hausmann, M., Ludwig, H., and Cremer, C. (1997) Nonenzymatic, low temperature fluorescence in situ hybridization of human chromosomes with a repetitive α-satellite probe. *Z. Naturforsch.* **52c,** 82–88.
43. Scherthan, H., Jarratsch, M., Li, B., et al. (2000) Mammalian meiotic telomeres: protein composition and redistribution in relation to nuclear pores. *Mol. Biol. Cell* **11,** 4189–4203.
44. Lambelet, P., Pfeffer, M., Sayah, A., and Marquis-Waible, F. (1998) Reduction of tip–sample interaction forces for scanning near-field optical microscopy in a liquid environment, *Ultramicroscopy* **71,** 117–121.
45. Mannelquist, A., Iwamoto, H., Szabo, G., and Shao, Z. (2001) Near-field optical microscopy with a vibrating probe in aqueous solution. *Appl. Phys. Lett.* **78,** 2076–2078.
46. Mannelquist, A., Iwamoto, H., Szabo, G., and Shao, Z. (2002) Near-field optical microscopy in aqueous solution: implementation and characterization of a vibrating probe. *J. Microsc.* **205,** 53–60.

47. Hecht, B., Bielefeldt, H., Inouye, Y., and Pohl, D. W. (1997) Facts and artifacts in near-field optical microscopy. *J. Appl. Phys.* **81,** 2492–2498.
48. Kalkbrenner, T., Graf, M., Durkan, C., Mlynek, J., and Sandoghdar, V. (2000) High-contrast topography-free sample for near-field optical microscopy. *Appl. Phys. Lett.* **76,** 1206–1208.
49. Beuthan, J., Hausmann, M., Minet, O., Perner, B., Dressler, C., and Eberle, H. G. (2001) Approximative determination of the modulation transfer function of the scanning near field microscope using biological samples. *Techn. Messen.* **3/2001,** 127–130.

15

Porosome

The Fusion Pore Revealed by Multiple Imaging Modalities

Bhanu P. Jena

Summary

Secretion occurs in all cells of multicellular organisms and involves the delivery of secretory products packaged in membrane-bound vesicles to the cell exterior. Specialized cells for neurotransmission, enzyme secretion, or hormone release utilize a highly regulated secretory process. Secretory vesicles are transported to specific sites at the plasma membrane, where they dock and fuse to release their contents. Similar to other cellular processes, cell secretion is found to be highly regulated and a precisely orchestrated event. It has been demonstrated that membrane-bound secretory vesicles dock and fuse at porosomes, which are specialized supramolecular structures at the cell plasma membrane. Swelling of secretory vesicles results in a buildup of pressure, allowing expulsion of intravesicular contents. The extent of secretory vesicle swelling dictates the amount of intravesicular contents expelled during secretion. The discovery of the porosome, its isolation, its structure and dynamics at nanometer resolution and in real time, and its biochemical composition and functional reconstitution into artificial lipid membrane have been determined. The molecular mechanism of secretory vesicle fusion at the base of porosomes and vesicle swelling have also been resolved. These findings reveal the molecular machinery and mechanism of cell secretion. In this chapter, the discovery of the porosome, its isolation, its structure and dynamics at nanometer resolution and in real time, and its biochemical composition and functional reconstitution into artificial lipid membrane are discussed.

Key Words: Fusion pore; atomic force microscopy; electron microscopy.

1. Introduction

Secretion and membrane fusion are fundamental cellular processes regulating endoplasmic reticulum-Golgi transport, plasma membrane recycling, cell division, sexual reproduction, acid secretion, and the release of enzymes, hormones, and neurotransmitters, to name just a few. It is, therefore, no surprise that defects in secretion and membrane fusion give rise to diseases like diabetes,

From: *Methods in Molecular Biology, vol. 319: Cell Imaging Techniques: Methods and Protocols*
Edited by: D. J. Taatjes and B. T. Mossman © Humana Press Inc., Totowa, NJ

Alzheimer's, Parkinson's, acute gastroduodenal diseases, gastroesophageal reflux disease, intestinal infections due to inhibition of gastric acid secretion, biliary diseases resulting from malfunction of secretion from hepatocytes, polycystic ovarian disease as a result of altered gonadotropin secretion, and Gitelman disease associated with growth hormone deficiency and disturbances in vasopressin secretion are only a few examples. Understanding cellular secretion and membrane fusion helps not only to advance our understanding of these vital cellular and physiological processes, but in the development of drugs also to help ameliorate secretory defects, provide insight into our understanding of cellular entry and exit of viruses and other pathogens, and in the development of smart drug delivery systems. Therefore, secretion and membrane fusion play an important role in health and disease. Studies ***(1–21)***, in the last decade demonstrate that membrane-bound secretory vesicles dock and transiently fuse at the base of specialized plasma membrane structures called porosomes or fusion pores, to expel vesicular contents. These studies further demonstrate that during secretion, secretory vesicles swell, enabling the expulsion of intravesicular contents through porosomes ***(16,19–21)***. With these findings ***(1–21)***, a new understanding of cell secretion has emerged and confirmed by a number of laboratories ***(22–27)***.

Throughout history, the development of new imaging tools has provided new insights into our perceptions of the living world and profoundly impacted human health. The invention of the light microscope almost 300 yr ago was the first catalyst, propelling us into the era of modern biology and medicine. Using the light microscope, a giant step into the gates of modern medicine was made by the discovery of the unit of life—the cell. The structure and morphology of normal and diseased cells and of disease-causing microorganisms were revealed for the first time using the light microscope. Then, in 1938, with the birth of the electron microscope (EM), dawned a new era in biology and medicine. Through the mid-1940s and 1950s, a number of subcellular organelles were discovered and their functions determined using the EM. Viruses, the new life-forms were discovered and observed for the first time and implicated in diseases ranging from the common cold to acquired immune disease (aquired immune deficiency syndrome [AIDS]). Despite the capability of the EM to image biological samples at near-nanometer resolution, sample processing (fixation, dehydration, staining) results in morphological alterations and was a major concern. Then, in the mid-1980s, scanning probe microscopy evolved ***(1,28)***, further extending our perception of the living world to the near-atomic realm. One such scanning probe microscope, the atomic force microscope (AFM), has helped overcome both limitations of light and electron microscopy, enabling determination of the structure and dynamics of single biomolecules and live cells in three dimensions, at

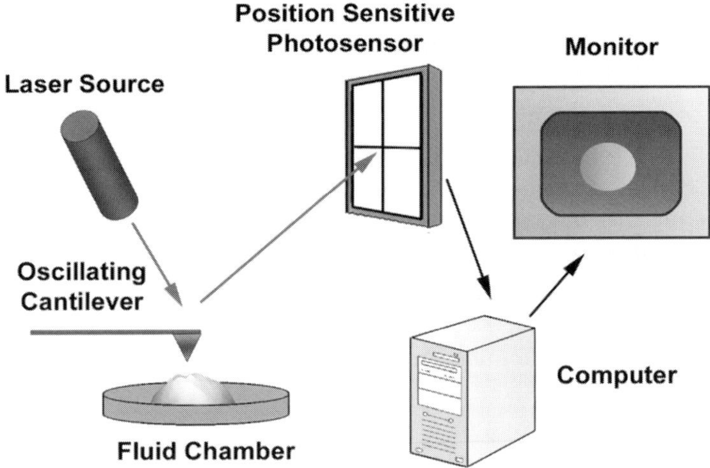

Fig. 1. Schematic diagram depicting key components of an atomic force microscope. (From **ref. *12***.)

near-angstrom resolution. This unique capability of the AFM has given rise to a new discipline of "nanobioscience," heralding a new era in biology and medicine. Using AFM in combination with conventional tools and techniques, this past decade has witnessed advances in our understanding of cell secretion ***(1–21)*** and membrane fusion ***(9,17,18,29)***, as noted earlier in the chapter.

The resolving power of the light microscope is dependent on the wavelength of the light used and, therefore, 250–300 nm in lateral and much less in depth resolution can be achieved at best. The porosome or fusion pore in live secretory cells are cup-shaped structures, measuring 100–150 nm at its opening and 15–30 nm in relative depth in the exocrime pancreas, and just 10 nm at the presynaptic membrane of the nerve terminal. As a result, it had evaded visual detection until its discovery using the AFM ***(3–8,15)***. The development of the AFM ***(28)*** has enabled the imaging of live cells in physiological buffer at nanometer to subnanometer resolution. In AFM, a probe tip microfabricated from silicon or silicon nitride and mounted on a cantilever spring is used to scan the surface of the sample at a constant force. Either the probe or the sample can be precisely moved in a raster pattern using a xyz piezotube to scan the surface of the sample (*see* **Fig. 1**). The deflection of the cantilever, measured optically, is used to generate an isoforce relief of the sample ***(30)***. Thus, force is used to image surface profiles of objects by the AFM, allowing imaging of live cells and subcellular structures submerged in physiological buffer solutions. To image live cells, the scanning probe of the AFM operates in physiological buffers and can do so under two modes: contact or tapping. In the contact mode, the probe is in direct contact with the sample surface as it scans at a constant vertical force. Although high-resolution AFM images can be obtained in this mode of AFM operation, sample height information generated might not be accurate because the vertical scanning force could depress the soft cell. However, information on the viscoelastic properties of the cell and the spring constant of the cantilever enables measurement of the cell height. In the tapping mode, on the other hand, the cantilever resonates and the tip makes brief contacts with the sample. In the tapping mode in fluid, lateral forces are virtually negligible. It is therefore important that the topology of living cells be obtained using both contact and tapping modes of AFM operation in fluid. The scanning rate of the tip over the sample also plays an important role on the quality of the image. Because cells are soft samples, a high scanning rate would influence its shape. Hence, a slow tip movement over the cell would be ideal and results in minimal distortion and better image resolution. Rapid cellular events might be further monitored by using section analysis. To examine isolated cells by the AFM, freshly cleaved mica coated with Cel-Tak has also been used with great success ***(3–8)***. Also, to obtain optimal resolution, the contents of the bathing medium as well as the cell surface to be scanned should be devoid of any debris.

2. Methods

2.1. Isolation of Pancreatic Acinar Cells

Acinar cells for secretion experiments, light microscopy, AFM, and EM were isolated using a minor modification of a published procedure. For each experiment, a male Sprague–Dawley rat weighing 80–100 g was euthanized by CO_2 inhalation. The pancreas was dissected and diced into 0.5-mm^3 sections with a razor blade, mildly agitated for 10 min at 37°C in a siliconized glass tube with 5 mL of oxygenated buffer A (98 m*M* NaCl, 4.8 m*M* KCl, 2 m*M* $CaCl_2$, 1.2 m*M* $MgCl_2$, 0.1% bovine serum albumin, 0.01% soybean trypsin inhibitor, 25 m*M* HEPES, pH 7.4) containing 1000 units of collagenase. The suspension of acini was filtered through a 224-μm Spectra-Mesh (Spectrum Laboratory Products, Rancho Dominguez, CA) polyethylene filter to remove large clumps of acini and undissociated tissue. The acini were washed six times, 50 mL per wash, with ice-cold buffer A. Isolated rat pancreatic acini and acinar cells were plated on Cell-Tak-coated (Collaborative Biomedical Products, Bedford, MA) glass cover slips. Two to three hours after plating, cells were imaged with the AFM before and during stimulation of secretion. Isolated acinar cells and hemiacinar preparations were used in the study because fusion of secretory vesicles at the protein membrane (PM) in these cells occurs at the apical region facing the acinar lumen.

2.2. Pancreatic Plasma Membrane Preparation

Rat pancreatic PM fractions were isolated using a modification of a published method. Male Sprague–Dawley rats weighing 70–100 g were euthanized by CO_2 inhalation. The pancreases were removed and placed in ice-cold phosphate-buffered saline (PBS), pH 7.5. Adipose tissue was removed and the pancreases were diced into 0.5-mm^3 pieces using a razor blade in a few drops of homogenization buffer A (1.25 *M* sucrose, 0.01% trypsin inhibitor, and 25 m*M* HEPES, pH 6.5). The diced tissue was homogenized in 15% (w/v) ice-cold homogenization buffer A using four strokes at maximum speed of a motor-driven pestle (Wheaton overhead stirrer). One and a half milliliters of the homogenate was layered over a 125-μL cushion of 2 *M* sucrose and 500 μL of 0.3 *M* sucrose was layered onto the homogenate in Beckman centrifuge tubes. After centrifugation at 145,000*g* for 90 min in a Sorvall AH-650 rotor, the material banding between the 1.2 *M* and 0.3 *M* sucrose interface was collected and the protein concentration was determined. For each experiment, fresh PM was prepared and used the same day in all AFM experiments.

2.2.1. Isolation of Synaptosomes, Synaptosomal Membrane, and Synaptic Vesicles

Synaptosomes, synaptosomal membrane, and synaptic vesicles were prepared from rat brains ***(31,32)***. Whole rat brains from Sprague–Dawley rats (100–150 g) were isolated and placed in ice-cold buffered sucrose solution (5 m*M* HEPES, pH 7.4, 0.32 *M* sucrose) supplemented with protease inhibitor cocktail (Sigma, St. Louis, MO) and homogenized using a Teflon–glass homogenizer (8–10 strokes). The total homogenate was centrifuged for 3 min at 2500*g*. The supernatant fraction was further centrifuged for 15 min at 14,500*g*, and the resultant pellet was resuspended in buffered sucrose solution, which was loaded onto 3-10-23% Percoll gradients. After centrifugation at 28,000*g* for 6 min, the enriched synaptosomal fraction was collected at the 10–23% Percoll gradient interface. To isolate synaptic vesicles and synaptosomal membrane ***(32)***, isolated synaptosomes were diluted with 9 vol of ice-cold H_2O (hypotonic lysis of synaptosomes to release synaptic vesicles) and immediately homogenized with three strokes in Dounce homogenizer, followed by a 30-min incubation on ice. The homogenate was centrifuged for 20 min at 25,500*g*, and the resultant pellet (enriched synaptosomal membrane preparation) and supernatant (enriched synaptic vesicles preparation) were used in our studies.

2.2.2. Preparation of Lipid Membrane on Mica and Porosome Reconstitution

To prepare lipid membrane on mica for AFM studies, freshly cleaved mica disks were placed in a fluid chamber. Two hundred microliters of the bilayer bath solution, containing 140 m*M* NaCl, 10 m*M* HEPES, and 1 m*M* $CaCl_2$, were placed at the center of the cleaved mica disk. Ten microliters of the brain lipid vesicles were added to the above bath solution. The mixture was then allowed to incubate for 60 min at room temperature, before washing (10X), using 100 µL bath solution/wash. The lipid membrane on mica was imaged by the AFM before and after the addition of immunoisolated porosomes.

2.3. Atomic Force Microscopy

"Pits" and fusion pores at the PM in live pancreatic acinar secreting cells in PBS, pH 7.5, were imaged with the AFM (Bioscope III, Digital Instruments) using both contact and tapping modes. All images presented in this chapter were obtained in the "tapping" mode in fluid, using silicon nitride tips with a spring constant of 0.06 Nm, and an imaging force of <200 pN. Images were obtained at line frequencies of 1 Hz, with 512 lines per image, and constant image gains. Topographical dimensions of "pits" and fusion pores at the cell PM were analyzed using the software nanoscopeIIIa 4.43r8 supplied by Digital Instruments.

2.4. ImmunoAFM on Live Cells

Immunogold localization in live pancreatic acinar cells was assessed after 5 min stimulation of secretion with 10 µ*M* of the secretagogue mastoparan. After stimulation of secretion, the live pancreatic acinar cells in buffer were exposed to a 1 : 200 dilution of α-amylase-specific antibody (Biomeda Corp., Foster City, CA) and 30 nm colloidal gold-conjugated secondary antibody for 1 min and were washed in PBS before AFM imaging in PBS at room temperature. "Pits" and fusion pores within and at the apical end of live pancreatic acinar cells in PBS, pH 7.5, were imaged by the AFM (Bioscope III, Digital Instruments) using both contact and tapping modes. All images presented were obtained in the "tapping" mode in fluid, using silicon nitride tips as described previously.

2.5. ImmunoAFM on Fixed Cells

After stimulation of secretion with 10 µ*M* mastoparan, the live pancreatic acinar cells were fixed for 30 min using ice-cold 2.5% paraformaldehyde in PBS. Cells were then washed in PBS, followed by labeling with 1 : 200 dilution of α-amylase-specific antibody (Biomeda Corp.) and 10 nm gold-conjugated secondary antibody for 15 min, fixed, washed in PBS, and imaged in PBS with AFM at room temperature.

2.6. Isolation of Zymogen Granules

Zymogen granules (ZGs) were isolated by using a modification of the method of our published procedure. Male Sprague–Dawley rats weighing 80–100 g were euthanized by CO_2 inhalation for each ZG preparation. The pancreas was dissected and diced into 0.5-mm^3 pieces. The diced pancreas was suspended in 15% (w/v) ice-cold homogenization buffer (0.3 *M* sucrose, 25 m*M* HEPES, pH 6.5, 1 m*M* benzamidine, 0.01% soybean trypsin inhibitor) and homogenized with a Teflon-glass homogenizer. The resultant homogenate was centrifuged for 5 min at 300*g* at 4°C to obtain a supernatant fraction. One volume of the supernatant fraction was mixed with 2 vol of a Percoll–sucrose–HEPES buffer (0.3 *M* sucrose, 25 m*M* HEPES, pH 6.5, 86% Percoll, 0.01% soybean trypsin inhibitor) and centrifuged for 30 min at 16,400*g* at 4°C. Pure ZGs were obtained as a loose white pellet at the bottom of the centrifuge tube.

2.7. Transmission Electron Microscopy

Isolated rat pancreatic acini and ZGs were fixed in 2.5% buffered paraformaldehyde (PFA) for 30 min, and the pellets were embedded in Unicryl resin and were sectioned at 40–70 nm. Thin sections were transferred to coated specimen transmission electron microscope (TEM) grids, dried in the presence of uranyl acetate and methyl cellulose and examined in a TEM.

2.8. Immunoprecipitation and Western Blot Analysis

Immunoblot analysis was performed on pancreatic PM and total homogenate fractions. Protein in the fractions was estimated by the Bradford method. Pancreatic fractions were boiled in Laemmli reducing sample preparation buffer for 5 min, cooled and used for sodium dodecyl sulfate–polyacrylamide gel electrophoresis (SDS-PAGE). PM proteins were resolved in a 12.5% SDS-PAGE and electrotransferred to 0.2-µm nitrocellulose sheets for immunoblot analysis with a SNAP-23 specific antibody. The nitrocellulose was incubated for 1 h at room temperature in blocking buffer (5% nonfat milk in PBS containing 0.1% Triton X-100 and 0.02% NaN_3) and immunoblotted for 2 h at room temperature with the SNAP-23 antibody (ABR, Golden, CO). The primary antibodies were used at a dilution of 1:10,000 in blocking buffer. The immunoblotted nitrocellulose sheets were washed in PBS containing 0.1% Triton X-100 and 0.02% NaN_3 and were incubated for 1 h at room temperature in horseradish peroxidase (HRP)-conjugated secondary antibody at a dilution of 1:2000 in blocking buffer. The immunoblots were then washed in the PBS buffer, processed for enhanced chemiluminescence, and exposed to X-OMAT-AR film. To isolate the fusion complex for immunoblot analysis, SNAP-23 specific antibody conjugated to protein A–Sepharose was used. One gram of total pancreatic homogenate solubilized in Triton/Lubrol solubilization buffer (0.5% Lubrol; 1 m*M* benzamidine; 5 m*M* ATP; 5 m*M* EDTA; 0.5 % Triton X-100, in PBS) supplemented with protease inhibitor mix (Sigma, St. Louis, MO) was used. SNAP-23 antibody conjugated to the protein A–Sepharose was incubated with the solubilized homogenate for 1 h at room temperature, followed by washing with wash buffer (500 m*M* NaCl, 10 m*M* Tris-HCl, 2 m*M* EDTA, pH 7.5). The immunoprecipitated sample attached to the immuno-Sepharose beads was incubated in Laemmli sample preparation buffer, prior to 12.5% SDS-PAGE, electrotransfer to nitrocellulose, and immunoblot analysis using specific antibodies to actin (Sigma), fodrin (Santa Cruz Biotechnology Inc., Santa Cruz, CA), vimentin (Sigma, St. Louis, MO), syntaxin 2 (Alomone Labs, Jerusalem, Israel), Ca^{2+}-β3 (Alomone Labs), and Ca^{2+}-α1c (Alomone Labs).

3. Porosome: A New Cellular Structure

Earlier electrophysiological studies on mast cells suggested the existence of fusion pores at the cell PM, which became continuous with the secretory vesicle membrane following stimulation of secretion *(33)*. AFM has confirmed the existence of the fusion pore or porosome and revealed its structure and dynamics in the exocrine pancreas *(3,4,7,8)*, neuroendocrine cells *(5,6)*, and neurons *(15)*, at near-nanometer resolution and in real time.

Isolated live pancreatic acinar cells in physiological buffer, when imaged with the AFM *(3,4,7,8)*, reveal at the apical PM, a group of circular "pits" measuring 0.4–1.2 μm in diameter that contain smaller "depressions" (*see* **Fig. 2**). Each depression averages between 100 and 150 nm in diameter, and, typically, three to four depressions are located within a pit. The basolateral membrane of acinar cells is devoid of either pits or depressions. High-resolution AFM images of depressions in live cells further reveal a cone-shaped morphology. The depth of each depression cone measures 15–30 nm. Similarly, growth hormone (GH)-secreting cells of the pituitary gland and chromaffin cells, β-cells of the exocrine pancreas, mast cells, and neurons, possess depressions at their PM, suggesting their universal presence in secretory cells. Exposure of pancreatic acinar cells to a secretagogue (mastoparan) results in a time-dependent increase (20–35%) in depression diameter, followed by a return to resting size on completion of secretion *(3,4,7,8)* (*see* **Fig. 3**; *see* Color Plate 16, following p. 274). No demonstrable change in pit size is detected following stimulation of secretion *(3)*. Enlargement of depression diameter and an increase in its relative depth after exposure to secretagogues correlated with increased secretion. Conversely, exposure of pancreatic acinar cells to cytochalasin B, a fungal toxin that inhibits actin polymerization, results in a 15–20% decrease in depression size and a consequent 50–60% loss in secretion *(3)*. Results from these studies suggested depressions to be the fusion pores in pancreatic acinar cells. Furthermore, these studies demonstrate the involvement of actin in regulation of both the structure and function of depressions. Analogous to pancreatic acinar cells, examination of resting GH-secreting cells of the pituitary *(5)* and

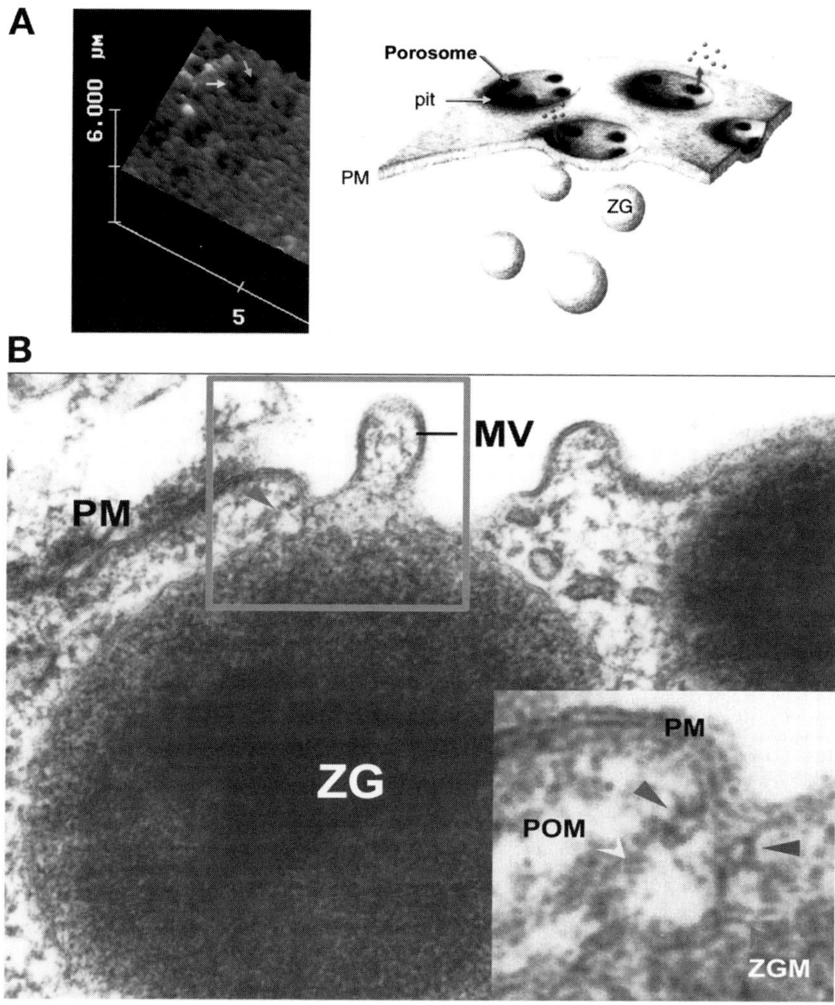

Fig. 2. **(A)** On the far left is an AFM depicting "pits" (arrow to left) and "depressions" within (right arrow), at the PM in live pancreatic acinar cells. On the right is a schematic drawing depicting depressions, at the cell PM, where membrane-bound secretory vesicles dock and fuse to release vesicular contents. **(B)** Electron micrograph depicting a porosome (arrowhead within box) close to a microvillus (MV) at the apical plasma membrane (PM) of a pancreatic acinar cell. Note association of the porosome membrane (POM) and the zymogen granule membrane (ZGM) (a gray arrowhead) of a docked zymogen granule (ZG), the membrane-bound secretory vesicle of exocrine pancreas. Also, a cross-section of the ring at the mouth of the porosome is seen (black arrowheads). (From **ref.** ***8***.)

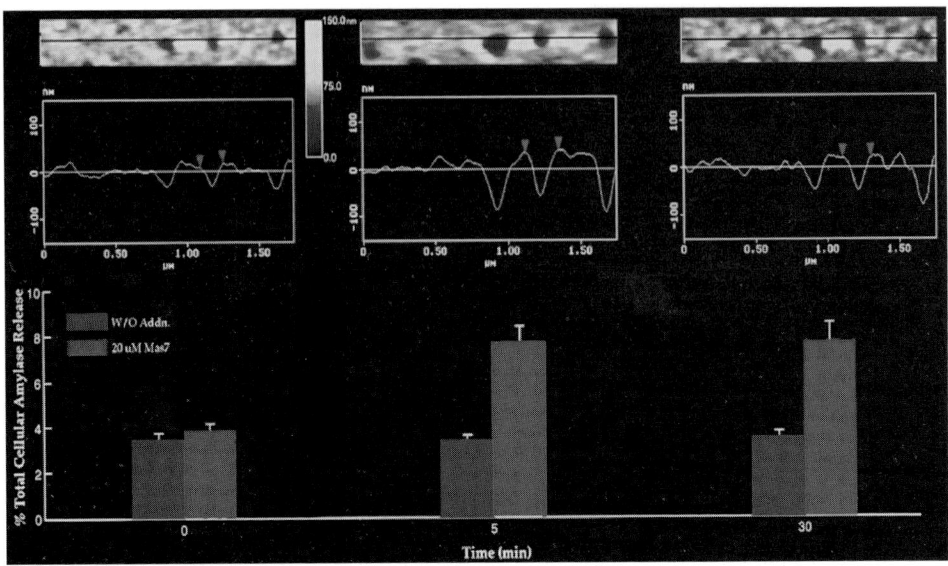

Fig. 3. Dynamics of depressions following stimulation of secretion. The top panel shows a number of depressions within a pit in a live pancreatic acinar cell. The scan line across three depressions in the top panel is represented graphically in the middle panel and defines the diameter and relative depth of the depressions; the middle depressions are represented by arrowheads. The bottom panel represents percent of total cellular amylase release in the presence and absence of the secretagogue Mas 7. Notice an increase in the diameter and depth of depressions, correlating with an increase in total cellular amylase release at 5 min after stimulation of secretion. At 30 min after stimulation of secretion, there is a decrease in diameter and depth of depressions, with no further increase in amylase release over the 5-min time-point. No significant increase in amylase secretion or diameter of depressions was observed in resting acini or those exposed to the nonstimulatory mastoparan analog Mas 17. (*See* Color Plate 16, following p. 274. From **refs.** ***3*** and ***11***.)

chromaffin cells of the adrenal medulla ***(6)*** also reveal the presence of pits and depressions at the cell PM (*see* **Fig. 4**). The presence of porosomes in neurons, β-cells of the endocrine pancreas, and in mast cells have also been demonstrated (*see* **Figs. 4** and **5** [*see* Color Plate 17, following p. 274]) ***(14,15)***. Depressions in resting GH cells measure 154 ± 4.5 nm (mean ± SE) in diameter. Exposure of GH cells to a secretagogue results in a 40% increase in depression diameter (215 ± 4.6 nm; $p < 0.01$) but no appreciable change in pit size. The enlargement of depression diameter during secretion and the known effect that actin depolymerizing agents decrease depression size and inhibit secretion ***(3)*** suggested depressions to be the fusion pores. However, a more direct determination of the function of depressions was required. This was achieved by immuno-

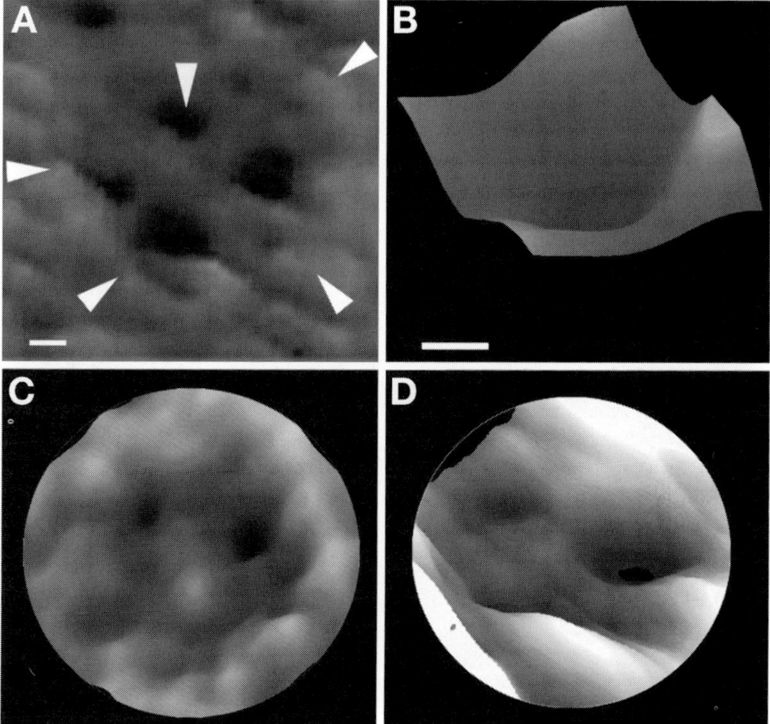

Fig. 4. AFM of depressions or porosomes or fusion pores in live secretory cell of the exocrine pancreas **(A, B)**, the GH-secreting cell of the pituitary **(C)**, and in the chromaffin cell **(D)**. Note the "pit" (arrowheads at the periphery of the structure) with four depressions (arrowhead at 12 o'clock). A high-resolution AFM of a single porosome is shown in B. Bars = 40 nm for A and B. Similarly, AFMs of pororsomes in β-cells of the endocrine pancreas have been demonstrated.

AFM studies. AFM localization at depressions of gold-conjugated antibody to a secretory protein demonstrated secretion to occur through depressions *(4,5)*. The membrane-bound secretory vesicles in exocrine pancreas called ZGs contain the starch-digesting enzyme amylase. Atomic force micrographs demonstrated localization of amylase-specific antibodies tagged with colloidal gold at depressions following stimulation of secretion *(4)* (*see* **Fig. 6**). These studies confirm depressions to be the fusion pores or porosomes in pancreatic acinar cells where membrane-bound secretory vesicles dock and fuse to release vesicular contents. Similarly, in somatotrophs of the pituitary, a gold-tagged GH-specific antibody is found to selectively localize at depressions following stimulation of secretion *(5)*, again identifying depressions in GH cells as fusion pores or porosomes. The porosome at the cytosolic side of the

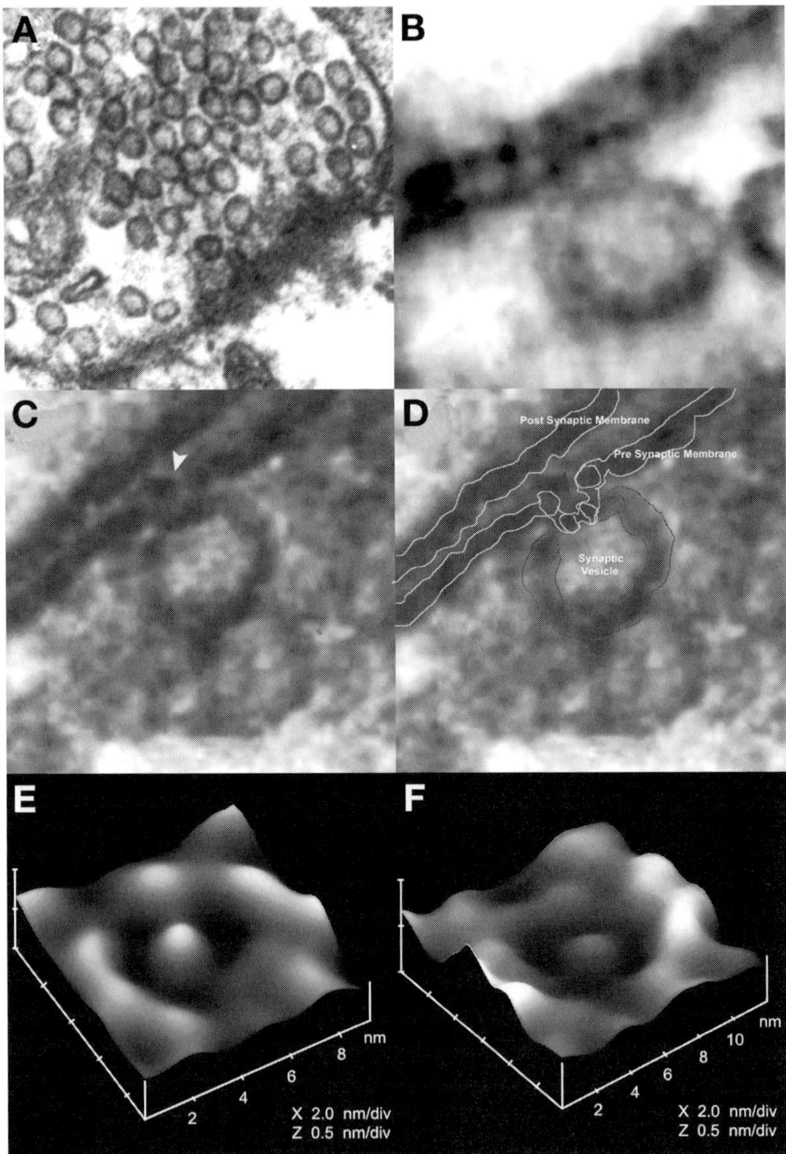

Fig. 5. Electron micrograph of porosomes in neurons. **(A)** Electron micrograph of a synaptosome demonstrating the presence of 40- to 50-nm synaptic vesicles. **(B–D)** Electron micrographs of neuronal porosomes that are 10- to 15-nm cup-shaped structures at the presynaptic membrane (yellow arrowhead), where synaptic vesicles transiently dock and fuse to release vesicular contents. **(E)** AFM of a fusion pore or porosome at the nerve terminal in live synaptosome. **(F)** AFM of isolated neuronal porosome, reconstituted into lipid membrane. (*See* Color Plate 17, following p. 274. From **ref. *15*.**)

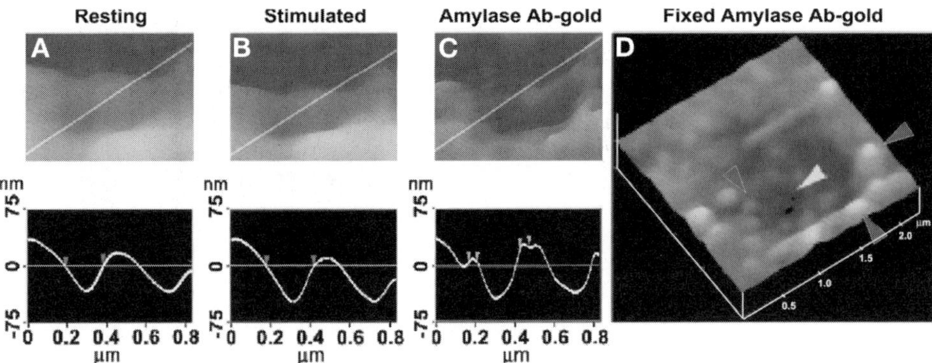

Fig. 6. Depressions are fusion pores or porosomes. Porosomes dilate to allow expulsion of vesicular contents. **(A, B)** AFM and section analysis of a pit and two out of the four fusion pores or porosomes, demonstrating enlargement following stimulation of secretion. Note the section analysis depicting the relative depth and width of the porosome in A through C. **(C)** Exposure of live cells to gold-conjugated-amylase antibody (Ab) results in specific localization of immunogold to the porosome opening. Amylase is one of the proteins within secretory vesicles of the exocrine pancreas. **(D)** AFM of a fixed pancreatic acinar cell, demonstrating a pit and porosomes within, and immunogold-labeling amylase at the site. The dark arrowhead points to immunogold clusters and the white arrowhead points to a porosome. (From **ref.** ***4.***)

plasma membrane in the exocrine pancreas *(7)* (*see* **Fig. 7**) and in neurons ***(15)*** has also been imaged at near-nanometer resolution in live tissue in buffer.

To determine the morphology of the porosome at the cytosolic side of the cell, pancreatic PM preparations were used. When placed on freshly cleaved mica, isolated PM in buffer tightly adheres to the mica surface, enabling imaging by AFM. The PM preparations reveal scattered circular disks measuring 0.5–1 µm in diameter, with inverted cup-shaped structures within *(7)*. The inverted cups range in height from 10 to 15 nm. On a number of occasions, ZGs ranging in size from 0.4 to 1 µm in diameter were found associated with one or more of the inverted cups. This suggested the circular disks to be pits and the inverted cups to be fusion pores or porosomes. To determine if the cup-shaped structures in isolated PM preparations are indeed porosomes, immuno-AFM studies were conducted. Because ZGs dock and fuse at the PM to release vesicular contents, it was hypothesized that if porosomes are at these sites, then PM-associated t-SNAREs should localize at the base of porosomes. The t-SNARE protein SNAP-23 has been identified and implicated in secretion from pancreatic acinar cells ***(34)***. A polyclonal monospecific SNAP-23 antibody recognizing a single 23-kDa band in Western blots of pancreatic PM fraction has been used in immuno-AFM studies. When the SNAP-23-specific antibody was

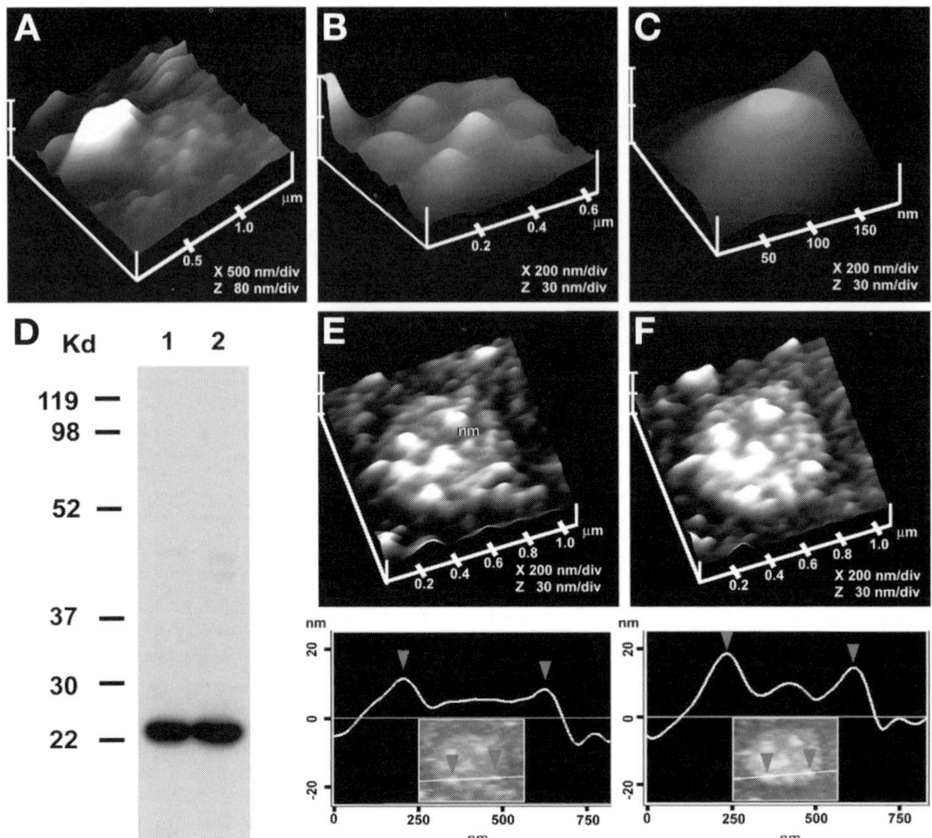

Fig. 7. Morphology of the cytosolic side of the porosome revealed in AFM studies on isolated pancreatic PM preparations. **(A)**. AFM of isolated PM preparation reveals the cytosolic end of a pit with inverted cup-shaped structures, the porosome. Note the 600-nm-diameter ZG at the left-hand corner of the pit. **(B)** Higher magnification of the same pit showing clearly the four to five porosomes within. **(C)** The cytosolic end of a single porosome is depicted in this AFM. **(D)** Immunoblot analysis of 10 μg and 20 μg of pancreatic PM preparations, using SNAP-23 antibody, demonstrates a single 23-kDa immunoreactive band. **(E, F)** The cytosolic side of the PM demonstrating the presence of a pit with a number of porosomes within, shown prior to **(E)** and following addition of the SNAP-23 antibody **(F)**. Note the increase in height of the porosome cone base revealed by section analysis (bottom pannel), demonstrating localization of SNAP-23 antibody at the base of the porosome (From **ref.** *7*).

added to the PM preparation during imaging with the AFM, the antibody selectively localized to the base of the cup-shaped structure, which is the tip of the inverted cup. These results demonstrate that the inverted cup-shaped structures in the isolated PM preparations are the porosomes observed from its cytosolic side *(7,8)*. Target membrane proteins SNAP-25 and syntaxin (t-SNARE) and secretory vesicle-associated membrane protein (v-SNARE) are part of the conserved protein complex involved in fusion of opposing bilayers ***(9,17,18,29)***. Since membrane-bounded secretory vesicles dock and fuse at porosomes to release vesicular contents, this suggested t-SNAREs associate at the porosome complex. It was, therefore, no surprise that the t-SNARE protein SNAP-23, implicated in secretion from pancreatic acinar cells, was located at the tip of the inverted cup (i.e., the base of the porosome) where secretory vesicles dock and fuse.

The structure of the porosome was further demonstrated using transmission electron microscopy (TEM) *(7,8)* (*see* **Fig. 4**). TEM studies confirm the fusion pore to have a cup-shaped structure, with similar dimensions as determined from AFM measurement. Additionally, TEM micrographs reveal porosomes to possess a basketlike morphology, with three lateral and a number of vertically arranged ridges. A ring at the base of the complex is also identified *(7)*. Because porosomes are found to be stable structures at the cell PM, it was hypothesized that if ZGs were to fuse at the base of the structure, it would be possible to isolate ZG-associated porosomes. Indeed, TEM of isolated ZG preparations reveal porosomes associated with docked vesicles *(7,8)*. As observed in whole cells, vertical structures were found to originate from within the porosome complex and appear attached to its membrane. As discussed later in this chapter, studies using full-length recombinant SNARE proteins and artificial lipid membranes demonstrated that t- and v-SNAREs located in opposing bilayers interact in a circular array to form conducting pores *(9)*. Because similar circular structures are observed at the base of the porosome and SNAP-23 immunoreactivity is found to localize at this site, this suggests that the t-SNAREs present at the base of porosomes are possibly arranged in a ring pattern.

3.1. Porosome: Isolation, Composition, and Reconstitution

In the last decade, a number of studies demonstrated the involvement of cytoskeletal proteins in secretion, and some studies implicated direct interaction of cytoskeleton protein with SNAREs ***(3,35–39)***. Furthermore, actin and microtubule-based cytoskeleton has been implicated in intracellular vesicle traffic ***(3,36)***. Fodrin, which was previously implicated in exocytosis ***(35)***, has recently been shown to directly interact with SNAREs ***(37)***. Studies demonstrate α-fodrin to regulate exocytosis via its interaction with the t-SNARE syntaxin family of proteins ***(37)***. The C-terminal coiled coil region of syntaxin interacts with α-fodrin, a major component of the submembranous cytoskeleton. Similarly, vimentin filaments interact with SNAP23/25 and, hence, are able to control the availability of free SNAP23/25 for assembly of the SNARE complex ***(36)***. All of these findings suggested that vimentin, α-fodrin, actin, and

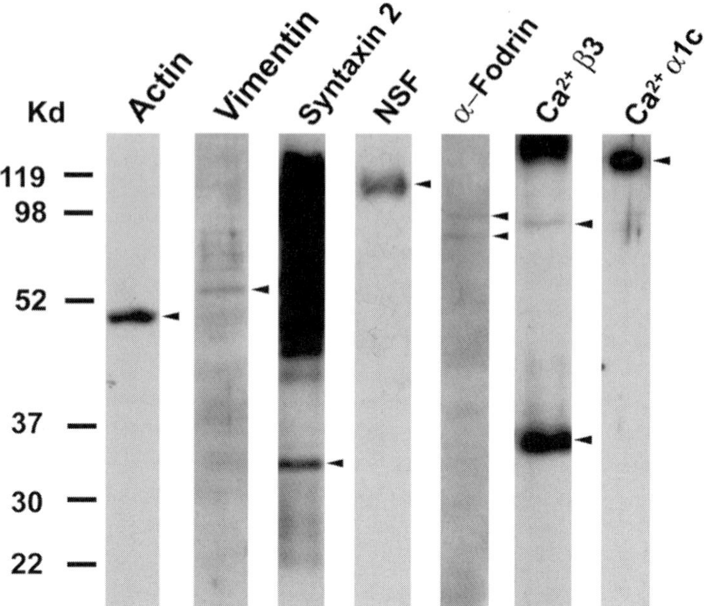

Fig. 8. SNAP-23-associated proteins in pancreatic acinar cells. Total pancreatic homogenate was immunoprecipitated using the SNAP-23-specific antibody. The precipitated material was resolved using 12.5% SDS-PAGE, electrotransferred to nitrocellulose membrane, and then probed using antibodies to a number of proteins. Association of SNAP-23 with syntaxin 2, with cytoskeletal proteins actin, α-fodrin, and vimentin, and calcium channels β3 and α1c, together with the SNARE regulatory protein NSF, is demonstrated (arrow heads). Lanes showing more than one arrowhead suggest the presence of isomers or possible proteolytic degradation of the specific protein. (From **ref.** *7.*)

SNAREs might be part of the porosome complex. Additional proteins such as v-SNARE (VAMP or synaptobrevin), synaptophysin, and myosin might associate when the porosome establishes continuity with the secretory vesicle membrane. The globular tail domain of myosin V contains a binding site for VAMP, which is bound in a calcium-independent manner ***(38)***. Further interaction of myosin V with syntaxin requires both calcium and calmodulin. It has been suggested that VAMP acts as a myosin V receptor on secretory vesicles and regulates formation of the SNARE complex ***(38)***. Interaction of VAMP with synaptophysin and myosin V has also been observed ***(39)***. In agreement with these earlier findings, recent studies demonstrated the association of SNAP-23, syntaxin 2, cytoskeletal proteins actin, α-fodrin, and vimentin, and calcium channels β3 and α1c, together with the SNARE regulatory protein NSF, in porosomes ***(7,8)*** (*see* **Fig. 8**). Additionally, chloride-ion channels ClC2 and

ClC3 were also identified as part of the porosome complex *(7,8)*. Isoforms of the various proteins identified in the porosome complex were subsequently demonstrated using 2D-BAC gel electrophoresis *(8,14)*. Three isoforms each of the calcium-ion channel and vimentin were found in porosomes *(8)*. Using yeast 2-hybrid analysis, recent studies confirm the presence and interaction of some of these proteins with t-SNAREs within the porosome complex *(14)*.

The size and shape of the immunoisolated porosome complex was determined in greater detail when examined using both negative staining EM and AFM *(8)*. The images of the immunoisolated porosome obtained by both EM and AFM were superimposable *(8)*. To further test whether the immunoisolated supramolecular complex was indeed the porosome, the complex was reconstituted into artificial liposomes, and the liposome-reconstituted complex examined using TEM *(8)*. Transmission electron micrographs reveal a 150–200-nm cup-shaped basketlike structure as observed for the porosome when coisolated with ZGs. To test if the reconstituted porosome complex was functional, purified porosomes were reconstituted into lipid membranes in an electrophysiological bilayer setup (EPC9) and challenged with isolated ZGs. Both the electrical activity of the reconstituted membrane as well as the transport of vesicular contents from the cis to the trans compartment were monitored. Results of these experiments demonstrate that the lipid membrane-reconstituted porosomes are functional supramolecular complexes (*see* **Figs. 9** and **10**) *(8)*. ZGs fused with the porosome-reconstituted bilayer, as demonstrated by an increase in capacitance and conductance and in a time-dependent release of the ZG enzyme amylase from the cis to the trans compartment of the bilayer chamber. Amylase is detected using immunoblot analysis of the buffer in the cis and trans chambers, using a previously characterized amylase-specific antibody *(4)*. As observed in immunoblot assays of isolated porosomes, chloride channel activities are also detected within the reconstituted porosome complex. Further, the chloride channel inhibitor DIDS was found to inhibit current activity in the porosome-reconstituted bilayer. In summary, these studies demonstrate that the porosome in the exocrine pancreas is a 100- to 150-nm-diameter supramolecular cup-shaped lipoprotein basket at the cell PM, where membrane-bound secretory vesicles dock and fuse to release vesicular contents. Similar studies have now been performed in neurons, demonstrating both the structural (*see* **Fig. 5E,F**) and functional reconstitution of the isolated neuronal porosome complex. The biochemical composition of the neuronal porosome has also been determined *(15)*.

3.2. Molecular Understanding of Cell Secretion

Fusion pores or porosomes are present in all secretory cells examined. From exocrine, to endocrine, to neuroendocrine, to neurons, where membrane-bound

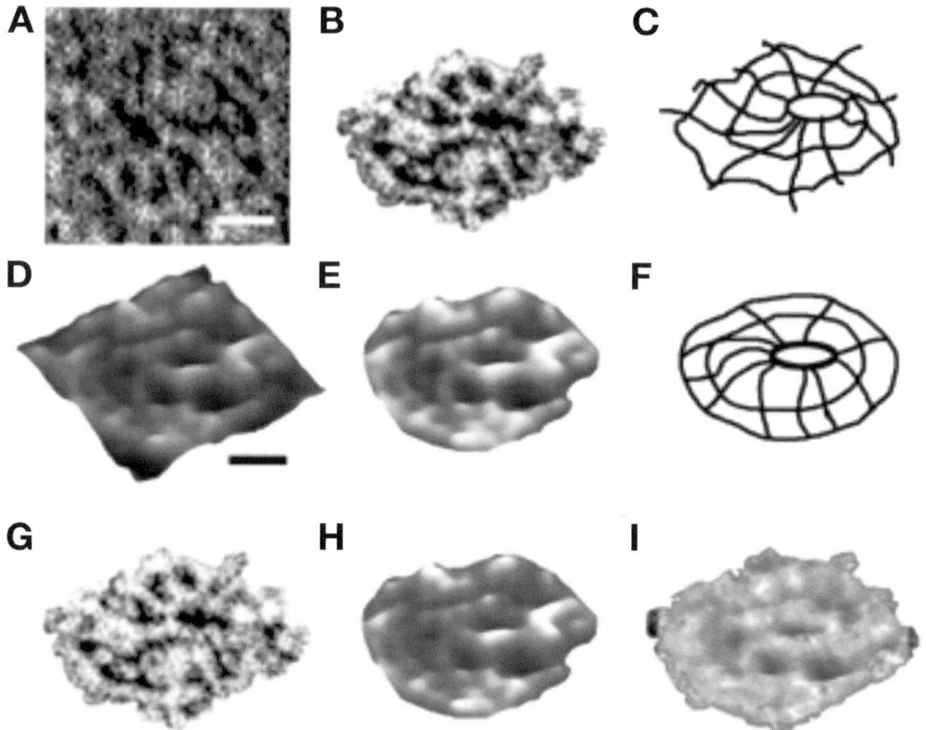

Fig. 9. Negatively stained EM and AFM of the immunoisolated porosome complex. **(A)** Negatively stained EM of an immunoisolated porosome complex from solubilized pancreatic plasma membrane preparations, using a SNAP-23-specific antibody. Note the 3 rings and the 10 spokes that originate from the inner smallest ring. This structure represents the protein backbone of the porosome complex, because the three rings and the vertical spikes are observed in EMs of cells and porosome coisolated with ZGs. Bar = 30 nm. **(B)** The EM of the fusion pore complex, cut out from **(A)**, and **(C)** an outline of the structure is presented for clarity. **(D–F)** AFM of the isolated porosome complex in near physiological buffer. Bar = 30 nm. Note the structural similarity of the complex, imaged both by EM **(G)** and AFM **(H)**. The EM and AFM micrographs are superimposable **(I)**.

secretory vesicles dock and transiently fuse to expel vesicular contents. Porosomes in pancreatic acinar or GH-secreting cells are cone-shaped structures at the plasma membrane, with a 100- to 150-nm-diameter opening. Membrane-bound secretory vesicles ranging in size from 0.2 to 1.2 μm in diameter dock and fuse at porosomes to release vesicular contents. Following fusion of secretory vesicles at porosomes, only a 20–35% increase in porosome diameter is demonstrated. It is therefore reasonable to conclude that secretory vesicles "transiently" dock and fuse at the site. In contrast to

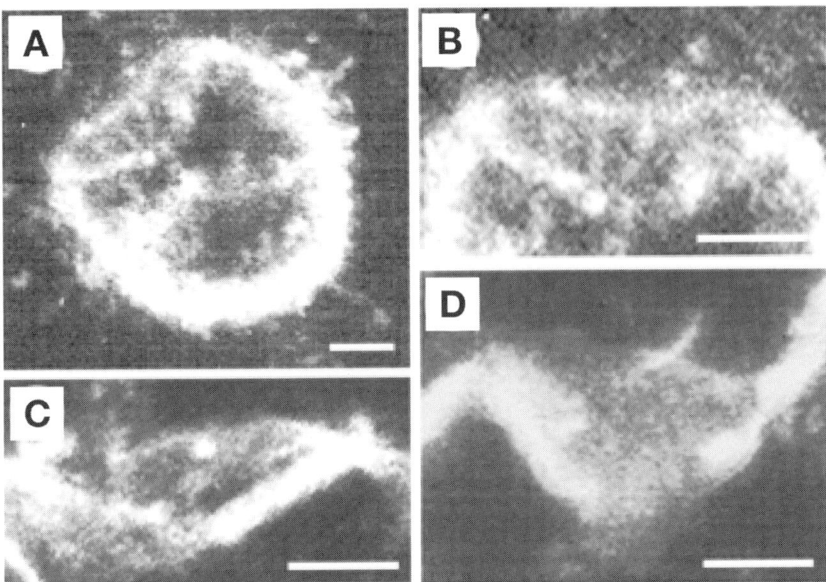

Fig. 10. Electron micrographs of liposome reconstituted porosome from pancreas demonstrating a cup-shaped basketlike morphology. (**A**) A 500-nm vesicle with an incorporated porosome is shown. Note the spokes in the complex. The reconstituted complex at greater magnification is shown in **B–D** Bar = 100 nm.

accepted belief, if secretory vesicles were to completely incorporate at porosomes, the PM structure would distend much wider than what is observed. Furthermore, if secretory vesicles were to completely fuse at the plasma membrane, there would be a loss in vesicle number following secretion. Examination of secretory vesicles within cells before and after secretion demonstrates that the total number of secretory vesicles remains unchanged following secretion *(10,26)*. However, the number of empty and partially empty vesicles increases significantly, supporting the occurrence of transient fusion. Earlier studies on mast cells also demonstrated an increase in the number of spent and partially spent vesicles following stimulation of secretion, without any demonstrable increase in cell size. Similarly, secretory granules are recaptured largely intact after stimulated exocytosis in cultured endocrine cells *(22)*. Other supporting evidence favoring transient fusion is the presence of neurotransmitter transporters at the synaptic vesicle membrane. These vesicle-associated transporters would be of little use if vesicles were to fuse completely at the plasma membrane to be compensatorily endocytosed at a later time. In further support, a recent study reports that single synaptic vesicles fuse transiently and successively without loss of vesicle identity *(23)*. Although the fusion of secretory vesicles at the cell PM occurs transiently, complete incorporation of membrane at the cell plasma membrane would occur when cells need to incorporate signaling molecules like receptors, second messengers, or ion channels. The discovery of the porosome and an understanding of the molecular mechanism of membrane fusion and the swelling of secretory vesicles required for expulsion of vesicular contents provide an understanding of secretion and membrane fusion in cells at the molecular level. These findings have prompted many laboratories to work in the area and further confirm these findings. Thus, the porosome is a supramolecular structure universally present in secretory cells, from the exocrine pancreas to the neurons, and in the endocrine to neuroendocrine cells, where membrane-bound secretory vesicles transiently dock and fuse to expel vesicular contents. Hence, the secretory process in cells is a highly regulated event, orchestrated by a number of ions and biomolecules.

Acknowledgment

This work was supported by grants DK-56212 and NS-39918 from the National Institutes of Health (BPJ).

References

1. Hörber, J. K. H. and Miles, M. J. (2003) Scanning probe evolution in biology. *Science* **302**, 1002–1005.
2. Anderson, L. L. (2004) Discovery of a new cellular structure—the porosome: elucidation of the molecular mechanism of secretion. *Cell Biol. Int.* **28**, 3–5.
3. Schneider, S. W., Sritharan, K. C., Geibel, J. P., Oberleithner, H., and Jena, B. P. (1997) Surface dynamics in living acinar cells imaged by atomic force microscopy: identification of plasma pembrane structures involved in exocytosis. *Proc. Natl. Acad. Sci. USA* **94**, 316–321.
4. Cho, S. J., Quinn, A. S., Stromer, M. H., et al. (2002) Structure and dynamics of the fusion pore in live cells. *Cell Biol. Int.* **26**, 35–42.
5. Cho, S. J., Jeftinija, K., Glavaski, A., Jeftinija, S., Jena, B. P., and Anderson, L. L. (2002) Structure and dynamics of the fusion pores in live GH-secreting cells revealed using atomic force microscopy. *Endocrinology* **143**, 1144–1148.
6. Cho, S. J., Wakade, A., Pappas, G. D., and Jena, B. P. (2002) New structure involved in transient membrane fusion and exocytosis. *Ann. New York Acad. Sci.* **971**, 254–256.
7. Jena, B. P., Cho, S. J., Jeremic, A., Stromer, M. H., and Abu-Hamdah, R. (2003) Structure and composition of the fusion pore. *Biophys. J.* **84**, 1337–1343.
8. Jeremic, A., Kelly, M., Cho, S. J., Stromer, M. H., and Jena, B. P. (2003) Reconstituted fusion pore. *Biophys. J.* **85**, 2035–2043.
9. Cho, S. J., Kelly, M., Rognlien, K. T., Cho, J., Hoerber, J. K. H., and Jena, B. P. (2002) SNAREs in opposing bilayers interact in a circular array to form conducting pores. *Biophys. J.* **83**, 2522–2527.

10. Cho, S. J., Cho, J., and Jena, B. P. (2002) The number of secretory vesicles remains unchanged following exocytosis. *Cell Biol. Int.* **26**, 29–33.
11. Jena, B. P. (2002) Fusion pore in live cells. *NIPS* **17**, 219–222.
12. Jena, B. P. (2003) Fusion pore: structure and dynamics. *J. Endocrinol.* **176**, 169–174.
13. Jena, B. P. (1997) Exocytotic fusion: total or transient. *Cell Biol. Int.* **21**, 257–259.
14. Jena, B. P. (2004) Discovery of the porosome: revealing the molecular mechanism of secretion and membrane fusion in cells. *J. Cell Mol. Med.* **8**, 1–21.
15. Cho, W. J., Jeremic, A., Rognlien, K. T., et al. (2004). Structure, isolation, composition and reconstitution of the neuronal fusion pore. *Cell Biol. Int.* **28**, 699–708. (published on-line August 25, 2004).
16. Kelly, M., Cho, W. J., Jeremic, A., Abu-Hamdah, R., and Jena, B. P. (2004). Vesicle swelling regulates content expulsion during secretion. *Cell Biol. Int.* **28**, 709–716 (published on-line August 25, 2004).
17. Jeremic, A., Kelly, M., Cho, W. J., Cho, S. J., Horber, J. K. H., and Jena, B. P. (2004) Calcium drives fusion of SNARE-apposed bilayers. *Cell Biol. Int.* **28**, 19–31 (published on-line 2003).
18. Jeremic, A., Cho, W. J., and Jena, B. P. (2004) Membrane fusion: what may transpire at the atomic level. *J. Biol. Phys. Chem.* **4**, 139–142.
19. Jena, B. P., Schneider, S. W., Geibel, J. P., Webster, P., Oberleithner, H., and Sritharan, K. C. (1997) G_i regulation of secretory vesicle swelling examined by atomic force microscopy. *Proc. Natl. Acad. Sci. USA* **94**, 13,317–13,322.
20. Cho, S. J., Sattar, A. K., Jeong, E.-H., et al. (2002) Aquaporin 1 regulates GTP-induced rapid gating of water in secretory vesicles. *Proc. Natl. Acad. Sci. USA* **99**, 4720–4724.
21. Abu-Hamdah, R., Cho, W. J., Cho, S. J., et al. (2004) Regulation of the water channel aquaporin-1: isolation and reconstitution of the regulatory complex. *Cell Biol. Int.* **28,** 7–17. (published on-line 2003).
22. Taraska, J. W., Perrais, D., Ohara-Imaizumi, M., Nagamatsu, S., and Almers, W. (2003) Secretory granules are recaptured largely intact after stimulated exocytosis in cultured endocrine cells. *Proc. Natl. Acad. Sci. USA* **100**, 2070–2075.
23. Aravanis, A. M., Pyle, J. L., and Tsien, R. W. (2003) Single synaptic vesicles fusing transiently and successively without loss of identity. *Nature* **423**, 643–647.
24. Tojima, T., Yamane, Y., Takagi, H., et al. (2000) Three-dimensional characterization of interior structures of exocytotic apertures of nerve cells using atomic force microscopy. *Neuroscience* **101**, 471–481.
25. Thorn, P., Fogarty, K. E., and Parker, I. (2004) Zymogen granule exocytosis is characterized by long fusion pore openings and preservation of vesicle lipid identity. *Proc. Natl. Acad. Sci. USA* **101**, 6774–6779.
26. Lee, J. S., Mayes, M. S., Stromer, M. H., Scanes, C. G., Jeftinija, S., and Anderson, L. L. (2004) Number of secretory vesicles in growth hormone cells of the pituitary remains unchanged after secretion. *Exp. Biol. Med.* **229**, 291–302.
27. Fix, M., Melia, T. J., Jaiswal, J. K., et al. (2004) Imaging single membrane fusion events mediated by SNARE proteins. *Proc. Natl. Acad. Sci. USA* **101**, 7311–7316.

28. Binnig, G., Quate, C. F., and Gerber, C. H. (1986) Atomic force microscope. *Phys. Rev. Lett.* **56**, 930–933.
29. Weber, T., Zemelman, B. V., McNew, J. A., et al. (1998) SNAREpins: minimal machinery for membrane fusion. *Cell* **92**, 759–772.
30. Alexander, S., Hellemans, L., Marti, O., Schneir, J., Elings, V., and Hansma, P. K. (1989) An atomic resolution atomic force microscope implemented using an optical lever. *J. Appl. Phys.* **65**, 164–167.
31. Jeong, E-H., Webster, P., Khuong, C. Q., Sattar, A. K. M. A., Satchi, M., and Jena, B. P. (1998) The native membrane fusion machinery in cells. *Cell Biol. Int.* **22**, 657–670.
32. Thoidis, G., Chen, P., Pushkin, A. V., et al. (1998) Two distinct populations of synaptic-like vesicles from rat brain. *Proc. Natl. Acad. Sci. USA* **95**, 183–188.
33. Monck, J. R., Oberhauser, A. F., and Fernandez, J. M. (1995) The exocytotic fusion pore interface: a model of the site of neurotransmitter release. *Mol. Membr. Biol.* **12**, 151–156.
34. Gaisano, H. Y., Sheu, L., Wong, P. P., Klip, A., and Trimble, W. S. (1997) SNAP-23 is located in the basolateral plasma membrane of rat pancreatic acinar cells. *FEBS Lett.* **414**, 298–302.
35. Bennett, V. (1990) Spectrin-based membrane skeleton: a multipotential adaptor between plasma membrane and cytoplasm. *Physiol. Rev.* **70**, 1029–1065.
36. Faigle, W., Colucci-Guyon, E., Louvard, D., Amigorena, S., and Galli, T. (2000) Vimentin filaments in fibroblasts are a reservoir for SNAP-23, a component of the membrane fusion machinery. *Mol. Biol. Cell.* **11**, 3485–3494.
37. Goodson, H. V., Valetti, C., and Kreis, T. E. (1997) Motors and membrane traffic. *Curr. Opin. Cell Biol.* **9**, 18–28.
38. Nakano, M., Nogami, S., Sato, S., Terano, A., and Shirataki, H. (2001) Interaction of syntaxin with α-fodrin, a major component of the submembranous cytoskeleton. *Biochem. Biophys. Res. Commun.* **288**, 468–475.
39. Ohyama, A., Komiya, Y., and Igarashi, M. (2001) Globular tail of myosin-V is bound to vamp/synaptobrevin. *Biochem. Biophys. Res. Commun.* **280**, 988–991.

16

Secretory Vesicle Swelling by Atomic Force Microscopy

Sang-Joon Cho and Bhanu P. Jena

Summary

The swelling of secretory vesicles has been implicated in exocytosis, but the underlying mechanism of vesicle swelling remained unknown. Earlier studies from our laboratory demonstrated the association of the α-subunit of heterotrimeric GTP-binding protein $G_{\alpha i3}$ with zymogen granule membrane and implicated its involvement in vesicle swelling. Mas7, an active mastoparan analog known to stimulate Gi proteins, was found to stimulate the GTPase activity of isolated zymogen granules and cause swelling. Increase in vesicle size in the presence of GTP, NaF, and Mas7 were irreversible and found to be KCl sensitive. However, Ca^{2+} had no effect on zymogen granule size. Taken together, these results indicated that zymogen granules, the membrane-bound secretory vesicles in exocrine pancreas, swell in response to GTP mediated by a $G_{\alpha i3}$ protein. Subsequently, our studies demonstrated that the water channel aquaporin-1 (AQP1) is also present at the zymogen granule membrane and participates in rapid GTP-induced and $G_{\alpha i3}$-mediated vesicular water gating and swelling. Isolated zymogen granules exhibit low basal water permeability. However, exposure of granules to GTP results in a marked potentiation of water entry. Treatment of zymogen granules with the known water channel inhibitor Hg^{2+} is accompanied by a reversible loss in both the basal and GTP-stimulable water entry and vesicle swelling. Introduction of AQP1-specific antibody raised against the carboxy-terminal domain of AQP1 blocked GTP-stimulable swelling of vesicles. Our results demonstrate that AQP1 associated at the zymogen granule membrane is involved in basal GTP-induced and $G_{\alpha i3}$-mediated rapid gating of water into zymogen granules of the exocrine pancreas.

Key Words: Secretory vesicle; AFM; swelling.

1. Introduction

Secretory vesicle swelling is critical for exocytosis *(1–4)*; however, the underlying mechanism of vesicle swelling remains unknown. In mast cells, increase in secretory vesicle volume following stimulation of secretion had been suggested from electrophysiological measurements *(5)*. The requirement of osmotic force in the fusion of phospholipid vesicles with a membrane bilayer

From: *Methods in Molecular Biology, vol. 319: Cell Imaging Techniques: Methods and Protocols*
Edited by: D. J. Taatjes and B. T. Mossman © Humana Press Inc., Totowa, NJ

had also been demonstrated *(4,6,7)*. Isolated zymogen granules (ZGs), the membrane-bound secretory vesicles in exocrine pancreas and parotid glands, posses Cl^- and ATP-sensitive K^+ selective ion channels at the vesicle membrane, whose activities are implicated in vesicle swelling *(8–16)*. Additionally, secretion of ZG contents in permeabilized pancreatic acinar cells requires the presence of both K^+ and Cl^- ions *(2)*. In vitro ZG–pancreatic plasma membrane fusion assays demonstrate potentiation of fusion in the presence of GTP and NaF (*2,17*, unpublished observation). Heterotrimeric $G_{\alpha i3}$ protein has been implicated in the regulation of both K^+ and Cl^- ion channels in a number of tissues *(18–22)*. Analogous to the regulation of K^+ and Cl^- ion channels at the cell membrane, the regulation of K^+ and Cl^- ion channels at the ZG membrane by a $G_{\alpha i}$ protein was demonstrated *(17)*. Isolated ZGs from exocrine pancreas swell rapidly in response to GTP and NaF *(17)*. These studies suggest the involvement of rapid water entry into ZGs following exposure to GTP. As opposed to osmotic swelling, membrane-associated water channels called aquaporins have been implicated in rapid volume changes in cells *(23,24)* and intracellular vesicles *(25,26)*. Therefore, the likely mechanism of ZG swelling via possible water channels at the ZG membrane was explored *(27)*. Our study demonstrates the presence of aquaporin-1 (AQP1) in ZG membranes and its participation in $G_{\alpha i3}$-mediated and GTP-induced vesicle water entry and swelling *(27)*.

2. Materials and Methods

2.1. Cell Fractions

Rat pancreatic fractions were prepared and their purity determined as described elsewhere *(28–30)*. Salt and detergent treatment of isolated ZG membrane preparations were performed at 4°C for 30 min. Following treatment, supernatant and particulate fractions were separated by centrifugation of the reaction mixture at 4°C for 1 h at 200,000*g*.

2.2. Immunoblot Analysis

Immunoblot analysis was performed on pancreatic fractions prepared as described in **ref.** ***26***. Protein in pancreatic fractions was estimated by the Bradford method *(31)*. Pancreatic fractions were boiled in Laemmli reducing sample preparation buffer *(32)* for 5 min. Equal loads of protein were resolved by 12.5% sodium dodecyl sulfate–polyacrylamide gel electrophoresis (SDS-PAGE), followed by electrotransfer to 0.2-μ*M* nitrocellulose sheets. The nitrocellulose was incubated for 1 h at room temperature in blocking buffer (5% nonfat milk in phosphate-buffered saline [PBS] containing 0.1% Triton X-100 and 0.02% NaN_3) and immunoblotted for 2 h at room temperature with either affinity-purified B-CHIP (AQP1) antibody raised against the holoprotein *(32)* (a gift from B. Baum and M.A. Knepper) or monoclonal vesicle-associated membrane protein (VAMP) antibody recognizing VAMP1 and VAMP2 from StressGen Biotechnologies Corp. Both primary antibodies were used at a dilution of 1 : 1000 in blocking buffer. The immunoblotted nitrocellulose sheets were washed in PBS containing 0.1% Triton X-100 and 0.02% NaN_3 prior to incubation for 1 h at room temperature in horseradish peroxidase (HRP)-conjugated secondary antibody at a dilution of 1:2000 in blocking buffer. The immunoblots were washed in PBS containing 0.1% Triton X-100 and 0.02% NaN_3, processed for enhanced chemiluminescence and exposure to X-OMAT-AR film. The exposed films were then developed and photographed.

2.3. Immunoelectron Microscopy

Sections were prepared from rat pancreatic lobules embedded in unicryl and double immunogold labeling performed *(33)* using affinity-purified polyclonal rabbit anti-$G_{\alpha i3}$ and an antibody raised to the carboxy terminal of AQP1. Sequential labeling and visualization involved blocking steps and the use of 10 nm gold-conjugated anti-rabbit and 5 nm gold-conjugated goat antibody.

2.4. Atomic Force Microscopy

Isolated ZG suspensions in 125 μ*M* KCl–Mes buffer, pH 6.5, were plated on Cell-Tak-coated glass cover slips. Ten minutes after plating, the cover slips were placed in a fluid chamber and washed with the KCl–Mes buffer to remove unattached granules, prior to imaging the attached ZG in KCl–Mes buffer or water, in the presence or absence of various concentrations of GTP, 25 μ*M* $HgCl_2$, and 100 μ*M* β-mercaptoethanol. AQP1 antibody raised against the carboxy-terminus domain of AQP1 (Santa Cruz) was introduced into the ZG lumen by incubating the ZG at 4°C for 5 min in KCl–Mes buffer containing streptolysin-O (SLO) and the antibody or vehicle, followed by a 2-min incubation at 30°C and three washes in ice-cold 125 m*M* KCl–Mes buffer, pH 6.5. At 4°C, the SLO binds to the cholesterol molecule at the ZG membrane, making approx 14-nm holes at 30°C, which allow antibodies (approx 8 nm) to enter the ZG. Washing ZG with ice-cold 125 m*M* KCl–Mes buffer reseals the holes and removes ZG-associated SLO. Following incubation, the ZGs containing AQP1 antibody or the vehicle were washed and imaged in the presence of 40 μ*M* GTP. ZG imaging and dynamics were performed using the Nanoscope IIIa, an AFM from Digital Instruments (Santa Barbera, CA). ZGs were imaged both in the "contact" and "tapping" mode in fluid *(34,35)*. However, all images presented in this manuscript were obtained in the "tapping" mode in fluid, using silicon nitride tips with a spring constant of 0.06 N/m and an imaging force of less than 200 nN. Images were obtained at line frequencies of 2 Hz, with 512 lines per image, and constant image gains. Time-dependent (resolution in seconds) morphological changes in ZG were obtained using section analysis. Topographical

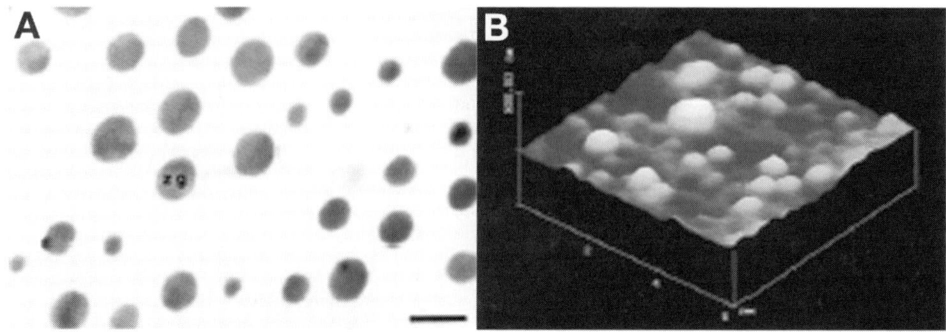

Fig. 1. Isolated ZGs ranging in size from 0.2 to 1.2 μ*M* obtained from rat pancreas as seen by electron microscopy and AFM. **(A)** An electron micrograph of the electron-dense ZGs. Note the purity of the ZG preparation. (bar = 1 μ*M*). **(B)** A three-dimensional AFM image of isolated ZGs adhering to a Cell-Tak-coated mica sheet. Note the size heterogeneity in the ZG population. (From **ref. *17*.**)

dimensions of ZG were analyzed using the software nanoscopeIIIa 4.43r8 supplied by Digital Instruments.

2.5. Measurement of Tritiated Water Entry

Isolated ZGs in 100 m*M* MES buffer, pH 6.5, containing 50,000 cpm of tritiated water were exposed to different stimulators, inhibitors, or their respective controls. The reaction mixture was incubated at room temperature for 30 min; ZGs were collected by centrifugation and tritiated water incorporation into them was estimated by liquid scintillation counting.

3. Results and Discussion

3.1. Purity and Size Heterogeneity of ZGs

The purity of ZGs was determined by imaging the isolated ZGs using a TEM **(Fig. 1A)**. **Figure 1A** is an electron micrograph of the purified electron-dense ZGs with no detectable subcellular contaminants. When isolated ZGs plated on Cell-Tak-coated mica sheets were placed in 125 m*M* Mes buffer (pH 6.5) and imaged by the AFM, three-dimensional images were obtained (*see* **Fig. 1B**) that demonstrated a size heterogeneity in the ZG population. Measurements of ZG images obtained by AFM demonstrated that the diameter of ZGs range from 0.2 to 1.2 μ*M*.

3.2. Association of Gi3 With ZGM

Immunoblot analysis of pancreatic total homogenate (TH) and zymogen granule membrane (ZGM) fractions using a Gi3-specific antibody after resolution by

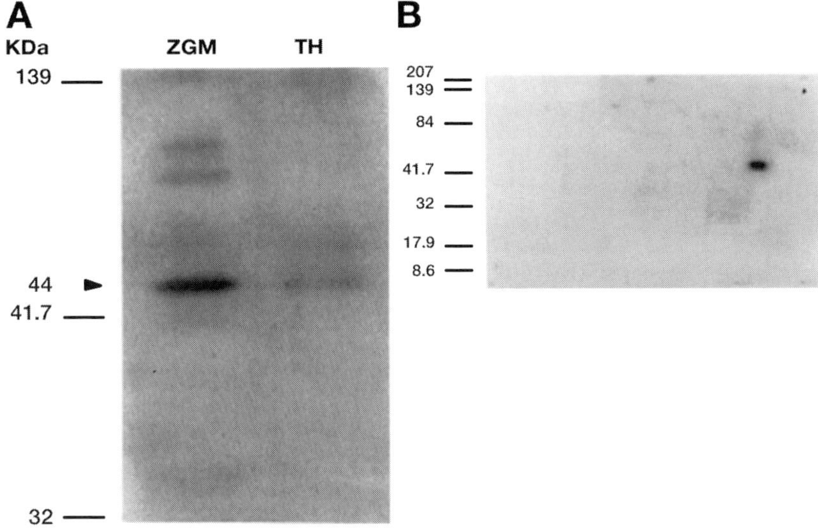

Fig. 2. Immunolocalization and biochemical detection of a $G_{\alpha i3}$ protein with ZGs from rat exocrine pancreas. **(A,B)** Immunoblot analysis. Identification of $G_{\alpha i3}$-immunoreactive antigen associated with ZGM. **(A)** Ten micrograms each of TH and ZGM was resolved using a single-dimension 12.5% SDS gel electrophoresis before electrotransfer onto nitrocellulose and immunoblotting. A 44-kDa immunoreactive band was detected in both fractions but enriched in the ZGM. **(B)** Twenty-five micrograms of ZGM protein was resolved on two-dimensional 16-BAC gel electrophoresis before electrotransfer and immunoblotting using the Gi3-specific antibody. Note a single spot at approx 44 kDa, suggesting that only one Gi3 isoform is present in the ZGM. (From **ref. *17*.**)

SDS/PAGE and electrotransfer to nitrocellulose membrane demonstrated a 44-kDa major immunoreactive band in both fractions. An increase in band intensity was found in the ZGM fraction over TH (*see* **Fig. 2A**). An affinity-purified rabbit polyclonal antibody (Santa Cruz Biotechnology) raised against a peptide corresponding to amino acid 345–354 mapping at the carboxy terminus of Gi3 of rat origin was used in the study. No immunoreactive bands were detected in the ZGM fraction when the nitrocellulose with resolved proteins was immunoblotted with affinity-purified Gs-, Gz-, and Gt-specific antibodies (not shown). To determine if more than one Gi isoform was present in ZGM, two-dimensional 16-BAC gel followed by immunoblotting using the Gi3- specific antibody was performed. Results from this study demonstrate a single spot at approximately the 44-kDa mark to be present in ZGM (*see* **Fig. 2B**). The 44-kDa immunoreactive band (single-dimension protein resolution) or spot (two-dimensional protein resolution) was absent in blots that were immunostained with peptide-neutralized antibody (not shown).

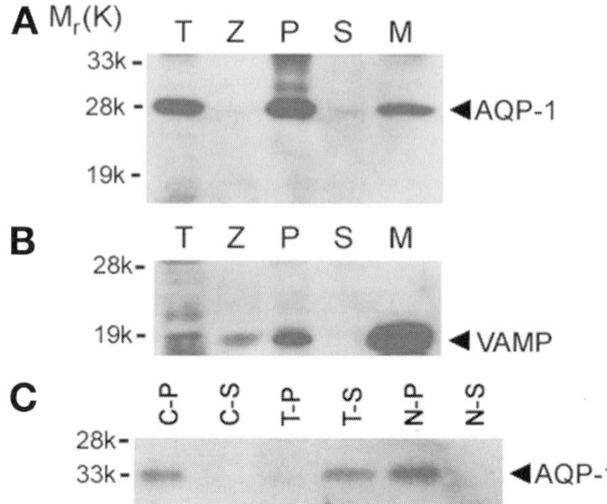

Fig. 3. Twenty micrograms of protein from each of the rat pancreatic fractions, total homogenate (T), zymogen granule (Z), 200,000*g* particulate (P) and supernatant (S) from total homogenates, and zymogen granule membrane (M), were resolved using 12.5% SDS-PAGE. The resolved proteins in each fraction were electrotransferred to nitrocellulose membrane and immunoblotted using AQP1 and VAMP-specific antibodies. The 28-kDa AQP1 immunoreactivity is associated with the M fraction **(A)**. The purity of isolated Z is demonstrated by the enriched presence of VAMP in the M fraction **(B)**. Zymogen granules are membrane-bound secretory vesicles packed with secretory proteins. The membrane fraction constitutes <1% of total granular proteins. Thus, very little AQP1 and VAMP immunoreactivity is detected in total granules, compared with several-fold enrichment in the granule membrane fraction. Affinity of AQP1 association with M was further examined by exposing 10 μg of granule membrane protein to buffer, salt, and detergent **(C)**. After treatment of the granule membrane, the particulate and supernatant fractions were separated by centrifugation at 200,000*g*. Examination of particulate (C-P) and supernatant (C-S) fractions prepared following PBS treatment, treatment with 1% Triton X-100 in PBS (T-P, T-S), or with 1 *M* KCl (N-P, N-S) demonstrates specific and tight association of AQP1 immunoreactivity with M. Exposure of the M to PBS or 1 *M* KCl failed to dislodge the AQP1 immunoreactivity; however, the nonionic detergent Triton X-100 was able to dissociate AQP1 from the granule membrane. (From **ref.** ***27***.)

3.3. Water Channel AQP1 at ZG Membrane

Pancreatic subcellular fractions ***(2,17,28,30)*** demonstrate the presence of AQP1 immunoreactivity in ZG membrane preparations (*see* **Fig. 3A**). A strong immunoreactive signal was also observed in the pancreatic particulate fraction. Because it is possible for ZG to have been contaminated with other subcellular components enriched in AQP1, it was necessary to determine the purity of the

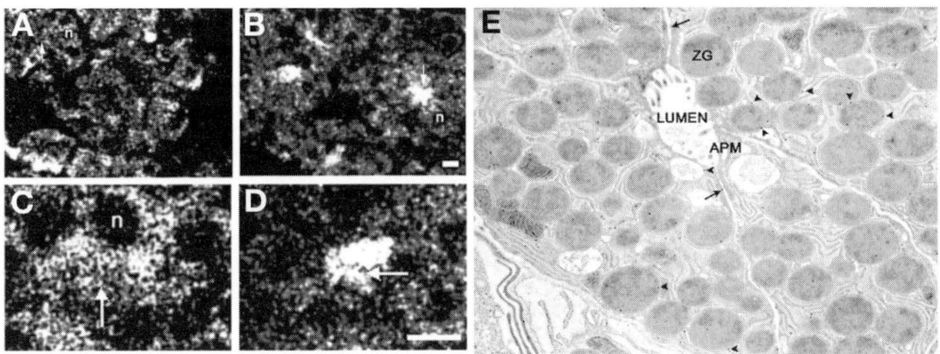

Fig. 4. AQP1 immunofluorescent confocal microscopy on stimulated and resting pancreatic lobules (**A–D**). Pancreatic lobules incubated for 1 h at 37°C in the absence (resting) or presence (stimulated) of the secretagogue carbamylcholine (1×10^{-5} *M*). A fourfold increase in amylase secretion from stimulated pancreatic lobules over resting was observed (data not shown). Following incubation, the lobules were fixed, cryosectioned, and processed for immunofluorescent confocal microscopy. Confocal images of AQP1 immunostained pancreatic tissue in resting **(A,C)** and stimulated **(B,D)** states indicate association of AQP1 with zymogen granules and its possible involvement in secretion. In resting acinar cells, AQP1 appears distributed throughout the cell including the apical region (arrow). The nucleus (n) located at the basolateral end of pancreatic acinar cells is devoid of AQP1-specific immunostaining. Following stimulation of secretion, much of the AQP1 immunoreactivity is localized at the apical region of pancreatic acinar cells where secretion is known to occur. (Bar = 10 μ*M*.) Immunogold electron microscopic (10-nm particles) localization of AQP1 to ZG (arrowheads) **(E)**. Note some AQP1 associated with the cell plasma membrane (arrow). Magnification: ×57,000 (reduced from original magnification). (From **ref.** ***27.***)

ZG preparation. The purity of ZG and the ZG membrane fractions were further determined by the enriched presence of VAMP immunoreactivity (*see* **Fig. 3B**). VAMP, also known as synaptobrevin, is a vesicle-associated membrane protein present in ZG membranes. A significant enrichment of VAMP immunoreactivity was observed in ZG membrane, compared to other pancreatic fractions. To determine the type of association of AQP1 with ZGs, the ZG membrane was treated with salt (1 *M* KCl) and a nonionic detergent (1% Triton X-100). Detergent treatment released the AQP1 immunoreactivity from ZG membranes, but salt had no effect (*see* **Fig. 3C**). These results show that AQP1, as its structure suggests ***(36,37)***, is tightly associated with the ZG membrane. To further determine the distribution of AQP1 in pancreatic acinar cells, AQP1 immunoreactivity was examined ***(38)***, using both light (*see* **Fig. 4A–D**) and electron microscopy (*see* **Fig. 4E**). Pancreatic acinar cells are polarized secretory cells which contain apically located ZGs. Abundant AQP1 immunoreactivity was

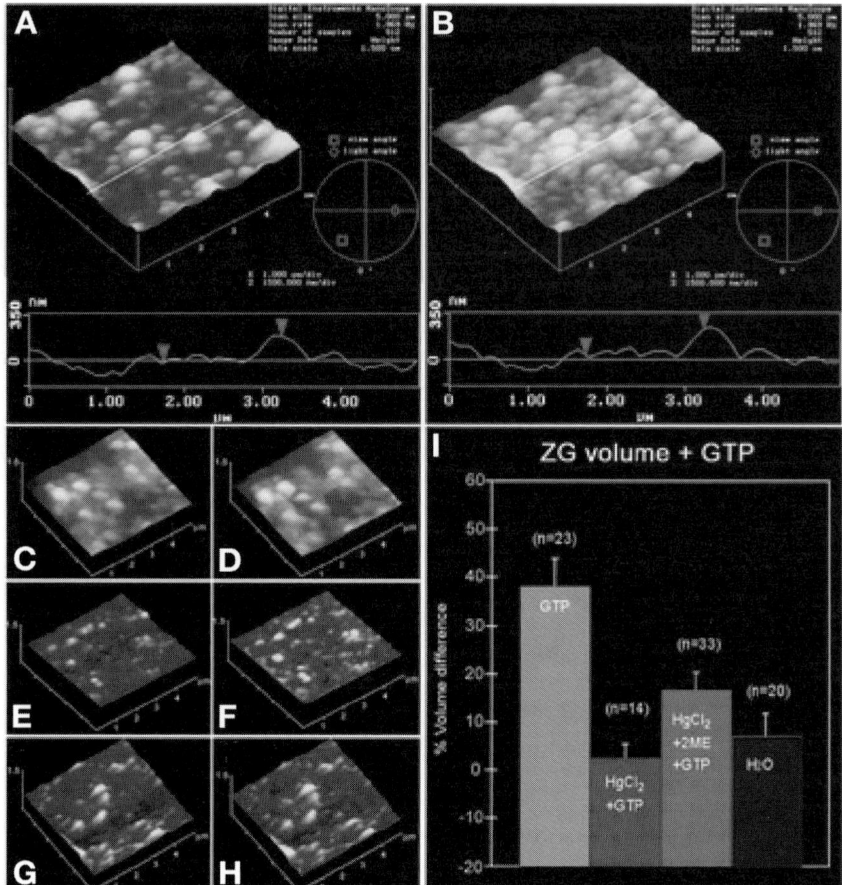

Fig. 5. Three-dimensional AFM images and section analysis of ZG, untreated **(A,B)** or pretreated with 25 μ*M* HgCl2 **(C,D)** and pretreated with 25 μ*M* $HgCl_2$ followed by 100 μ*M* β-mercaptoethanol **(E,F)** before **(A,C,E)** and after **(B,D,F)** exposure to 40 μ*M* GTP. Notice the increase in size of ZG (from A to B and I) in the three-dimensional array, section analysis, and bar graph, following exposure to GTP. Pretreatment of ZG with $HgCl_2$ abolishes the GTP effect (from C to D and I). However, the inhibitory effect of $HgCl_2$ is partially reversed in the presence of β-mercaptoethanol (from E to F and I). Incubation of ZG in water instead of 125 m*M* KCl–Mes buffer, pH 6.5, has little effect on ZG size (time 0 min G to time 30 min H and I). (From **ref.** ***27.***)

detected at the apical region of these cells. In resting acini (*see* **Fig. 4A–D**), AQP1 immunoreactivity was detectable throughout the cell, but its primary localization was at the apical region. Electron microscopy further confirms the presence of AQP1 at the ZG membrane (*see* **Fig. 4E**). It is well established that following a secretory stimulus, there is release of granular contents at the

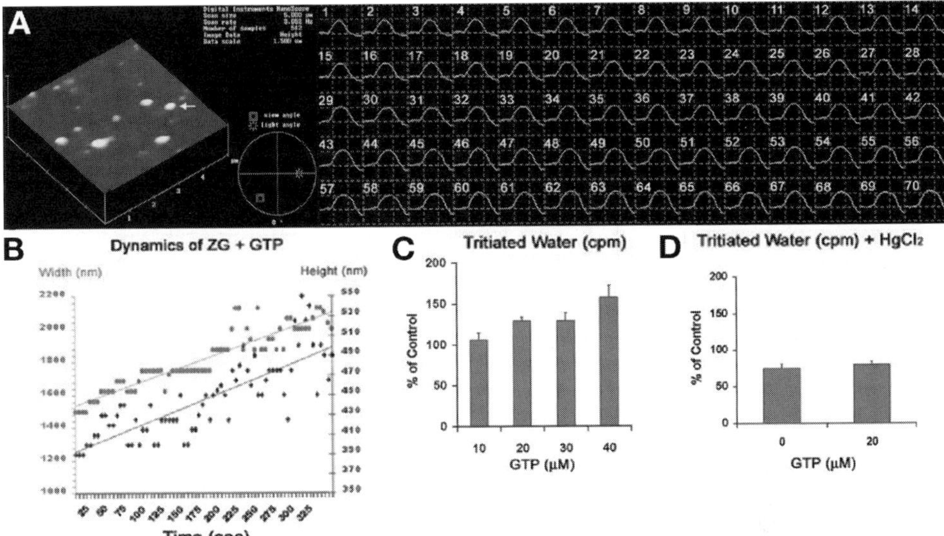

Fig. 6. The height and width of a single ZG (arrow) is monitored in seconds, following exposure to 40 μ*M* GTP (**A,B**). Notice the linear time-dependent increase in both height and width of the ZG following exposure to GTP A GTP dose-dependent increase in tritiated water entry is observed in ZG (**C**). Exposure of ZG to $HgCl_2$ inhibits both basal and GTP-induced tritiated water entry into ZG (**D**). (*See* Color Plate 18 following p. 274. From **ref.** *27*.)

apical lumen of pancreatic acini. Immunocytochemistry performed using light microscopy demonstrated intense AQP1-specific labeling at the apical region of the cell (*see* **Fig. 4B,D**), suggesting the involvement of AQP1 in secretion.

3.4. GTP-Induced ZG Swelling Is $HgCl_2$ Sensitive

As previously demonstrated, exposure to GTP results in swelling (percent volume increase = 38.22 ± 5.87; mean ± S.E.) of isolated ZG preparations (*see* **Figs. 5A,B,I** and **6A–C**). Within seconds, following exposure to 40 μ*M* GTP, a marked increase in water entry into ZG was demonstrated (*see* **Fig. 6A,B**). Water permeability of ZG was GTP dose dependent, as assayed by using the entry of tritiated water (*see* **Fig. 6C**). The increased entry of tritiated water was 6.5% over control in the presence of 10 μ*M* GTP, 29.6% in the presence of 20 μ*M* GTP, and 58% in the presence of 40 μ*M* GTP. However, exposure of ZG to $HgCl_2$, a known inhibitor of AQP1 ***(18,39)***, results in inhibition of both the basal (75.1 ± 6.17% of control) as well as the GTP-mediated (80.43 ± 3.55% of control) water permeability (*see* **Figs. 5C,D,I** and **6D**). The effect of $HgCl_2$ on ZG was reversed in the presence of 1 m*M* β-mercaptoethanol (*see* **Fig. 5E,F,I**), as demonstrated by a 16.83 ± 3.53% GTP-induced increase in volume of $HgCl_2$-treated vesicles following exposure to 1 m*M* β-mercaptoethanol.

3.5. Functional AQP1 at ZG Membrane

To determine if the GTP-induced water entry was the result of the AQP1 water channel function or osmosis driven, isolated ZGs were incubated in water and their size was monitored. Isolated ZG incubated in water had little effect (*see* **Fig. 5G,H,I**), as shown by no significant change in percent ZG volume over control (7.32 ± 4.51%). The presence of functional AQP1 in ZG membranes was confirmed from tritiated water (^3H_2O) permeability experiments on isolated ZG. ZGs were incubated in buffer containing ^3H_2O, in the presence and absence of GTP and/or $HgCl_2$. GTP is known to induce swelling of isolated ZGs ***(17)***. Exposure of isolated ZGs to GTP (10–40 μ*M*) results in an increase in water entry in a dose-dependent manner (*see* **Fig. 6C**). In the presence of $HgCl_2$, GTP-induced water entry was inhibited (*see* **Fig. 6D**). Isolated ZGs are stable for hours in low-pH (pH 6–6.5) buffers. At neutral or alkaline pH, isolated ZGs rapidly lyse. All studies were therefore performed in pH 6.5 buffer. These results, however, do not exclude the possibility that yet unidentified mercury-sensitive water channels might contribute to GTP-stimulable water entry into ZGs. This issue was addressed by introducing, into ZGs, AQP1-specific antibody raised against the carboxyl domain of the water channel.

3.6. AQP1 Regulates ZG Swelling

The introduction of AQP1-specific antibody into isolated ZG was carried out by permeabilizing ZG using SLO. We next carried out experiments to determine whether the AQP1 antibody actually entered the SLO-treated ZGs and, if so to determine their distribution within the ZG. Both Western blot assay and immuno-electron microscopy demonstrate the entry of AQP1 antibody into ZGs (*see* **Fig. 7A–E**; *see* Color Plate 19 following p. 274). When intact and permeabilized ZGs were separately exposed to the AQP1 antibody, resolved using SDS-PAGE, followed by transfer to nitrocellular membrane and immunoblot probed using a secondary antibody conjugated to a chemiluminescent marker, the AQP antibody was detected in the SLO-treated ZG, demonstrating entry of the antibody. No detectable signal was seen in the intact ZG fraction exposed to the AQP1 antibody (*see* **Fig. 7A**). Electron microscopy performed in the intact and SLO-treated ZG demonstrated intense immunogold localization in the SLO-treated ZGs, confirming AQP1 entry into the ZGs. Specific immunogold labeling at the luminal side of the ZG membrane in the SLO-treated batch further confirms binding of the antibody to the carboxy domain of AQP1. When AQP1-specific antibody was introduced into SLO permeabilized ZGs (*see* **Fig. 7A–F**),

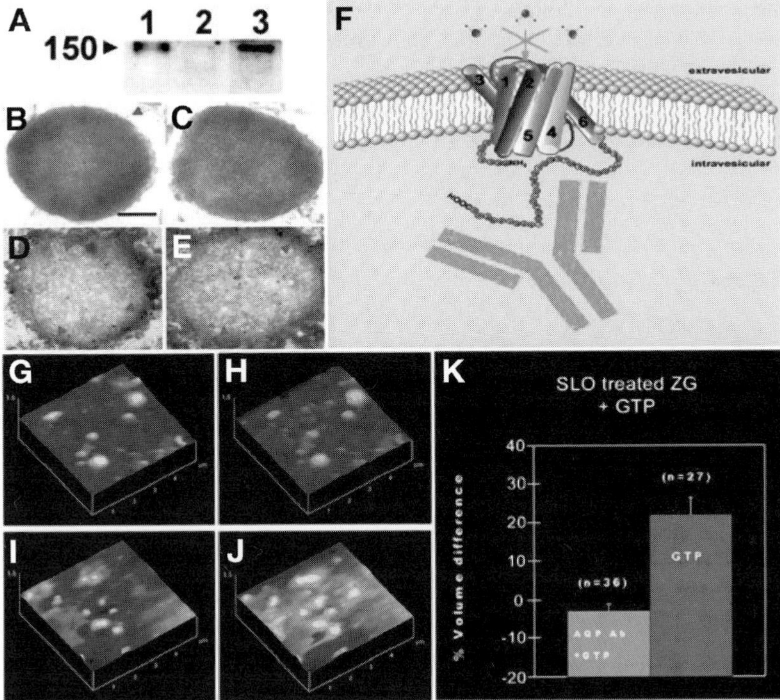

Fig. 7. Immunoblot assay demonstrating the presence of AQP1 antibody in SLO permeabilized ZG (**A**). Lane 1 is AQP1 antibody alone; lane 2 is nonpermeable ZG exposed to antibody; lane 3 is permeable ZG exposed to AQP1 antibody. Immuno-electron micrographs of intact ZGs exposed to AQP1 antibody demonstrate little labeling **(B,C)**. (Bar = 200 nm.) Contrarily, SLO-treated ZG demonstrate intense gold labeling at the luminal side of the ZG membrane **(D,E)**. AQP1 regulates GTP-induced water entry in ZG. Schematic diagram of ZG membrane depicting AQP1-specific antibody binding to the carboxy domain of AQP1 at the intragranular side, to block water gating **(F)**. AQP1 antibody introduced into ZG blocks GTP-induced water entry and swelling (from **G** to **H,** following GTP exposure; **G,H,K**). However, only vehicle introduced into ZG, retains the GTP-stimulable effect (from **I** to **J**, following GTP exposure; **I–K**). (*See* Color Plate 19 following p. 274. From **ref.** ***27***.)

GTP-induced water entry was inhibited (**Fig. 7G,H,K**). However, in the absence of AQP1-specific antibody, SLO permeabilized ZGs elicit strong GTP-induced swelling (% volume increase of 21.95 ± 4.42; *see* **Fig. 7I–K**). These studies demonstrate that at pH 6.5, AQP1 participates in GTP-induced gating of water in ZG of pancreatic acinar cells. Previous studies demonstrated that the carboxyl domain of AQP1 does not participate in water entry ***(40)***. However, the inhibition of AQP1 by the antibody could result from the structural distortion of AQP1 by binding of the comparatively larger antibody molecule. To understand the detailed pathway of GTP-induced signaling in the regulation of AQP1 in ZG membrane, further studies are being conducted. Studies using liposomes reconstituted with AQP1, the $G_{\alpha i3}$ protein, and other signaling molecules will help elucidate the molecular mechanism of $G_{\alpha i3}$-mediated and GTP-induced AQP1-involvement in water entry into ZG.

Acknowledgments

This work was supported in part by research grants DK56212 and NS39918 to B.P.J. from the National Institutes of Health. S.J.C. is a recipient of an NIH postdoctoral fellowship award (DK60368).

References

1. Alvarez de Toledo, G., Fernandez-Chacon, R., and Fernandez, J. M. (1993) Release of secretory products during transient vesicle fusion. *Nature* **363,** 554–558.
2. Curran, M. J. and Brodwick, M. S. (1991) Ionic control of the size of the vesicle matrix of beige mouse mast cells. *J. Gen. Physiol.* **98,** 771–790.
3. Monck, J. R., Oberhauser, A. F., Alvarez de Toledo, G., and Fernandez, J. M. (1991) Is swelling of the secretory granule matrix the force that dilates the exocytotic fusion pore? *Biophys. J.* **59,** 39–47.
4. Finkelstein, A., Zimmerberg, J., and Cohen, F. S. (1986) Osmotic swelling of vesicles: its role in the fusion of vesicles with planar phospholipid bilayer membranes and its possible role in exocytosis. *Annu. Rev. Physiol.* **48,** 163–174.
5. Fernandez, J. M., Villalon, M., and Verdugo, P. (1991) Reversible condensation of mast cell secretory products in vitro. *Biophys. J.* **59,** 1022–1027.
6. Holz, R. W. (1986) The role of osmotic forces in exocytosis from adrenal chromaffin cells. *Annu. Rev. Physiol.* **48,** 175–189.
7. Almers, W. (1990) Exocytosis. *Annu. Rev. Physiol.* **52,** 607–624.
8. Gasser, K. W., DiDomenico, J., and Hopfer, U. (1988) Secretagogues activate chloride transport pathways in pancreatic zymogen granules. *Am. J. Physiol.* **254,** G93–G99.
9. Fuller, C. M., Deetjen, H. H., Piiper, A., and Schulz, I. (1989) Secretagogue and second messenger-activated Cl^- permeabilities in isolated pancreatic zymogen granules. *Pflugers Arch.* **415,** 29–36.
10. Fuller, C. M., Eckhardt, L., and Schulz, I. (1989) Ionic and osmotic dependence of secretion from permeabilised acini of the rat pancreas. *Pflugers Arch.* **413,** 385–394.
11. Gasser, K. W. and Hopfer, U. (1990) Chloride transport across the membrane of parotid secretory granules. *Am. J. Physiol.* **259,** 413–420.
12. Thevenod, F., Gasser, K. W., and Hopfer, U. (1990) Dual modulation of chloride conductance by nucleotides in pancreatic and parotid zymogen granules. *Biochem. J.* **272,** 119–126.
13. Piiper, A., Plusczyk, T., Eckhardt, L., and Schulz, I. (1991) Effects of cholecystokinin, cholecystokinin JMV-180 and GTP analogs on enzyme secretion from

permeabilized acini and chloride conductance in isolated zymogen granules of the rat pancreas. *Eur. J. Biochem.* **197,** 391–398.

14. Thevenod, F., Chathadi, K. V., and Hopfer, U. (1992) ATP-sensitive K^+ conductance in pancreatic zymogen granules: block by glyburide and activation by diazoxide. *J. Membr. Biol.* **129,** 253–266.
15. Takuma, T., Ichida, T., Okumura, K., Sasaki, Y., and Kanazawa, M. (1993) Effects of valinomycin on osmotic lysis of zymogen granules and amylase exocytosis from parotid acini. *Am. J. Physiol.* **264,** G895-G901.
16. Thevenod, F., Hildebrandt, J-P., Striessnig, J., de Jong, H. R., and Schulz, I. (1996) Chloride and potassium conductances of mouse pancreatic zymogen granules are inversely regulated by a approximately 80-kDa mdr1a gene product. *J. Biol. Chem.* **271,** 3300–3305.
17. Jena, B. P., Schneider, S. W., Geibel, J. P., Webster, P., Oberleithner, H., and Sritharan, K. C. (1997) Gi regulation of secretory vesicle swelling examined by atomic force microscopy. *Proc. Natl. Acad. Sci. USA* **94,** 13,317–13,322.
18. Preston, G. M., Carroll, T. P., Guggino, W. B., and Agre, P. (1992) Appearance of water channels in *Xenopus* oocytes expressing red cell CHIP28 protein. *Science* **256,** 385–387.
19. Ito, H., Tung, R. T., Sugimoto, T., et al. (1992) On the mechanism of G protein beta gamma subunit activation of the muscarinic K+ channel in guinea pig atrial cell membrane. Comparison with the ATP-sensitive K+ channel. *J. Gen. Physiol.* **99,** 961–983.
20. Schwiebert, E. M., Kizer, N., Gruenert, D. C., and Stanton, B. A. (1992) GTP-binding proteins inhibit cAMP activation of chloride channels in cystic fibrosis airway epithelial cells. *Proc. Natl. Acad. Sci. USA* **89,** 10,623–10,627.
21. Kirsch, G. E., Codina, J., Birnbaumer, L., and Brown, A. M. (1990) Coupling of ATP-sensitive K^+ channels to A1 receptors by G proteins in rat ventricular myocytes. *J. Am. J. Physiol.* **259,** H820–H826.
22. Schwiebert, E. M., Light, D. B., Fejes-Toth, G., Naray-Fejes-Toth, A., and Stanton, B. A. (1990) A GTP-binding protein activates chloride channels in a renal epithelium. *J. Biol. Chem.* **265,** 7725–7728.
23. Marinelli, R. A., Pham, L., Agre, P., and LaRusso, N. F. (1997) Secretin promotes osmotic water transport in rat cholangiocytes by increasing aquaporin-1 water channels in plasma membrane. Evidence for a secretin-induced vesicular translocation of aquaporin-1. *J. Biol. Chem.* **272,** 12,984–12,988.
24. Knepper, M. A. (1994) The aquaporin family of molecular water channels. *Proc. Natl. Acad. Sci. USA* **91,** 6255–6258.
25. Yasui, M., Hazama, A., Kwon, T. H., Nielsen, S., Guggino, W. B., and Agre, P. (1999) Rapid gating and anion permeability of an intracellular aquaporin. *Nature* **402,** 184–187.
26. Knepper, M. A. and Inoue, T. (1997) Regulation of aquaporin-2 water channel trafficking by vasopressin. *Curr. Opin. Cell Biol.* **9,** 560–564.
27. Cho, S. J., Abdus Sattar, A. K., Jeong, E. H., et al. (2002) Aquaporin 1 regulates GTP-induced rapid gating of water in secretory vesicles. *Proc. Natl. Acad. Sci. USA* **99,** 4720–4724.

28. Jena, B. P., Padfield, P. J., Ingebritsen, T. S., and Jamieson, J. D. (1991) Protein tyrosine phosphatase stimulates Ca(2+)-dependent amylase secretion from pancreatic acini. *J. Biol. Chem.* **266,** 17,744–17,746.
29. Cameron, R. S., Cameron, P. L., and Castle, J. D. (1986) A common spectrum of polypeptides occurs in secretion granule membranes of different exocrine glands. *J. Cell Biol.* **103,** 1299–1313.
30. Rosenzweig, S. A., Miller, L. J., and Jamieson, J. D. (1983) Identification and localization of cholecystokinin-binding sites on rat pancreatic plasma membranes and acinar cells: a biochemical and autoradiographic study. *J. Cell Biol.* **96,** 1288–1297.
31. Bradford, M. M. (1976) A rapid and sensitive method for the quantitation of microgram quantities of protein utilizing the principle of protein-dye binding. *Anal. Biochem.* **72,** 248–254.
32. Laemmli, U. K. (1970) Cleavage of structural proteins during the assembly of the head of bacteriophage. T4. *Nature* **227,** 680–685.
33. Bendayan, M. (1984) Protein A-gold immunocytochemistry: technical approach, applications and limitations. *J. Electron Microsc. Tech.* **1,** 243–270.
34. Schneider, S. W., Sritharan, K. C., Geibel, J. P., Oberleithner, H., and Jena, B. P. (1997) Surface dynamics in living acinar cells imaged by atomic force microscopy: identification of plasma membrane structures involved in exocytosis. *Proc. Natl. Acad. Sci. USA* **94,** 316–321.
35. Henderson, R. M., Schneider, S., Li, Q., Hornby, D., White, S. J., and Oberleithner, H. (1996) Imaging ROMK1 inwardly rectifying ATP-sensitive K^+ channel protein using atomic force microscopy. *Proc. Natl. Acad. Sci. USA* **93,** 8756–8760.
36. Walz, T., Hirai, T., Murata, K., et al. (1997) The three-dimensional structure of aquaporin-1. *Nature* **387,** 624–627.
37. Cheng, A., van Hoek, A. N., Yeager, M., Verkman, A. S., and Mitra, A. K. (1997) Three-dimensional organization of a human water channel. *Nature* **387,** 627–630.
38. Jena, B. P., Gumkowski, F. D., Konieczko, E. M., von Mollard, G. F., Jahn, R., and Jamieson, J. D. (1994) Redistribution of a rab3-like GTP-binding protein from secretory granules to the Golgi complex in pancreatic acinar cells during regulated exocytosis. *J. Cell Biol.* **124,** 43–53.
39. Agre, P., Bonhivers, M., and Borgnia, M. J. (1998) The aquaporins, blueprints for cellular plumbing systems. *J. Biol. Chem.* **273,** 14,659–14,662.
40. Murata, K., Mitsuoka, K., Hirai, T., et al. (2000) Structural determinants of water permeation through aquaporin-1. *Nature* **407,** 599–605.

17

Imaging and Probing Cell Mechanical Properties With the Atomic Force Microscope

Kevin D. Costa

Summary

This chapter describes the use of the atomic force microscope (AFM) to probe and map out regional variations in apparent elastic properties of living cells. The importance of mechanics in the field of cell biology is becoming more widely appreciated, and the AFM has unique advantages for cell mechanics applications. However, care must be taken in the acquisition, analysis, and interpretation of AFM indentation data. To help make this powerful technique accessible to a broad range of investigators, detailed procedures are provided for all stages of the AFM experiment from sample preparation through data analysis and visualization.

Key Words: AFM; scanning probe microscope; nanoindentation; elastography; micromechanics; elastic modulus; stiffness; viscoelasticity; cytoskeleton.

1. Introduction

1.1. Role of Mechanics in Cell Biology

The cell is the basic building block of all living organisms. Every tissue in the body is comprised of cells, and the specialized function of various tissues and organs is fundamentally related to differences in mechano-electro-chemical characteristics of the individual constituent cells and their organization within the tissues. Cell mechanics deals with the mechanical properties and functions of cells *(1)*, including how such properties relate to individual components of the internal cytoskeleton *(2–5)*, how cells respond to external mechanical stimulation *(6–8)*, and how they, in turn, remodel and alter the extracellular matrix *(9–12)*. Indeed, many fundamental aspects of cellular function, including shape, deformability, motility, division, and adhesion, are critically dependent on the mechanical properties (e.g., stiffness, nonlinearity, anisotropy, and heterogeneity) of the cell. There is also growing evidence that

From: *Methods in Molecular Biology, vol. 319: Cell Imaging Techniques: Methods and Protocols*
Edited by: D. J. Taatjes and B. T. Mossman © Humana Press Inc., Totowa, NJ

the state of health or disease of a cell might be detectable as changes in cell mechanical properties *(13)*.

Although micropipet aspiration has become an established method for characterizing the mechanical properties of cells floating in suspension *(14–18)*, a number of alternative techniques (including magnetic twisting cytometry *(19)*, laser tracking microrheology *(20)*, magnetic tweezers *(21)*, the optical stretcher *(22)*, and various cell indenters *(23–25)* have been developed to study mechanical properties of adherent cells with a well-defined cytoskeleton. In particular, following its invention in 1986 as a high-resolution imaging tool *(26)*, the atomic force microscope (AFM) has rapidly become a popular method for studying mechanical properties of living cells in culture *(13,27)*.

1.2. Principle of Operation of AFM

The AFM is well suited for cell mechanics applications because of its high sensitivity (subnanonewton), high spatial resolution (submicrometer), and the ability to be used for real-time measurements in a physiologic aqueous cell culture environment. An important advantage of AFM over other cell mechanics techniques is the ability to combine high-resolution scanning with nano-indentation, which allows direct correlation of local mechanical properties with underlying cytoskeletal structures *(4,28)*. Furthermore, commercial availability of the AFM makes it accessible to a broad range of investigators.

In principle, the AFM is a relatively simple instrument that involves laser tracking of the deflection of a microscopic cantilever probe as its tip scans, indents, or otherwise interacts with the cell (*see* **Fig. 1**). The AFM probe is the transducer of the instrument and typically consists of a rectangular or V-shaped cantilever beam about 100–300 μm long and about 0.5 μm thick, microfabricated of silicon or silicon nitride (*see* **Fig. 2**). The standard AFM probe has an integrated pyramid-shaped tip with a blunted point having a radius of curvature in the 50- to 100-nm range. It is this tip that physically contacts the cell, whereas the cantilever serves as a soft spring to measure the contact force. The AFM sensor uses a laser beam reflected off the end of the cantilever probe and onto a four-quadrant photodetector to monitor vertical and lateral deflections of the probe due to forces at the tip. The distance between the probe and the photodetector amplifies the reflected laser motion such that movements of the AFM tip on the order of 0.1 nm can be reliably detected *(29)*.

The actuator that moves the AFM probe relative to the sample is a piezoelectric ceramic that deforms in response to applied voltages. The typical *z*-range is about 6 μm, although custom configurations have achieved a *z*-range up to about 20 μm to better accommodate irregular biological samples *(30)*. Piezoelectric positioners

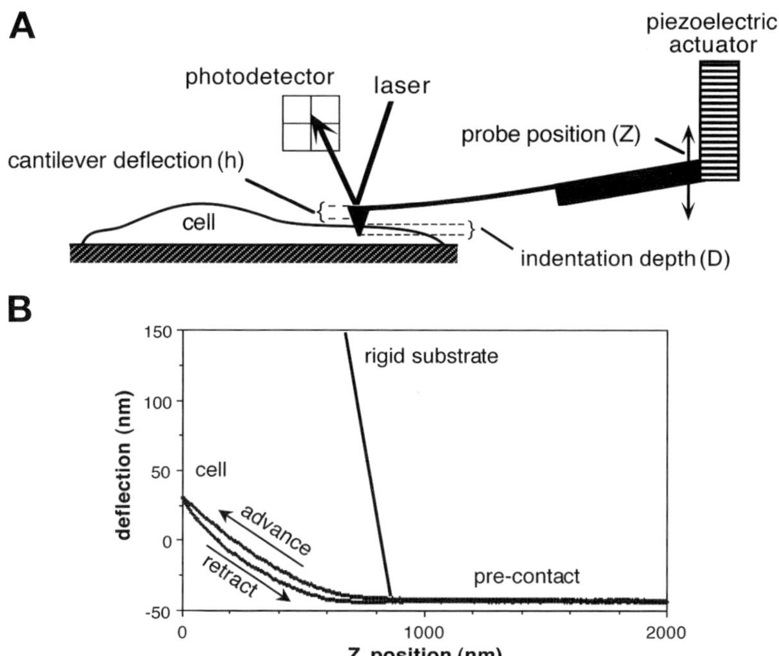

Fig. 1. Schematic of AFM cell indentation experiment. **(A)** Deflection of the AFM cantilever probe is sensed from the reflection of a laser onto a four-quadrant photodetector, and the position of the probe is controlled by a piezoelectric ceramic actuator. As the probe approaches and eventually contacts the cell, further changes in probe position, *Z*, result in a combination of cantilever deflection, *h*, and cell indentation, *D*, dependent on the spring constant of the probe, the geometry of the tip, and the mechanical properties of the cell. **(B)** The resulting measurements of cantilever deflection vs probe position during advancement and retraction of the probe yield the so-called "force curve." On a rigid substrate, the force curve has a linear postcontact region, whereas on a soft sample such as a cell, the postcontact region is more complex.

are used to control movement in the *x-y*-plane as well, with a maximum scan range typically around 100 × 100 µm. Although piezoelectric materials are inherently nonlinear and hysteretic, these effects can be overcome by software compensation (open-loop design) or direct strain-gage monitoring (closed-loop design) for precise positioning of the AFM tip. In the standard AFM configuration, the sample is moved relative to a stationary probe. However, for cell biology applications, it is more convenient to place the entire AFM on the stage of an inverted light microscope to allow simultaneous visualization, including fluorescence microscopy, of the cells ***(30)***. In this configuration, the AFM probe is moved relative to a stationary sample.

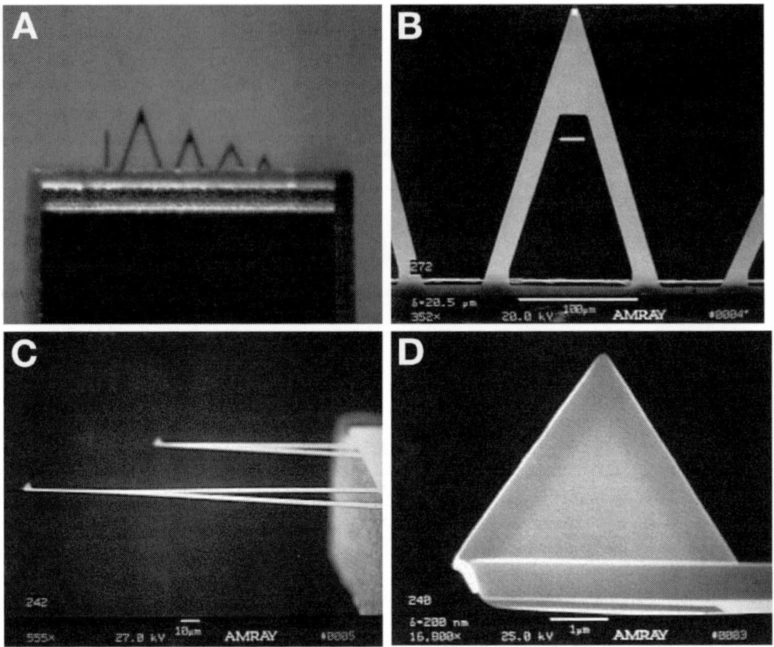

Fig. 2. **(A)** Bright-field image of silicon nitride Microlever® probes on a chip. **(B–D)** Scanning electron microscope images of various AFM probes with standard, unsharpened, integrated pyramidal tips. Scale bars: 100 µm **(B)**, 10 µm **(C)**, 1 µm **(D)**.

The standard AFM controller allows the tip to be raster-scanned over the surface of the cell while monitoring the tip deflection, yielding a high-resolution image of the cell topography. It is also possible to extend and retract the probe perpendicular to the cell surface, thus performing a nano-indentation experiment that can be used to measure the contact force and extract cell mechanical properties.

1.3. Overview

As a wide variety of studies demonstrate ***(4,28,31–36)***, the AFM can provide novel information relating cellular structure, mechanical properties, and functions. To help make such tests accessible to a broad range of investigators, this chapter focuses on procedures required to prepare cells for AFM analysis, to characterize the AFM probe, to conduct the AFM indentation experiment, and to analyze and visualize the resulting data. The recent method of analyzing AFM indentation data to extract a pointwise apparent modulus ***(37)*** is highlighted. As the AFM continues to be developed for cell mechanics applications, other testing procedures and methods of analysis might emerge ***(28,38,39)***, but the standard indentation experiment is likely to remain an important protocol for characterizing cell material properties.

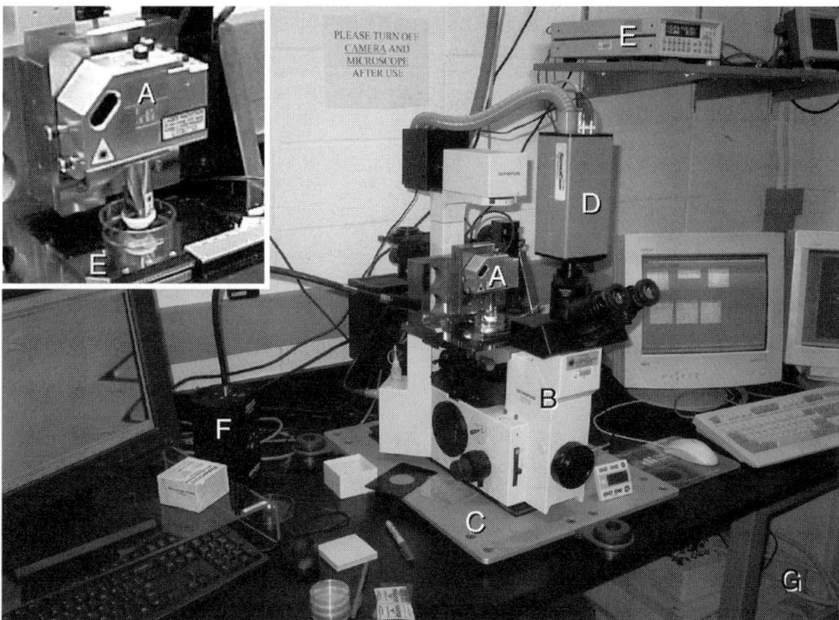

Fig. 3. Typical AFM setup for cell biology applications includes **(A)** AFM scanner housing the laser, photodetector, and piezoelectric actuator, with the cantilever probe mounted at the end exposed to the Petri dish contents, **(B)** inverted light microscope with custom stage, **(C)** vibration isolation platform, **(D)** cooled charge-coupled device camera with vacuum hose leading to remote fan, **(E)** fluid sample heater and controller, **(F)** fiberoptic lamp, and **(G)** AFM controller electronics and computer.

The AFM setup used in the author's laboratory and described herein is a Bioscope® from Digital Instruments (Veeco Metrology Group, Santa Barbara, CA), and a number of accessories and options were also obtained from Digital Instruments for compatibility and convenience (*see* **Fig. 3** and **Note 1**). However, the materials and methods detailed below are intended to apply to any AFM that can be used in a fluid environment to study living cells in monolayer cultures.

2. Materials

2.1. Cell Culture

Beyond the standard solutions and materials used for culturing the cell types of interest, there are two specially recommended cell culture supplies.

1. Gibco CO_2-independent medium (cat. no. 18045, Invitrogen Corp, Carlsbad, CA) supplemented with 4 m*M* L-glutamine according to the manufacturer's instructions (*see* **Note 2**).

2. A 60 × 15-mm polystyrene tissue culture dish (cat. no. 353002, Becton Dickinson, Franklin Lakes, NJ) or other small open-top culture substrate of comparable stiffness that fits under the AFM scanner.

2.2. Probe Setup and Characterization

1. Low-power (40× to 80× max) dissecting microscope (e.g., model SZ-60, Olympus, Woodbury, NY) with external light source.
2. Several microscopic, flat-tipped, stainless-steel tweezers, such as EREM's broad straight tip (cat. no. 2ASA), fine straight tip (cat. no. 3CSA), and 10° offset tip (cat. no. 5ASA), available from Cooper Hand Tools (Raleigh, NC).
3. 30G Hypodermic needles (cat. no. 305128, Becton Dickinson) and 3-mL disposable syringes (cat. no. 309585, Becton Dickinson).
4. Cotton-tipped swabs (cat. no. 14-960-3N, Fisher Scientific, Pittsburgh, PA) and small delicate task Kimwipes® (cat. no. 06-666, Fisher).
5. Glass microscope slides (e.g., cat. no. 12-550C, Fisher) or mica sheets (cat. no. 56, Ted Pella, Redding, CA).
6. Bottled compressed air duster (e.g., cat. no. 23-022523, Fisher).
7. Double-sided cellophane tape (Scotch 137, 3M Corp.) and small Post-it® notes (cat. no. 653, 3M Corp.).
8. Unsharpened, gold-coated, silicon-nitride (Si_3N_4) AFM cantilever probes with integrated pyramid tips (model no. MLCT-AUHW, Veeco Metrology Group, Santa Barbara, CA) (*see* **Notes 3** and **4**).
9. Silicone-bottom plastic containers for storing and organizing individual AFM probes (e.g., TipBox®; BioForce Nanosciences, Ames, IA) (*see* **Fig. 4C**).
10. Ultrasharp® silicon calibration grating with a regular pattern of sharp points having a cone angle <10° and tip radius <10 nm (cat. no. TGT01, Silicon-MDT, Moscow, Russia, available from K-TEK International, Portland, OR) for characterizing the shape, angle, and radius of curvature of AFM probe tips.
11. Micromachined silicon force calibration cantilevers (model no. CLFC-NOBO, Veeco Metrology) for calibrating the AFM probe spring constant.
12. Optional ozone/UV (ultraviolet) decontamination chamber (e.g., Ozone PSD-II; Novascan Technologies, Ames, IA) for cleaning and reusing expensive custom AFM probes.

2.3. AFM System

1. Atomic force microscope compatible with operation in fluid (*see* **Note 1**), including controller electronics and driver software.
2. Inverted light microscope (model IX-70, Olympus) with 10×, 20×, and 40× long working distance objective lenses and trinocular head adapter for video camera mount (*see* **Note 5**). The microscope stage is typically a custom part designed to support a particular model of AFM and supplied by the AFM manufacturer. The stage itself should be adjustable to facilitate alignment of the AFM probe over the objective lens and should also allow separate *x*-*y* translation of the sample relative to the AFM probe.

3. A gooseneck fiberoptic light source (Fiber-Lite 190; Dolan-Jenner, Lawrence, MA) provides illumination of the sample by shining light into the AFM head, which has a built-in mirror for directing the light downward toward the objective lens on the inverted microscope. Manual adjustment of the fiberoptic lamp position can dramatically alter shadowing and contrast for visualizing the cell culture sample.
4. Bioscope fluid sample heater (Digital Instruments) for maintaining cells at a physiological temperature of 37°C. To minimize evaporation of fluid and resulting changes in osmolarity of the cell culture medium and to avoid condensation of fluid on the inner electronics of the AFM scanner, this heater is supplied with special silicone covers that fit over a standard 60-mm culture dish during AFM scanning (*see* **Note 6**).
5. A vibration isolation platform (e.g., Stable Table; VayTek, Fairfield, IA) to minimize vertical and horizontal mechanical vibration of the microscope. Placing the AFM on a slab of solid marble and suspending it from the ceiling using bungee cords can be an effective alternative ***(4)***.
6. A video camera mounted on the trinocular head of the microscope and connected to a screen and VCR provides a convenient means of visualizing and recording AFM experiments and monitoring the condition and stability of the cell culture. For low-light fluorescence applications, we use a cooled digital video camera (Sensicam, Cooke Corp, Auburn Hills, MI) with an optional remote fan to minimize mechanical vibration on the microscope. The camera is connected to the microscope using a 0.63× coupler lens (model no. D63IXC; Diagnostic Instruments, Sterling Heights, MI) so that the magnification and field of view in the camera match that in the oculars.

3. Methods

Before beginning these procedures, the AFM controller, video monitors and computers, vacuum pump, lamps, and other major electronics should be turned on and warmed up for at least 60–90 min to achieve a stable temperature and to minimize drift of electrical signals during the experiment.

3.1. Preparing and Mounting the AFM Probe

The following methods for preparing and mounting the AFM cantilever probe on the scanner head will help minimize damage to the cells, the probes, and the scanner.

1. Carefully separate a row of rectangular chips from the wafer of cantilevers (*see* **Fig. 4**) by lifting one end, then the other, with a pair of broad-tipped microscopic tweezers. Holding the row of chips with one pair of tweezers, use a second pair to separate individual chips by carefully snapping them apart at the score line. Arrange the individual chips in a TipBox® for subsequent use.
2. Using microscopic tweezers, carefully grasp one chip from the side and examine under the dissecting microscope to inspect for bent, broken, or dirty probes. Stabilizing the chip against the dissecting scope stage with tweezers, use a 30G needle in the other hand to carefully break off any probes longer than the one selected

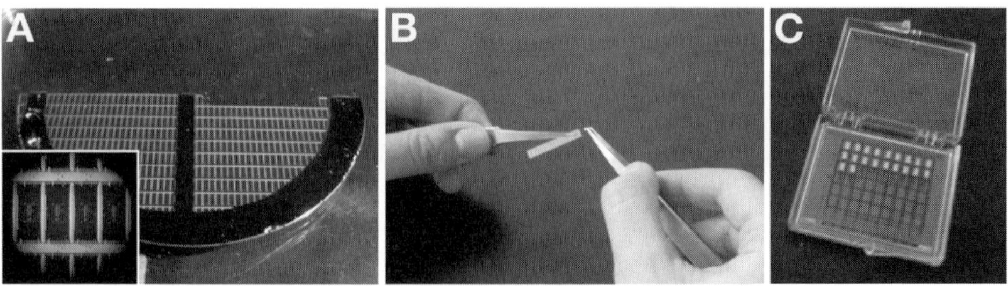

Fig. 4. **(A)** A half-wafer of Microlever probes with one row that has been removed. Inset shows magnified view of a region of the wafer. **(B)** Individual chips must be snapped off the end of a connected row. **(C)** Separated chips are arranged in a silicone-bottom storage box.

for the experiment (*see* **Note 3**). Only this one probe will be monitored by the AFM laser, and longer probes could contact and potentially damage the cell undetected.

3. Carefully mount the chip on a holder appropriate for use in fluids and inspect once more to ensure that the probe is undamaged.
4. Mount the holder on the AFM scanner and then hold the scanner over a Kimwipe® at about a 20° angle with the probe facing down. Using a 30G needle on a 3-mL syringe, carefully apply a few drops of culture medium alongside the AFM probe, allowing fluid to engulf the probe without introducing air bubbles and without dripping down into the scanner (*see* **Note 7**).
5. Right the scanner, move it onto the inverted microscope stage, and plug it in; then use the stepper motor to lower it into a small volume of culture medium on a glass slide (or mica sheet). Use just enough fluid (approx 2 mL) to form a meniscus between the probe holder and the glass slide. Adjust the fiberoptic light and focus to view the AFM probe using the inverted microscope. If any air bubbles are observed, remove the scanner, remove the probe holder and carefully blot it dry with the corner of a Kimwipe® or cotton swab and compressed air; repeat **step 4**.
6. Adjust the laser spot so that it reflects off the tip of the cantilever (the optical microscope is very helpful for this step) and provides an adequate voltage signal centered on the photodetector. The photodetector voltage should be monitored for several minutes to ensure that the system has stabilized, as the probe might deflect as it is heated by the laser.
7. Use the stepper motor to carefully move the AFM toward the glass slide (*see* **Note 8**). Adjust the set-point reference voltage to 0 V, the vertical deflection voltage to about –1 V, and the horizontal deflection voltage to 0 V. Also set the scan size to zero so that the probe remains stationary when it initially contacts the glass surface (and later the cell). Finally, use the AFM control software to automatically advance the probe until it makes contact, or engages, on the glass slide with a force proportional to the difference between the vertical deflection voltage and the set-point reference voltage.
8. In contact force mode, perform an indentation test on the glass slide. The *z*-range and scan rate can be adjusted by the user and should be set to values comparable

to what will be used with the cell indentation experiments. We have found that typical values of 3 μm for *z*-range and 1 Hz for scan rate work well for many cell types. Note that because the slide is effectively rigid compared to the stiffness of the cantilever, there is no indentation, so for every nanometer that the probe advances, the tip deflection should increase by exactly 1 nm. The resulting measurements of tip deflection vs probe position (the "force curve") should ideally have a flat precontact region with an abrupt transition to a linear postcontact region (*see* **Fig. 1B**).

9. A manual trace of the linear postcontact curve provides a conversion factor, or detector sensitivity, in nanometers per volt that is used to convert the photodetector signal into probe deflection.

For standard AFM imaging applications, the shape of the precontact region of the force curve is of little consequence. However, for cell indentation studies, the transition from precontact to postcontact is often subtle and gradual. Determination of this contact point is most accurate and reliable when the precontact portion of the curve is ideally flat and linear with minimal noise. We, therefore, make an extra effort to optimize the shape of the precontact region as follows:

1. Withdraw the AFM probe approx 1 mm from the glass slide. Readjust the set point and horizontal deflection voltages to 0 V, and set the vertical deflection voltage to about +1 V; then engage the AFM tip. Setting the vertical deflection above the set point will cause the AFM controller software to think the tip is immediately in contact with the sample even though it is not (note that the control-F key sequence can also be used with Digital Instruments Nanoscope software to perform such a "false engage").
2. In the contact force mode, perform a series of repeated excursions of the AFM probe over the full *z*-range of the scanner. Because the probe is far from any surface, the resulting force curve should ideally be a straight flat line indicating zero probe deflection. We have found that by adjusting the location of the laser spot on the AFM probe, it is possible to drastically alter the shape of this noncontact curve: The curve can be nonlinear and can exhibit oscillatory reflection artifacts, and the overall slope can switch from positive to negative (*see* **Fig. 5**). We typically aim for a force curve in which the deviation from a flat straight line is less than 4 nm over the full *z*-range of the scanner, at times accepting some nonlinearity to obtain the maximum possible smoothness.
3. Once the laser position has been optimized, it should not be adjusted for the remainder of the experiment. Withdraw the AFM probe and repeat **steps 7–9** above to determine the sensitivity of the photodetector with the laser in its new optimized location on the probe. If the laser position is accidentally altered during an experiment, this process should be repeated before force measurements are resumed.
4. Withdraw the AFM probe several millimeters using the stepper motor, and then lift the scanner to ensure that nothing bumps the probe when the glass slide is removed and exchanged for another sample.

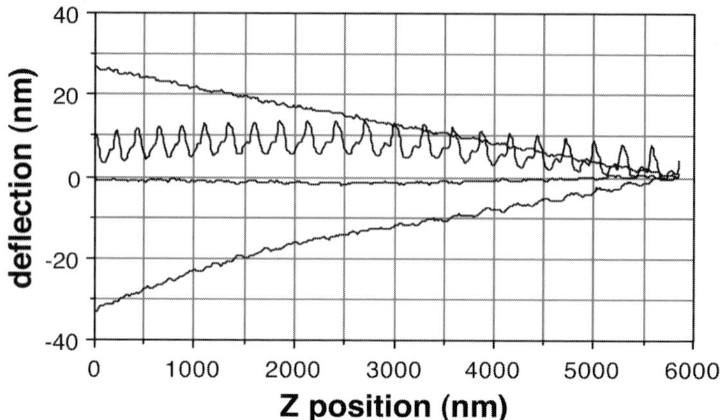

Fig. 5. Four examples of noncontact force curves in water (retraction data omitted for clarity) obtained using a single cantilever undergoing repeated 6-μm excursions at 1 Hz. By altering the position of the laser reflection near the end of the cantilever, substantial deviations from the ideal zero-deflection noncontact curve could be observed.

3.2. AFM Probe Characterization

The two main characteristics of the AFM cantilever probe that must be accurately determined for each experiment are (1) the spring constant, κ, which is required to compute contact force from probe deflection, and (2) the tip geometry, which is required to accurately extract the elastic modulus from the force–depth indentation response. Although nominal values of spring constant and tip radius might be adequate for some purposes such as mapping of relative stiffness on a single cell, accurate evaluation of cell mechanical properties and comparison from one experiment to another requires more careful probe characterization.

3.2.1. Spring Constant

The spring constant of the AFM probe is required to compute the contact force from the measured probe deflection (*see* **Note 9**). Several alternative methods have been proposed for experimentally determining the spring constant of an AFM cantilever from measurements of its resonance frequency ***(40–42)*** or thermal noise behavior ***(29)***. We prefer another simple method using a reference cantilever for calibration ***(43,44)***. This method requires no assumptions about the geometry of the AFM probe, it can be performed in an aqueous environment, and the probe undergoes a bending mode of deformation consistent with cell indentation experiments. The calibration process is as follows:

1. Carefully mount a calibration cantilever on a clean dry glass slide using a small piece of double-stick tape.
2. Add a few drops of deionized water alongside the probe using a 3-mL syringe with a 30G needle, again using caution to avoid introducing air bubbles as the liquid

engulfs the chip. Place this slide on the microscope stage and lower the AFM scanner until a meniscus forms between the AFM probe and the calibration cantilever, carefully adding drops of liquid with the syringe as needed (*see* **Note 10**).

3. Engage the AFM tip on the glass as described in **step 7** of **Subheading 3.1.** for determining the photodetector sensitivity.
4. Obtain a force curve on the glass slide as described in **step 8** of **Subheading 3.1.** and save this measurement in a data file.
5. Withdraw the scanner several millimeters using the stepper motor.
6. Engage the AFM tip on the end of the calibration cantilever (*see* **Note 11**).
7. Obtain a force curve on the calibration cantilever using the same scan settings as in **step 4** and save this measurement in a data file. In this case, the force curve should also have a linear postcontact region, but the slope should appear less than on glass because of bending of the calibration cantilever by the AFM probe.
8. Withdraw the AFM probe several millimeters using the stepper motor and then lift the scanner to ensure that nothing bumps the probe when the calibration cantilever is removed and exchanged for another sample.
9. Export the two above data files and fit a straight line to the postcontact data to obtain the slope of this region. The slope on glass, m_g, should ideally be unity (i.e., 1000 nm/μm). The slope on the calibration cantilever, m_c, should have a smaller value. From these two values and the known spring constant of the calibration cantilever, κ_c, one can compute the spring constant of the AFM probe, κ_p, using the formula $\kappa_p = \kappa_c(m_g - m_c)/(m_c \cos\theta)$, where θ is the acute angle between the AFM probe and the calibration cantilever (typically about 10°, depending on how the probe is mounted on the particular model of AFM scanner).

3.2.2. Tip Geometry

As will be described in **Subheading 3.5.2.**, quantification of cell mechanical properties from AFM indentation measurements requires accurate knowledge of the geometry of the AFM probe tip (*see* **Note 12**). To characterize tip geometry, one can use scanning electron microscopy to obtain a high-resolution image of the tip (*see* **Fig. 2D**). However, this requires special facilities and is often inconvenient and costly to perform on a regular basis. A simple alternative is tip self-imaging *(45)*, in which the AFM probe is scanned over a sample with very sharp features compared to the tip geometry, yielding a reversed image of the tip itself (*see* **Fig. 6**).

1. Carefully mount the Ultrasharp® silicon calibration grating on a clean dry glass slide using a small piece of double-stick tape. Then add fluid and transfer to the AFM stage as described in **step 2** of **Subheading 3.2.1.**
2. Engage the AFM tip on the calibration grating using the procedure described in **step 7** of **Subheading 3.1.**, using caution to not crash the probe and damage the calibration grating.
3. In the contact imaging mode, increase the *x-y* scan size to about 5 μm, using a scan speed of about 1 Hz and a low pixel resolution (128 pixels) to get a quick image of the tip. Use the *x*-axis and *y*-axis offset controls to center the probe tip in the

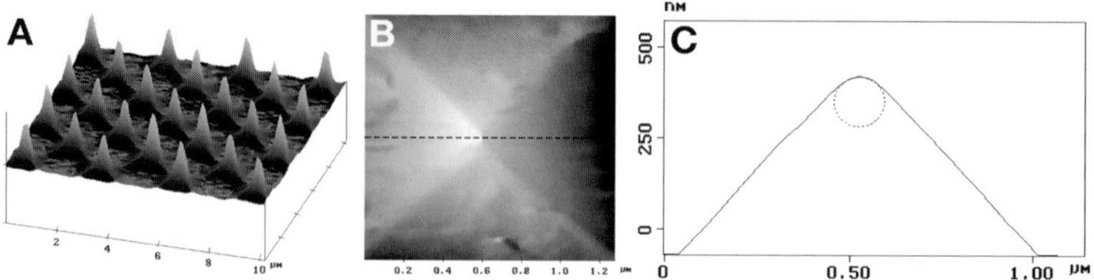

Fig. 6. **(A)** 10 × 10-µm contact mode scan of an Ultrasharp® calibration grid with expanded *z* scale. **(B)** Magnified 1.27-µm scan of a single sharp feature (top view) reveals a reverse image of the pyramidal Si_3N_4 AFM probe tip. **(C)** A 1:1 aspect ratio cross-section through the peak of the height image (broken line in B) reveals a blunt tip with a 70-nm radius of curvature indicated by the dotted circle.

image and adjust the scan size as needed to capture the tip geometry. Then perform a final scan at high pixel resolution and save this image file.

4. The file can then be exported and analyzed using image processing software to determine the radius of curvature of the probe tip (*see* **Fig. 6C**), or the AFM software might have this capability built in it.

3.3. Cell Indentation

3.3.1. Cell Culture Preparation

Analysis of mechanical properties by AFM indentation is compatible with standard cell culture techniques for most adherent cell types using confluent or subconfluent monolayer cultures in a small (e.g., 60 mm) culture dish or similar flat, stiff substrate (*see* **Note 13**). Prior to moving the cells onto the stage of the AFM, standard cell culture medium should be removed and replaced with a minimal volume (e.g., 2–3 mL in a 60-mm dish) of CO_2-independent medium (*see* **Notes 2** and **7**) and the cells should be allowed to equilibrate to the temperature at which the AFM experiment will be conducted.

3.3.2. Single-Force Curves

The most fundamental cell mechanics experiment performed using the AFM is the simple indentation test described as follows:

1. After placing the cell culture dish on the microscope stage, use the stepper motor to lower the AFM until it makes contact with the surface of the cell culture medium. Center the laser reflection on the photodetector and monitor its position for several minutes, as it will tend to drift until the probe and sample become thermally equilibrated.
2. Adjust the set-point reference voltage to 0 V, the vertical deflection voltage to about –1 V, and the horizontal deflection voltage to 0 V. Also set the *x*-axis and *y*-axis

offsets to zero and the scan size to zero so that the probe remains stationary when it initially contacts the cell. Finally, align the probe tip with a region of interest as described in **Note 11** and use the AFM control software to automatically advance the probe until it engages on the cell at the desired location.

3. In the contact force mode, perform an indentation test on the cell. The z-range and scan rate should be adjusted by the user to obtain a desirable force curve. We have found typical values of 3 μm for z-range and 1 Hz for scan rate to work well for many cell types. The ideal force curve should have a flat precontact region that transitions around one-half to three-quarters of the way through the z-excursion to a nonlinear postcontact region where the AFM tip is indenting the soft cell (*see* **Fig. 1B**). Hysteresis between the advancing and retracting portions of the curve might also be observed (*see* **Notes 14** and **15**).
4. Save these measurements of tip deflection vs probe position in a data file. Performing multiple indentations at the same location is one way to evaluate stability of the cells to characterize potential changes in cell properties in response to probing.

3.3.3. Arrays of Force Curves (Force Mapping)

Considering the highly localized nature of AFM indentation measurements obtained using a standard probe and the substantial regional variability of the cytoskeleton and other intracellular structures, single-indentation tests might be inadequate to characterize cell mechanical properties for some applications. Another approach is to perform an array of indentations covering some region of interest on the cell, which then allows evaluation of the variability in the measurement and also can be used to map the elastic properties of the cell ***(13,46)***. This can be combined with contact mode imaging (and fluorescence microscopy) to correlate elastic properties with specific cytoskeletal structures.

1. A simple assessment of regional variability of cell mechanical properties can be obtained by performing several indentations in a small array in the region of interest. A square array of up to 4 × 4 indentations covering up to a 5 × 5-μm region of the cell provides a relatively rapid (<30 s) assessment of local mechanical heterogeneity, with the variability tending to decrease as the size of the region probed also decreases. This is easiest when the AFM controller software allows such indentation arrays to be executed automatically, although such an experiment could also be performed by adjusting the x-axis and y-axis offsets manually.
2. Cell indentation can also be used to obtain a more detailed map of mechanical properties. Such "force mapping" techniques generally involve performing a much larger array of indentations (16 × 16 to 128 × 128 locations) covering a substantial region (up to 100 × 100 μm) of one or more cells (*see* **Note 16**). The actual indentation force, or some derived measure of elastic properties, is then rendered at each location to form an image with each pixel representing one of the indentation sites (see **Subheading 3.6.** for further details). Note that the acquisition time for such arrays of force curves can be quite long (hours), during which time it is

important to maintain a stable culture environment and have some independent verification of cell viability (such as video microscopy).

3. To obtain structurally correlated information on mechanical properties with more rapid scan times, it is possible to combine indentation arrays with contact mode imaging (*see* **Note 17**). With the AFM probe withdrawn a few hundred microns above the cell, adjust the set-point reference and the horizontal deflection voltages to 0 V and the vertical deflection voltage to about –0.5 V to engage at a relatively low contact force. Also set the scan size and *x*- and *y*-axis offsets to zero. Engage the AFM tip at the desired location on the cell (*see* **Note 11**). Set the *x-y* scan speed to 1 Hz or less and increase the scan size to obtain a contact mode image of the desired region of interest (*see* **Note 18**). Based on this image, use the mouse to graphically zoom in on a region of the cell with features of interest. With the new scan size and *x*-axis and *y*-axis offsets so determined, switch to the force mapping mode to obtain an array of indentations as described in **step 2.**

3.4. Cleaning Up

1. At the end of the experiment, withdraw the AFM probe and use the stepper motor to back up the scanner nearly to the limit of the focus range of the optical microscope. This keeps the scanner distant enough to avoid damage when exchanging samples, yet close enough to use the optical microscope to view the AFM probe during setup for the next experiment (*see* **Subheading 3.1.**).
2. Unplug the scanner and remove the probe holder. Use microscopic tweezers to remove the probe and set it aside. Rinse the holder thoroughly with deionized water, dry with the corner of a Kimwipe® or cotton swab, and blow dry with the compressed air duster. Alcohol and other solvents can be used for cleaning calibration standards, but they should not be used on the AFM probe holder to avoid damaging delicate parts.
3. The probe can be disposed of or be rinsed and saved for ozone/UV cleaning and reuse. Note that ozone cleaning can erode and alter the AFM tip, so the tip geometry of reused probes should be characterized as in **Subheading 3.2.2.**

3.5. Data Analysis

The most prevalent method of analyzing AFM indentation data has been application of the so-called "Hertz model" of contact between a rigid indenter and an elastic sample ***(47,48)***. However, it is important to appreciate that the Hertz model makes a number of simplifying assumptions that do not readily apply to the cell indentation problem; these include infinitesimal deformation of a semi-infinitely thick and flat sample, having homogeneous isotropic linear elastic material properties, and using an axisymmetric indenter ***(37,49)***. We have developed a more generalized analysis method whereby computation of an apparent elastic modulus as a function of indentation depth can reveal when deviations from the standard Hertz model become important ***(37)*** (*see* **Note 19**).

The two main steps in the analysis are identifying the initial point of contact with the cell and calculating the pointwise elastic modulus.

3.5.1. Identifying the Contact Point

When indenting a stiff sample, the contact point is readily apparent as an abrupt transition from the precontact region to the postcontact region, as described in **step 8** of **Subheading 3.1.** However, when indenting a cell, the initial contact event occurs when the smallest part of the tip contacts the soft cell membrane, resulting in a smooth and continuous transition from the precontact to postcontact region, as described in **step 3** of **Subheading 3.3.2.** Accurately identifying the initial point of contact, which impacts most subsequent analyses of cell mechanical properties, is not a trivial matter. A number of procedures have been proposed to identify the contact point by fitting some portion of the postcontact data to the Hertz model ***(27,50,51)***. However, when the data deviate from the expected theory, the estimated contact point can become clearly inaccurate ***(35)***. We have therefore developed a two-part polynomial model-fitting algorithm that utilizes both precontact and postcontact data to identify the contact point (*see* **Fig. 7**). This iterative fitting procedure takes just a few seconds per force curve when implemented in Matlab on a personal computer.

1. Using the advance portion of the force curve, select a subset of the data clearly in the precontact region and calculate the average precontact deflection. Also, the z-position data are negated so that the postcontact regime has a positive slope (i.e., increases to the right, rather than to the left, as it is acquired in the standard force curve).
2. Specify a maximum deflection that will be included in the postcontact fit, capturing the earliest portion of the indentation response. A value of 15 or 20 nm beyond the precontact deflection is typical. Begin by assuming a trial contact point, z^{trial}, that corresponds to the data point at half this maximum deflection.
3. Use a standard Levenberg–Marquardt least squares minimization algorithm ***(52)*** to fit the following two-part polynomial model of probe deflection to the data in the specified neighborhood of the trial contact point, z^{trial},

$$h(z) = \begin{cases} a_1 + b_1 z & \text{for } z \le z^{\text{trial}} \\ a_2 + b_2(z - z^{\text{trial}}) + c_2(z - z^{\text{trial}})^2 & \text{for } z > z^{\text{trial}} \end{cases}$$

   where a_1, b_1, b_2, and c_2 are model parameters adjusted in the fitting procedure, and the constraint $a_2 = a_1 + b_1 z^{\text{trail}}$ is enforced to ensure continuity between the precontact and postcontact regimes.
4. Record the resulting value of the minimization error function for this value of z^{trial}.
5. Moving one data point at a time toward the beginning of the precontact region, repeat **steps 3** and **4** using each data point sequentially as the trial contact point for the fitting procedure (*see* **Fig. 7A**).

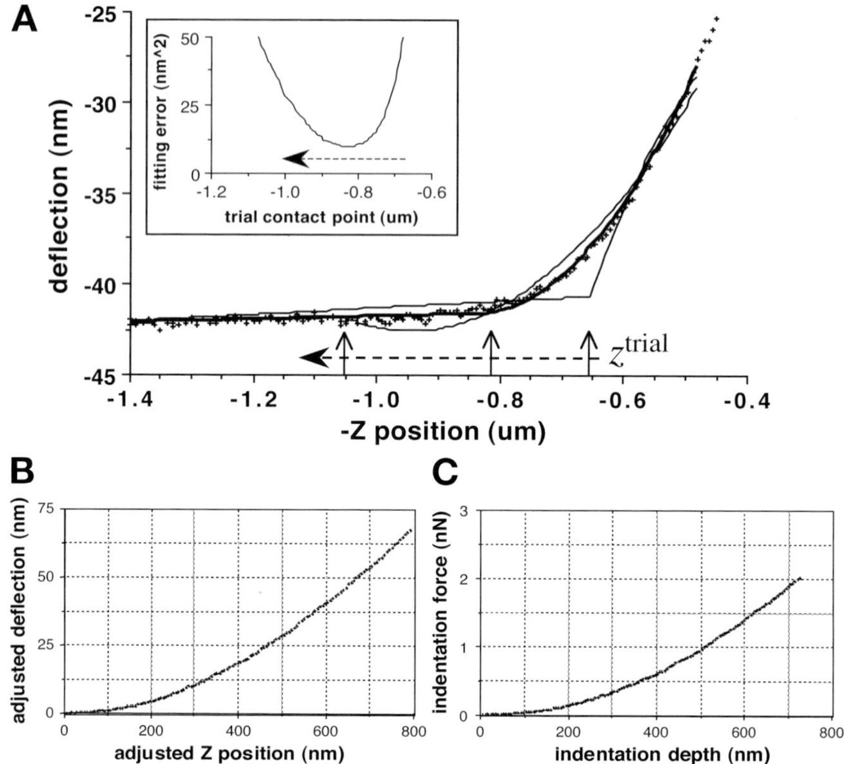

Fig. 7. Analysis of indentation data obtained from the advance portion of the force curve in **Fig. 1B**. **(A)** Procedure for identifying the contact point using a two-part polynomial model-fitting algorithm (*see* text for details). The data scale is expanded to highlight the region of initial contact. Inset shows the fitting error function vs the trial contact point, the minimum of which is used to determine the actual contact point. **(B)** The adjusted postcontact deflection vs z-position with the origin at the contact point. **(C)** The resulting indentation force-vs-depth curve.

6. A plot of the minimization error vs z^{trial} will exhibit a minimum value for the data point at which a best fit is obtained between the linear precontact region and the quadratic (or other polynomial) initial postcontact region (*see* **Fig. 7A,** inset). This data point provides an estimate of the contact point.
7. Finally, the fitting routine is repeated one last time, with the contact point itself included as a fitted model parameter in case the actual point of contact, z^*, occurred somewhere in between two acquired data points.
8. Subtract the (z^*, h^*) coordinates of the contact point from the (z, h) values of each data point, i, in the force curve so that the contact point is shifted to become the origin of the curve using $(z', h')_i = (z, h)_i - (z^*, h^*)$. Precontact data are discarded, leaving the postcontact data (*see* **Fig. 7B**).

9. For each data point, the cantilever spring constant is used to calculate the indentation force, $F = \kappa \times h'$. With κ in newtons/meter (N/m) and h' in nanometers (nm), F has units of nanonewtons (nN). The indentation depth (nm) is calculated using $D = z' - h'$. The resulting plot of indentation force vs depth (*see* **Fig. 7C**) is used to determine the elastic modulus of the cell.

3.5.2. Calculating the Pointwise Modulus

In theory, the indentation force and depth are related by an equation of the form ***(37)*** $F = 2\pi \times \tilde{E} \times \phi(D)$, where $\tilde{E}$ is an apparent elastic modulus that is equivalent to $E/2(1-\nu^2)$ in classical linear elasticity theory, with E being the Young's modulus and $\nu = 0.5$ being the Poisson's ratio for an incompressible material. The term $\phi(D)$ is an expression determined by the geometry of the AFM probe tip. For a blunt-tipped cone with tip angle, 2α, that transitions at radius b into a spherical tip of radius R, the geometric term is determined as follows ***(37,53)***.

For indentations smaller than the tip defect ($D < b^2/R$), the equation is that for a simple spherical indenter:

$$\phi(D) = \frac{4}{3\pi}\sqrt{RD^3}$$

For indentations larger than the tip defect ($D \geq b^2/R$),

$$\phi(D) = \frac{2}{\pi}\left\{aD - \frac{a^2}{2\tan\alpha}\left[\frac{\pi}{2} - \arcsin\left(\frac{b}{a}\right)\right] - \frac{a^3}{3R} + \sqrt{a^2 - b^2}\left[\frac{b}{2\tan\alpha} + \frac{a^2 - b^2}{3R}\right]\right\}$$

where the radius of contact, a, is determined from the following expression:

$$D - \frac{a}{R}\left(a - \sqrt{a^2 - b^2}\right) - \frac{a}{\tan\alpha}\left[\frac{\pi}{2} - \arcsin\left(\frac{b}{a}\right)\right] = 0$$

A smooth transition from the cone to the spherical tip is obtained when $b = R\cos\alpha$ The indentation response for a standard pyramidal silicon nitride probe is equivalent to that for a cone with $\alpha = 39.5°$ ***(37)***; a tip defect with R in the range 50–100 nm is typical (*see* **Subheading 3.2.2.**).

In the general equation $F = 2\pi \times \tilde{E} \times \phi(D)$, the indentation force and depth are measured in the AFM experiment and $\phi(D)$ is determined from the tip geometry as above (*see* **Note 20**). Therefore, the only unknown is $\tilde{E}$. Rather

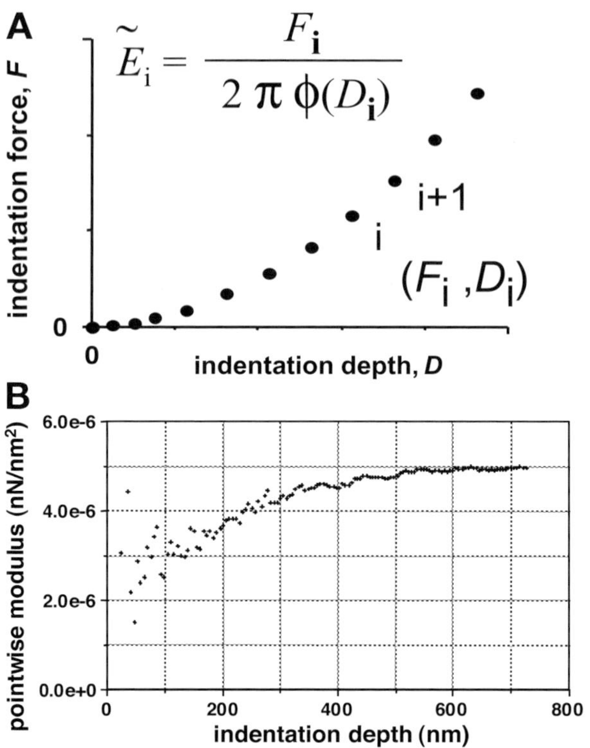

Fig. 8. **(A)** Schematic procedure for calculating the pointwise elastic modulus $\tilde{E}_i$, from individual indentation force–depth data points. **(B)** Plot of pointwise modulus vs indentation depth obtained from data in Fig. 7C.

than assume that $\tilde{E}$ is constant *a priori* (*see* **Note 21**), we solve the equation at each data point, *i*, to obtain a pointwise apparent modulus, $\tilde{E}_i = F_i/(2\pi\phi(D_i))$, as illustrated in **Fig. 8**. With F in nN and D in nm, $\tilde{E}$ has units of nN/nm^2, where 10^{-6} nN/nm^2 = 1000 N/m^2 = 1 kPa.

The elastic properties of the cell are determined from a plot of the pointwise modulus vs indentation depth *(37)*. Changes in the value of $\tilde{E}_i$ indicate deviations from the classical linear elastic Hertz model.

3.6. Data Presentation

The above pointwise modulus analysis can reveal details of local cell mechanical properties that are simply ignored in more traditional analysis methods. However, the increased complexity of the resulting data impacts its subsequent presentation and interpretation. We have found the following

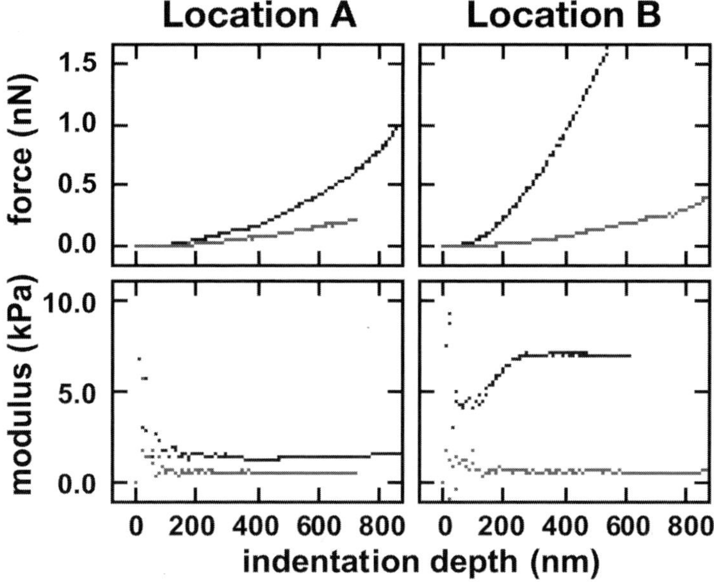

Fig. 9. Indentation force and corresponding pointwise modulus vs indentation depth at two locations 2 μm apart on a living NIH 3T3 fibroblast cell. The two curves in each panel were obtained under control conditions (upper curve) and after 30 min treatment with cytochalasin-B (lower curve).

methods useful for conveying regional variations in the pointwise modulus and correlating these with local cytoskeletal structures.

3.6.1. Pointwise Modulus Curves

A plot of pointwise modulus vs indentation depth directly reveals how the apparent stiffness of the cell varies as it is deformed. For example, measurements on a control NIH 3T3 fibroblast cell (*see* **Fig. 9**) show that $\tilde{E}_i$ can be nearly constant with indentation depth at one location, whereas it can exhibit a strongly increasing depth dependence at a neighboring location just 2 μm away. The increased apparent stiffness with depth might reflect nonlinear elastic properties (sometimes incorrectly called "strain hardening") typical of macroscopic biological tissues or it might reveal through-thickness heterogeneity of material properties, with the AFM interacting with stiffer structures as it probes deeper into the cell. AFM indentation of the same cell following treatment with cytochalasin-B (4 μ*M* for 30 min) results in smaller values of $\tilde{E}_i$ that are essentially independent of indentation depth or location on the cell, suggesting that the previous nonlinear behavior could have been primarily the result of the filamentous actin cytoskeleton, and not simply an artifact of the underlying stiff substrate.

Because the value of $\tilde{E}_i$ can vary substantially with indentation depth, it is necessary to specify the indentation depth when comparing modulus values. Linear interpolation between nearest-neighbor data points can be used to easily obtain modulus values at specific indentation depths for performing descriptive and comparative statistics.

3.6.2. Arrays of Modulus Curves

By performing a more detailed force mapping of a region of a cell, it is possible to correlate the mechanical behavior with underlying cytoskeletal structures. **Figure 10A** shows a contact mode AFM image of a 15 × 15-µm square region of the periphery of a human aortic endothelial cell cultured on a polystyrene dish at room temperature, showing linear structures representing actin stress fibers within the cell. The black square indicates a 3.5 × 3.5-µm subregion where the AFM was zoomed in on an area of interest with a prominent diagonal stress fiber. An 8 × 8 array of 64 indentations was performed in this subregion at an indentation rate of 3 µm/s, and the resulting pointwise modulus curves are plotted in a matrix corresponding to their relative location on the cell (*see* **Fig. 10B**). Strongly nonlinear behavior is exhibited in a diagonal zone coincident with the prominent stress fiber, whereas adjacent regions appear softer with less depth variation.

3.6.3. Images of Regional Stiffness

Arrays of modulus curves as described above contain a tremendous amount of information about the detailed regional and depth-dependent mechanical properties of living cells. However, presented as such, the data can be overwhelming and not ideally suited for presentation in a figure or to an audience. One alternative is to extract the pointwise modulus at a specified indentation depth and create an image in which this modulus value is represented on a gray scale, (i.e., image contrast is based on the local cell stiffness). The resulting image of elastic properties (an "elastogram" *[13]*) provides a convenient visual representation of the regional cell stiffness, although it must be emphasized that with nonlinear elastic behavior, the appearance of the map can vary substantially at different indentation depths. **Figure 10C** shows an example of such an elastogram, with $\tilde{E}_i$ extracted at $D = 200$ nm for the same region represented by the array of modulus curves described earlier. This representation makes it clear that the actin stress fibers are stiffer than the surrounding cytoplasm. However, only by examining the individual modulus curves can one also appreciate that the stress fibers selectively give rise to nonlinear elastic properties, whereas the cytoplasm behaves like a nearly linear elastic material.

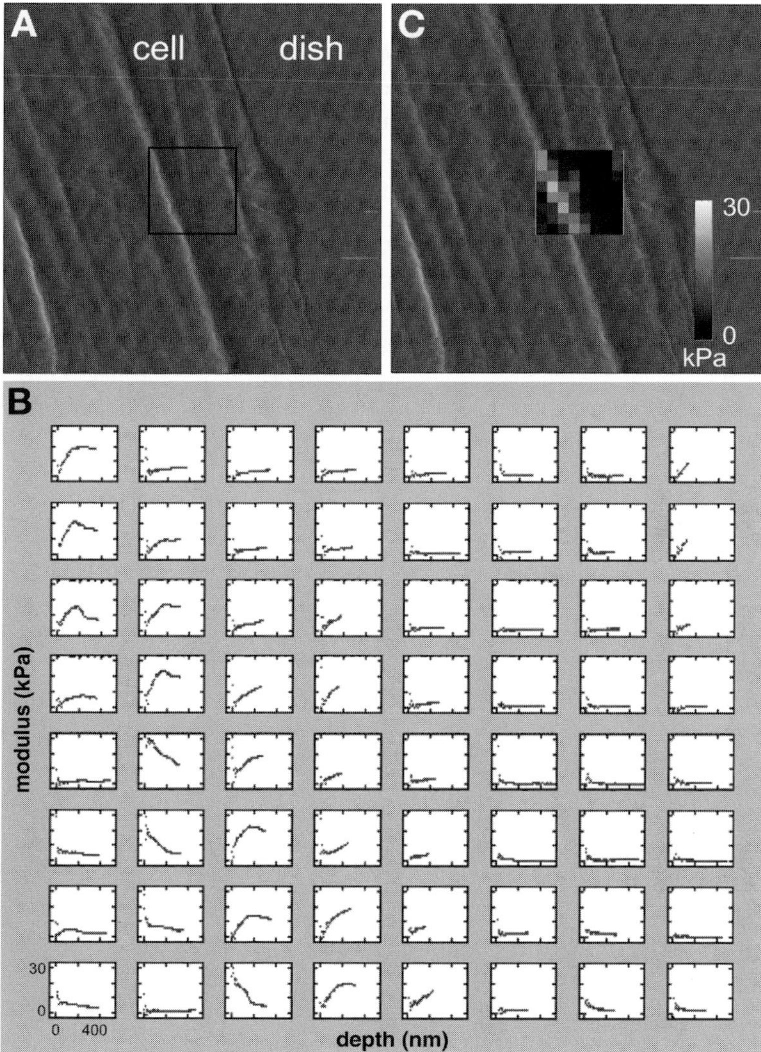

Fig. 10. **(A)** Contact mode AFM image (15 × 15 μm) of the peripheral region of a human aortic endothelial cell, revealing cytoskeletal actin stress fibers. The black square indicates a 3.5 × 3.5-μm region where an 8 × 8 array of 64 indentations was performed to measure local cell mechanical properties. **(B)** Pointwise modulus vs indentation depth from the 8 × 8 array of indentations described in **A**. All curves are shown on the same scale of 0–30 kPa for modulus and 0–400 nm for depth. **(C)** The same contact mode image as **A**, with a superposed gray-scale map of the pointwise modulus from **B** extracted at 200 nm depth.

4. Notes

1. Most commercial AFMs are supported on a three-point frame, with the instrument placed on top of the sample. Alternatively, the Bioscope (Digital Instruments, Santa Barbara, CA) is supported from behind such that the end of the scanner (where the AFM probe is mounted) remains exposed. This permits easy access to the sample and direct visualization of the probe and the fluid volume. It also facilitates compatibility with fluid-exchange systems, electrical stimulators, micropipet manipulators, and other techniques and devices useful for cell biology experiments.
2. When used according to the manufacturer's instructions, this special CO_2 independent culture medium allows cells to be maintained in a normal room-air environment for several hours while maintaining a constant pH, unlike normal bicarbonate-buffered culture media. Naturally, if the AFM is equipped with an environmental chamber that maintains a physiologic 5 to 10% CO_2 environment, this special media is not necessary.
3. A broad range of AFM cantilevers are commercially available, offering variations in shape, stiffness, material composition, surface coating, and tip geometry to satisfy a wide range of AFM applications. Here, we recommend unsharpened, gold-coated, silicon nitride Microlevers® (*see* **Fig. 2**), which are well suited for cell indentation studies for a number of reasons. Unsharpened pyramidal probe tips minimize cell damage and help maintain viability for extended scans ***(4,54,55)***. Sharpened tips offer higher resolution for some AFM imaging applications but tend to penetrate the cell membrane and damage living cells ***(56)***. Gold coating enhances reflectivity for use with cell culture media containing phenol red, which attenuates AFM laser light. Silicon nitride (Si_3N_4) Microlevers® are fabricated with six cantilevers per chip, having spring constants ranging from 0.01 to 0.5 N/m. It is important that the spring constant of the probe be well matched to the sample stiffness. If the probe is too stiff, it will not deflect during contact and could damage the cell; if the probe is too soft, it will not indent the cell sufficiently to obtain reliable material properties. Typically, a spring constant of 0.02–0.1 N/m works well with most biological cells.
4. Commercial AFM probes are microfabricated in wafers of several hundred at a time ***(57,58)***. Whereas stoichiometry, thickness, and tip characteristics can vary from one wafer to another, AFM probes from a single wafer are often quite consistent ***(40,59,60)***. Therefore, it is advisable to purchase AFM probes in wafers or half-wafers to maximize consistency and minimize cost to less than $10 per probe, making it practical to use a new probe for each experiment.
5. If the inverted microscope is to be used exclusively with the AFM, then the standard microscope lamp and condenser components are not required because the AFM blocks this light path. The AFM is also compatible with (standard or confocal) fluorescence microscopy ***(30,61)***, offering a powerful tool for mechanotransduction studies relating specific molecular responses to targeted force simtuli. As such, the light microscope and video monitoring system are very useful but not absolutely necessary. Some investigators have used a standard AFM configuration without a light microscope to probe confluent monolayer cell cultures without

direct visualization ***(28)***, although this approach is obviously limited for targeting and monitoring individual cells.

6. Various custom enclosures and heating systems have been built for the purpose of maintaining cells at body temperature ***(4)***, including placing the entire AFM in a warm room ***(28)***. Such systems must be very stable to minimize thermal fluctuations of the cantilever. The advantage is a more physiologic environment during the experiment. Alternatively, performing experiments at room temperature could slow metabolic processes and make the cells more stable during lengthy protocols.
7. Dangerously high voltages are required to drive the piezoelectric elements controlling the probe position. It is especially critical when working in saline solutions that no liquid comes in contact with the scanner electronics, as this could severely damage the instrument. The scanner should be regularly inspected, cleaned, and calibrated according to the manufacturer's instructions to ensure that it is operating properly. If any fluid leakage is suspected during an experiment, unplug the scanner immediately and dry off all components with compressed air and cotton swabs, inspecting it under the dissecting microscope.
8. Adjusting the laser spot and making gross adjustments of the scanner position relative to a transparent sample are both greatly simplified when the AFM is mounted on an inverted light microscope for direct visualization. To avoid accidentally crashing the AFM tip into the clean transparent glass slide, it is helpful to make an ink spot on the slide with a laboratory marker that can be used to unambiguously identify the focal plane of the top surface of the glass (note the cells themselves serve this purpose in the Petri dish). The stepper motor can then be used to lower the AFM scanner until the probe is just out of focus, bringing it close to the surface without accidentally bumping into it and breaking the AFM probe or damaging the scanner.
9. According to classical elasticity theory ***(62)***, for a simple cantilever beam with a thin rectangular cross-section, the spring constant relating a point force at the end of the beam to its maximum deflection is given by $\kappa = Et^3w/4L^3$, where w is the width, L is the length, t is the thickness, and E is the elastic Young's modulus of the beam material. This basic formula has been modified to correct for the fact that the silicon nitride AFM probes commonly used for cell mechanics applications typically have a more complicated V- or A-shaped geometry and that the load on the probe tip is not applied exactly at the end of the cantilever ***(63,64)***. However, the greatest limitation when using such formulas is uncertainty in the values of E and t (especially because of the strong nonlinear dependence on t), which are inherently imprecise as a result of the vapor-deposition process used in fabricating wafers of AFM probes ***(57,59)***. Although variability in the spring constant of cantilevers from a given wafer is typically about 10% ***(40,60)***, measured values of κ can vary by 50% or more between wafers and can differ substantially from nominal values supplied by the manufacturer ***(40,60,65)***, necessitating calibration for accurate force measurements.
10. When the AFM is temporarily moved out of a liquid bath, as when exchanging samples, it sometimes happens that the voltage signal on the photodetector suddenly drops very low. Do not readjust the laser! This can occur when the volume of fluid on the scanner tip is small enough that the cantilever itself becomes bent by the

surface tension of the drop, so the laser reflection moves off the photodetector. When the voltage on the photodetector falls below some minimum value, software-controlled lowering of the AFM is prevented in order to protect the probe and scanner from accidentally contacting a surface undetected. To increase the photodetector signal and restore software control of the scanner position, the user must increase the volume of fluid on the scanner tip, either by manually lifting a dish of liquid up into contact with the scanner tip or by adding a few drops of liquid onto the scanner using the 3-mL syringe and 30G needle. When the cantilever is released from the surface tension of the drop, the photodetector signal will suddenly be restored.

11. Accurately engaging the tip of the AFM probe at a particular location on a sample can be challenging, even with the aid of a light microscope. However, this process can be greatly simplified when the microscope is equipped with a video camera and monitor. First, focus on the sample and use the stepper motor to bring the AFM within a few hundred micrometers of the sample. Next, adjust the microscope focus so the tip of the AFM probe comes clearly into focus on the video monitor. Mark the exact location of the probe tip on the monitor using the adhesive corner of a small Post-it note. Focus back on the sample and use the microscope's *x-y* stage to position the sample so the region of interest coincides with the tip location marked on the video monitor. Proceed to engage the AFM tip, which will slowly come into focus at the exact location marked on the video monitor, and hence at the desired location on the sample.
12. Standard silicon nitride AFM probes have a four-sided pyramidal tip geometry, with an inclination angle relative to the cantilever of 54.7°, defined by the crystal structure of the etched silicon molds on which they are fabricated ***(57,66)***. The tip of the pyramid is not an ideal point, but, instead, has a variably blunted profile ***(66,67)*** that can be approximated as a sector of a sphere. Some investigators have attached glass or polystyrene microspheres to AFM probes to obtain a simplified tip geometry ***(68–70)***, but this decreases spatial resolution and requires additional technical expertise. Custom probe modification can also be performed by some commercial vendors (e.g., Novascan Technologies, Ames, IA).
13. For accurate interpretation of indentation data, it is necessary for the cell to remain stationary during the measurement. Although typically not a problem for cells attached to a substrate, this becomes an issue for studying cells in suspension, as they tend to roll or move during indentation. In such cases, it might be possible to allow the cell to incubate for a brief period to initiate attachment, to use a non-specific binding agent such as poly-L-lysine (cat. no. P4707, Sigma, St. Louis, MO) to secure the cell to the substrate, or to use a micropipet or other device to hold the cell in place during the indentation experiment.
14. Adhesion between the AFM probe and the cell might be observed as negative deflections during probe retraction, especially near the point of initial contact with the cell. However, by the time the probe reaches its fully retracted position, the retraction curve should coincide with the flat precontact region of the advance curve. Sometimes long-range adhesion between the cell and the probe can be observed as downward deflection at the start of the advance curve and the end of the retract

curve. If this occurs, the user should extend the *z*-range or back up the tip (withdrawing completely if necessary) until the precontact region becomes flat again.

15. It is important to inspect the probe tip regularly under the microscope to ensure that no debris has accumulated on the probe. If contamination is suspected, withdraw the AFM and perform indentations on a clean glass slide (as described in **steps 7** and **8** in **Subheading 3.1.**) to ensure that the tip is clean and the postcontact curve is linear with unity slope. If this is not the case, the probe can sometimes be cleaned by removing the scanner and carefully rinsing the probe with fresh liquid using the 3-mL syringe and 30G needle or by carefully drying off the probe holder (without removing it from the scanner) with the corner of a Kimwipe® or cotton swab and gentle compressed air and rewetting the probe. Check the photodetector sensitivity (*see* **Subheading 3.1.**) to make sure that the laser and probe alignment have not changed during this process; if they have, simply record the new sensitivity and proceed with experiments.
16. Because cell height can vary by several micrometers over such large regions, it might be helpful to acquire force curves using a "trigger mode," in which the probe position is automatically adjusted to achieve a set maximum deflection (usually relative to the initial precontact deflection) for each indentation. This helps ensure consistency of force curves over a highly variable sample and also minimizes cell damage from excessively deep indentations on higher parts of the cell, although softer parts must be indented more deeply to achieve a given trigger deflection. For such applications, it is advantageous to have a scanner with an extended *z*-range of at least 10 μm.
17. Tapping mode imaging (in which feedback is used to maintain a constant amplitude of oscillation of the probe tip) is also well suited for imaging delicate samples such as cells ***(71)***. However, switching between tapping mode and force mode for indentations requires resetting of a number of control parameters. "Phase imaging" is a form of tapping mode in which changes in the phase of the tip oscillation relative to the probe input signal are interpreted to reflect the viscoelastic properties of the cell ***(72,73)***. One advantage of this technique is a rapid scan time (about 4 min for a 128 × 128 pixel image), but the physical interpretation of the data remains unclear and the resulting map is primarily of qualitative value.
18. A number of studies have documented stability and viability of cultured cells during contact mode imaging ***(4,65,74)***. However, it is also quite possible to damage a cell during this process ***(55,75,76)***. Therefore, it is advisable to perform a few test images for a given cell type and AFM probe to determine the best values of the most important scan parameters. In general, cells are more easily damaged at higher contact forces, so the set-point reference voltage should be decreased to the minimum value that still provides an acceptable image. Higher tip velocities also tend to cause damage, which means that a lower scan rate (in Hz) should be used with a larger scan size. It can be helpful to adjust the scan angle so that the fast-scan direction coincides with the direction of least change in cell height, such as moving along the length of actin stress fibers rather than repeatedly traversing the fibers to minimize disruption. A pixel resolution of 128 or 256 scan lines is often

adequate to visualize features of interest without unnecessarily lengthy scan times. Finally, a simple method to estimate the desired scan size on a particular cell without trial-and-error adjustments while scanning the cell is as follows. Immediately after engaging on the cell, use the stepper motor to back up the AFM probe about 10–20 μm out of contact with the cell while the software is still "engaged." Then increase the scan size until the probe is visibly moving over the region of the cell you wish to scan; again, Post-it® notes placed on a video monitor can help with this process. An AFM image will not be obtained during this time, of course, because the probe is not in contact with the cell. Once the scan size is determined, the probe can again be engaged on the cell to obtain a contact mode image with minimal further parameter adjustment.

19. In recognition of the limitations of the Hertz model for cell mechanics applications, a number of alternate approaches for analyzing AFM indentation data have also been developed. The method of force integration to equal limits overcomes several practical difficulties with standard AFM indentation, such as contact point uncertainty, to yield regional maps of relative cell stiffness ***(28)***, but the analysis is ultimately founded on the same assumptions as the Hertz model. Dimitriadis et al. have recently developed an analytic correction for the Hertz model when applied to very thin samples, such as cells ***(51)***. McElfresh and co-workers developed an analysis that explicitly accounts for surface interactions with the cell membrane for AFM indentation of sperm cells, but the approach makes other assumptions, such as single-point contact, that limit applicability to more general cell types ***(77)***. In AFM "force modulation," small perturbations are applied upon a larger indentation ***(78)***, for which Mahaffy and co-workers have recently developed a theoretical framework for extracting quantitative frequency-dependent elastic and viscous moduli of biological cells ***(38,68)***.
20. The Hertz equation for a simple ideal cone is readily obtained for the case $b = 0$. However, it must be emphasized that ignoring (or underestimating) the blunt tip defect will result in overestimation of the elastic modulus, especially when based on data from the early response at small indentation depths ***(37,53)***.
21. The standard approach is to perform a least squares fit of the general Hertz contact equation to the measured force–depth data set and extract a single value of $\tilde{E}$ (or a related modulus) from each indentation test. This might be reasonable if the cell truly behaves like a simple linear elastic material (i.e., if the pointwise modulus, $\tilde{E}_i$, is indeed independent of indentation depth). However, if the cell is actually a nonlinear material, then such a fitting procedure can lead to gross errors in the estimated value of elastic modulus ***(37)***. Other attempts to determine a depth-dependent modulus have been made by fitting subsets of the postcontact data ***(35,79)***. However, the resulting values of elastic modulus are difficult to interpret because the estimated contact point was also allowed to vary.

Acknowledgment

This work was funded in part by a CAREER Award from the NSF (BES-0239138).

References

1. Zhu, C., Bao, G., and Wang, N. (2000) Cell mechanics: Mechanical response, cell adhesion, and molecular deformation. *Annu. Rev. Biomed. Eng.* **2,** 189–226.
2. Elson, E. L. (1988) Cellular mechanics as an indicator of cytoskeletal structure and function. *Annu. Rev. Biophys. Biophys. Chem.* **17,** 397–430.
3. Pourati, J., Maniotis, A., Spiegel, D., et al. (1998) Is cytoskeletal tension a major determinant of cell deformability in adherent endothelial cells? *Am. J. Physiol.* **247,** C1283–C1289.
4. Rotsch, C. and Radmacher, M. (2000) Drug-induced changes of cytoskeletal structure and mechanics in fibroblasts: an atomic force microscopy study. *Biophys. J.* **78,** 520–535.
5. Trickey, W. R., Vail, T. P., and Guilak, F. (2004) The role of the cytoskeleton in the viscoelastic properties of human articular chondrocytes. *J. Orthop. Res.* **22,** 131–139.
6. Heidemann, S. R., Kaech, S., Buxbaum, R. E., and Matus, A. (1999) Direct observations of the mechanical behaviors of the cytoskeleton in living fibroblasts. *J. Cell Biol.* **145,** 109–122.
7. Sato, M., Nagayama, K., Kataoka, N., Sasaki, M., and Hane, K. (2000) Local mechanical properties measured by atomic force microscopy for cultured bovine endothelial cells exposed to shear stress. *J. Biomech.* **33,** 127–135.
8. Costa, K. D., Lee, E. J., and Holmes, J. W. (2003) Creating alignment and anisotropy in engineered heart tissue: role of boundary conditions in a model three-dimensional culture system. *Tissue Eng.* **9,** 567–577.
9. Wakatsuki, T. and Elson, E. L. (2003) Reciprocal interactions between cells and extracellular matrix during remodeling of tissue constructs. *Biophys. Chem.* **100,** 593–605.
10. Swartz, M. A., Tschumperlin, D. J., Kamm, R. D., and Drazen, J. M. (2001) Mechanical stress is communicated between different cell types to elicit matrix remodeling. *Proc. Natl. Acad. Sci. USA* **98,** 6180–6185.
11. Tamariz, E. and Grinnell, F. (2002) Modulation of fibroblast morphology and adhesion during collagen matrix remodeling. *Mol. Biol. Cell* **13,** 3915–3929.
12. Harris, A. K. (1994) Multicellular mechanics in the creation of anatomical structures, in Biomechanics of Active Movement and Division of Cells, *Volume H-84* (Akkas, N., ed.), Springer-Verlag, Berlin, pp. 87–129.
13. Costa, K. D. (2004) Single-cell elastography: probing for disease with the atomic force microscope. *Dis. Markers* **19,** 139–154.
14. Hochmuth, R. M. and Waugh, R. E. (1987) Erythrocyte membrane elasticity and viscosity. *Annu. Rev. Physiol.* **49,** 209–219.
15. Schmid-Schonbein, G. W., Sung, K. -L. P., Tozeren, H., Skalak, R., and Chien, S. (1981) Passive mechanical properties of human leukocytes. *Biophys. J.* **36,** 243–256.
16. Evans, E. and Yeung, A. (1989) Apparent viscosity and cortical tension of blood granulocytes determined by micropipet aspiration. *Biophys. J.* **56,** 151–160.
17. Jones, W. R., Ting-Beall, H. P., Lee, G. M., Kelley, S. S., Hochmuth, R. M., and Guilak, F. (1999) Alterations in the Young's modulus and volumetric properties of

chondrocytes isolated from normal and osteoarthritic human cartilage. *J. Biomech.* **32,** 119–127.
18. Miyazaki, H., Hasegawa, Y., and Hayashi, K. (2000) A newly designed tensile tester for cells and its application to fibroblasts. *J. Biomech.* **33,** 97–104.
19. Wang, N., Butler, J. P., and Ingber, D. E. (1993) Mechanotransduction across the cell surface and through the cytoskeleton. *Science* **260,** 1124–1127.
20. Yamada, S., Wirtz, D., and Kuo, S. C. (2000) Mechanics of living cells measured by laser tracking microrheology. *Biophys. J.* **78,** 1736–1747.
21. Alenghat, F. J., Fabry, B., Tsai, K. Y., Goldmann, W. H., and Ingber, D. E. (2000) Analysis of cell mechanics in single vinculin-deficient cells using a magnetic tweezer. *Biochem. Biophys. Res. Commun.* **277,** 93–99.
22. Guck, J., Ananthakrishnan, R., Mahmood, H., Moon, T. J., Cunningham, C. C., and Kas, J. (2001) The optical stretcher: a novel laser tool to micromanipulate cells. *Biophys. J.* **81,** 767–784.
23. Petersen, N. O., McConnaughey, W. B., and Elson, E. L. (1982) Dependence of locally measured cellular deformability on position on the cell, temperature, and cytochalasin B. *Proc. Natl. Acad. Sci. USA* **79,** 5327–5331.
24. Felder, S. and Elson, E. L. (1990) Mechanics of fibroblast locomotion: quantitative analysis of forces and motion at the leading lamellas of fibroblasts. *J. Cell Biol.* **111,** 2513–2526.
25. Koay, E. J., Shieh, A. C., and Athanasiou, K. A. (2003) Creep indentation of single cells. *J. Biomech. Eng.* **125,** 334–341.
26. Binnig, G., Quate, C. F., and Gerber, C. (1986) Atomic force microscope. *Phys. Rev. Lett.* **56,** 930–933.
27. Radmacher, M. (2002) Measuring the elastic properties of living cells by the atomic force microscope. *Methods Cell Biol.* **68,** 67–90.
28. A-Hassan, E., Heinz, W. F., Antonik, M. D., et al. (1998) Relative microelastic mapping of living cells by atomic force microscopy. *Biophys. J.* **74,** 1564–1578.
29. Butt, H. -J. and Jaschke, M. (1995) Calculation of thermal noise in atomic force microscopy. *Nanotechnology* **6,** 1–7.
30. Lehenkari, P. P., Charras, G. T., Nykanen, A., and Horton, M. A. (2000) Adapting atomic force microscopy for cell biology. *Ultramicroscopy* **82,** 289–295.
31. Lekka, M., Laidler, P., Gil, D., Lekki, J., Stachura, Z., and Hrynkiewicz, A. Z. (1999) Elasticity of normal and cancerous human bladder cells studied by scanning force microscopy. *Eur. Biophys. J.* **28,** 312–316.
32. Matzke, R., Jacobson, K., and Radmacher, M. (2001) Direct, high-resolution measurement of furrow stiffening during division of adherent cells. *Nature Cell Biol.* **3,** 607–610.
33. Mathur, A. B., Collinsworth, A. M., Reichert, W. M., Kraus, W. E., and Truskey, G. A. (2001) Endothelial, cardiac muscle and skeletal muscle exhibit different viscous and elastic properties as determined by atomic force microscopy. *J. Biomech.* **34,** 1545–1553.
34. Charras, G. T., Lehenkari, P. P., and Horton, M. A. (2001) Atomic force microscopy can be used to mechanically stimulate osteoblasts and evaluate cellular strain distributions. *Ultramicroscopy* **86,** 85–95.

35. Rotsch, C., Jacobson, K., and Radmacher, M. (1999) Dimensional and mechanical dynamics of active and stable edges in motile fibroblasts investigated by using atomic force microscopy. *Proc. Natl. Acad. Sci. USA* **96,** 921–926.
36. Domke, J., Parak, W. J., George, M., Gaub, H. E., and Radmacher, M. (1999) Mapping the mechanical pulse of single cardiomyocytes with the atomic force microscope. *Eur. Biophys. J.* **28,** 179–186.
37. Costa, K. D. and Yin, F. C. (1999) Analysis of indentation: implications for measuring mechanical properties with atomic force microscopy. *J. Biomech. Eng.* **121,** 462–471.
38. Mahaffy, R. E., Shih, C. K., MacKintosh, F. C., and Kas, J. (2000) Scanning probe-based frequency-dependent microrheology of polymer gels and biological cells. *Phys. Rev. Lett.* **85,** 880–883.
39. Mathur, A. B., Truskey, G. A., and Reichert, W. M. (2000) Atomic force and total internal reflection fluorescence microscopy for the study of force transmission in endothelial cells. *Biophys. J.* **78,** 1725–1735.
40. Cleveland, J. P., Manne, S., Bocek, D., and Hansma, P. K. (1993) A nondestructive method for determining the spring constant of cantilevers for scanning force microscopy. *Rev. Sci. Instrum.* **64,** 403–405.
41. Sader, J. E., Larson, I., Mulvaney, P., and White, L. R. (1995) Method for the calibration of atomic force microscope cantilevers. *Rev. Sci. Instrum.* **66,** 3789–3798.
42. Sader, J. E., Chon, J. W. M., and Mulvaney, P. (1999) Calibration of rectangular atomic force microscope cantilevers. *Rev. Sci. Instrum.* **70,** 3967–3969.
43. Ruan, J. -A. and Bhushan, B. (1994) Atomic-scale friction measurements using friction force microscopy: Part I—General principles and new measurement techniques. *ASME J. Tribol.* **116,** 378–388.
44. Tortonese, M. and Kirk, M. (1997) Characterization of application specific probes for SPMs. *Micromach. Imag. SPIE* **3009,** 53–60.
45. Jensen, F. (1993) Z calibration of the atomic force microscope by means of a pyramidal tip. *Rev. Sci. Instrum.* **64,** 2595–2597.
46. Heinz, W. F. and Hoh, J. H. (1999) Spatially resolved force spectroscopy of biological surfaces using the atomic force microscope. *Trends Biotechnol.* **17,** 143–150.
47. Weisenhorn, A. L., Khorsandi, M., Kasas, S., Gotzos, V., and Butt, H. -J. (1993) Deformation and height anomaly of soft surfaces studied with an AFM. *Nanotechniques* **4,** 106–113.
48. Sneddon, I. N. (1965) The relation between load and penetration in the axisymmetric Boussinesq problem for a punch of arbitrary profile. *Int. J. Eng. Sci.* **3,** 47–57.
49. Johnson, K. L. (1985) Contact Mechanics, Cambridge University Press, New York.
50. Domke, J. and Radmacher, M. (1998) Measuring the elastic properties of thin polymer films with the atomic force microscope. *Langmuir* **14,** 3320–3325.
51. Dimitriadis, E. K., Horkay, F., Maresca, J., Kachar, B., and Chadwick, R. S. (2002) Determination of elastic moduli of thin layers of soft material using the atomic force microscope. *Biophys. J.* **82,** 2798–2810.
52. Press, W. H., Teukolsky, S. A., Vetterling, W. T., and Flannery, B. P. (1992) *Numerical Recipes in FORTRAN 77: The Art of Scientific Computing*, Cambridge University Press, New York.

53. Briscoe, B. J., Sebastian, K. S., and Adams, M. J. (1994) The effect of indenter geometry on the elastic response to indentation. *J. Phys. D: Appl. Phys.* **27,** 1156–1162.
54. Hoh, J. H. and Schoenenberger, C. -A. (1994) Surface morphology and mechanical properties of MDCK monolayers by atomic force microscopy. *J. Cell Sci.* **107,** 1105–1114.
55. Schaus, S. S. and Henderson, E. R. (1997) Cell viability and probe–cell membrane interactions of XR1 glial cells imaged by atomic force microscopy. *Biophys. J.* **73,** 1205–1214.
56. Haydon, P. G., Lartius, R., Parpura, V., and Marchese-Ragona, S. P. (1996) Membrane deformation of living glial cells using atomic force microscopy. *J. Microsc.* **182,** 114–120.
57. Albrecht, T. R., Akamine, S., Carver, T. E., and Quate, C. F. (1990) Microfabrication of cantilever styli for the atomic force microscope. *J. Vac. Sci. Technol. A* **8,** 3386–3396.
58. Tortonese, M. (1997) Cantilevers and tips for atomic force microscopy. *IEEE Eng. Med. Biol. Mag.* **16,** 28–33.
59. Weisenhorn, A. L., Maivald, P., Butt, H. -J., and Hansma, P. K. (1992) Measuring adhesion, attraction, and repulsion between surfaces in liquids with an atomic force microscope. *Phys. Rev. B* **45,** 11,226–11,232.
60. Senden, T. J. and Ducker, W. A. (1994) Experimental determination of spring constants in atomic force microscopy. *Langmuir* **10,** 1003–1004.
61. Horton, M. A., Charras, G., Ballestrem, C., and Lehenkari, P. (2000) Integration of atomic force and confocal microscopy. *Single Mol.* **1,** 135–137.
62. Landau, L. D. and Lifshitz, E. M. (1970) *Theory of Elasticity,* Pergamon, Oxford.
63. Sader, J. E. and White, L. (1993) Theoretical analysis of the static deflection of plates for atomic force microscope applications. *J. Appl. Phys.* **74,** 1–9.
64. Sader, J. E. (1995) Parallel beam approximation for V-shaped atomic force microscope cantilevers. *Rev. Sci. Instrum.* **66,** 4583–4587.
65. Le Grimellec, C., Lesniewska, E., Giocondi, M. C., Finot, E., Vie, V., and Goudonnet, J. P. (1998) Imaging of the surface of living cells by low-force contact-mode atomic force microscopy. *Biophys. J.* **75,** 695–703.
66. Schwarz, U. D., Haefke, H., Reimann, P., and Güntherodt, H. J. (1994) Tip artefacts in scanning force microscopy. *J. Microsc.* **173,** 183–197.
67. Grütter, P., Zimmermann-Edling, W., and Brodbeck, D. (1992) Tip artifacts of microfabricated force sensors for atomic force microscopy. *Appl. Phys. Lett.* **60,** 2741–2743.
68. Mahaffy, R. E., Park, S., Gerde, E., Kas, J., and Shih, C. K. (2004) Quantitative analysis of the viscoelastic properties of thin regions of fibroblasts using atomic force microscopy. *Biophys. J.* **86,** 1777–1793.
69. Charras, G., Lehenkari, P., and Horton, M. (2002) Biotechnological applications of atomic force microscopy. *Methods Cell Biol.* **68,** 171–191.
70. Benoit, M. (2002) Cell adhesion measured by force spectroscopy on living cells. *Methods Cell Biol.* **68,** 91–114.

71. Vie, V., Giocondi, M. C., Lesniewska, E., Finot, E., Goudonnet, J. P., and Le Grimellec, C. (2000) Tapping-mode atomic force microscopy on intact cells: optimal adjustment of tapping conditions by using the deflection signal. *Ultramicroscopy* **82,** 279–288.
72. Nagao, E. and Dvorak, J. A. (1999) Phase imaging by atomic force microscopy: analysis of living homoiothermic vertebrate cells. *Biophys. J.* **76,** 3289–3297.
73. Hansma, H. G., Kim, K. J., Laney, D. E., et al. (1997) Properties of biomolecules measured from atomic force microscope images: a review. *J. Struct. Biol.* **119,** 99–108.
74. Rotsch, C., Braet, F., Wisse, E., and Radmacher, M. (1997) AFM imaging and elasticity measurements on living rat liver macrophages. *Cell Biol. Int.* **21,** 685–696.
75. You, H. X., Lau, J. M., Zhang, S., and Yu, L. (2000) Atomic force microscopy of living cells: a preliminary study of the disruptive effect of cantilever tip on cell morphology. *Ultramicroscopy* **82,** 297–305.
76. Braet, F., de Zanger, R., Seynaeve, C., Baekeland, M., and Wisse, E. (2001) A comparative atomic force microscopy study on living skin fibroblasts and liver endothelial cells. *J. Electron Microsc.* **50,** 283–290.
77. McElfresh, M., Baesu, E., Balhorn, R., Belak, J., Allen, M. J., and Rudd, R. E. (2002) Combining constitutive materials modeling with atomic force microscopy to understand the mechanical properties of living cells. *Proc. Natl. Acad. Sci. USA* **99(Suppl. 2),** 6493–6497.
78. Radmacher, M., Tillmann, R., and Gaub, H. (1993) Imaging viscoelasticity by force modulation with the atomic force microscope. *Biophys. J.* **64,** 735–742.
79. Radmacher, M., Fritz, M., Kacher, C. M., Cleveland, J. P., and Hansma, P. K. (1996) Measuring the viscoelastic properties of human platelets with the atomic force microscope. *Biophys. J.* **70,** 556–567.

18

Reflection Contrast Microscopy

The Bridge Between Light and Electron Microscopy

F. A. Prins, I. Cornelese-ten Velde, and E. de Heer

Summary

Reflection contrast microscopy (RCM) is a light microscopic method to image cells at high definition and enhanced sensitivity compared to conventional bright-field microscopy. RCM images have very high contrast, which makes them easily applicable for digital image analysis. Because ultrathin sections are mostly used in this method, RCM also functions by bridging light with electron microscopy: the combination of ultrastructural with histochemical studies. RCM can also replace electron microscopy for rapid and simple screening of large quantities of samples for immunocytochemical staining. Special attention is paid to small biological objects, which have to be processed for RCM. If you encounter the limits of brightfield microscopy, in resolution, sensitivity or handling of the specimen, RCM will be a feasible option.

Reflection contrast microscopy methods use only slightly adjusted electron microscopy methods for specimen preparation. Therefore, many familiar techniques for ultrathin specimen preparation can be applied. It is essential that only refractive index differences exist in those areas that are of interest and that the further specimen is as optically homogenic as possible, with a refractive index as close to that of glass as possible. Therefore, plastic embedding is recommended.

Key Words: Ultrathin section; immunocytochemistry; high definition; specimen preparation; reflection mode of CLSM; reflection contrast microscopy; enhanced detection sensitivity; image analysis; bright-field microscopy; transmission electron microscopy; high-resolution; high-contrast microscopy; combining light and electron microscopy.

1. Introduction

Reflection contrast microscopy (RCM) can be used as a bridge between light microscopy (LM) and electron microscopy (EM) because it visualizes ultrathin sections, normally used in transmission electron microscopy (TEM), at the light microscopic level (*see* **Fig. 1A**). The main advantages of RCM (properties) (compare **Fig. 1B–D**) are increased sensitivity and image definition compared to bright-field microscopy; the theoretical resolution (200 nm) can actually be

From: *Methods in Molecular Biology, vol. 319: Cell Imaging Techniques: Methods and Protocols*
Edited by: D. J. Taatjes and B. T. Mossman © Humana Press Inc., Totowa, NJ

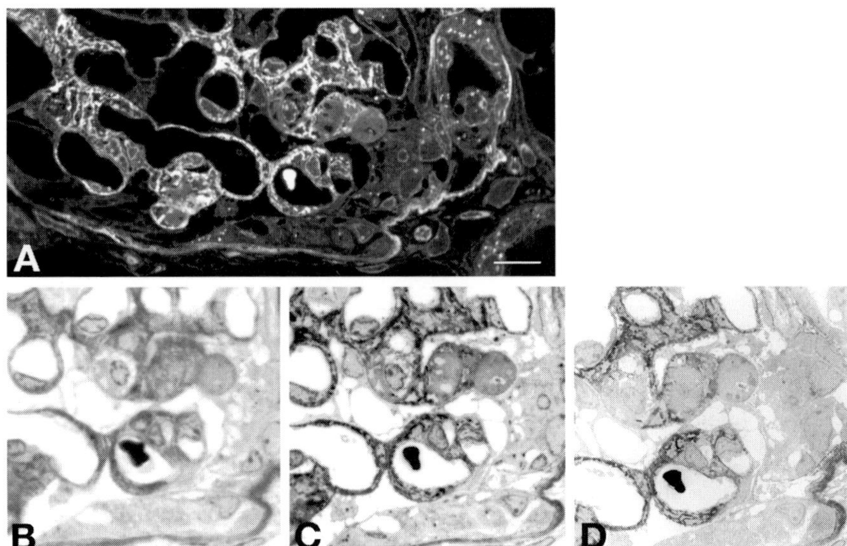

Fig. 1. High-power images of sequential sections of the distribution of laminin in glomeruli during development of glomerulosclerosis in mice with experimental lupus nephritis. These glomeruli demonstrate typical properties of RCM imaging: increased definition, enhanced contrast, and increased sensitivity in detecting the immunoperoxidase product of diaminobenzidine. **(A)** RCM image of ultrathin Epon-embedded section counterstained with toluidine blue. Using reflection, there is a larger overview of the whole section than by using TEM. **(B)** Bright-field microscopic (BFM) image of 1-µm-thick section counterstained with toluidine blue. **(C)** The same as **A**, but digitally inverted, for ease of comparison with BFM and TEM image. **(D)** TEM image of a sequential ultrathin section, which shows the detection of the DAB-ox with high resolution (only lightly contrasted with lead citrate). Bar = 20 µm.

obtained in RCM. Compared to TEM, RCM provides fast information at an overview. For instance, immunolabel can be detected at such a low magnification (630–1000×) that it facilitates finding the labeled sites in TEM to study its ultrastructure ***(1,2)***. RCM is a suitable method for even the tiniest objects (morphological study using a simple light microscope), for which you initially do not know anything else than to subject them to TEM analysis (*see* **Note 1**), or to perform immunofluorescence (e.g., leptospires in **ref. *3***). With RCM, one can make images of small to extremely small objects, biopsies, and single cells without the need for any immunocytochemical staining because you see all there is, which have a refractive index different from glass. In **Tables 1** and **2** many types of application and discipline are listed in which RCM has been used. A special feature of RCM technology is the use of sequential ultrathin sections, so that almost the same tissue can be studied by RCM and TEM; the distance between the sections is 50–100 nm (*see* **Fig. 1C,D**).

Table 1
Type of RCM Applications Using RCM

Immunolocalization studies at (sub) cellular level in cells and tissues
Comparison of sequential ultrathin sections for multiple labels
In situ hybridization
Autoradiographic detection
Tracer studies of molecules conjugated with a label (radioactive, peroxidase [PO], gold, or indirect with a labeled antibody)
Detection of small inclusions, fragmented foreign material (different refractive index)
Detection at LM level of objects below the resolution of the LM, as a result of a bright, isolated signal against a dark background
Previewing of electron microscopic ultrathin sectioning of tiny or thin objects
Quality control of cell and tissue culture
Attachment and movement of living cells

Table 2
Type of Discipline Using RCM

Discipline	Method	Resin	Label	Ref.
Nephrology	Post-embedding; Pre-embedding	Lowicryl K4M; EPON	IGS/IGSS, Po-DAB	***17–20***
Oncology	Pre-embedding	EPON	Po-DAB	***15,21,22***
Pathology	Pre-embedding	EPON	Po-DAB	***3,18,22***
Cell biology	Post-embedding; living cells	Lowicryl K4M	IGSS	***7,8,15,23–25***
Pharmacology	Auto radiography	Spurr	Silver	***26***
Neurology	Post-embedding	Spurr	IGSS	***1,2,5,6,27***
Plant biology			IGSS	***28***
Transplantation	Post-embedding	Lowicryl K4M	IGS/IGSS	***29***
Molecular biology	*In situ* hybridization		APh/Po-DAB	***16,30–32***
Overview of study's RCM				***3,6,7,11,33–40***

Abbreviations: IGS, immunogold staining; IGSS, immunogold silver staining; Po-DAB, peroxidase–diaminobenzidine; Aph, alkaline phosphatase.

All types of TEM specimen preparation can be used for RCM, so that the book by Nasser Hajibagheri on electron microscopy is very useful ***(4)***. The main modification for RCM use is that ultrathin sections are placed on glass slides. Also, less rigorous fixation can be used, omitting osmium tetroxide (which influences images formation because of strong reflection) or sometimes glutaraldehyde to maintain antigenicity in case of immunocytochemistry. Fewer specimens are useful for RCM from the LM specimen preparation field—only plastic-embedded and whole-cell (chromosomes, etc.) preparations. Generally, the LM section is too thick, which yields blurred images. However, optically quite homogenic objects with a strong reflecting label (like silver-stained neurons) can be used with RCM to visualize the label in depth through the object ***(5,6)***. Thus, with good understanding of the image formation in RCM (*see* **Subheading 1.1.**) many new objects, specimens, and applications can be discovered or developed.

Living cells also can be studied by RCM. Actually, this was the first application of RCM ***(7)***. In an aqueous environment, cell contacts with the glass are less reflective (darker) and can be studied ***(8)***.

1.1. Principles of Reflection Contrast Microscopy

Reflection contrast microscopy is a light microscopic technique where image formation is based on reflected light rays. It uses a microscope with incident light, in contrast to transmitted light, which is used for bright-field microscopy. The RCM images look like fluorescence images, because generally the background is dark (*see* **Note 2**). They are also made on an adapted fluorescence microscope, which is equipped with devices so that only the light reflected at the object forms the image. This is extremely difficult for biological objects, for which the reflected light intensities are generally much less than 1% of the incident light, which is very low compared to the 4% reflection intensity at a glass–air interface (of which there are many in a microscope); therefore, an optical trick is applied so that only the light reflected at the object forms the image. This omittance of disturbing reflections, which would decrease contrast in the image, is indicated by the additive "contrast" in RCM.

The microscope conformation of RCM consists of a combination of two contrast-enhancing principles: the antiflex principle ***(9)*** and the central stop system ***(10)***. The latter is located in front of the lamp house to create access to an aperture plane conjugate with the upper entrance pupil of the objective, and it consists of central black circles and aperture diaphragms (the size depends on the objective lens). Passing this combination makes the light beam ring shaped. The central stop prevents the central part of the light beam from reflecting at the backside of the objective (in a RCM, it prevents the existence of rays, which otherwise would reflect against the underside of the glass slide, (the amount of light at this glass–air interface is about 4%) and blurs the object image (*see* **Note 3**). The antiflex principle consists of a set of crossed polarizers (a pol-block) and a quarter-wavelength plate (located at the front lens of the objective). It functions as follows. The light is linearly polarized by a 0° polarizer,

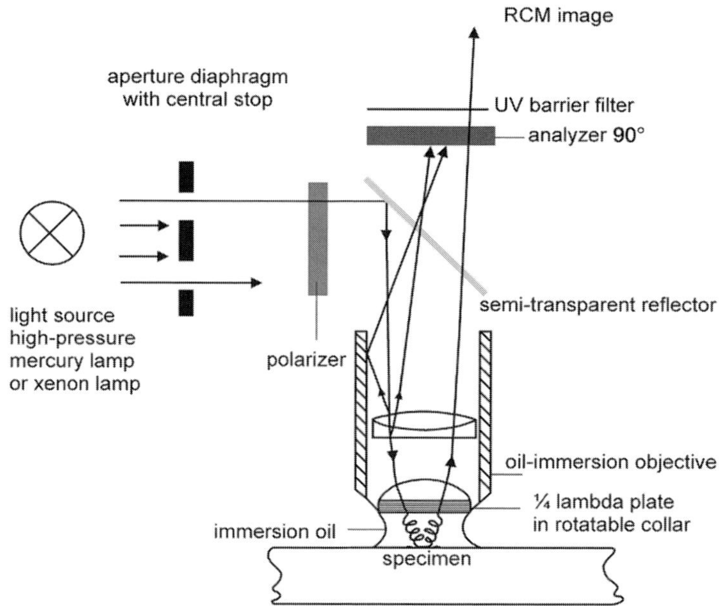

Fig. 2. Light path of the RCM.

deflected by the 50/50 reflector in the vertical light path, and upon passing the quarter-wavelength plate, the light is circularly polarized. After reflection at the object, the light passes the quarter-wavelength plate again and is rotated another 45°. Related to the incident ray, the polarization state has turned 90°, so that this light (reflected at the object) can pass the analyzer, which is crossed to the polarizer. All light rays that reflect inside the microscope are stopped by the analyzer. The light path is depicted schematically in **Fig. 2**.

Image formation in RCM is based on both reflection and interference of reflected light rays ***(11)***. What we see of our world is based on reflection of light. Oil on a water surface is seen as colors because of interference of the reflected light waves at the surface oil–air interface with those reflected light waves at the oil–water interface. The colors seen also depend on the thickness of the oil layer and on the angle of reflection. Translated to a microscope image, the thickness of an object, the numerical aperture (NA) of the objective, and the size of the aperture diaphragm (outer ring of central stop) determine the image formed next to the properties of the object (i.e., the refractive index, absorption coefficient, and surface texture). Reflection of light occurs at an interface of refractive index differences, and the intensity of reflection is proportional to the difference in refractive indices. Interference of reflected light occurs in thin layers, where, at the first interface, a portion of the light ray is reflected and the

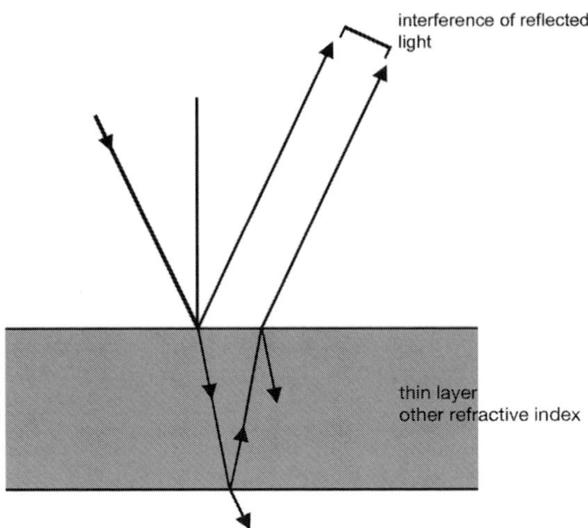

Fig. 3. Principle of interference of reflected light waves in thin layers.

remainder is transmitted; this transmitted light ray reflects at the next interface, transmits the former interface, and can interfere with the directly reflected part of its ray because they have the same frequency (*see* **Fig. 3**).

1.2. Introduction to the Methods

The methods are given for the whole procedure from tissue to RCM image. This is a very large field and many books even in this series have been written about this. Thus, we present our main routine for the preparation of specimens. Moreover, methods in the TEM field for preparing ultrathin sections can be used, including immunocytochemistry, with pre-embedding and postembedding labeling, different plastics, ultracryomicrotomy, and so forth. Methods for specimen preparation from the fields of parasitology, microbiology, or cytogenetics can also be useful.

Because ultrathin sections are needed (*see* **Note 4**), in general the tissue is cut in small pieces (2 mm^2) or one starts with smaller objects, which are fixed (lightly for maintaining antigenicity in case of immunocytochemistry) and embedded. Two methods are given: one for the morphological study and the other for immunocytochemical study. For immunocytochemical staining, there are two possibilities: either the immuno-incubation is performed before (pre-) or after (post-) embedding of the tissue. In pre-embedding immunocytochemistry, the tissue is mostly sliced with a vibratome in sections of about 70–90 µm thickness, which are immunostained, followed by embedding in plastic and cut in ultrathin sections. In postembedding immunocytochemistry, the ultrathin sections are stained, and depending on the type of embedding media used (affecting accessibility of the antigenic sites for the antibodies), further etching steps might be needed (e.g., EPON™). However, whole cells, cultures, suspensions or micro-organisms, chromosomes, or other isolated cell-constituent specimens can be made, with or without embedding and ultrathin sectioning. Even living cells can be studied (*see* **Subheading 3.2.**).

Reflection contrast microscopy can be easily performed if a fluorescence microscope stand is available. The adaptation is provided in a method, followed by operating the RCM. If a confocal laser scanning microscope is available, we will describe how to make RCM images with it. In the last part, some methods about recording and interpretation of RCM images are presented.

2. Materials

2.1. Specimen Preparation

2.1.1. Major Equipment and Supplies

1. Razor blade (single edge), wax plate, pair of tweezers, rubber policeman.
2. Stereo microscope with light source.
3. Ultramicrotome (Leica UCT) or a routine, motorized microtome (Leica RM2165).
4. Vibrating blade microtome (Leica VT 1000 S).
5. Small freezer with top lid (–20°/–30°C).
6. Eppendorf capsules, 1.5 mL.
7. Flat embedding molds (Agar Scientific Ltd., UK, cat. no. G3531).
8. Small Beem capsules (size 3 and 00) (Polaron, England).
9. Beem capsules with conical tip (Agar Scientific Ltd., UK, cat. no. G361-1).
10. Glass, cryo-, or regular diamond knife (Diatome-Switzerland or Drukker, The Netherlands) (*see* **Fig. 4**).
11. Small wire loop (3 mm in diameter) (Agar Scientific Ltd., UK, cat. no. T5011) (*see* **Fig. 4**).
12. Coated glass slides (*see* **Subheading 3.1.1.**).
13. Transwell filters (Costar, Cambridge, UK, cat. no. 3414).
14. Immunofluorescence slides, 10 wells (Polysciences Europe, cat. no.18357).
15. Hot plate (30–80°C).
16. Ultraviolet polymerization lamp (Polysciences Europe, cat. no. 8778).
17. Moist chamber.
18. Dako pen (cat. no. S 2002) and a water-resistant ink pen.
19. Glass writing diamond marker (A. J. Cope Ltd., UK, code LA530).
20. Polyethylene gloves.
21. Burner for ethanol (*see* **Fig. 4**).

2.1.2. Major Chemicals and Solutions

1. Hank's culture wash medium.
2. Phosphate buffer: Solution A: 0.2 *M* sodium di-hydrogen ortho-phosphate (NaH_2PO_4), 0.497 g $NaH_2PO_4{\cdot}2H_2O$ in each 100 mL. Solution B: 0.2 *M* di-sodium

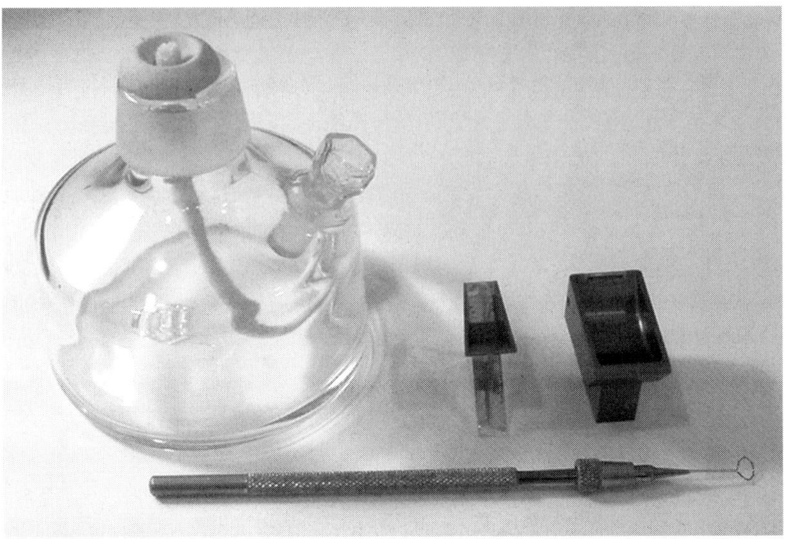

Fig. 4. Tools for ultrathin sectioning. Left to right: burner for ethanol, glass knife, diamond knife, and wire loop.

hydrogen ortho-phosphate (Na_2HPO_4), 2.328 g Na_2HPO_4 in 100 mL.To prepare 0.2 *M* phosphate buffer, pH 7.4, mix 20 mL solution A and 80 mL solution B.

3. 0.2 *M* Sodium cacodylate buffer: 21.4 g sodium cacodylate in 400 mL distilled water, adjust pH to 7.4 with 1 *M* HCl and fill up to final volume of 500 mL (storage: at 4°C; stability: a few weeks; hazards: sodium cacodylate is toxic by inhalation, ingestion, and skin contact; irritation to skin, eyes, and mucous membranes; inhalation of small concentrations over long period will cause poisoning and should be treated as a suspected carcinogen; and could cause adverse mutagenic or teratogenic effects; toxicity data: LD_{50} = 2600 mg/kg oral, rat).
4. Store buffer morphology: mix 0.2 *M* sodium cacodylate buffer with the same volume of distilled water (1:1) and add 3% sucrose.
5. Store buffer immunohistochemistry: 2% paraformaldehyde in 0.1 *M* phosphate buffer, pH 7.4.
6. Phosphate-buffered saline (PBS): 150 m*M* NaCl, 10 m*M* phosphate buffer, pH 7.4.
7. Incubation buffer: 1% bovine serum albumin (BSA), (Sigma, cat. no. A-9647), 0.1% gelatin (Merck, cat. no. 070), in PBS, mix together on a hot plate at 60°C.
8. 2% BSA in PBS (put 100 mL PBS in a glass vial and add 2 g BSA on top of it; do not stir until it is dissolved).
9. 0.2% Aurion BSA-C in PBS (Aurion, The Netherlands).
10. Aldehyde quench buffer: 0.05 *M* glycine in PBS, or 0.5 *M* ammonium chloride in PBS.
11. Sodium azide.

12. Protein blocking buffer: Add 5% normal serum (same species as secondary antibody) to PBS.
13. DAB: freshly prepared 0.05% 3,3´-diaminobenzidine tetra-hydrochloride (Fluka) in 0.05 *M* Tris- HCl, pH 6.0.
14. 30% Hydrogen peroxide (Merck, cat. no. 7209). *Note:* hazard.
15. 0.2% Light green in distilled water (BAH/Gurr, cat. no. 34204).
16. 0.2% Mayer's hematoxylin (Merck, cat. no. 15938) in distilled water.
17. 1% Toluidine blue in distilled water.
18. Aminosilane glass-coating solution: 2% Amino-propylethoxylane (Sigma, cat. no. A-3648) in acetone (storage: at 2–8°C; hazards: corrosive, causes burns; harmful by inhalation, in contact with skin, and if swallowed; wear suitable protective clothing, gloves, and eye and face protection).
19. Paraformaldehyde (Merck, cat. no. 4005) (storage as a solution at 4°C; stability: best to prepare fresh for each fixation, but can be stored for a few weeks; hazards: paraformaldehyde harmful by ingestion and irritant to the skin, eyes, and respiratory system; toxicity data: LD_{50} = 800 mg/kg oral, rat; LC_{50} = 0.9 mg/L inhalation, rat; should be handled in a fume hood with gloves).
20. 25% Glutaraldehyde (BDH, cat. no. 360802F): "EM" grade should be used (storage at 4°C; stability: several years; hazards: harmful by ingestion and inhalation; extremely irritating to the eyes; no prolonged skin contact; toxicity data: LD_{50} = 134 mg/kg oral, rat. Should be handled in a fume hood with gloves).
21. Osmium tetroxide (BDH, cat. no. 36219). (Should be handled in a fume hood with gloves.)
22. Lowicryl K4M kit (PolySciences Ltd.). (Storage: at 25°C; hazards: harmful by ingestion and inhalation; irritating to skin; eyes, and respiratory system. Should be handled in a fume hood with gloves).
23. Unicryl (Storage: at 4°C; hazards: harmful by ingestion and inhalation; irritating to skin, eyes and respiratory system; toxicity data: LD_{50} = 5000 mg/kg oral, rat. Should be handled in a fume hood with gloves).
24. Epon (Poly/Bed 812 kit 21959, Polysciences, Germany). *Note:* hazard.
25. Spurr's resin (resin kit R1032, AGAR Scientific, UK) *Note:* hazard.
26. Immunogold reagents: Either pA-, pG-, or IgG-bound gold particles (size 1–40 nm) (Amersham, UK or Aurion, The Netherlands).
27. Primary antibodies.
28. Rapid mounting medium ("Entellan"; Merck, cat. no. 7961).
29. Immersion oil for fluorescence microscopy; refractory index (n_e) = 1.518 (23°C) (Zeiss).
30. 1% Periodic acid in distilled water.
31. 10% Gelatin in PBS.

2.1.3. Glass Slide, Cleaning, and Coating

1. 96% Ethanol.
2. Acetone (Pre analyze [PA] quality).

3. Aminosilane solution: 2% aminopropylethoxylane (Sigma, cat. no. A-3648) in acetone.
4. Glass slides, all types of ready-to-use microscope slides.

2.1.4. Fixatives

1. For immunohistochemistry: 4% paraformaldehyde in 0.1 *M* phosphate buffer. Dissolve 8 g of paraformaldehyde in 100 mL distilled water by heating to 60°C. Add a few drops of 1 *N* sodium hydroxide until the solution clears, cool to room temperature, and filter. Mix 10 mL of 8% paraformaldehyde with 10 mL of 0.2 *M* phosphate buffer and add glutaraldehyde (25%) to a final concentration of 0.05 to 0.2%.
2. For morphology: 1.5% glutaraldehyde and 1% paraformaldehyde in 0.1 *M* cacodylate buffer adjusted to pH 7.4 for the primary fixation (mix 6 mL of 25% glutaraldehyde, 12.5 mL of 8% paraformaldehyde, and 50 mL of 0.2 *M* cacodylate buffer, then add 31.5 mL of distilled water). If EM is also used, postfix with 1% osmium tetroxide in 0.1 *M* cacodylate (mix 2 mL of 2% aqueous osmium tetroxide with 2 mL of 0.2 *M* cacodylate buffer, pH 7.4).

2.1.5. Pre-Embedding

1. Vibrating blade microtome (Leica VT 1000 S).
2. Beem capsules (size 3 and 00) (Polaron, England).

2.1.6. Embedding Resins

1. Epon (Poly/Bed 812 kit, cat. no. 21959; Polysciences, Germany). Mix 45 mL nadic methyl anhydride (NMA), 50 mL dodecinyl succinic anhydride (DDSA), and 80 mL Poly/Bed 812. Divide into 10-mL portions in 20-mL glass vials and store in a freezer (several months). Before using, add 0.15 mL 2,4,6-{tri(dimethylaminoethyl)phenol} (DMP-30) to 10 mL Epon mix and stir well.
2. Spurr's resin (resin kit, R1032; AGAR scientific, UK). *Note:* hazard. Mix 52 mL nonenyl succinic anhydride (NSA), 20 mL vinyl cyclohexene dioxide (ERL 4206), 14 mL diglycidyl ether of polypropylene glycol (DER 736), 1.2 mL dimethylaminoethanol (DMAE) or equal amounts just before using.
3. Lowicryl K4M (resin Polar Kit, cat. no. 15923; Polysciences Inc., Germany). *Note:* hazard. Mix 2.25 g crosslinker (A) and 12.75 g monomer (B), and 0.075 g initiator (C).
4. Unicryl (resin kit, cat. no. BA250; BB International). *Note:* hazard.

2.1.7. Ultracryo-sectioning

1. Ultramicrotome equipped with cryo-attachment (Leica Germany).
2. Diamond cryoknife or a glass knife.
3. Copper or aluminum cryo-specimen mounting pins (Leica, Germany).
4. Small wire loop (3 mm in diameter) made from platinum wire (0.2 mm) and mounted on a wooden stick (self-made).
5. 2.3 *M* Sucrose in PBS.
6. Liquid nitrogen Thermos flask.
7. Liquid nitrogen storage tank (for specimen storage).
8. Aminosilane coated glass slides (*see* **Subheading 3.1.1.**).

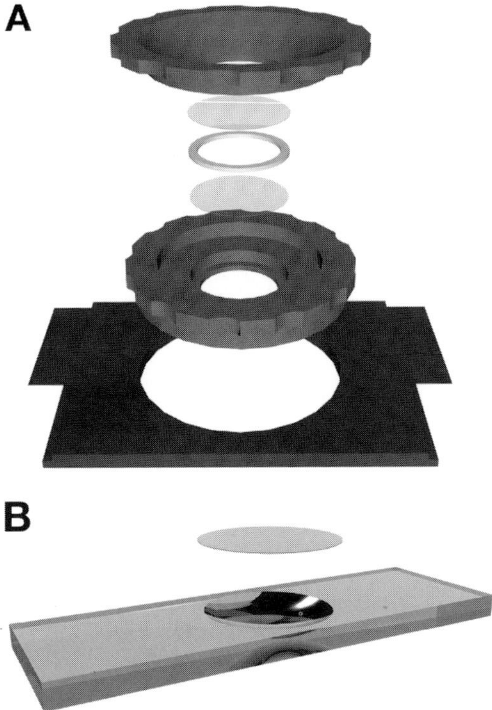

Fig. 5. Cell culture/observation chambers: **(A)** specially designed chamber; **(B)** glass slide with cavity.

2.2. Observation of Living Cells

1. Round cover glasses (25 mm in diameter) (Omnilabo International, The Netherlands; cat. no. 211045).
2. Glass slide with cavity (*see* **Fig. 5B**) (Omnilabo International, The Netherlands; cat. no. 408501).
3. Cell culture chambers (Agar Scientific Ltd., UK cat. no. G3541) or special chambers, with a silicon-ring between two round cover glasses (*see* **Fig. 5A**).
4. Nail polish.

2.3. RCM Adaptation Equipment

1. An incident light microscope, such as a fluorescence microscope, Leica DMR, or Zeiss Axioskop.
2. RCM oil-immersion objective equipped with a quarter lambda plate in the front lens system (*see* **Fig. 6D**): Plan-neofluor 63×/1.25 oil Antiflex from Zeiss, or the HCX PL Fluotar 100×s/1.30 oil RC from Leica.
3. A lamphouse with a high-pressure mercury or xenon lamp as used in fluorescence microscopy.

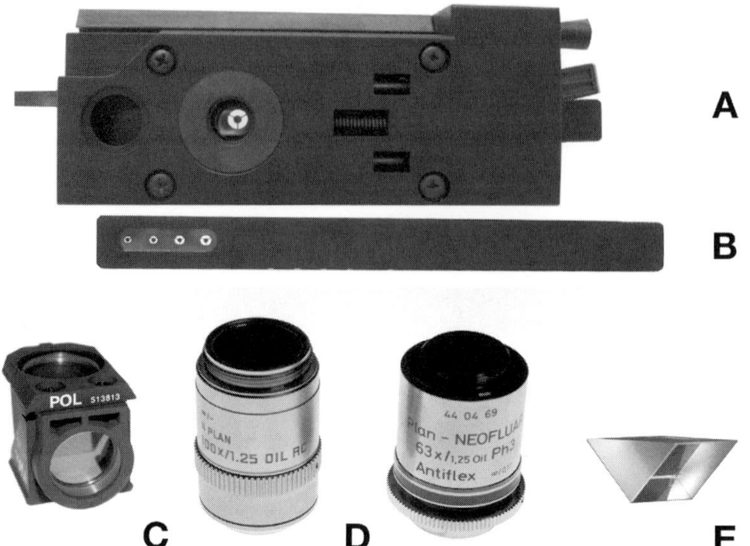

Fig. 6. Equipment for adaptation of an incident/fluorescence microscope: **(A)** RCM module containing, the following: **(B)** slider with different central stops, **(C)** POL-block, **(D)** RCM objectives, and **(E)** glass prism.

4. A RCM module, Leica (*see* **Fig. 6A**) adapting one slider with different sets of central stops (*see* **Fig. 6B**).
5. A POL- block containing a polarizer, a neutral beam splitter, and an analyzer (*see* **Fig. 6C**), a 10-mm Right-Angle Prism (Melles Griot, code: 01PRB 009 http://www.mellesgriot.com) (*see* **Fig. 6E**), instead of the central stop.
6. Protection filter (preventing heat, infrared) for the polarizer.

2.4. Confocal Laser Scanning Microscope in the Reflection Mode

1. Confocal laser scanning microscope (LSM-510, AxioPlan-2 from Zeiss, Jena Germany, or TCS SP2, DM-I RBE from Leica, Heidelberg, Germany). An inverted or upright microscope can be used. Laser lines: argon (range: 450–514 nm; 200 mW); helium–neon (He–Ne1; 543 nm; 4 mW); helium–neon (He–Ne2; 633 nm; 15 mW).
2. Important components for confocal RCM are oil-immersion objectives and a neutral beam splitter (always present in a confocal laser scanning microscope [CLSM]).
3. 5× or 10× Dry objective.

2.5. Recording of RCM Images

1. Conventional analog camera in combination with Dia positive film of 100 to 200 ASA.

2. Examples of digital cameras are Leica DC 200, and DC 300 and the AxioCam from Zeiss.

3. Methods

3.1. Specimen Preparation From Tissues and Cells

The protocol is as follows:

- Fixation of tissue/cells to a small cubic form;
- Pre-embedding immunostaining*;
- Dehydration and embedding of blocks;
- Sectioning of blocks;
- Ultracryosectioning*;
- Staining of sections (immuno)-histochemistry.

Very thin objects yield RCM images with the highest image definition. To visualize or immunolabel the inside of a cell (intracellular labeling), it has to be sectioned. To section tissue/cells ultrathin, they must be made firm by embedding in plastic or by freezing. Before infiltration with the plastic or freezing, tissue/cells need to be fixed: with a mild fixative for immunocytochemical studies so that antigenicity is maintained, and with a strong fixative for a morphological study (*see* **Note 5**). Ultrathin sectioning (50–250 nm) of the embedded or frozen tissue/cells is then performed using an ultra-/motorized conventional ultramicrotome or an ultracryomicrotome. The sections are put on cleaned and coated glass slides and marked by a circle at the underside of the glass slide (for ease of finding the sections). Thereafter, the ultrathin sections can be immunostained (postembedding). Depending on the accessibility of the antigenic sites, further etching steps might be required (e.g., Epon and Spurr's). Ultrathin cryosections have the highest accessibility, followed by hydrophilic resins such as LR-White, Lowicryl, and Unicryl, in which the antigenic sites are accessible in the top layer, and, finally, Epon and Spurr's, which are least accessible. The cryo-method is the best option for studies with new antibodies, of which antigenic sites are unknown or the preservation of antigenicity upon fixation is unknown.

For inter-(extra)cellular labeling, or for instance extracellular matrix components in tissue, the pre-embedding method can also be chosen. A strong fixation is applied after the immunostaining step. This extra fixation is often applied in RCM methods.

Staining of antigens at the surface of (single) cells grown on glass can be visualized directly (i.e., without embedding and sectioning) with RCM or confocal laser scanning (CLS-RCM). However, when, for instance, internalization

*This step also comprises other steps from this protocol.

studies are performed with probes conjugated to reflective markers (colloidal gold, peroxidase), then in order to study the inside of the cells, they have to be sectioned and the appropriate method is preembedding.

For single-cell/small biopsy studies with these RCM methods, there are many ways to section the cells (e.g., parallel or perpendicular to the attachment substrate of the cells). Another advantage of the method is that because of the thickness (50–100 nm) of ultrathin sections, many sections can be made of even single cells. They can also be collected in their sequential order. Sequential (serial) sections can be collected consecutively on glass slides, or one on glass and the next on grid and respectively be used for localization of different antigens on sequential sections (multilabeling studies) or in combination with EM (subcellular localization).

A central theme in methods for preparing RCM specimens is to adhere to clean working conditions; because of the sensitivity of RCM, any dirt or debris is visible. Therefore, all glass slides used must be cleaned, and some solutions millipore-filtered (*see* **Note 6**). For immunohistochemistry, coated slides are used.

3.1.1. Glass Slides, Cleaning, and Coating

For morphological studies:

1. Clean slides by washing them with 96% ethanol for 1 min and with acetone for 1 min.
2. Wipe dry with tissue in one movement (avoid dust) and store in a dust-free box.

For immunostaining:

1. Wash slides with acetone for 2X 10 min.
2. Coat overnight with 2% aminosilane in acetone in a covered jar.
3. Wash with acetone for 2X 10 min.
4. Wash with distilled water for 5 min.
5. Wipe dry with tissue in one movement (avoid dust) and store in a dust-free box.

Alternatively, use glass slides with adhesive coating (Star-Frost Adhäsiv-Objektträger; Knittel-Gläser, Germany).

3.1.2. Fixation of Tissue/Cells to a Small Cubic Form

3.1.2.1. Tissues, Biopsies

After removal of the tissue, cut into small cubes (size <2 mm^3) on a wax plate with a sharp, new, single-edged razor blade and submerge the cubes in the appropriate fixative (*see* **Subheading 2.1.2.**) (*see* **Note 7**).

The fixation time for tissue should be at least 4 h at 4°C. After sufficient fixation, tissues or cells can be stored in the following: 3% Sucrose in cacodylate buffer 4 h or overnight at 4°C for morphology processing; 2% Paraformaldehyde/PBS 4 h or overnight at 4°C for immunoprocessing.

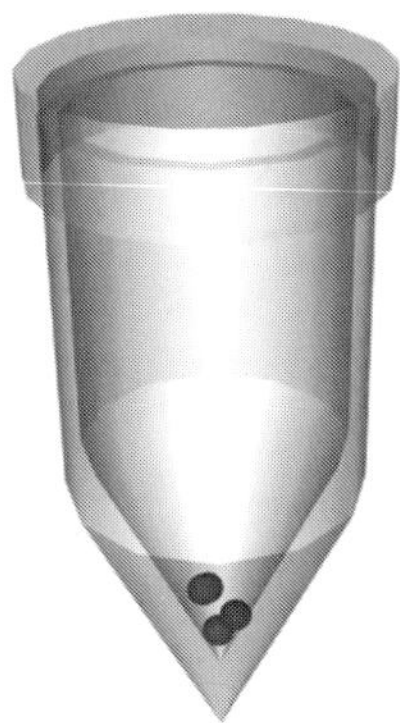

Fig. 7. Beem capsule with a conical tip with small samples.

3.1.2.2. Small Biopsies

Biopsies smaller than 2 mm^3 include spheroids, cell aggregates, needle biopsies, isolated glomeruli, or islets of Langerhans.

1. Fixation in Eppendorf capsules with conical tips (*see* **Note 7**).
2. Wash biopsies with wash buffer (do not centrifuge); carefully remove buffer after allowing the tissue pieces to settle by gravity.

For morphology, fixation is with osmium tetroxide (*see* **Note 8**); further processing (*see* **Subheading 3.1.4.**) is performed in Beem capsules with conical tips (*see* **Fig. 7**).

For immunohistochemistry, blocks are made of biopsies in gelatin:

3. Add 100 µL of 10% prewarmed gelatin/PBS (37°C) and mix.
4. Centrifuge quickly when gelatin is still liquid at 825*g* for 5 min.
5. Allow the gelatin to solidify in a refrigerator at 4°C for 10 min.
6. Cut off the bottom of the tube with a sharp, new, single-edged razor blade just above and beneath the gelatin. Also, carefully remove the surrounding remains of the tube.
7. The gelatin block(s) are postfixed in stored buffer for immunohistochemistry, at 4°C for 4–16 h.
8. Cut away any gelatin not containing cells and divide in cubes of 1–2 mm^3 under a stereomicroscope on a wax plate with a razor blade.
9. Process the pieces as described in the embedding procedure for tissue (*see* **Subheading 3.1.4.2.**) or ultrathin cryosectioning (*see* **Subheading 3.1.5.2.**, **step 3**).

3.1.2.3. Single-Cell Suspensions

1. Harvest cells after trypsinization and wash in culture medium with serum in 10-mL tube.
2. Centrifuge at 300*g* for 5 min.

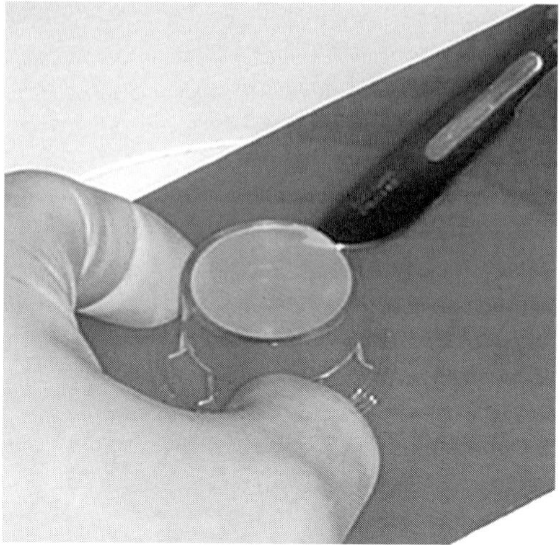

Fig. 8. Cutting away of the transwell filter.

3. Remove medium and dissolve pellet in fixation fluid.
   - For morphology, fix in 1.5% glutaraldehyde for 30 min at room temperature;
   - For immunohistochemistry, fix in 4% paraformaldehyde with 0.05–0.2% glutaraldehyde 30 min at room temperature (*see* **Note 7**).
4. Centrifuge at 300*g* for 5 min.
5. Remove the fixative and add 1 mL of PBS to the pellet.
6. Put the cells in an Eppendorf tube (1.5 mL).
7. Centrifuge again, remove the PBS, and gently loosen the pellet by tapping the tube. For further steps, see **Subheading 3.1.2.2.**, after **step 2**.

3.1.2.4. Monolayers (on Plastic or Glass)

1. Wash the cells with Hank's medium.
2. Fix the cell layer for 30 min at 4°C (*see* **Note 7**).
3. Remove the cells from the layer with a rubber policeman and collect in a centrifuge tube.
4. Proceed from **Subheadings 3.1.2.3.** and **3.1.3.4.**

3.1.2.5. Cells Growing on Transwells

1. Wash the cells with Hank's medium.
2. Fix the cell layer for 30 min at 4°C (*see* **Note 7**).
3. Carefully remove the filter from the well with a razor blade under a stereomicroscope and put this on a wax plate, with the cells facing upward (*see* **Fig. 8**).
4. Carefully cut the filter in slices approx 2 mm in width and place two or three slices on top of each other in a rubber embedding mold or place them upright in an Eppendorf tube.

5. Add 10% preheated gelatin/PBS (37°C) into the module or tube.
6. Allow the gelatin to solidify in a refrigerator at 4°C, or on ice.
7. Remove from the mold or cut off the bottom of the tube with a sharp, new, single-edged razor blade just above and beneath the gelatine. Also, carefully remove the surrounding remains of the tube.
8. Fix the gelatin with filters with the stored buffer overnight, also at 4°C.
9. Process the pieces as in the embedding procedure for tissue, described in **Subheading 3.1.4.** or ultrathin cryosectioning (*see* **Subheading 3.1.5.2., step 4**).

3.1.3. Tissue Processing for Pre-Embedding (See ***Note 8****)*

All steps, including the glutaraldehyde fixation, are performed at 4°C to avoid bacterial growth; thereafter, steps are performed at room temperature.

1. Fix tissue blocks (4–10 mm^3) briefly after perfusion or by immersion with fixative for 16 h (*see* **Note 7**).
2. Cut vibratome sections (70–100 μm thick) and transfer sections with a small paintbrush into PBS in a 10-mL glass vial.
3. Proceed with all further incubation steps by gently stirring on a rotor.
4. Incubate sections with quench buffer for 1 h (*see* **Note 9**).
5. Permeabilize with 0.05% Triton X-100 in PBS for 30 min (*see* **Note 10**).
6. Incubate with protein blocking buffer for 30 min.
7. Incubate with specific primary antibody (1–5 μg/mL) in 1% BSA with 20 m*M* NaN_3/PBS (*see* **Note 11**) for 16 h (*see* **Note 12**).
8. Wash with PBS 5X 10 min.
9. Incubate in peroxidase-conjugated antibody in 1% BSA/PBS for 2 h. or incubate in appropriate colloidal gold (1 nm or 5 nm)-conjugated reagent in 1% BSA/PBS for 2 h.
10. Wash with PBS for 5X 10 min.
11. Fix with 1% glutaraldehyde in PBS for 20 min at room temperature.
12. Wash with PBS for 3X 10 min at room temperature.
    a. For Peroxidase–DAB Staining. Develop with freshly prepared diaminobenzidine medium (add 10 μL of 30% H_2O_2 to the filtered DAB solution) for 20–40 min in the dark. Wash peroxidase–DAB-stained sections with PBS for 3X 20 min.
    b. For Immunogold Staining (*see* **Note 13**) Wash sections thoroughly with distilled water 4X 15 min. Develop gold with silver reagent in the dark with the Danscher method for 30 min to 1h at room temperature (*see* **Subheading 3.1.6.4.**). Wash thoroughly with distilled water 4X 15 min.
    c. For both staining procedures, continue with **step 13** (*see* **Note 14**).
13. Postfix with 0.5% osmium tetroxide in 0.1 *M* cacodylate buffer, pH 7.4, at room temperature for 20 min.
14. Further processing (*see* **Subheading 3.1.4.1., step 2**) is performed in Beem capsules size 00 with the tip cut off. The vibratome sections must be embedded in the cover of the Beem capsules by placing the capsule upside down (*see* **Fig. 9**). It is then possible to cut sections parallel to the surface of the vibratome sections.

Fig. 9. Beem capsule with vibratome section embedded in plastic.

Fig. 10. Example of embedding in gelatin capsules of cells grown on glass.

3.1.4. Dehydration and Embedding

3.1.4.1. Dehydration and Embedding (of Tissue/Cells/Immunostained Vibratome Sections) in EPON

1. Postfix with 1% osmium tetroxide in cacodylate buffer (tissue for 1 h and cells for 10 min). If tissue must be studied by TEM and RCM, postfixation with osmium tetroxide is necessary (*see* **Note 14**). Wash with distilled water for 2X 10 min.
2. Dehydrate through an ethanol series, from 50%, 70%, 80%, 90%, and 2X 100% ethanol. All steps last 45 min for tissue samples and 5 min for single cells.
3. Embedding: remove the final absolute ethanol and add EPON/ethanol mix 1:1 overnight.
4. Infiltrate for 2 h with EPON or Spurr's/ethanol mix 2:1 followed by pure EPON or Spurr's resin for 2 h.
5. Finally, embed tissue blocks in Beem capsules or mold.

5a. For monolayer cells on glass or petri dishes (*see* **Note 15**), put EPON-filled gelatin capsules on top of them (*see* **Fig. 10**).

6. Polymerize the blocks at 60°C for 24–48 h.

Remove capsules with the embedded monolayer from the glass slide: Remove the excess EPON around the capsules with a small hand drill or with a single-edged razor-blade. Dip the whole object glass in liquid nitrogen. Control the broken surface for glass particles under a stereomicroscope and remove them (*see* **Note 16**).

3.1.4.2. Dehydration and Embedding in Lowicryl K4M or Unicryl, for Immunohistochemistry

Cells and tissue samples are prepared for Lowicryl K4M or Unicryl resin embedding with minor modifications according to the progressive lowering-temperatures technique ***(12)***.

Dehydration:

1. Dehydrate the samples in 30% (v/v) ethanol at 4°C for 30 min.
2. 50% ethanol at –4°C for 30 min.
3. 70% ethanol at –20°C for 30 min.
4. 80% ethanol at –20°C for 30 min.
5. 90% ethanol at –20°C for 30 min.
6. 100% ethanol at –20°C for 30 min.

Infiltration:

7. Ethanol/resin (1:1) at –20°C for 60 min.
8. Ethanol/resin (1:2) at –20°C for 60 min.
9. Immerse in pure resin at –20°C for 60 min.
10. Immerse in pure resin at –20°C overnight.
11. Optionally, add fresh resin (60 min, –20°C) before final embedding of the samples in Beem capsules (*see* **Note 17**).

Polymerization:

12. Capsules are placed in a 96-well microtiter plate under ultraviolet (UV) lamps for 1 d at –20°C (*see* **Note 18**).
13. Polymerization is continued for 2 d at room temperature under UV lamps in a fume hood.

3.1.5. Sectioning

3.1.5.1. Sectioning for RCM and TEM, Collecting, and Storing

Sections can be cut on an ultramicrotome (100 nm) (*see* **Note 19**) or a routine motorized microtome (250 nm) using a glass or histo-diamond knife.

1. Clamp the specimen into the specimen holder and trim the excess resin from the block face and from the edges of the block using a single-edged razor blade.
2. Cut a semithin section of 1 μm, collect onto water, and use a preparation needle to transfer the section to a drop of water on a microscope slide.
3. Dry on a hot plate at approx 60°C and stain with toluidine blue for 1 min on a hot plate. Wash briefly with distilled water and dry on the hot plate for 20 min.

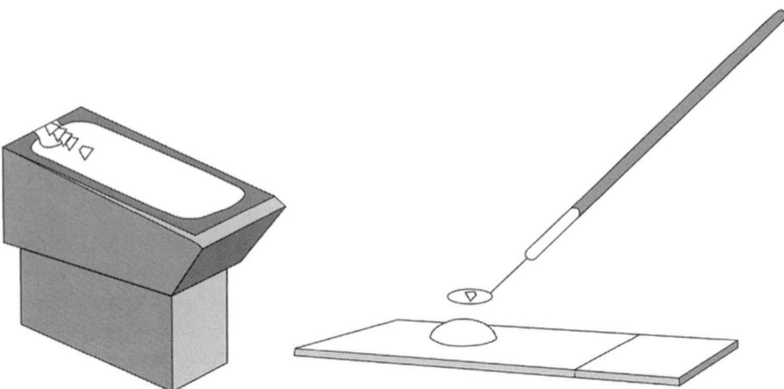

Fig. 11. Transfer of thin sections from a knife to a glass slide using a droplet in a wire loop.

4. Select an area using the microscope and trim the face to a trapezoid shape with a histo-diamond knife).
5. Fill the boat with water (*see* **Note 20**) and cut the ultrathin sections.

Collecting the Sections for TEM and RCM, Sequential Sectioning

For TEM: If you collect Lowicryl, Unicryl, or cryosections, use grids with a formvar or carbon film. For sequential TEM/RCM, one section from a small ribbon is picked up with a wire loop and the next one with a grid.

1. Cut a small ribbon; isolate the sections in pairs using an eyelash mounted on a cocktail stick with nail polish, which is also used for moving the sections on the water.
2. Separate the pair in single sections and place the grid in the water beneath one section and raise it at a slight angle so that the first section is in the center of the grid (*see* **Note 21**).
3. Slowly raise the grid out of the water and remove the water with filter paper; place the grid in a Petri dish with round filter paper to dry.
4. Move the wire loop into the water beneath the second section and raise it so that the second section is in the center of the loop; place it on a water drop on a clean glass slide (coated if immunocytochemistry will be performed) (under a dissecting microscope) (*see* **Fig. 11**). Clean wire loop after use in an ethanol flame (*see* **Fig. 4**).
5. Dry on a hot plate at approx 60°C for morphology and 40°C for immunohistochemistry.
6. Under a stereomicroscope, encircle the sections on the reverse side of the slide with water-resistant ink or, if a cap/prism will be used for microscopy, with a diamond writer (*see also* **Subheading 3.4.1., step 4**).

Only for morphology: Counterstain with 1% toluidine blue for 1 min on a hot plate and wash with distilled water and dry on a hot plate. Cover with a small amount of normal mounting medium and use a clean cover glass. You can also use immersion oil as mounting medium; then fix the cover glass with nail polish. If you leave the section with oil without a cover glass, clean it with ethanol if it becomes dirty. Store slides in a dust-free slide container.

3.1.5.2. Ultrathin Cryosectioning for RCM, Starting From Fresh Material*

1. Fix fresh, living material with 4% paraformaldehyde and 0.1–0.2% glutaraldehyde in 0.1 *M* phosphate buffer, pH 7.4, for up to 2 h (*see* **Note 7**).
2. For cells; fix for 30 min and carefully scrape with a rubber policeman from the solid substrate, pellet by centrifugation in an Eppendorf tube (1.5 mL), and embed in gelatin as described in **Subheadings 3.1.2.2.** and **3.1.2.3**.
3. Cut sample into slabs and then into small blocks no larger than 2 mm^3 (*see* **Note 22**).
4. Transfer blocks into 2.3 *M* sucrose and leave overnight at 4°C (*see* **Note 23**).
5. Mount individual blocks onto specimen pins (supplied with cryo-ultramicrotome) and freeze by immersion in liquid nitrogen.
6. Cool cryo-chamber of ultramicrotome to –110°C and lock the knife into the holder.
7. Transfer the specimen pin to the cryo-chamber of the ultramicrotome.
8. Adjust thickness setting to section at 500 nm and cut into the block until a flat surface is produced.
9. Advance knife and specimen block and cut sections automatically to 100 or 150 nm.
10. Pick up sections on a drop of 2.3 *M* sucrose held in a wire loop (*see* **Fig. 11**).
11. Warm up the drop to which the sections are attached and place on a dry glass slide.
12. Store under sucrose at 4°C, or after marking around the drop of sucrose with a Dako pen, remove the sucrose with PBS and incubate directly for immunostaining (*see* **Subheading 3.1.6.1.** or **3.1.6.2.**).

3.1.6. Immunostaining

3.1.6.1. Immuno-Peroxidase Staining of Cryosections or Lowicryl/Unicryl Sections (*See* Note 24)

1. Perform all incubations in a moist chamber at room temperature.
2. Quench free aldehyde groups with quench buffer for 15 min (*see* **Note 9**).
3. Preincubate with protein blocking buffer for 10 min.
4. Incubate for 1 h at room temperature (or 16 h at 4°C) with specific antibody in incubation buffer.
5. Wash with PBS 4X 5 min in a jar.
6. Incubate with a peroxidase-conjugated secondary antibody in a suitable dilution in incubation buffer for 1 h.
7. Wash with PBS 4X 5 min.
8. Fix with 1% glutaraldehyde in PBS for 5 min (*see* **Note 25**).
9. Wash with PBS 2X 5 min.
10. Add 10 µL of H_2O_2 to 100 mL DAB solution. Develop in the dark with DAB + H_2O_2 solution for 10 min and wash with distilled water 2X 5 min.

11. Counterstain with light green for 2 min (*see* **Note 26**).
12. Wash with distilled water for 1 min.
13. Dry on a hot plate at 40°C for 10 min.
14. Cover the sections with a very small droplet of mounting medium and a clean cover glass.

3.1.6.2. Immunogold-Silver Staining of Cryosections or Lowicryl/Unicryl Sections

1. Start with **Subheading 3.1.6.1., steps 1–5** (immuno-peroxidase staining); then preincubate with 0.2% solution of BSA-C in PBS for 10 min.
2. Incubate with a gold-conjugated (1–10 nm) secondary antibody in a suitable dilution in incubation buffer for 2 h.
3. Wash with PBS 4X 5 min.
4. Fix for 5 min with 2% glutaraldehyde in PBS.
5. Wash with PBS 2X 5 min.
6. Wash with distilled water for 4X 5 min.
7. Develop immunogold in the dark with the silver enhancer (mix just prior to incubation) for 20–40 min (*see* **Subheading 3.1.6.4.**).
8. Wash with distilled water for 4X 5 min.
9. Counterstain with toluidine blue for 2 min, with hematoxylin for 1 min, or with 1 µg/mL propidium iodide for 5 min (*see* **Note 26**).
10. Wash with distilled water for 1 min.
11. Dry on a hot plate at 40°C for 10 min.
12. Cover the sections with a small amount of mounting medium and a clean cover glass.

3.1.6.3. Immunogold-Silver Staining of EPON or Spurr's Sections

Immunoperoxidase can be applied on these sections as well.

1. Perform all incubations in a moist chamber at room temperature.
2. Etch with 10% H_2O_2 for 5 min; then wash with distilled water 2X 5 min.
3. Treat with 1% periodic acid solution for 10 min and wash with distilled water 2X 5 min.
4. Preincubate with protein blocking buffer for 10 min.
5. Incubate with specific antibody in PBS, 1% BSA, and 1% normal serum of second antibody for 1 h at room temperature (or 16 h at 4°C).
6. Wash with PBS for 5X 5 min.
7. Preincubate with 2% BSA in PBS for 10 min.
8. Incubate with a gold-conjugated (1–5 nm for RCM and 10 nm for TEM) secondary antibody in a suitable dilution in 2% BSA in PBS for 2 h.
9. Wash with PBS for 5X 5 min.
10. Fix with 2% glutaraldehyde in PBS for 5 min.
11. Wash with PBS for 2X 5 min, and wash with distilled water for 4X 5 min.
12. Develop immunogold in the dark with the silver enhancer (mix just prior to incubation for 20–40 min) (*see* **Subheading 3.1.6.4.**).

13. Wash with distilled water for 4X 5 min.
14. Counterstain EPON or Spurr's sections with toluidine blue for 2 min or Spurr's sections with hematoxylin for 1 min.
15. Wash with distilled water for 1 min.
16. Dry on a hot plate at 40°C for 10 min.
17. Cover the sections with a small amount of mounting medium and a clean cover glass.

3.1.6.4. Silver Enhancement of Gold *(13)*

Only enhance with silver if the colloidal gold particles are smaller than 15 nm, because 15-nm gold particles are visible in RCM.

Solution A: Mix 25 g arabic gum in 50 mL distilled water for 1 d and filter through a gauze bandage. Dispense in 3-mL portions and store in freezer.
Solution B: Dissolve 25.5 g citric acid and 23.5 g sodium-citrate and check that pH = 3.5.
Solution C: Dissolve 0.85 g hydroquinone in 14 mL distilled water and filter through 0.2-µm filter. Prepare solution shortly (store for no more than 1 d) before use.
Solution D: Dissolve 0.11 g silver lactate in 14 mL distilled water and filter through 0.2-µm filter (store in the dark).

Before use, add 0.5 mL solution B, 0.75 mL solution C, and 0.75 mL solution D to 3.0 mL solution A, mix well, and drip with a Pasteur pipet on the sections and develop in the dark (e.g., incubation box or light-tight box). Make sure sections do not dry before the silver enhancement.

3.1.7. Single-Cell Experiments

3.1.7.1. Staining the Surface of Cells With Immunogold*

This is a nice specimen for demonstration of reflection with CLSM (*see* **Fig. 12**) (*see* **Note 27**). All steps were carried out at room temperature and incubation was performed in 24-well plates.

1. Culture cells on 10-mm-diameter cover glass in a Petri dish.
2. Wash with Hank's medium for 2X 1 min.
3. Fix with 1% paraformaldehyde in 0.1 *M* phosphate buffer, pH 7.4, for 10 min.
4. Treat with aldehyde quench buffer for 20 min.
5. Preincubate with 2% BSA in PBS for 10 min.
6. Incubate with specific primary antibody diluted in 1% BSA in PBS for 1 h. Alternatively, use a colloidal gold (5–40 nm)-conjugated antibody and go to **step 10.**
7. Wash with PBS 4X 5 min.
8. Preincubate with 2% BSA in PBS for 10 min.
9. Incubate with gold (1–10 nm)-conjugated antibody diluted in 1% BSA in PBS for 2 h.

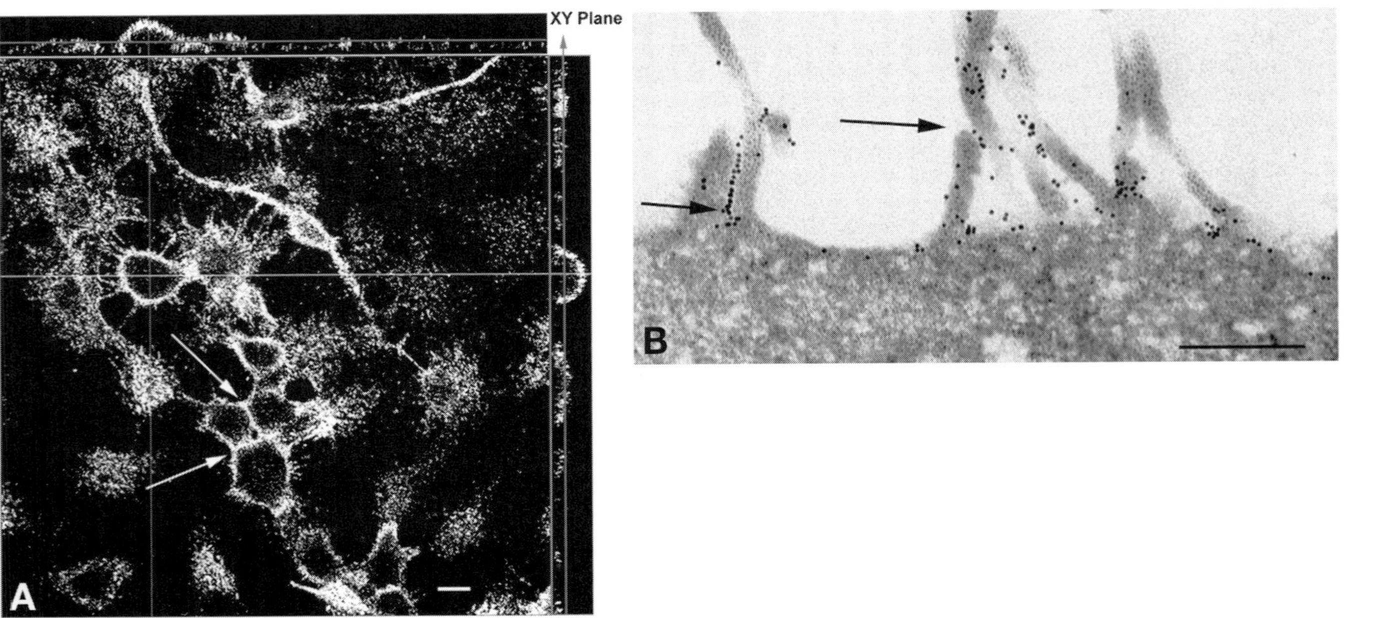

Fig. 12. Reflection contrast image of a surface labeling with immunogold-silver showing the small attachments and detecting small amounts of an antibody on fine structures. These structures as focal adhesions are visible when using 10 nm gold with silver enhancement. Scanning was performed with the 488-nm laser line in the reflectance mode and a pinhole size of 10 μm. U2 cells after mild fixation and labeled with anti-Ep-CAM, and gold (10 nm)-conjugated antibody and a short silver enhancement. **(A)** Typical CLSM image showing an ortho display of a z-scan. The line in the *XZ* and *YZ* plane show the *Z*-height of the *XY* image (large image) (bar = 10 μm). **(B)** High-magnification TEM image of cell membrane staining without silver enhancement after embedding in Epon (*see* **Note 27**) (bar = 0.2 μm).

10. Wash with PBS 4X 5 min.
11. Fix with 2% glutaraldehyde in PBS for 10 min.
12. Wash with PBS for 2X 5 min and in distilled water 4X 5 min.
13. Enhance with silver in the dark for 20–40 min, as described in **Subheading 3.1.6.4.**
14. Counterstain with 1 µg/mL propidium iodide (fluorescent stain).

Mounting single cells or very small objects in immersion oil (*see* **Note 24**):

15. Wash with distilled water.
16. Dehydrate through 40%, 75%, 95%, and 2X 100% ethanol for 2 min (each step) (*see* **Note 28**).
17. Incubate in a mix of 1:1 ethanol/immersion oil (refractive index = 1.518) for 10 min.
18. Incubate in a mix of 1:2 ethanol/immersion oil for 10 min.
19. Incubate in pure immersion oil for 10 min.
20. Take the cover glasses carefully out of the Petri dish/wells (using a fine forceps).
21. Maintain immersion oil on cells.
22. Put a small drop of immersion oil on the object slide.
23. Place the cover glass (cell side up) on the drop of immersion oil and seal with a small amount of nail polish.
24. Store the slides horizontally in a dry box so that cells are maintained in and covered with oil.
25. Scan the slides with a CLSM in the reflection mode, in combination with differential interference contrast or with fluorescence using the multitrack method. Be careful with the objectives, because cells can be damaged easily, or cover with cover glass and seal with nail polish.

3.1.7.2. Internalization of Endocytic Tracers (Gold)

1. Culture cells on 10-mm-diameter cover glass in a Petri dish.
2. Cells were pulsed in a medium containing BSA/5 nm gold or a gold-conjugated antibody for 1 h at 37°C (*see* **Note 29**).
3. Wash with medium and culture for 16 h at 37°C.
4. Stop endocytosis by washing with cold Hank's medium and fix with 1% glutaraldehyde in 0.1 *M* cacodylate buffer, pH 7.4, for 10 min.
5. Wash with Hank's medium for 1X 2 min and in distilled water for 4X 5 min.
6. Continue from **Subheading 3.1.7.1., step 13** and/or for other possibilities (*see* **Note 27**).

3.1.7.3. Internalization of Peroxidase-Conjugated Antibody

1. Culture cells in a 96-well plate.
2. Place the plate on ice, add peroxidase-conjugated antibody (type of antibody that binds to the cell and will internalize) to the medium, and incubate for 1 h.
3. Wash with fresh medium and place the plate at 37°C.

4. Wash with cold Hank's medium after different time intervals (1, 4, and 24 h) for 2X 2 min.
5. Fix after each time interval with 1% glutaraldehyde in 0.1 *M* cacodylate buffer, pH 7.4, for 10 min at 4°C.
6. Wash with 0.1 *M* cacodylate buffer (pH 7.4) 2X 10 min at room temperature.
7. Develop with freshly prepared DAB (add 10 µL H_2O_2 to 100 mL DAB solution) for 30 min in the dark and wash with PBS.
8. Postfix with1% osmium tetroxide in 0.1 *M* cacodylate buffer, pH 7.4, for 10 min.
9. Wash 4X 5 min with distilled water.
10. Dehydrate with ethanol from 50%, 70%, 80%, to 95% (each step for 5 min) and 100% ethanol for 2X 10 min.
11. Infiltrate with 1:1 ethanol/EPON for 2 h or overnight.
12. Fill the wells with pure EPON and polymerize at 60°C for 24–48 h.
13. Use a minisawing device to remove the wells from the plate and remove the remaining plastic from the embedded EPON blocks. Proceed with sectioning (**Subheading 3.1.5.1., step 4**) (*see* **Note 16**).

3.2. Observation of Living, Unstained Cells

Living cells can be studied by RCM through the cover glass on which they are growing. A cover glass is thin enough to obtain a focused image of the cells. The underside of the cells is studied (e.g., processes of adhesion, migration, and locomotion). The contact sites of the cell with the glass are dark in RCM, whereas the aqueous surrounding is bright. *(8,14)*. Special chambers were designed and an inverted RCM was used to visualize living cells *(7)*.

1. Grow the cells on round cover glasses (25 mm in diameter) in a Petri dish.
2. Heat the stage of the microscope (e.g., with temp control 37- 2 [Zeiss, cat. no. 1052-320]) to 37°C. Check the temperature with a small thermometer in a Petri dish with water on the stage.
3. Place the cover glasses into the holders or on object glass with cavity, with the cells facing down (*see* **Fig. 5A or B**) and add some medium from the Petri dish, avoiding inclusion of air bubbles. They will influence the quality of the images and will damage the cells. First, dry off the outside of the cover glasses with tissue, and in case of glass slides with a cavity, encircle the edge of the cover glass with nail polish, to avoid drifting of the glass and drying of the cells.
4. Place the holder onto the microscope stage and put a drop of oil on the cover glass.
5. Gently approach the cover glass with the objective, focus (if necessary, use phase-contrast or differential interference contrast microscopy to locate the cells), and collect images.

3.3. How to Adapt a Fluorescence Microscope for RCM

A microscope stand with incident illumination either upright or inverted, equipped with a strong light source like a mercury high-pressure or xenon lamp (*see* **Note 30**) is the basis for RCM. The three extra parts needed for RCM are an oil-immersion objective with a quarter-lambda plate in the front lens, a polarization block, and a central stop system (diaphragm module) or cap/prism.

Only Leica has a complete adaptation set for RCM available (Leica reflection contrast device RC; Wetzlar, Germany) for a DAS Mikroskop Type R (DMR) microscope. An antiflex objective (also with a rotateable quarter-lambda plate on the front lens) and polarization block is available from Zeiss. The adaptation of a Zeiss Axioskop provides a good RCM also. However, in principle, all incident light microscopes can be adapted for RCM; a polarization block is available from all microscope firms (sometimes named immunogold-block or epi-polarization block), three objectives (50×, 63×, and 100×) exist, a central stop system can be developed, or a cap/prism can be used instead.

3.3.1. RCM on a Leica DMR

1. Position a protection filter for the polarizer in front of the lamp (the polarizer can be damaged by long exposure to intensive light); remove the lamp housing (Leica 105z or 106z) and slot the protection filter into the light exit window of the lamp housing from the front.
2. Position the RCM module into the space reserved for it, in the horizontal light path between lamp and illuminator.
3. Exchange a filter block in the epi-fluorescence illuminator for a polarization block.
4. Rotate the RC objective into the light path.

3.3.2. Zeiss Axioskop

An alternative for the module is the use of a cap (glass lens) or prism, linked by oil immersion to the bottom of the glass slide, which improves the image contrast in a manner similar to the central stop (*see* **Fig. 13**) (*see* **Subheading 3.3.1., steps 1, 3,** and **4**).

Link the cap or prism with a very small droplet of immersion oil to the underside of the glass slide at the marked circle. Next, place the slide on the microscope stage (*see* **Note 31**).

3.4. Microscopy

3.4.1. Operating the RCM

Optimize microscope settings

1. Center the incident illumination carefully by using the adjusting screws on the lamp housing (*see* **Note 32**).
2. Center the field diaphragm.

Fig. 13. If no central stop system is available, the use of a prism/cap makes RCM of ultrathin sections possible. The prism is mounted with oil to the glass slide. It functions by directing the light outside of the objective; no reflected light at another place than at the specimen can enter the objective.

3. Choose the size of central stop in the RCM module in accordance with the RCM objective and center its projected image using the Bertrand lens of the microscope in the middle of the image of the arc of the lamp. Find and focus the ultrathin specimen on slide.
4. Circle the sections under a stereomicroscope on the back of the slide with a pen, closely around the specimen, or with a diamond writer if you use a cap/prism. Alternatively, if no stereomicroscope is available, unstained ultrathin sections are detectable by reflecting bright light (e.g., of a lamp) at the section upon tilting the glass slide.
5. Put a drop of immersion oil on top of the sections (the marked area) of the slide.
6. Use bright-field microscopy to find and center the circle with oil (thin sections) with a 5× or 10× dry objective (avoid getting immersion oil on the specimen).
7. Swing in the RCM oil-immersion objective into the observing position and close the field diaphragm slightly and focus. The field diaphragm is simultaneously visible in the field of view. Use the fine focus control and carefully focus. If no image of the field diaphragm is observed, lower the microscope stage (but without losing contact with the immersion oil) until an image of the field stop is obtained.
8. You are now in the plane of the specimen and within the marked area of the slide and should find the specimen by moving the stage slowly in various directions.

Optimize image

9. Rotate the front lens (quarter-wave plate) until the highest reflection is obtained. If the image is not satisfactory (*see* **Note 33**), consider **steps 10** and **11**.
10. For uneven distribution of light, (a) center and align the lamp better, (b) locate the air bubble in the immersion oil between specimen and objective (remove the bubble by swinging objective in and out of the oil), and (c) center the central stop or diaphragm.
11. For poor contrast, (a) rotate the front lens until the highest reflection is observed, (b) change the stop and or aperture diaphragm, (c) improve the contrast by closing down the field diaphragm, (d) control the presence of gray filters in the light path, and (e) improve the lamp alignment by adjustment of the collector of the lamp housing. Finally, collect the images (*see* **Subheading 3.5.**).
12. After examination with the microscope, the slides should be stored vertically in a microscope slide box, to allow the oil to drain away from the slide (except for cells immersed in oil, which should be stored horizontally).
13. For microscopic re-examination of stored dirty slides, first remove the oil with 100% ethanol and allow to dry (*see* **Note 34**).

3.4.2. How to Use the Confocal Laser Scanning Microscope as a Reflection Contrast Microscope

First, find the specimen site for RCM imaging using bright-field microscopy or phase-contrast microscopy. Next, use CLSM in the reflection mode by inserting the NT splitter and removing all dichroic mirrors. To mimic the RCM image with xenon light, choose laser light intensities that combine with white light and reduce the size of pinhole. Refinement of gain(s), amplification offset(s), and focus must be optimized before CLRSM images can be scanned.

1. Put the slide with the circle centered in the beam of the light that is in the center of the condenser front lens. Use bright-field microscopy with low light intensity to find the thin section with a 5× or 10× dry objective. Find and focus the section, or use phase-contrast or differential interference contrast microscopy to find and focus the cells. Use stained (toluidine blue) sections for good visibility; unstained ultrathin sections are visible because of refractions on the edge of sections, in case no circle is set after putting the ultrathin sections on glass (*see also* **Subheading 3.4.1., step 4**).
2. Only lower the stage to change the objective; do not move the stage laterally. Place a drop of oil on the slide and insert the 25× or 63× oil objective and focus again using bright-field microscopy.
3. Locate part of the section of interest in the middle of the field of view and shut the bright-field light down; move the condenser out of position if not needed for differential interference contrast microscopy.
4. Turn on the lasers (argon, He–Ne1, and He–Ne2).

Configuration of the confocal laser scanning microscope:

5. For the LSM-510 (Zeiss), insert the NT 80/20/543 or NT 80/20 neutral beam splitter. For the TCS SP2 (Leica), insert the RT 70/30 neutral beam splitter. For other CLSMs, insert the neutral beam splitter (reflection mode) (*see* **Note 35**).

6. Use one or three detection channel(s) (photomultiplier tubes [PMTs]) (*see* **Note 36**).
7. Code each laser line with its own color in one or each detection channel (488 nm, blue; 543 nm, green; 633 nm, red).
8. Optimize the laser intensity values as when using white visible light. For ultrathin stained sections, use the following illumination intensity values for the LSM510 (Zeiss): 6% of the 488-nm line, 15% of the 543-nm line, 2% of the 633-nm line.
9. Optimize the detector pinhole diameter for the PMT in use. Use 25–100 μm for weakly stained ultrathin sections and use a smaller pinhole (10–30 μm) for gold or gold/silver staining (*see* **Note 37**).

Scan control:

10. Set detector gain to 50%. Use the fast-scan mode and focus with the fine-focus knob untill maximum reflection intensity (focused image) is achieved.
11. Optimize the detector gain of the PMT in use during continuous scanning until there is a bright and contrast-rich image of all known structures that are seen separately (optimized detector gain, "gain").
12. Split into RGB images. Optimize the amplification offset for each color so that the background (reflection of the glass) is black or dark for each laser line (optimized amplification offset, "black level").

Scan image:

13. Make a single image with a low-scan speed and choose the minimal frame size of 1000 × 1000 and minimal 8-bit pixel depth for high definition.

3.5. RCM Images

3.5.1. Conventional Photographic Recording

1. Focus the reflected graticule (square with double lines in the middle of the overlay) by adjusting the eye lenses of the observation eyepieces until the double lines of the focusing aid are sharply and separately focused.
2. Adjust the image focus with the coarse/fine controls of the microscope.
3. Close the field diaphragm just around the format outline.
4. For exposure time, use spot measuring if there is high contrast in the image (in most cases) or use integral measuring. Point spot on the highest reflection part; only if a part of the spot covers highly reflective gold/silver particles, adjust for the percentage that is covered.
5. Take picture (*see* **Note 38**).
6. Dia positive film can be easily digitized on a high-resolution slide scanner.

3.5.2. Digital Recording

Reflecting contrast microscopy images show a high contrast that should be kept in the recording; therefore, use a high dynamic range. Also, RCM images have very high definition, so that a high-resolution camera is needed (charge-coupled devise basic resolution of 1.3 megapixels [8-bit color]). The purpose of the recording is to obtain an image as good as the one observed by eye through the oculars, with the same colors and dynamic range.

1. Start with a desired image in focus in the eyepieces.
2. Insert the camera in the light path (parfocal) and direct all of the light to the camera.
3. Operate the camera according to the manual; set automatic white balance.
4. Fine-focus the image on the screen during real-time imaging.
5. Set the black level, gamma for brightness according to the real observed image.
6. Capture the image.
7. For a bright-field-microscopy-like image, invert the image. For an EM-like image, invert the image and convert to black and white (*see* **Fig. 1A,C**).
8. Because of high contrast, image analysis of RCM images is easily performed; there are fewer steps for image improvement. Begin, perhaps, with segmentation.

3.5.3. Image Interpretation

The color of the tissue is complementary to the bright-field color because the wavelengths that are strongly absorbed are also strongly reflected. The effect of osmium tetroxide fixation on the RCM image is an enhancement of contrast of membranes and the peroxidase–DAB product, resembling more that of an inverted TEM image.

Interpretation of immunocytochemical staining in a RCM image is best done by comparing with a RCM image of a control section; namely all incubation steps performed but without the specific primary antibody.

In this chapter, we did not show colored images; however, the colors in the RCM-image facilitate interpretation. For example; the gold color can be easily detected in an immunogold (15–40 nm)-stained section. Undecalcified bone, in an ultrathin Epon section counterstained with toluidine blue, is light blue and cells are orange. These examples in black and white do not have a high-contrast difference but a good color distinction.

4. Notes

1. To study ultrastructure, EM is the best way, because RCM is a light microscopic method and the resolution depends on the wavelength of the imaging rays/particles and on the numerical aperture (NA). The NA of the RC objective should be 1.4 for highest resolution. However, CLSM is even better with the same objective, because as a result of the scanning principle, the resolution is $\sqrt{2}$ times higher.
2. Reflection contrast microscopic images generally show the background of the tissue as black and the tissue itself as white, resembling a fluorescent image or an inverted TEM image. If the background of the image is bright and the object dark, it looks like a bright-field image; then the reflection at the interface of the glass slide is the strongest. With gelatin-coated glass slides, this effect is obtained.
3. A cap can be used instead of a central stop system. It can be a prism or a small lens, and its function is to prevent light that reflects beneath the object from entering the

objective. The main interface that causes these reflections is the one between the bottom of the glass slide and air (*see also* **Fig. 13** and **Subheading 3.3.2.**).

4. For visualization of morphology at the highest definition, the specimen must be ultrathin and the refractive index as close to glass as possible. Between 50 and 100 nm, the strongest, brightest reflection is observed, namely white-light yellowish. This color is the same as the thin section has when it is floating on the water bath of the ultramicrotome. Submerging an ultrathin plastic section or immersing an ultrathin cryosection (sectioned after freezing the tissue) in oil (*see* **Subheading 3.1.7.1., steps 15–23**) or an embedding medium with the same refractive index as glass provides good RCM images of all material present for morphology. Enhancement of contrast of these kinds of image can be obtained by overall staining (e.g., toluidine blue) or specific enhancement of certain structures (e.g., DNA with acridine orange). However, even thicker sections (0.25 μm) cut on a motorized routine microtome provide reasonable RCM images.
5. Mild and reversible fixatives (formaldehyde, acetone, methanol, etc.) as used for paraffin embedding destroy ultrastructure normally not visible with bright-field microscopy, but visible with RCM ***(15)***.
6. All solutions are made with Milli–Q water (i.e., double-distilled filtered water) Blocking buffers and solutions for silver enhancement (*see* **Subheading 3.1.6.4.**) are filtered, except the solutions containing gold particles or antibodies.
7. Fixation can vary and strongly depends on the tissues or cells under study. However, the general rule is to use freshly made paraformaldehyde in the range of 1–4% in 0.1 *M* phosphate or cacodylate buffer, adjusted to pH 7.4. Glutaraldehyde is subsequently added to the proper concentration (0.05–0.2%) for further improvement of the ultrastructural morphology without loss of antigenicity.
   a. *Fixing of tissues*: if perfusion is not possible, then immersion fixation is an option. However, it is important to transfer the sample to the fixative as rapidly as possible after removal and to avoid all forms of mechanical trauma. If necessary, the tissue can be carefully sliced into small slabs as soon as it has been immersed in fixative.
   b. *Fixing of cells*: for preservation of morphology, 10-min fixation is sufficient, and a lower concentration of 0.1–1% glutaraldehyde in buffer can be used. For immunocytochemistry, prolonged fixation in fixative containing glutaraldehyde is not recommended. For cryosectioning, longer than 2 h in fixative containing glutaraldehyde causes the samples to become brittle and difficult to cut. After incubation in fixative, store in store buffer (*see* **Subheading 2.1.1.2.**).
8. Pre-embedding immunogold or peroxidase–DAB labeling is an approach that applies to the localization of both intracellular and extracellular antigens. It also can be applied for the localization of extracellular antigens in unfixed material. Using in vivo, bound, peroxidase-conjugated antibodies, fixation with glutaraldehyde can be performed after the immuno-incubations.
9. The blocking of free aldehydes can be achieved in various ways. PBS–50 m*M* NH_4Cl seems to be the most efficient method. This step reduces immunocytochemical background.

10. When antibody is bound in vivo, the use of detergents is not always necessary, especially when a small amount immunogold (1–5 nm) is used as a label. With a mild fixation, such as 2% paraformaldehyde, the penetration of the Fab fragments of an antibody is 5–7 μm. This is enough to cut several ultrathin sections parallel to the vibratome section.
11. Azide is used to stop bacterial growth and to prevent internalization of antibodies by unfixed cells.
12. When performing pre-embedding labeling, the time involved with incubation and washing might have to be prolonged to warrant complete penetration of reagents to internal antigens and removal of unreacted reagents!
13. Silver enhancement of gold can be applied either before or after osmium tetroxide fixation. Wash extensively after osmium tetroxide fixation because traces of this strong oxidizing agent will interfere with adequate silver enhancement. Osmium tetraoxide could remove the metallic silver deposits; therefore, apply it briefly (20 min) and at a lower concentration (0.5%) at 4°C.
14. For observation with RCM, osmium tetroxide fixation is not necessary. A light counterstaining of the ultrathin sections with toluidine blue or hematoxylin provides RCM images with high contrast and much morphological detail.
15. For example, steps done before starting this protocol: For morphology, fix cells with 1% glutaraldehyde in 0.1 *M* cacodylate buffer, pH 7.4, for 10 min; for immunocytochemistry, see **Subheading 3.1.7.1., steps 1–13** or **Subheading 3.1.7.2., steps 1–6** (*see also* **Note 13**).
16. Start sectioning directly with ultrathin cutting, parallel to the monolayer.
17. Avoid introducing oxygen from air, as it interferes with polymerization. Mixing the components is simply achieved by gently stirring with a glass Pasteur pipet. Before using, let the mixture cool in the freezer (–20°C). When using Unicryl, there is a ready-to-use mix, and it is not necessary to exclude air from this resin during polymerization.
18. It is important to use indirect UV irradiation on all sides of the capsules. The illumination system is put in a freezer whose walls are covered with aluminum foil, thus permitting the radiation of UV light to all sides of the capsules.
19. For new users, it is best to ask a person who has done this before, because this is a technique that you have to learn and practice for a while.
20. For hydrophilic embedding medium (Lowicryl K4M, Unicryl), use a low water level and a higher cutting speed of 3–5 mm/s. For Spurr's resin and EPON, use a speed of 1 mm/s.
21. Stabilize your hand on the microtome during manipulation of sections.
22. Preparing small blocks of specimen: Ideally, the pieces of tissue or cell pellets should be trimmed to a pyramidal shape so that when mounted onto the specimen stub, the base is wider than the top. If possible, the block face should be flat and resemble a square, rectangle, or trapezoid.
23. Sucrose infiltration: Although an overnight infiltration in 2.3 *M* sucrose is recommended, it is possible to leave tissue pieces and cell pellets in the sucrose for much shorter times (1–2 h). However, uneven infiltration could occur that will make the sample more difficult to section.

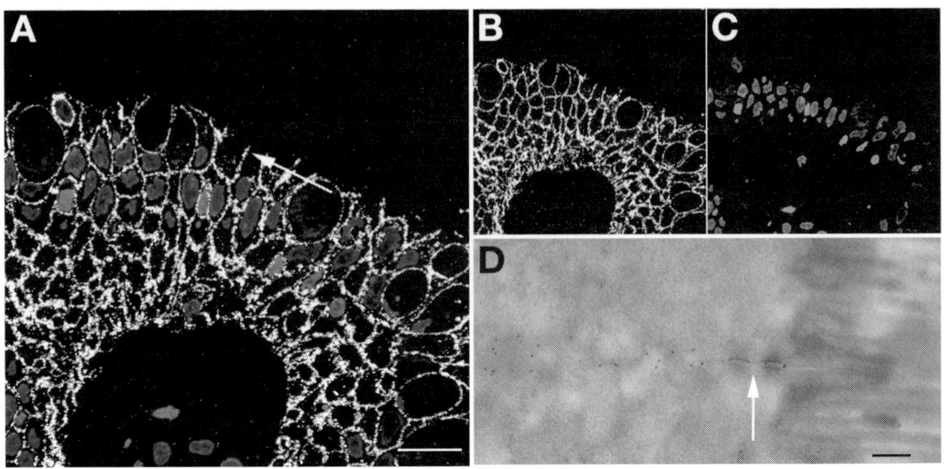

Fig. 14. The fluorochrome propidium iodide shows no reflection at all. Therefore, this fluorescence staining is useful as a counterstain in combination with reflection in the multitracking confocal laser scanning microscope. Lowicryl K4M section (250 nm) of human colonic epithelium stained with MoAb against Ep-CAM conjugated to 10 nm gold and silver-enhanced (post-embedding immunogold–silver method). Propidium iodide was used as a counterstain for DNA, showing nuclei. **(A)** Overlay image of reflection and fluorescence (bar = 20 μm); **(B)** only reflection image; **(C)** only fluorescence image; **(D)** high-magnification TEM image of cell membrane immunogold staining (arrow) without silver enhancement (bar = 0.1 μm).

24. Speel et al. ***(16)*** described other substrates for use with alkaline phosphatase, peroxidase, or glucose oxidase that also show colored reflection in RCM. For most substrates, fixation of the enzyme precipitates in a protein matrix is essential; thereafter, they can be mounted in immersion oil. For mounting sections in immersion oil, use a very small amount of oil, cover with a glass slip, press between filter paper, and seal with nail polish.
25. The peroxidase is still active at this glutaraldehyde concentration, and H_2O_2 will cause less damage because of better fixation.
26. Light green is less reflective and, therefore, a better counterstain with DABox reflection. Toluidine and hematoxylin are strong reflecting counterstains, especially useful with silver-staining. Propidium iodide does not reflect but is suitable in multitrack confocal reflection–fluorescence combination (*see* **Fig. 14**). It is useful for counterstaining nuclei in Lowicryl/Unicryl and cryosections, but it does not stain EPON sections properly. Acridin orange both reflects and fluoresces and provides a strong reflection on thin sections.
27. More possibilities by combining different protocols. For example, after **Subheading 3.1.7.2.**, take cells back to ethanol, wash twice to remove immersion oil, and then embed in Epon **(Subheading 3.1.4.1., step 4)**, without post-osmium tetroxide fixation, until ultrathin sectioning (*see* **Fig. 15**).

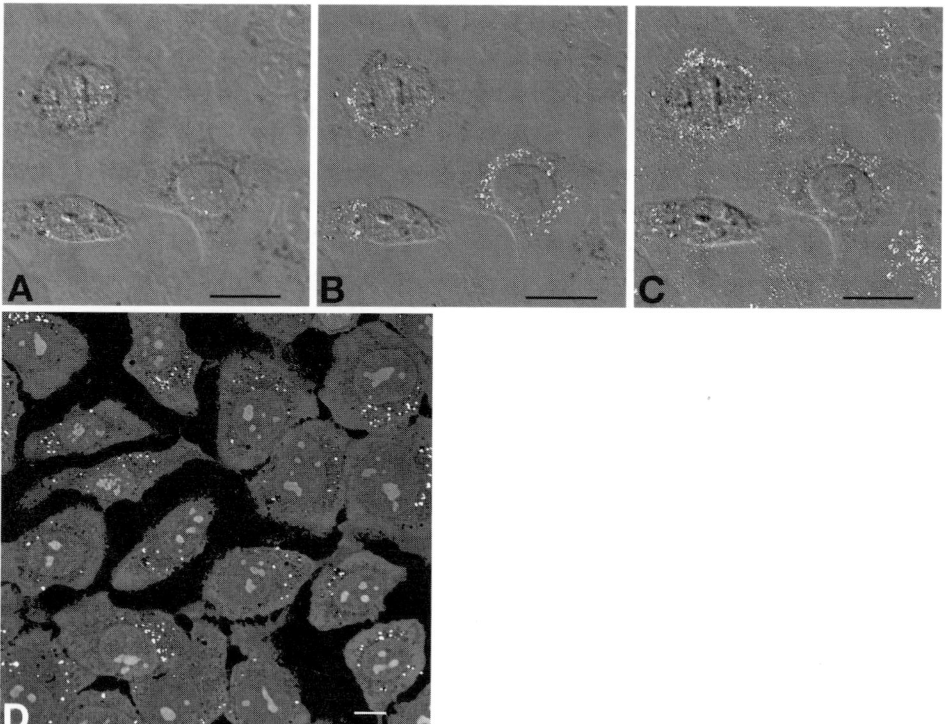

Fig. 15. Example of whole-cell experiment combined with a pre-embedding method. Internalization of BSA–5 nm gold and silver enhancement on whole squamous carcinoma cells (U2). **(A–C)** At three different heights from top to bottom of cells, overlay images of a RCM–confocal slice on differential interference contrast microscopical image (not confocal!). The reflection images are made with a small pinhole (10 μm) (*see* **Note 39**). **(D)** CLS-RCM image of a 90-nm Epon section of the same cells as shown in **(A–C)** after embedding in Epon, and counterstained with 0.01% toluidine blue. CLS-RCM image is made using three laser lines (633 nm, 543 nm, and 488 nm) and detected with pseudocolor gradation RGB in one channel in the reflection mode. (bar = 10 μm)

28. If water is still included in the specimen, then very strong white reflections will occur, which disturb the image. Therefore, dehydrate with 100% ethanol.
29. Small gold particles do not influence the capping process. For observation of strongly stained sections (e.g., toluidine blue-stained ultrathin sections), use a halogen lamp of 3200 K.
30. Mercury high-pressure lamps (150 or 100 W) emit strong emission lines dominating the color of the reflection by the specimen. Xenon lamps (75 W) emit an almost continuous spectrum and provide "true" colors. Upon reflection, the wavelength does not change, so the color of the incident light determines which colors can be seen in the image. The reflectance of the specimen determines what wavelength and in what amount will be reflected.

31. When using this cap/prism method, the specimen must be circled with a diamond writer pen, otherwise the ink will dissolve.
32. Realign the arc of the lamp and the central stop every time the lamp is replaced.
33. How do you know you have a proper/real RCM image? Inverting the image should yield a familiar bright-field microscopic or TEM image. A reference image can be made of a familiar (osmicated) ultrathin TEM section placed on glass and visualized directly with oil on the section.
34. Specimens can be embedded in mounting medium and covered with cover glasses, but they will fade away after months. They can also be embedded in immersion oil, covered with cover glasses, and sealed with nail polish.
35. The NT 80/20/543 has a neutral splitting function. About 20% of the laser energy is reflected toward the sample. From the reflected light, about 80% will pass this same splitter toward the photomultiplier tubes (PMTs).
36. One PMT is used to avoid pixel shift between the different channels.
37. For the gold or gold/silver markers, a use higher laser power and smaller pinhole. This configuration is also optimal for z-scan on "gold-stained" whole cells (*see* **Fig. 12**).
38. To acquire a satisfactory picture, take more pictures with different exposure times. Use the automatic measured exposure time and half and double it.
39. A very small pinhole can be used when there is strong reflection in the specimen. This allows also for z-scanning with more optical slices; however, be aware of the fact that sometimes very strong signals can be detected in more than one optical section.

References

1. Yamaguchi, H., Maat-Schieman, M. L. C., van Duinen, S. G., et al. (2000) Amyloid β protein (Aβ) starts to deposit as plasma membrane-bound form in diffuse plaques of brains from hereditary cerebral hemorrhage with amyloidosis-dutch type, alzheimer disease, and nondemented age subjects. *J. Neuropath. Exp. Neurol.* **59**, 723–732.
2. Natté, R., Yamaguchi, H., Maat-Schieman, M. L. C., et al. (1999) Ultrastructural evidence of early non-fibrillar Aβ42 in capillary basement membrane of patients with hereditary cerebral hemorrhage with amyloidosis, dutch type. *Acta Neuropathol.* **98**, 577–582.
3. Ploem, J. S., Cornelese-ten Velde, I., Prins, F. A., Bonnet, J., and de Heer, E. (1997) Reflection contrast microscopy. *Sci. Tech. Inf.* **XI(4)**, 98–113.
4. Nasser Hajibagheri, M. A. (1999) *Electron Microscopy: Methods and Protocols,* Humana, Totowa, NJ.
5. Filler, T. J., Rickert, C. H., Fassnacht, U. K., and Pera, F. (1994) Reflection contrast microscopy within chrome–alum hematoxylin stained thick tissue–sections. *Histochemistry* **101**, 375–378.
6. Filler, T. J. and Peuker, E. T. (2000) Reflection contrast microscopy (RCM): a forgotten technique? *J. Pathol.* **190**, 635–638.
7. Ploem, J. S. (1975) Reflection-contrast microscopy as a tool for investigation of the attachment of living cells to a glass surface, in *Mononuclear Phagocytes in Immunity* (van Furth, R., ed.), Blackwell, Oxford, pp. 405–421.

8. Verscheuren, H. (1985) Interference reflection microscopy in cell biology: methodology and applications. *J. Cell Sci.* **75**, 279–301.
9. Piller, H. (1959) Enhancement of contrast in reflected-light microscopy. *Zeiss-Werkschrift* **34**, 87–90.
10. Stach, E. (1949) Verkommungen de Kohlen-Auflichtmikroskopie. *Glückauf* **85**, 117–122.
11. Cornelese-ten Velde, I., Bonnet, J., Tanke, H. J., and Ploem, J. S. (1988) Reflection contrast microscopy. Visualization of (peroxidase-generated) diaminobenzidine polymer products and its underlying optical phenomena. *Histochemistry* **89**, 141–150.
12. Carlemalm, E., Villiger, W., Hobot, J. A., Acetarin, J. D., and Kellenberger, E. (1985) Low temperature embedding with Lowicryl resins: two new formulations and some applications. *J. Microsc.* **140**, 55–72.
13. Danscher, G. (1981) Localization of gold in biological tissue. A photochemical method for light and electronmicroscopy. *Histochemistry* **71**, 81–88.
14. Pluta, M. (1989) *Advanced Light Microscopy. Specialized Methods.* Elsevier PWN–Polish Scientific Publishers, Warszawa, pp. 197–210.
15. Hoetelmans, R. W., Prins, F. A., Cornelese-ten Velde, I., van der Meer, I., van de Velde, C. J., and van Dierendonck, J. H. (2001) Effects of acetone, methanol, or paraformaldehyde on cellular structure, visualized by reflection contrast microscopy and transmission and scanning electron microscopy. *Appl. Immunohistochem. Mol. Morphol.* **9**, 346–351.
16. Speel, E. J. M., Kramps, M., Bonnet, J., Ramaekers, F. C. S., and Hopman, A. H. (1993) Multicolour preparations for in situ hybridization using precipitating enzyme cytochemistry in combination with reflection contrast microscopy. *Histochemistry* **100**, 357–366.
17. Prins, F. A., Bruijn, J. A., and de Heer, E. (1996) Applications in renal immunopathology of reflection contrast microscopy, a novel superior light microscopical technique. *Kidney Int.* **49**, 261–266.
18. Kootstra, C. J., Bergijk, E. C., Veninga, A., et al. (1995) Qualitative alterations in laminin expression in experimental lupus nephritis. *Am. J. Pathol.* **147**, 476–488.
19. Bergijk, E. C., Munaut, C., Baelde, J. J., et al. (1992) A histologic study of the extracellular matrix during the development of glomerulosclerosis in murine chronic graft-versus host disease. *Am. J. Pathol.* **140**, 1147–1156.
20. Leer van, E. H. G., Ronco, P., Verroust, P., van der Wal, A. M., Hoedemaeker, P. J., and de Heer, E. (1993) Epitope specificity of anti-gp330 autoantibodies determines the development of proteinuria in active Heymann nephritis. *Am. J. Pathol.* **141**, 821–829.
21. Cornelese-ten Velde, I. and Prins, F.A. (1990) New sensitive light microscopical detection of colloidal gold on ultrathin sections by RCM. Combination of reflection contrast and electron microscopy in post-embedding Immunogold. *Histochemistry* **94**, 61–71.
22. de Jong, D., Prins, F. A., Mason, D. Y., Reed, J. C., van Ommen, G. B., and Kluin, P. M. (1994) Subcellular localization of bcl-2 protein in malignant and normal lymphoid cells. *Cancer Res.* **54**, 256–260.

23. Balzar, M. and Prins F. A. (1999) The structural analysis of adhesions mediated by Ep-CAM. *Exp. Cell Res.* **246**, 108–121.
24. Curtis, A. S. G. (1964) The mechanism of adhesion of cells to glass. A study by interference reflection microscopy. *J. Cell. Biol.* **20**, 199–215.
25. Kruidering, M., Maasdam, D. H., Prins F. A., de Heer, E., Mulder, G. J., and Nagelkerke, J. F. (1994) Evaluation of nephrotoxicity in vitro using a suspension of highly purified porcine proximal tubular cells and characterization of the cells in primary culture. *Exp. Nephrol.* **2**, 334–344.
26. Neelissen, J. A. M., Arth, C., Wolff, M., Schrijvers, A. H. G. J., Jungingen, H. E., and Bode, H. E. (2000) Visualization of percutaneous ^{3}H-estradiol and ^{3}H-norethindrone acetate transport across human epidermis as a function of time. *Acta Derm. Venereol.* **208**, 36–43.
27. Rickert, C. H., Filler, T. J., and Gullotta, F. (1997) Reflection contrast microscopy on silver-stained amyloid deposits in Alzheimer's disease. *Sci. Tech. Inf.* **XI(5)**, 136–142.
28. Diaz, C. L., Melchers, P. J. J., Hooykaas, P. J. J., Lugtenberg, B. J. J., and Kijne, J. W. (1989) Root lectin as a determinant of host-plant specificity in the Rhizobium–legume symbiosis. *Nature* **338**, 579–581.
29. van der Burg, M. P. M., Guicherit, O. R., Frölich, M., et al. (1994) Assessment of islet isolation efficacy in dogs. *Cell Transplant.* **3**, 91–101.
30. Landegent, J. E., Jansen in de Wal, N., van Ommen, G. B., et al. (1984) Chromosomal localization of a unique gene by non-autoradiographic in situ hybridization. *Nature* **317**, 175–177.
31. Landegent, J. E., Jansen in de Wal, N., Ploem, J. S., and van der Ploeg, M. (1985) Sensitive detection of hybridocytochemical results by means of reflection-contrast microscopy. *J. Histochem. Cytochem.* **33**, 1241–1246.
32. Macville, M. V. E., van Dorp, A. M., Wiesmeijer, K. C., Dirks, R. W., Fransen, A. M., and Raap, A. (1995) Monitoring morphology and signal during non-radioactive in situ hybridization procedures by reflection-contrast microscopy and transmission electron microscopy. *J. Histochem. Cytochem.* **43**, 665–674.
33. Cornelese-ten Velde, I., Bonnet, J., Tanke, H. J., and Ploem, J. S. (1990) Reflection contrast microscopy performed on epi-illumination microscope stands: comparison of reflection contrast- and epi-polarization microscopy. *J. Microsc.* **159**, 1–13.
34. Cornelese-ten Velde, I. and Prins, F. A. (1990) New sensitive light microscopical detection of colloidal gold on ultrathin sections by RCM. Combination of reflection contrast and electron microscopy in post-embedding immunogold histochemistry. *Histochemistry* **94**, 61–71.
35. Cornelese-ten Velde, I., Wiegant, J., Tanke, H. J., and Ploem, J. S. (1989) Improved detection and quantitation of the (immuno) peroxidase product using reflection contrast microscopy. *Histochemistry* **92**, 153–160.
36. Ploem, J. S., Cornelese-ten Velde, I., Prins, F. A., and Bonnet, J. (1995) Reflection-contrast microscopy: an overview. *RMS* **30/3**, 185–192.
37. Ploem, J. S., Prins, F. A., and Cornelese-ten Velde, I. (1998) Reflection-contrast microscopy for high definition images with light microscopy: practical tips using new Leica DM R equipment for RCM. *Sci. Tech. Inf.* **CDR 1**, 47–86.

38. Prins, F. A., van Diemen-Steenvoorde, R., Bonnet, J., and Cornelese-ten Velde, I. (1993) Reflection contrast microscopy of ultrathin sections in immunocytochemical localization studies: a versatile technique bridging electron microscopy with light microscopy. *Histochemistry* **99**, 417–425.
39. Ploem, J. S., Prins, F. A., and Cornelese-ten Velde, I. (1999) Reflection-contrast microscopy, in *Light Microscopy in Biology: A Practical Approach*, 2nd ed., (Lacey, A.J., ed.), Oxford University Press, Oxford, pp. 275–310.
40. Ploem, J. S. (1975) General introduction. *Ann. NY Acad. Sci.* **254**, 40.

19

Three-Dimensional Analysis of Single Particles by Electron Microscopy

Sample Preparation and Data Acquisition

Teresa Ruiz and Michael Radermacher

Summary

Electron microscopy of single particles has recently become a very popular field in both biological and material sciences. It might be difficult for a novice researcher new to this field to know how to start tackling a new project. This chapter is designed to serve as a guideline for anyone starting a new project to determine a three-dimensional structure using single-particle techniques. The chapter describes the basic techniques necessary to prepare the samples and acquire the data to calculate a three-dimensional reconstruction in easy-to-understand, step-by-step instructions. It starts with the basic preparation of support films and the usage of a variety of staining techniques needed to assess the quality of the sample and the viability of the project. It ends with a detailed description of vitreous ice preparations designed to acquire high-resolution structural information. Guidelines and tips are given on how to record the best images with an electron microscope. Although this chapter is geared to researchers new to the field, experts might find it not only useful as a reference but also valuable because of the number of practical tips included.

Key Words: Three-dimensional analysis; single particles; cryo-electron microscopy; sample preparation; data acquisition; data collection; deep staining; methyl amine tungstate; ammonium molybdate; uranyl acetate; vitreous ice; random conical; electron microscope alignment; improved coherence; point mode.

1. Introduction

Three-dimensional cryo-electron microscopy of single particles has matured into a well-established biophysical technique *(1)*. Outstanding are the reconstructions of the ribosome in different functional states *(2–7)*. Recently, the reconstruction of phosphofructokinase from *Saccharomyces cerevisiae* at better than 11 Å resolution *(8)* in combination with molecular replacement has led to an initial model to phase native X-ray data, in a situation where heavy-atom derivative data was not available. During the past 15 yr, X-ray structures of subdomains have been

From: *Methods in Molecular Biology, vol. 319: Cell Imaging Techniques: Methods and Protocols*
Edited by: D. J. Taatjes and B. T. Mossman © Humana Press Inc., Totowa, NJ

fitted into cryo-electron microscopy structures to either localize them in the whole macromolecule or understand conformational changes in different states ***(9–12)***. At the resolutions currently achieved by cryo-electron microscopy of single particles, we would not be surprised if using electron microscopy data to provide starting phases for X-ray data when heavy-atom derivatives cannot be obtained would become a standard technique and not a singular event as for phosphofructokinase.

There is no general technique for preparing and analyzing single particles that can be applied to all samples. Each sample has its own characteristics and requires tailoring of the preparation and image processing techniques to accommodate them. It is true that our final goal is to determine the structure in its most native environment from vitrified samples ***(13–15)***; however, mastering a variety of staining techniques is a necessity. Any new molecule should be first analyzed in negative-stain preparations, carefully selecting the stain that least affects the structure. Stained specimens are highly visible and the effort in data collection and image analysis is much less than for cryo-specimens. The analysis of stained samples allows one to optimize the biochemical purification and is of invaluable help to reach a first understanding of the molecular architecture and variability of the sample ***(16)***. Moreover, the correct image analysis strategy can be more easily established in these preparations. For example, in the study of phosphofructokinase (an elongated molecule), the superiority of using Radon transform algorithms became evident ***(17)***. In particular, the possibility to perform simultaneous rotational/translational alignments presented a great advantage in this study and it should benefit all studies of nonglobular molecules. Only after a thorough analysis using negative-stain techniques has been carried out can the quality of frozen hydrated samples be properly judged. Imaging a new specimen in ice without prior analysis in negative stain can lead to many weeks of lost work and frustration.

A very important fact to remember is that all electron microscopy preparations are radiation sensitive ***(18–20)***. This is most visible when examining vitrified or glucose-embedded samples where bubbling of the media can easily be observed ***(21,22)***. However, negatively stained specimens also suffer radiation damage. Stain redistribution and microcrystal formation occur, which prevents the imaging of high-resolution details. This loss of high resolution might often pass unnoticed. By burning samples needlessly you might miss important features of your specimen that otherwise would have been obvious. Therefore, minimal or low-dose settings should always be used in the microscope when collecting data ***(22,23)***.

This chapter was designed for researchers with some basic knowledge of electron microscopy techniques. Depending on your background, some topics might seem either too basic or too advanced; we hope that, on average, you find them useful. We will first describe the preparation of support films best suited for single-particle imaging and a variety of different staining techniques. This will be followed by detailed procedures for the preparation of frozen-hydrated samples on holey grids coated with a thin carbon layer. The settings to obtain high-resolution results using a microscope equipped with a LaB_6 cathode and the data collection procedures optimum for imaging single particles are described.

2. Materials

Necessary materials and small tools for general electron microscopy and negative stain preparations are as follows:

1. Precision tweezers (either with an O-ring for closure or with reversible action).
2. Grids (the material depends on the sample; for our samples, copper is acceptable).
3. Carbon-coated grids.
4. Holey grids.
5. Carbon-coated holey grids.
6. Negative-staining solutions or the powder to make the solutions.
7. Filter paper for blotting: Whatman no. 40 or no. 1 (medium) or Whatman no. 41 or no. 4 (fast).
8. Parafilm.
9. Pasteur pipet (glass, quartz, or plastic).
10. Automatic micropipettor.
11. Pipet tips.
12. Water.

Many of the materials (staining solutions, carbon-coated grids, holey grids) needed for sample preparation can be purchased from electron microscopy supply companies:

Ladd Research (www.laddresearch.com); Ernest F. Fullam (www.fullam.com); Structure Probe (www.2spi.com); Ted Pella (www.tedpella.com); Electron Microscopy Sciences (www.emsdiasum.com); Nanoprobes (www.nanoprobes.com); Quantifoil (www.quantifoil.com).

However, in most cases, and to have a tight and better control on the quality, the materials are prepared in the laboratory. The same applies to certain small tools not available commercially. We will recommend to the real beginners to purchase the materials off the shelf from any of the electron microscopy supply companies. Those with certain experience in electron microscopy and with a minimally equipped laboratory for electron microscopy might benefit from the following instructions for the in-house preparation of the materials and small tools described here.

2.1. Support Films

Many substrates have been assayed for their viability to be used as supports for electron microscopy samples ***(24,25)***. A good substrate should be highly transparent to electrons, have minimal intrinsic structure, and be stable upon electron irradiation. For cryo-electron microscopy preparations researchers started favoring grids coated with perforated carbon films (holey grids)—the optimum holes having diameters between 1.5 and 10 μm. Images were taken of the sample in the vitrified water layer spanning the holes (suspensions) where the background noise coming from the support film was zero. A further advantage was, as a general accepted rule, that particles adopted many different orientations in suspension, unlike particles supported on a thin carbon film, which lie in preferred orientations in response to their interaction with the support film. However, the effect of the air–water interface was neglected and, as it has been observed in several cases, also gives rise to preferred orientations. In our laboratory, we favor the use of holey grids coated with a thin carbon film (5–10 nm); the holey film providing a strong support for the thin carbon. Images are collected over the holes with a minimal background arising from the thin carbon. This approach presents two general advantages. First, once the optimum sample concentration has been found using staining techniques, the same concentration is good for vitreous ice preparations (usually below 100 μg/mL). Second, the later correction of the transfer function of the microscope will be more accurate using the thin carbon on which the sample lies rather than the thick carbon that makes the holes. Moreover, it facilitates the preparation of membrane protein samples, avoiding the meniscus problems encountered in suspensions attributable to the presence of detergent in the buffers and also the need for using very high concentrations (approx 10 mg/mL) to have a significant amount of sample in the bare holes.

2.1.1. Holey Grids

Several techniques have been used to produce perforated films ***(26)***. Grids with perforated films are available from many manufacturers, in particular Quantifoil, which is a perforated film with holes very well defined in size, shape, and arrangement ***(27)***. Here, we present a method that produces reliable holes independent of the ambient conditions (e.g., humidity) and was passed on to us by the group of J. Frank in Albany, NY.

2.1.1.1. Cleaning of the Slides

1. Select approx 20 microscope glass slides without scratches.
2. Rinse the slides thoroughly with distilled water. Many suppliers coat the slides with a thin film of detergent that must be removed.
3. Dry the slides with lint-free Kimwipes.
4. Coat each slide very lightly with nose grease or, if you prefer, spread some light facial cream on the top of your hand and use this instead (Yves Rocher's cream for mixed skin types works well; however, the common Nivea cream does not).
5. Remove the grease from the slides using lint-free Kimwipes. Wipe always in the same direction along the long axis of the slide. This helps the production of holes that are aligned along the long axis of the slide.
6. Put the treated slides in a Petri dish and cover them to avoid dust deposition.

2.1.1.2. Preparation of the Formvar Solution

1. Prepare 1.2 mL of 90% solution of glycerol (anhydrous of high purity) in double-distilled water (ddH_2O).
2. Put 45 mL of chloroform (99.0–99.5% purity) in a 100 mL bottle.
3. Weigh 0.2 g of Formvar (polyvinyl formal). First add some of the Formvar to the surface of the chloroform and dissolve it gently (mimic a centrifuge motion with your wrist; do not shake).
4. Keep adding the Formvar slowly until all of it is dissolved.
5. Add 45 mL of acetone (99.5% purity); the bottle will become slightly warm.
6. Add 0.92 mL of the 90% glycerol solution to the bottle.
7. Close the bottle tightly, cover it with aluminum foil, and shake energetically for 10 min.
8. Remove the foil, open the bottle, and sonicate with a sonic probe for 4 min without cooling. Dip the sonicator tip approx 1 cm below the liquid surface level. After sonication, the bottle will be lukewarm and the solution clear.
9. Close the bottle, cover it with the aluminum foil, and wait 15 min before preparing the Formvar films.

2.1.1.3. Preparation of Formvar Holey Films

1. Fill a 50-mL beaker with the Formvar solution and a 100-mL beaker with acetone.
2. Dip the slide into the 50-mL beaker; let it sit for 10 s.
3. Slowly withdraw the slide at an angle from the solution. The surface that makes an acute angle with the surface of the solution (the bottom surface) is the one that will be used.
4. Let the slide dry. You can lean it against a rack in an almost vertical position to dry.
5. Dip the slide in the beaker with acetone and stir it gently for 5 s. Keep the slide vertical. Let it dry as in **step 4** before (*see* **Note 1**).
6. Observe the slide with a phase-contrast microscope to asses the amount of holes obtained and their size. If the holes are too small, wait a few minutes before dipping a new slide in the Formvar solution (*see* **Notes 2–4**).
7. Prepare as many slides as you can before the Formvar solution is exhausted or the holes become too large (usually 1 h after sonication).

2.1.1.4. Preparation of Formvar Holey Grids

1. Fill a crystallizing dish to the rim with ddH_2O.
2. Cut the edges of the plastic film with a diamond knife so that it will strip easily off the glass slide.
3. Breath gently on the edges of the slide to facilitate striping and dip the slide at an angle into the crystallizer. The plastic film should float off on the surface of the water.
4. Place 300 or 400 mesh grids on top of the film, try to always put the same side facing the plastic.
5. Pick the film with a piece of parafilm supported by a glass slide. Lower the parafilm perpendicular to the water surface touching the plastic film, keep lowering it into the water until all the grids are attached to it, and then lift the parafilm out of the water at a shallow angle.

6. Trim slightly the parafilm and let the grids dry on a Petri dish.
7. Once the grids are dry, remove them from the parafilm with a pair of tweezers and put them on a glass slide for carbon coating. At this point, you can proceed to a preselection of your grids by looking at them in the phase-contrast microscope.
8. Coat the grids with a very thick layer of carbon (80–100 nm).

2.1.2. Holey Grids With Thin Carbon

1. Select some holey grids. Grids with 2- to 3-µm holes and very few broken squares are good.
2. Put holey grids, carbon side up, on a Petri dish with filter paper (Whatman #4 is good) and gently add approx 2.5 mL of chloroform at the periphery of the paper to wet it slowly (use the same solvent to remove the plastic as was used to prepare the plastic films).
3. Keep the Petri dish closed for 30 min or overnight if you prefer.
4. Open the Petri dish for at least 10 min to let the rest of the chloroform evaporate. If the bottom of the Petri dish is cold, the chloroform has not finished evaporating.
5. Coat the holey grids with a thin layer of carbon (5–10 nm) either using a Smith carbon coating trough (from Ladd Research Inc. *[28]*) or a homemade device based on the same principle.
6. Let them dry for about 30 min before using them.

2.2. Stains

Many articles and books have been written describing stains for electron microscopy ***(24,25,29)***. In this chapter, we will focus on the most commonly used in our laboratory: uranyl acetate, methylamine tungstate, ammonium molybdate, and Nanovan (sold as a 2% solution by Nanoprobes ***[30,31]***).

2.2.1. Uranyl Acetate

Uranyl acetate can be purchased from Ted Pella (ref. no.: 19481). This stain can be used at different concentrations (0.5–4%); however, one important thing to remember is that at lower concentrations, the grain is finer. A solution of approx 4% uranyl acetate is saturated, the exact value depends on the manufacturer. The pH of the staining solution should be around 4.0; attempts to raise the pH might result in the precipitation of uranyl acetate.

1. To prepare a 1% solution of uranyl acetate, put 200 mg of uranyl acetate in 20 mL of ddH_2O.
2. Vortex until all the stain is dissolved.
3. Filter the solution with a 0.22-µm filter.
4. Store the solution either covered with aluminum foil or in a dark container.

2.2.2. 2% Ammonium Molybdate

Ammonium molybdate can be purchased from Agar Scientific Ltd (ref. no.: R1156) or from Ted Pella (ref. no.: 19482).

1. To prepare 5 mL of stain, put 100 mg of ammonium molybdate in 5 mL of ddH_2O and vortex until it is dissolved.
2. Check the pH with pH paper; the starting value is acidic.
3. Adjust the pH to 7.0 with NH_4OH (250 µL or 300 µL of a 5% solution should be enough). Be careful because when pH 6.5 is reached, the pH seems to stabilize for a while, and adding a little more makes the pH jump to 7.5 or 8.0. Sometimes it is safer to stay at pH 6.5. Do not titrate back; if the pH gets to high, start again.
4. Filter the solution with a 0.22-µm filter, discarding the first few drops, and store the filtered solution in the refrigerator until use.
5. The staining solution is good for 1 wk if stored at 4°C, but the best staining is achieved with stain prepared fresh.

2.2.3. 0.25% Methylamine Tungstate With 10 µg Bacitracin

Methylamine tungstate can be purchased from Agar Scientific Ltd. (ref. no.: R1219) or from Ted Pella (ref. no.: 18353). Bacitracin can be purchased from Sigma. A stock solution of bacitracin (1 mg/mL) in water is stable for 1–3 mo in the refrigerator.

1. To prepare 20 mL of stain, put 50 mg of methylamine tungstate in 20 mL of ddH_2O and vortex until it is dissolved (the solution might still be slightly cloudy).
2. Check the pH with pH paper; it should be around 5.5 or 6.0.
3. Adjust the pH to 7.0 with NH_4OH (5 µL or 6 µL of a 5% solution should be enough). Methylamine tungstate is not a good buffer; therefore, the pH will change very fast. Do not titrate back; if the pH gets too high, start again.
4. Filter the solution with a 0.45-µm filter, discarding the first few drops.
5. To the filtrate, add bacitracin from the stock solution to a final concentration of 10 µg/mL.
6. Cover the stain with aluminum foil and keep in the refrigerator until use. To achieve the best results, prepare the solution fresh the same day or the night before because the nice staining capabilities diminish the longer the solution is stored. Only in extreme cases can a 1-wk-old solution that has been stored at 4°C in the dark should be used, with less than optimal results.

2.3. Vitreous Ice

Necessary materials and small equipment specific for vitreous ice preparations are as follows:

1. Safety glasses.
2. Liquid nitrogen.
3. Ethane gas.

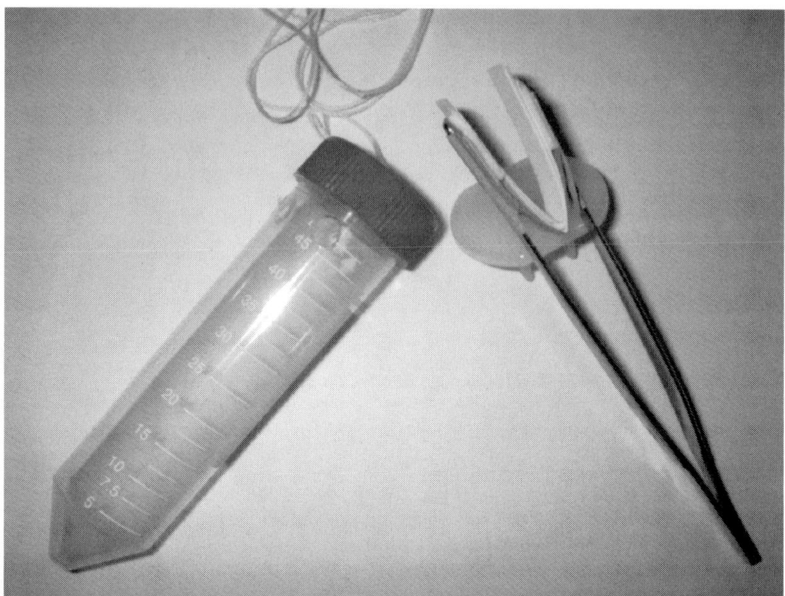

Fig. 1. **Left**: 50-mL Falcon tube modified to be used for long-term storage of cryo-samples; **right**: blotting tweezers for double-sided blotting.

4. Liquid nitrogen container with a small neck (4 L).
5. Small nitrogen container for storing the falcon tube with the frozen grids (1 L).
6. Plunger ***(32)***.
7. Blotting tweezers to perform blotting from both sides of the grid (*see* **Fig. 1**); to make them, proceed as follows:
   a. Take a pair of tweezers with round tips (12/8).
   b. Take a thin piece of cardboard (1 cm wide and 6 cm long), bend it to a V-shape, and glue it with tape to the tweezers bridging the tips.
   c. Stick a piece of soft padding (1 cm wide and 1.5 cm long) to each side. Either a thin piece of styrofoam or a piece of mounting tape is good.
   d. Put a double-stick tape over the soft padding.
   e. Cut a piece of filter paper (1 cm wide and 5 cm long), bend it to a V-shape, and bridge the tips as in **step b**. This will prevent the blotting paper from slipping off the blotting tweezers.
   f. The blotting paper also will be bent to a V-shape and will be placed in the gap between the tips.
8. Two long tweezers with round tips (18/8 is a good size).
9. Short tweezers with round tips (10/8 is a good size).
10. 200-µL pipet tips with the nose slightly cut (to serve as the nozzle of the tubing used to pour the ethane).
11. Filter paper for blotting: Whatman no. 40 (medium) or Whatman no. 41 (fast).
12. Three to six precision tweezers for plunging (e.g., Dumont N5 biology).

13. Millipore tweezers for closing and transferring cryo-boxes (or tweezers and screwdriver).
14. Cryo-boxes to store the frozen grids.
15. A 30- or 50-mL Falcon tube for long-term storage of the grid-boxes in nitrogen with appropriate holes for airing and a string to retrieve them from the liquid nitrogen refrigerator (*see* **Fig. 1**).
16. Automatic micro-pipettor.
17. 10 μl pipette tips.
18. Holey carbon grids coated with a thin carbon film.

3. Methods

3.1. Staining Techniques

3.1.1. Uranyl Acetate Staining

This is the first step that one uses when analyzing a new sample. There are many different ways of staining with uranyl acetate, and many readers probably know one or two of them; therefore, we will not describe them in this chapter. However, we will take the time to describe here a method not often used in our field, which we have called the drop method. This method is especially useful when samples are in buffers containing either detergents or phosphate. Moreover, it is very successful for removing unwanted precipitates that form on the grid and that otherwise would not permit the correct visualization of the sample.

1. Coat the outer surface of a large Petri dish with parafilm (approx 15 cm in diameter).
2. Put 3 drops of stain (approx 150 μL each) on the parafilm.
3. Add 5 μL of sample to the grid and wait 30–60 s.
4. Place the grid on top of the first drop and move it gently over the surface. Be careful not to allow it to go inside of the stain drop.
5. Move the grid to the next drop and do the same as before.
6. Move the grid to the last drop and let it float on the surface for 30 s.
7. Pick up the grid, remove the excess liquid by touching the edge of the grid with a filter paper (wick), and dry it fast.
8. Remove stain drops with a pipet.
9. Set up new stain drops for the next grid in a clean area of the Parafilm.

3.1.2. Deep Staining

This procedure, first described by Stoops et al. for methylamine tungstate ***(33)***, is used in our laboratory with either methylamine tungstate, ammonium molybdate, or Nanovan. We have found that most soluble proteins stain well with methylamine tungstate, whereas membrane proteins stain better with ammonium molybdate. Nanovan is our favorite stain when looking at large complexes or vesicles coated with membrane proteins (*see* **Notes 5–10**).

1. Add 6 μL of sample to the grid.
2. Wait 30–60 s.

3. Wick the sample with filter paper (Whatman no. 4 or no. 41, fast papers). A small layer of liquid should be left on the surface of the grid. The grid should never be allowed to dry. Before wicking, you should have ready the pipet with the next drop of stain.
4. Add 6 µL of the staining solution.
5. Wick the stain.
6. Add 6 µL of the staining solution.
7. Wick the stain.
8. Add 6 µL of the staining solution.
9. Let it stand for a few seconds, wick the stain, and air-dry fast.

3.2. Vitreous Ice Preparations

We carry out all our vitreous ice preparations in the cold room to minimize evaporation. The evaporation rate is proportional to the surface/volume ratio of the drop; therefore, the most critical point in the preparation is the time between blotting the grid and vitrification of the sample in liquid ethane ***(13,32,34)***.

1. Clean cryo-boxes with ethanol and dry them well with a lint-free Kimwipe. Make sure that the holes are not wet. Outline with a permanent marker the outside of the lid and also the slit of the screw (this will allow you to see them easily under liquid nitrogen). Check that all the boxes open/close easily because things get tighter at liquid nitrogen temperatures.
2. Cut Whatman no. 40 paper to size (1 cm wide and 5 cm long), fold into a V-shape, and store it in Petri dishes. For each grid that you plan to make, you need two pieces of paper (*see* **Note 11**).
3. Clean the tips of the plunging tweezers with ethanol. Make sure that all the tweezers hold the grids properly and check that they do not show capillarity. Normally, we add 5 µL of sample; if your sample buffer contains detergent, test with the buffer to determine the maximum amount that you should use to avoid capillary effects. This amount depends on the buffer; 3–4 µL might be fine.
4. Fill the 4-L container with liquid nitrogen.
5. To prepare ethane nozzles, cut off 1 mm of the 200-µL pipet tips and store them in a Petri dish. As the tips are sold, the hole is very small and it might take a long time to fill the ethane pot or, worse, it might clog, preventing the ethane from coming out.
6. Put the 30- to 50-mL Falcon tube inside of the small liquid nitrogen container and fill it with nitrogen.
7. Transport the ethane bottle to the cold room (*see* **Note 12**).
8. Transfer all of the necessary equipment to the cold room.
9. Fill the styrofoam plunger box with liquid nitrogen. Make sure that the level of nitrogen is high enough to liquefy the ethane gas. A few refills will be necessary to allow all of the large surfaces to cool down (*see* **Notes 13** and **14**).
10. Open the main valve of the ethane bottle.
11. Use a new, and therefore clean of water contamination, ethane tip. Every time that you need to pour ethane, take a new tip. In the cold room, the cold tips get a lot of condensation, and if you reuse them, this condensation will end up in your ethane pot and later on your grids.

12. Put on the safety glasses.
13. Filling the ethane pot for the first time. First, make sure that the level of nitrogen is high enough; if the level is low, the ethane gas will not liquify. Hold the tubing approx 10 cm from the tip to prevent getting your hands too cold or your fingers in contact with the liquid ethane. Position the tip inside of the ethane pot approximately at the center and about 0.5 cm from the top. Open the fine valve of the bottle to allow the ethane gas to flow. Increase the flow until steam forms at the tip. At this moment, lower the tip all the way to the bottom of the ethane pot, making sure that it is touching one of the walls. You can increase the flow of ethane a little more. Now you should see that the steaming has stopped and you will start hearing a sizzling noise. This means that the ethane is starting to liquify. If you look carefully inside of the pot, you should see how it is starting to fill up with liquid. As the pot fills, raise the tip along the pot wall with the liquid level, always keeping the tip about 1–2 mm below the liquid surface while filling the pot. Once the pot is filled to the rim, slightly decrease the gas flow, take the tip away from the pot, and close the fine valve completely. The main valve will remain open for the duration of the plunging (*see* **Notes 15–23**).
14. Take one of the cryo-boxes. Open it slightly so that the lid moves freely. Pick it up with the small tweezers and put it inside of the liquid nitrogen in the plunger box. With a pair of tweezers, move the lid of the box so that one of the holes is ready to accept a grid (*see* **Fig. 2**).
15. Set a pair of tweezers with a grid in the guillotine; tighten them well but not excessively.
16. Put a drop of sample (size depends on your test for capillarity [*see* **step 3**]) and wait for 20–30 s.
17. Place a fresh set of blotting paper in the blotting tweezers. Turn the guillotine to the position that best fits your blotting technique (*see* **Fig. 3**). Hold the blotting tweezers with one hand and the manual guillotine release with the other. Blot the grid for 4–5 s, quickly remove the blotting tweezers out of the way, and release the guillotine to plunge the grid into the ethane.
18. Loosen the tweezers from the guillotine, always keeping the grid inside of the ethane.
19. Transfer the grid from the ethane pot into the cryo-box. Once the grid is inside of the hole, move the lid over the grid and make another hole available for the next grid (*see* **Notes 24–28**).
20. Repeat **steps 15–19** for the other grids in this box.
21. Take the Millipore tweezers, the large tweezers, and the nitrogen container with the Falcon tube. Cool down the Millipore tweezers. Close the lid of the cryo-box using just one of the legs of the Millipore tweezers as a screwdriver. Keep the Millipore tweezers inside of the nitrogen. Take the Falcon tube out of the nitrogen container, holding it with the large tweezers along the wall and place it as close to the plunger box as possible. Take the cryo-box with the Millipore tweezers, holding it from the screw, and drop it very fast into the Falcon tube. Place the Falcon tube back into the nitrogen container and close the lid.
22. Refilling the ethane pot. If the ethane level has dropped low, you have to refill it. If the ethane is frozen, you have to reliquify it (*see* **Notes 15–23**). Put a new tip on

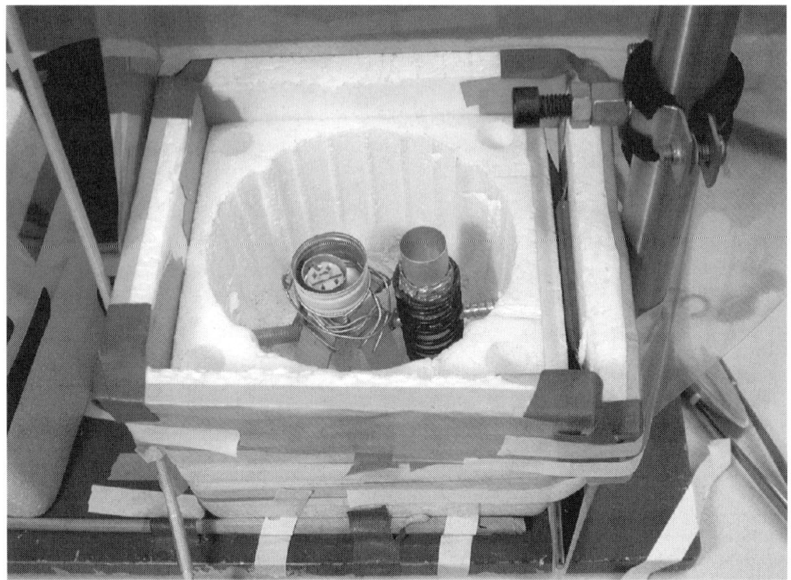

Fig. 2. Bottom part of the plunger device showing the disposition of the ethane pot and the cryo-box.

the tubing. Hold the tubing carefully, as before. Position the tip just above the ethane. Open the fine valve on the bottle to start the flow of ethane gas. Increase the flow until you see steam coming out from the tip. At this moment, lower the tip 1–2 mm below the liquid level, keeping it close to the cold wall. Again, once the pot is filled to the rim, slightly decrease the gas flow, take the tip away from the pot, and close the fine valve completely.

23. Use the same procedure (repeat **steps 14–21**) to plunge a new set of grids.
24. When you have finished all of the plunging, close the Falcon tube that contains all of the cryo-boxes and transfer it to a liquid nitrogen refrigerator (approx 35 L).
25. Close the main valve of the ethane bottle and take the bottle close to a fume hood. With the main valve closed, open the fine valve to release the pressure in the line. When the meter shows zero pressure, close the fine valve again and take the bottle to its storage place.
26. Clean all the plunging tools (dry the moisture with a lint-free Kimwipe).

3.3. Electron Microscope Alignment

3.3.1. Standard Alignment

Align the microscope following the manufacturer's instruction manual. The procedure described here should be taken as a guideline for the order in which the different optical elements should be aligned, as it is based on the alignment of the FEI CM microscope series with which we are most familiar.

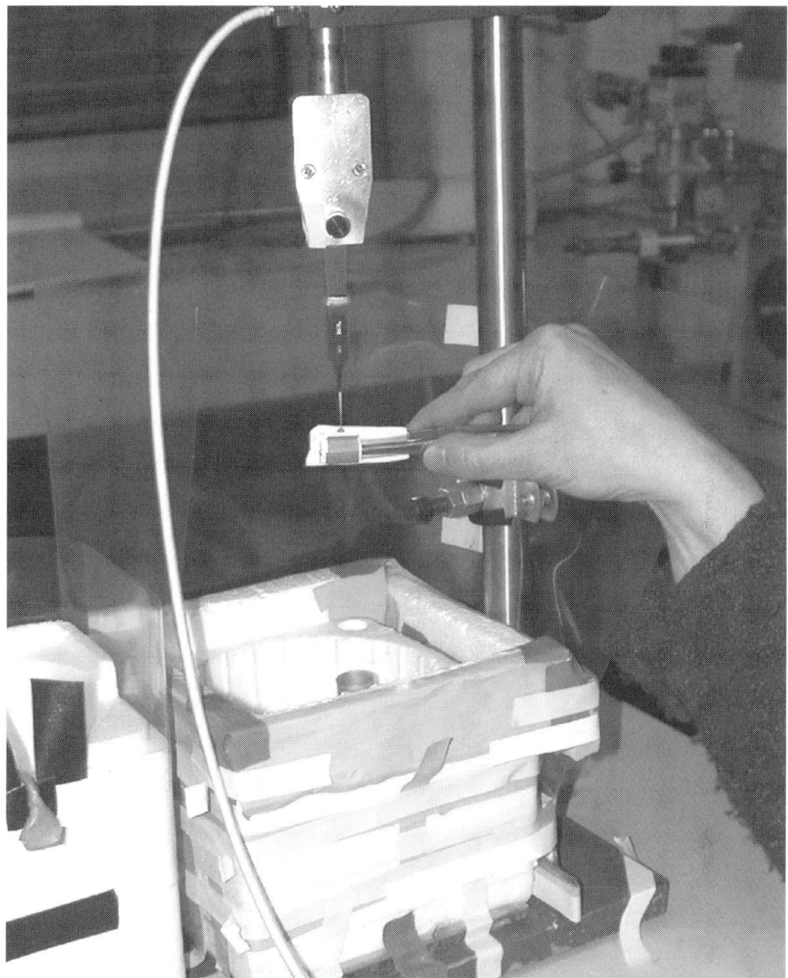

Fig. 3. Double-blotting technique.

1. Insert a known specimen into the microscope.
2. If the voltage was changed to a new value, wait 30–60 min to allow the lenses to equilibrate.
3. Adjust the eucentricity of the grid at low and high magnification (×3000 to ×60,000).
4. In the mode where the image will be acquired, start fine-tuning the alignment for the specific conditions where the data will be collected (e.g., magnification, C_1 lens excitation, bias, C_2 aperture, and objective aperture). Remove the objective and selected area or diffraction apertures.
5. Correct first gun tilt and gun shift (*see* **Note 29**).

6. Center the condenser aperture. Focus the beam to a small spot. Center the spot using the beam deflection coils. Spread the beam and center it using the condenser aperture adjustments. Repeat a few times until the beam is well centered.
7. Correct condenser astigmatism.
8. Check again the eucentricity of the grid and focus the specimen.
9. Align the beam deflection coils, the objective lens, and the image deflection coils according to the manufacturer's instructions. For the CM microscope series, align the beamcoils pivot points first, then the rotation center alignment. Loop around these two alignments until they are both correct. Finally, align the image shift (*see* **Notes 30** and **31**).
10. Center the objective aperture. Insert the objective aperture, focus the specimen, and focus the beam to a small spot. Go to diffraction mode and center the aperture. Focus the aperture (make it sharp) with the imaging lens system. Make sure that under these conditions that the aperture is larger than the diffraction spot. If the aperture is smaller than the diffraction spot, you will get distortions in the negatives. To solve this problem, either select a larger objective aperture or change the C_1 lens excitation to get a smaller spot. Go back to the imaging mode.
11. Go to the center of the grid square and correct the objective astigmatism.
12. Align properly all of the modes of your minimal-dose or low-dose kit (search, focus, and exposure).
13. Select an area of the grid without carbon to set up the illumination. Go to the exposure mode. At a magnification of approx ×60,000, (good for data collection of single particles), spread the beam to cover at least the large viewing screen. Adjust the illumination to obtain a meter reading of 1.5–1.7 s. When the sample is in the beam path, the meter reading will show increased exposure times. Set the manual plate exposure mode to 1 s.

3.3.2. Improved Illumination Conditions for High Coherence

The beam coherence can be improved by adjusting the C_2 lens settings and the gun settings and by finding the optimal combination of C_1 lens excitation and C_2 aperture ***(8)***.

3.3.2.1. C_2 Lens Settings

The C_2 lens should be over focused and the illumination should cover at least the whole area of the large viewing screen. A beam focused to a small spot on the screen worsens the effects of spherical aberration because of the large aperture angle. A focused beam is almost incoherent because of the large difference in the electron path.

3.3.2.2. Gun Settings

1. Mechanical settings of the gun.
   a. Use a LaB_6 cathode.
   b. Use the smallest Wehnelt aperture available (in our case, for the CM120 FEI microscope, it was 0.3 mm).

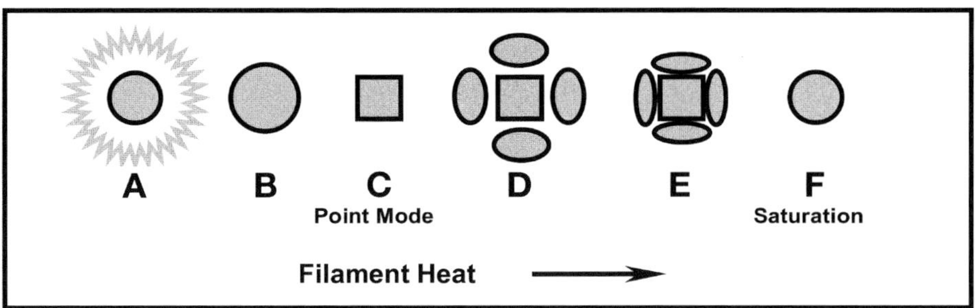

Fig. 4. Drawing of the different stages of heating of a LaB_6 cathode.

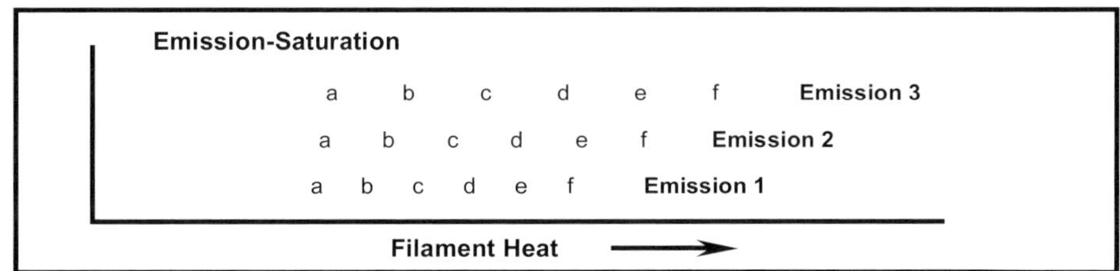

Fig. 5. Relationship between the different stages of heating of a LaB_6 cathode and the filament emission.

   c. Use a small distance between the tip and the Wehnelt aperture (in our case, for the CM120 FEI microscope, it was 0.12–0.13 mm).

2. Electronic settings of the gun.
   a. In **Fig. 4** are represented the different stages through which a LaB_6 cathode passes while it is heated. Most laboratories collect their images with the filament saturated (stage f). We use the filament slightly unsaturated in the stage c or point mode. This provides the highest spatial coherence.
   b. The Wehnelt is set at a voltage slightly more negative than the cathode, therefore preventing the free flow of electrons toward the anode. The voltage difference between the cathode and the Wehnelt can be set at different values via the bias or emission controls. As the voltage difference is decreased, more electrons pass through the Wehnelt aperture and the emission increases. Another effect of reducing the bias or increasing the emission is that, as is shown in **Fig. 5**, the filament saturation is achieved at higher currents or higher filament heat values.

3.3.2.3. Optimal Combination of C_1 Lens Excitation and C_2 Aperture

1. The beam coherence is adjusted by the spot size (C_1) and condenser aperture (C_2).
2. Increasing the spot size increases the coherence.
3. Decreasing the size of the C_2 aperture increases the coherence.
4. As a guide, in **Table 1** we present some of the combinations tested for coherence in a CM120 FEI microscope equipped with a LaB_6 cathode (after spreading the

Table 1
Different Combinations of C_1 Lens Excitation and C_2 Aperture Tested for Coherence

	Spot size			
	3	4	5	6
	(500 nm)	(300 nm)	(200 nm)	(150 nm)
C_2 aperture	100 µm	100 µm	200 µm	200 µm

beam as described earlier, and still being able to collect low-dose images in 1 s). The best conditions were found for spot size 4 (300 nm) and 100-µm C_2 aperture.

3.3.2.4. Simple Protocol to Obtain Point Mode at 100 KV

1. Preparatory settings at 120 kV where the filament is saturated. We have not determined the optimum values for point mode at this voltage because we collect our data at 100 kV.
   a. Settings: Emission 1; C_2 aperture 100 µm; spot size 4.
   b. Saturate the filament (standard procedure).
   c. At ×60,000, expand the illumination to cover the large screen; check that the meter reading is 1 s or less.
   d. If not adjust, it to 1 s by increasing the emission (decreasing the bias).
   e. If necessary, increase the filament limit by one click.
   f. This determines the final filament limit, which should not be exceeded.
2. Final settings at 100 kV for point mode operation.
   a. Wait at least 30 min for the lenses to equilibrate at 100 kV.
   b. Start with the final conditions obtained previously for 120 kV.
   c. Decrease filament heat to obtain the point mode.
   d. At ×60,000, expand the illumination to cover the large screen; check that the exposure time is 1 s or less.
   e. If not, increase the emission setting by 1.
   f. To obtain the point mode again, it might be necessary to increase the heat (never exceed the limit!).

3.4. Data Collection

This section has been divided into two parts. The first one describes the basic collection procedure for untilted images and the second one describes the collection of Random Conical datasets (*see* **Notes 32–38**).

3.4.1. Untilted Images or 0° Images

1. Align the microscope using the normal holder and a negative-stain grid. When finished, reset the stage to the zero position and remove the holder.

2. Switch the filament down all the way. Cover the lead glass window if you are doing cryo to prevent liquid nitrogen from falling on the glass window.
3. Transfer either the negative-stained grid or the frozen grid to the holder and insert the holder in the microscope.
4. Wait for the microscope to recover a good vacuum Ion getter pump ([IGP]=6, on FEI microscopes). If you are using a room-temperature holder, wait a few minutes for it to stabilize. If you are using a cryo-holder, wait 30–45 min for the temperature of the holder to stabilize. Make sure that the shutter in the cryo-holder is closed and that the beam is blanked. During this time, you can monitor the temperature of the holder with the Gatan temperature controller unit. When the cryo-holder has reached its minimum temperature, remove the plug of the temperature controller unit.
5. Switch on the filament. Remove the cover from the lead glass window. Release the beam blank, and for the cryo-holder open, also its shutter.
6. Go to search mode and find a visible area on the grid. Adjust the eucentric height at ×3000. Go back to the exposure mode and make a finer adjustment of the height at higher magnifications. This step is very important if you want to collect a Random Conical dataset or a tomography dataset.
7. Go back to the search mode, remove the objective aperture, and select a low magnification (×100 to ×200). Move around the grid to select areas with either the right thickness of ice or a good stain distribution.
8. Select all usable areas. Blank the beam. Insert the objective aperture. Go back to your normal search magnification.
9. Move to one of the selected grid squares. Because grids are not completely flat, go to a neighbor grid square that is not good (if you have many good squares in an area, just give one up), fine adjust the eucentric height, and make sure that the minimal-dose kit is still well aligned.
10. Go back to the good square and select a good area.
11. Focus the specimen and reset the defocus to zero. At a magnification of ×100,000 and dialing a small defocus, you can check if there is drift. Stare at the grain and make sure that the holder is stable.
12. Blank the beam and lift the screens.
13. Dial the desired defocus value. Because there is less contrast, frozen hydrated specimens will require larger defoci than negatively stained ones. The necessary defocus value depends on the voltage and the size of the specimen. At higher voltages, larger defoci will be necessary. The smaller the particle, the larger the necessary defocus. The first time that you look at your sample in a frozen hydrated state, it is recommended that you take pictures at different defoci to know what would be a good range later.
14. Acquire an image.
15. After recording the image, go to the exposure mode, put down both screens, depress the beam blank, and look at the meter to get a more quantitative idea of the thickness of the sample. If you have adjusted the intensity of the light to be 1.72 s in an area without carbon; good areas (for 0° images) should show meter readings of about 2.16–2.52 s.

16. Keep taking as many images as you can.
17. When you have finished your data collection, remove the holder (before removing the holder, empty the prevacuum buffer tank to avoid having the vacuum collapse while you remove the holder).
18. Place the holder in the transfer station and check if your grid was up or down in the holder.
19. If this was the last frozen hydrated grid, empty the nitrogen of the cryo-holder and warm it up by gently blowing nitrogen gas in the Dewar and also on the tip. If you remove the clip-ring to take your grid out, put it back in its place. It is cleaner than having it in the clip-ring tool.

3.4.2. Tilt-pairs for Random Conical Reconstructions

1. Do **steps 1–9** of **Subheading 3.4.1.** When selecting areas for tilt-pairs at low magnification, remember that close to the center of the grid you will be able to achieve higher tilts.
2. Go to the maximum possible tilt; 55°–60° would be a good value. Be careful when tilting.

   a. When you see the shadow of either the holder or the grid bar coming into the field of view, stop tilting and reduce the tilt angle by approx 5° from this value, otherwise you will get a pseudoastigmatism in your image.
   b. Tilt slowly, at least for the last 15° to keep the specimen drift to a minimum.
   c. Always approach the final tilt angle from the same direction. If you overshoot, go back several degrees and try again.
3. On the tilted specimen, select an area for imaging using the search mode.
4. Focus in the focus mode. If two positions are available for focusing, use them both.

   a. Make sure that all the focus positions are on the tilt axis. If not, there will be a difference between the focus and the exposure mode, even if the grid is 100% flat.
   b. On the FEI CM series microscopes, the focusing positions are defined by a radius and an angle. The angle of the two positions is rarely at 0° and 180°. Determine the correct angle for your microscope beforehand (see the microscope manual). Typical values for the CM series of microscopes are 145° and 325°.

5. Observe that there is no drift, dial the desired defocus, and record an image.
6. Go to 0° tilt.
7. In the search mode, select the same area as before. This has to be done precisely; the best is to center a small feature on a mark on the small viewing screen. If it is not possible to achieve a good eucentricity of the stage, you can observe the specimen in focus mode during tilting; tilt slowly and correct the position as you tilt. Be very careful to keep the position so that you do not accidentally expose the imaging area.
8. Focus specimen as in **step 4**.
9. Observe that there is no drift; dial the desired defocus and record an image.
10. If you have a slow-scan charge-coupled device (CCD) camera, record a second image with the camera. Observe that the area you photographed was good. On a TVIPS CCD camera, you should be able to observe not only the ice/stain

thickness but also the preservation of fine details in your particles. If you do not have a CCD camera, go to exposure mode and check the area you photographed. Also check that all imaging conditions were good. (illumination, defocus, etc.)

11. Repeat **steps 2–10** to collect more data within the same grid square. Make sure to avoid exposed areas.
12. Proceed to the next preselected grid square. You should check the eucentricity when you move far away. Repeat **steps 2–10** for each image. Continue until either all good areas are exhausted or the film in the camera is finished.

4. Notes

1. Do not keep slides immersed in acetone for long times because the plastic film will dissolve.
2. The quantity and size of the holes depends on the amount of glycerol added to the solution. It is a parameter that you can vary; at a higher concentration of glycerol, the holes are larger and more numerous.
3. The Formvar solution should be used after sonication; if the solution is left standing for a long time, it is better to make a new one.
4. The films cast on the glass slides should be transferred to the grids as soon as possible. The longer they sit around, the more difficult it is to remove them from the slide.
5. If the surface of the grid is allowed to dry when using methylamine tungstate, the chances of having deep stain are minimal. With ammonium molybdate, it is not as critical.
6. Deep-stained areas seem to have a preferential direction. This might be avoided by using a filter paper with a cutout of the shape of the grid. Cut triangles of paper and then using a hole-punch, cut the tip of the triangle. You cannot get the whole perimeter of the grid but at least one-half or two-thirds of it.
7. Avoid getting the solutions up the tips of the tweezers as a result of capillary action. This will produce undesirable staining artifacts.
8. Grids prepared with methylamine tungstate are only good for about 1 wk and grids prepared with ammonium molybdate are good for about 2 wk. Images taken after these periods will be of lesser quality.
9. The purpose of the technique is to obtain areas on the grid that show deep stain. These areas can be described as rivers or lakes of stain that extend many grid squares. Areas surrounding them are usually dry. Another useful tip for recognizing these good areas is by looking at the change that occurs upon irradiation. Before electron irradiation, these regions look very smooth; after irradiation (the dose to collect and image is enough), the particles are visible as white dots at the low magnifications used in search mode of the minimal-dose kit.
10. All of the images should be taken under low-dose conditions. These stains are very radiation sensitive.
11. Do not make the papers too wide or blotting will be a problem; look at the blotting tweezers to get the right size. Make some spares just in case.
12. Always transport the gas bottles using a bottle trolley. Tighten the bottle properly with ropes or belts to the trolley. When you tip the trolley to move the bottle, put your hand over the bottle for safety.

13. Do not pour nitrogen in the empty ethane pot or it will take a long time to evaporate. Do not pour nitrogen over ethane.
14. Always check that the plunger is well adjusted. Put a pair of tweezers with a grid in the guillotine. Plunge it gently by hand and make sure that the tip of the tweezers is 3–5 mm below the rim of the ethane pot and that it falls in the center.
15. Ethane is a flammable gas. Do not use close to open flames or sparks. Do not use close to electrical equipment during on/off operation unless they are explosion-proof.
16. Ethane is used for vitrifying specimens because it freezes them faster that liquid nitrogen. If for some reason you get in contact with liquid ethane, you will have a cold burn. Stop what you are doing safely and as fast as possible and pour cold water over the burn for 5–10 min. Afterward, treat it as a high-degree burn. If aloe is available, put some over the burned skin.
17. If there is no steam coming out of the tip when you start filling up the ethane pot, either the tip is clogged, the main valve of the bottle is closed, or the bottle is empty.
18. Frozen ethane: If the level of nitrogen in the plunger box is too high, the ethane will freeze first at the walls and at the bottom of the pot. Frozen ethane occupies a larger volume than liquid ethane and, as a result, you will observe a bulging of the ethane surface. To melt the frozen ethane, you can add some fresh ethane gas. Put a new tip on the tubing. Place the tip just above the ethane surface, open the fine valve, and when the steam starts coming out, dip the tip inside of the ethane pot. Open the fine valve a little more and direct the tip toward the frozen areas. As you do this, the ethane will melt. Stir gently the ethane with the tip, making sure that no frozen ethane remains at the bottom of the pot. When all of the ethane is liquid again, raise gently the tip toward the edge of the pot, lower the gas flow a little, remove the tip from the pot, and close the fine valve completely.
19. If you remove the tip from the ethane pot still having a large flow of gas, you will get a lot of steam on the plunger box.
20. If you close the fine valve before removing the tip from the ethane pot, some ethane will flow back into the tip and into the tubing. Try to avoid it.
21. If the ethane gas flow is too strong when you are either refilling the pot or melting the frozen ethane, some of the liquid ethane will splash out of the pot. Try to avoid having a strong flow of gas.
22. If the level of liquid ethane stopped rising and there is some steam coming out, the pot is not cold enough. Stop the flow of gas and add more liquid nitrogen to the plunger box. Be careful not to put the nitrogen inside of the ethane pot.
23. If the level of liquid ethane stops rising and there is no steam coming out, either the tip is clogged or the bottle is empty.
24. If you transfer the grid very fast, you will observe some white scum on it; do not worry because this is just frozen ethane and it will disappear after a few hours of exposure to liquid nitrogen. However, if you transfer the grid very slowly, then you risk getting it too warm and you might have initiated the transformation of amorphous ice into cubic ice. As always, a speed in between is the best.
25. If putting the grid inside of the cryo-box is taking too long, be careful because cold metal surfaces will burn your fingers. The tweezers will be getting colder, so

search for a new pair of clean tweezers, cool the tips down in the deepest part of the cryo-box pot, and try again.

26. Always cool down the tools in liquid nitrogen before you touch the frozen grids.
27. If a cold tool has been taken out of the liquid nitrogen, do not reuse it until it is warmed up and you have dried it with a Kimwipe.
28. Air currents are the main source of contamination on your grids. Try to avoid them. Your movements close to the liquid nitrogen should not be too rough, to minimize air currents.
29. If the gun is not well aligned, you will have a reduced illumination and also a reduced coherence in your beam.
30. If the beam deflection coils or the objective lens are not well aligned, either the beam, the image, or both will show displacements when focusing.
31. In the FEI CM microscope series, if the beam moves while focusing, you have to realign the beamcoils pivot points. However, if the image moves while focusing, then it is the rotation center alignment that requires adjustments.
32. Lots of drift: Check the nitrogen level in both the anticontaminator and the cryo-holder. The nitrogen inside the cryo-holder should not boil. Refill if necessary. If the nitrogen level was acceptable, move the holder slowly to another area and check the drift again. If it is still the same, it is best to remove the grid and put in a new one. Either the grid does not make a good contact with the holder, it is not sitting properly in its place, or it is too bent. In the end, you will save time if you put in a new grid.
33. For both 0° and tilted images, if there is a large difference in focus between the two focal positions of the minimal-dose kit, the grid might be warped. Another possibility in the case of tilted images is that the focal positions are not positioned on the tilt axis.
34. If there is contamination during transfer of the grid to the holder and the transfer was smooth, check whether there are air currents where you are doing your transfers.
35. Be careful when you collect tilt-pairs. Grid bars have a given thickness. If you look at your grid in search, you will observe that as you tilt from 0° to a high tilt, certain areas of the grid square start becoming visible while on the opposite side some areas become obscured by the grid bar. Select areas that are visible on both.
36. When specimens are tilted, you will observe that the value shown on the exposure meter reading increases. This is caused by the larger specimen thickness that the electrons have to penetrate (easy to calculate with simple trigonometric formulas). Do not increase the beam intensity to compensate for this effect or you will be irradiating your sample with a higher electron dose.
37. Images of tilted specimens can have defocus gradients in the micrometer range (1–3 μm). Select the defocus that ensures that the whole image is underfocused and where the least defocus area is defocused enough so that your particles are still visible.
38. Always record in which orientation the grid was placed on the holder to know which side of the grid was facing the electron beam. Please note that some standard or room-temperature holders turn the grid upside down when inserted in the microscope. This is important to be certain of the handedness of the structure and to make sure that enantiomorphic structures are not mixed in the reconstruction (*see* Chapter 20).

References

1. Baumeister, W. and Steven, A. C. (2000) Macromolecular electron microscopy in the era of structural genomics. *Trends Biochem. Sci.* **25,** 624–631.
2. Rawat, U. B., Zavialov, A. V., Sengupta, J., et al. (2003) A cryo-electron microscopic study of ribosome-bound termination factor RF2. *Nature* **421,** 87–90.
3. Frank, J. (2001) Cryo-electron microscopy as an investigative tool: the ribosome as an example. *Bioessays* **23,** 725–732.
4. Spahn, C. M., Grassucci, R. A., Penczek, P., and Frank, J. (1999) Direct three-dimensional localization and positive identification of RNA helices within the ribosome by means of genetic tagging and cryo-electron microscopy. *Structure* **7,** 1567–1573.
5. Mueller, F., Sommer, I., Baranov, P., et al. (2000) The 3D arrangement of the 23 S and 5 S rRNA in the *Escherichia coli* 50 S ribosomal subunit based on a cryo-electron microscopic reconstruction at 7.5 A resolution. *J. Mol. Biol.* **298,** 35–59.
6. Harms, J., Tocilj, A., Levin, I., et al. (1999) Elucidating the medium-resolution structure of ribosomal particles: an interplay between electron cryo-microscopy and X-ray crystallograhy. *Structure* **7,** 931–941.
7. Stark, H., Rodnina, M. V., Wieden, H. J., Zemlin, F., Wintermeyer, W., and van Heel, M. (2002) Ribosome interactions of aminoacyl-tRNA and elongation factor Tu in the codon-recognition complex. *Nature Struct. Biol.* **9,** 849–854.
8. Ruiz, T., Mechin, I., Bar, J., Rypniewski, W., Kopperschlager, G., and Radermacher, M. (2003) The 10.8–A structure of *Saccharomyces cerevisiae* phosphofructokinase determined by cryoelectron microscopy: localization of the putative fructose 6-phosphate binding sites. *J. Struct. Biol.* **143,** 124–134.
9. Roseman, A. M., Ranson, N. A., Gowen, B., Fuller, S. D., and Saibil, H. R. (2001) Structures of unliganded and ATP-bound states of the *Escherichia coli* chaperonin GroEL by cryoelectron microscopy. *J. Struct. Biol.* **135,** 115–125.
10. Ranson, N. A., Farr, G. W., Roseman, A. M., et al. (2001) ATP-bound states of GroEL captured by cryo-electron microscopy. *Cell* **107,** 869–879.
11. Klaholz, B. P., Pape, T., Zavialov, A. V., et al. (2003) Structure of the *Escherichia coli* ribosomal termination complex with release factor 2. *Nature* **421,** 90–94.
12. Agrawal, R. K., Linde, J., Sengupta, J., Nierhaus, K. H., and Frank, J. (2001) Localization of L11 protein on the ribosome and elucidation of its involvement in EF-G-dependent translocation. *J. Mol. Biol.* **311,** 777–787.
13. Dubochet, J., Adrian, M., Chang, J. J., et al. (1988) Cryo-electron microscopy of vitrified specimens. *Q. Rev. Biophys.* **21,** 129–228.
14. Dubochet, J., Booy, F. P., Freeman, R., Jones, A. V., and Walter, C. A. (1981) Low temperature electron microscopy. *Annu. Rev. Biophys. Bioeng.* **10,** 133–149.
15. McDowall, A. W., Smith, J. M., and Dubochet, J. (1986) Cryo-electron microscopy of vitrified chromosomes in situ. *EMBO J.* **5,** 1395–1402.
16. Radermacher, M., Ruiz, T., Wieczorek, H., and Grüber, G. (2001) The structure of the V1-ATPase determined by three-dimensional electron microscopy of single particles. *J. Struct. Biol.* **135,** 26–37.
17. Ruiz, T., Kopperschläger, G., and Radermacher, M. (2001) The first three-dimensional structure of phosphofructokinase from *Saccharomyces cerevisiae* determined by electron microscopy of single particles. *J. Struct. Biol.* **136,** 167–180.

18. Unwin, P. N. (1974) Electron microscopy of the stacked disk aggregate of tobacco mosaic virus protein. II. The influence of electron irradiation of the stain distribution. *J. Mol. Biol.* **87,** 657–670.
19. Glaeser, R. M. (1971) Limitations to significant information in biological electron microscopy as a result of radiation damage. *J. Ultrastruct.* Res. **36,** 466–482.
20. Cosslett, V. E. (1978) Radiation damage in the high resolution electron microscopy of biological materials: a review. *J. Microsc.* **113,** 113–129.
21. Glaeser, R. M. and Taylor, K. A. (1978) Radiation damage relative to transmission electron microscopy of biological specimens at low temperature: a review. *J. Microsc.* **112,** 127–138.
22. Unwin, P. N. and Henderson, R. (1975) Molecular structure determination by electron microscopy of unstained crystalline specimens. *J. Mol. Biol.* **94,** 425–440.
23. Knauer, V., Schramm, H. J., and Hoppe, W. (1978), in Proceedings of the 9th International Congress on Electron Microscopy, Vol. 2.
24. Hayat, M. A. (1989) *Principles and Techniques of Electron Microscopy.* Biological Applications, CRC, Boca Raton, FL.
25. Hayat, M. A. and Miller, S. E. (1990) Negative Staining, McGraw-Hill, New York.
26. Baumeister, W. and Hahn, M. (1978) Specimen supports in principles and techniques of electron microscopy, in *Biological Applications* (Hayat, M. A., ed.), Van Nostrand Reinhold, New York, Vol. 8, pp. 1–113.
27. Ermantraut, E., Wolhfart, K., and Tichelaar, W. (1998) Perforated support foils with predefined hole size, shape a nd arrangement. *Ultramicroscopy* **74,** 75–81.
28. Smith, P. R. (1981) A trough designed to facilitate the coating of electron mciroscope grids. *Philips Electron Opt. Bull.* **115,** 13.
29. Harris, J. R. (1996) Negative Staining and Cryoelectron Microscopy: The Thin Film Techniques, Springer-Verlag, New York.
30. Tracz, E., Dickson, D. W., Hainfeld, J. F., and Ksiezak-Reding, H. (1997) Paired helical filaments in corticobasal degeneration: the fine fibrillary structure with NanoVan. *Brain Res.* **73,** 33–44.
31. Gregori, L., Hainfeld, J. F., Simon, M. N., and Goldgaber, D. (1997) Binding of amyloid beta protein to the 20 S proteasome. *J. Biol. Chem.* **272,** 58–62.
32. Trinick, J. and Cooper, J. (1990) Concentration of solutes during preparation of aqueous suspensions for cryo-electron microscopy. *J Microsc.* **159,** 215–222.
33. Stoops, J. K., Momany, C., Ernst, S. R., et al. (1991) Comparisons of the low-resolution structures of ornithine decarboxylase by electron microscopy and X-ray crystallography: the utility of methylamine tungstate stain and Butvar support film in the study of macromolecules by transmission electron microscopy. *J. Electron. Microsc. Tech.* **18,** 157–166.
34. Cyrklaff, M., Adrian, M., and Dubochet, J. (1990). Evaporation during preparation of unsupported thin vitrified aqueous layers for cryo-electron microscopy. *J. Electron Microsc. Tech.* **16,** 351–355

20

Three-Dimensional Reconstruction of Single Particles in Electron Microscopy

Image Processing

Michael Radermacher and Teresa Ruiz

Summary

Three-dimensional electron microscopy of single macromolecular assemblies has made large strides forward over the last decade. A large number of image processing techniques have been developed and many have found general distribution. For the proper usage of the wide range of available techniques, a clear concept of all processing steps is essential. This chapter provides step-by-step instruction for the three-dimensional reconstruction of an unknown macromolecule. Where possible, the limitations of the techniques are explained. The chapter attempts to be sufficiently general such so as not to adhere to a single image processing system. Described are alignment techniques for two and three dimensions, classification procedures, and the usage of three-dimensional reconstruction algorithms.

Key Words: Electron microscopy; image processing; three-dimensional reconstruction; Radon transform; image alignment; image classification; cross-correlation; single particles; random conical; angular refinement.

1. Introduction

The term "single particles" throughout this chapter is used in its traditional meaning and designates macromolecules without symmetry or only a low degree of symmetry. Specifically not included are techniques specialized in exploiting structural repeats such as those found in helical or icosahedral particles or two-dimensional crystals.

There are a large number of different reconstruction techniques, and within each technique, different algorithms are being used. However, all of these methods follow similar principles. Therefore, a detailed example of one technique opens the door to all other techniques. The most comprehensive book about image processing in electron microscopy is Frank *(1)*.

From: *Methods in Molecular Biology, vol. 319: Cell Imaging Techniques: Methods and Protocols*
Edited by: D. J. Taatjes and B. T. Mossman © Humana Press Inc., Totowa, NJ

The theory of three-dimensional reconstruction algorithms is based on the projection theorem ***(2–4)***, which states that the two-dimensional Fourier transform of the projection of a three-dimensional object is a central section through the three-dimensional Fourier transform of the object. For example, the two-dimensional Fourier transform of a projection provides the measurement of the Fourier coefficients in a plane that passes through the origin of the three-dimensional transform. The direction of this plane is the same as the direction of the projection relative to the three-dimensional object. Thus, by collecting projections from a variety of directions, eventually the complete three-dimensional Fourier transform of the object can be determined, and, in principle, the inverse Fourier transform of these data provides the three-dimensional structure. The reconstruction is rarely done by actual Fourier interpolation and inverse transforms, but a wide variety of algorithms exist that achieve a reconstruction from projections, each with its own set of advantages and disadvantages. The projection theorem exists not only for Fourier transforms but also for Radon transforms ***(5,6)***. In recent years, several groups have taken advantage of the mathematical properties of Radon transforms and developed new alignment and reconstruction algorithms based on its theory ***(7–11)***.

Three basic approaches are used to determine the structure of single particles, their main difference being the data collection strategy. These approaches are: tomography ***(12–15)***, the random conical reconstruction technique ***(16–18)***, and angular reconstitution ***(7,19,20)***.

Tomography of single particles is based on the same principles as medical tomography ***(21–40)***. A large number of images are collected from the same specimen area at different angles. A typical tilt series would start at –60° and proceed in 1° or 2° steps up to +60°. The limited tilt range of the microscope stage does not allow sampling of the full three-dimensional Fourier transform and a double wedge in the three-dimensional transform is undetermined. Remedies that alleviate this problem are the combination of two tilt series recorded with the tilt axis rotated by 90° between them ***(41)*** or the collection of a conical tilt series ***(42,43)***. In a conical tilt series, the specimen is tilted by a fixed high angle, typically 60°, and then rotated within this tilted plane by small angular increments, recording an image for each position. This geometry allows measuring the three-dimensional Fourier transform with a better angular coverage than does a single axis tilt series, and only a double cone is undetermined. At 60°, this double cone encompasses 13.4% of the volume in Fourier space in contrast to the 33% missing for single axis tilting. The two tomographic approaches, single axis and conical tilting, allow the reconstruction of individual particles; however, the multiple exposures required can lead to extensive radiation damage, in many cases preventing a reconstruction at high resolution. These techniques have found, as their main application, the three-dimensional reconstruction of cellular organelles like mitochondria. However, combined with averaging techniques, they can be extremely powerful also for the reconstruction of frozen hydrated particles.

Random conical reconstruction ***(16–18)*** and angular reconstitution ***(7)*** both overcome the problem of radiation damage through averaging over identical particles. The dataset for a random conical reconstruction consists of pairs of micrographs; one image in each pair is recorded at a high tilt angle of the specimen, typically 45° to 60°, and a second image of the same specimen area is recorded at 0°. If the specimen consists of identical particles attached in preferred in-plane orientations relative to the supporting carbon foil, then one tilt image provides a conical tilt series, yet with random azimuthal angles, in contrast to a series of images of one particle recorded in a conical geometry with equal angular increments. Every particle in the tilt image provides one view in a conical tilt series. Because the azimuthal angles are random, the technique is called the random conical reconstruction technique. The 0° image serves two purposes: one is the finding of the in-plane orientation of each particle, and the second purpose is to classify the particles into groups of identical projections to ensure that the reconstruction originates from projections of identical particles. Differences between the classes can have two major reasons. One reason is that the particles might have more than one orientation on the grid but are otherwise identical, the other is that the differences are caused by true differences among the molecules on the grid, a situation that is rather common. Which of the two is the reason for the class differences can only be decided after separate reconstructions for each class have been carried out. The images that are being used for the actual three-dimensional reconstruction are those extracted from the tilt image, and these particles have only been exposed once to the electron beam. Thus, the radiation damage can be kept to a minimum.

Angular reconstitution is based on the assumption that particles assume a wide variety of orientations on the specimen support. The images are again aligned and classified as is done for the 0° image in the random conical reconstruction technique. However, in angular reconstitution, the assumption is made that the differences between the classes are caused solely by differences in particle orientations. Under these conditions, the class averages can be uniquely related to each other and a projection angle relative to a common structure can be assigned to each particle projection. After a first low-resolution reconstruction, the classification is relaxed, and in the final step, each image is aligned to the common three-dimensional model.

The image processing in this chapter will focus on the reconstruction from a random conical dataset followed by a reference-based three-dimensional projection alignment, which uses the three-dimensional model calculated using random conical techniques as reference to align new projections in space. All of the techniques used to analyze the 0° image are the same as those used in any two-dimensional study. A two-dimensional study should precede any three-dimensional reconstruction, because it can serve to get familiar with the gross structure of the molecules. A two-dimensional study allows a fast assessment of the quality of preparation and often the results influence the biochemistry of the project. Not rarely, in the first attempts to solve the structure of an unknown molecule, the researcher is faced with a specimen that shows a wide variety of shapes, even though all biochemical techniques indicate a highly pure preparation. A critical interpretation of both biochemical and structural results will lead, in most cases, to a preparation that shows a reasonable homogeneity. The extent of the homogeneity varies from specimen to specimen. Not only is the quality influenced by the original preparation but, obviously, also by the age of the preparation and, not to be neglected, by the dilution before being applied to the grid. It has been observed that the stability of a macromolecule can be highly dependent on its concentration ***(44)***. The main reason for focusing on the random conical technique is the particle variability mentioned earlier. Random conical and tomographic techniques are currently the only techniques that, if used properly, do not lead to artificial three-dimensional structures when specimen homogeneity cannot be guaranteed. In fact, these techniques can be used to determine the variability of the sample.

The description of the processing is based on our experience using the image processing system SPIDER and the associated display program WEB ***(45)***. However, other systems can be used if they have certain basic features implemented. Essential are an interactive display program for the selection of images from tilt-pairs, Fourier transforms and Fourier filters, correlation functions for translational and rotational alignments, and good image classification algorithms. For example, correspondence analysis followed by classification is one of the most powerful techniques and is available in most systems used in electron microscopy. It is also advantageous that the system has a good scripting language so that the different algorithms can be combined in a flexible manner. For most systems, you should be able to obtain example scripts from colleagues and these examples will give you a good start with your project. Even if scripting at first seems more tedious than if you had a single program that does everything, in the end it will give you the flexibility required to obtain the optimum results from your data.

Image processing systems for electron microscopy include XMIPP, SUPRIM, MDPP, EMAN, IMAGIC, SPARK, and the EM-System. Information about most of the systems can be found on the em-outreach website http://em-outreach.sdsc.edu/community-codes/MicroscopySoftware.html. SPIDER and WEB are available from the Wadsworth Center (Albany, NY), SPARK is developed at the Dipartimento di Chimica Strutturale e Stereochimica

Inorganica (Milano, Italy), and XMIPP is available from the Biocomputing Unit at the Centre National de Biotechnologia CNB (Madrid, Spain). The EM-System, one of the oldest image processing systems developed for three-dimensional electron microscopy, is available from the Max-Planck-Institute for Biochemistry (Martinsried/Munich). An overview over many image processing systems can be found in a special issue of the *Journal of Structural Biology* (Vol. 116, 1996).

2. Random Conical Reconstruction

2.1. Image Recording and Negative Selection

1. Record pairs of micrographs, the first image with high tilt (~50° to 60°), and the second image of the same specimen area at 0° (see Chapter 19) (*see* **Note 1**).
2. Inspect the tilt-pairs visually. Make sure that the specimen area visible in the 0° image is the same as in the tilt image (*see* **Note 2**).
3. Inspect the negatives in the optical diffractometer. Thon rings should be visible out to the resolution that is intended. The complete area in the tilt image must be in underfocus. If not, the area in overfocus should be indicated (best with a felt-tip marker at the edge of the micrograph) such that it can be excluded during scanning. Alternatively, the overfocused area can be marked on a print copy of the negative and then excluded either during scanning or during particle picking. If a sufficiently large number of tilt-pairs is available, those that show overfocused areas should not be used.

2.2. Image Processing: Basic Considerations

Any image processing requires a reasonable choice of a wide range of parameters. Frequently used parameters are the values of low-pass and high-pass filters, the search ranges and increments, the number of classes in classification algorithms, and many others. Guidelines to the proper selection of these parameters will be given for every step; these can only be guidelines because every specimen will behave differently. It is worthwhile to spend some time at certain stages of the processing to determine the parameters for the best performance of the algorithm and fine-tune them to the data. Make sure that the image processing system that you use gives you access to these parameters where needed. Improper selection of parameters can lead to a reconstruction of lower quality than the data allow or, in the worst case, to wrong results.

Image processing of large datasets require a reasonable organization of the data such that it is possible to retrieve any step of the processing at any later step in the procedure. When a sample is analyzed for the first time, some steps might need multiple runs until satisfactory results are obtained and it is essential to be able to tell them apart. Image names might sound suggestive when you create them, but 1 mo later, they might not be clear anymore.

You will create different types of file during the processing of your data:

Image files are sometimes stored as series of images in one file; other times, each image has its own file. In most systems, the name of an image file consists of a name followed by a number that uniquely identifies the image within an image series.

Data files contain numbers, like coordinates, angles, magnifications, and so forth, which will be referred to as "document files" in accordance with the nomenclature used in SPIDER.

Printable files contain the descriptions and results of the processing, which will be referred to as "results files," also using the nomenclature of SPIDER.

Command files and scripts contain the sequences of image processing commands. Scripts can normally be reused for several applications without changes ("procedures" in SPIDER), whereas command files are used to run one specific job ("batch files" in SPIDER).

Other binary files that do not belong to any of the above categories are occasionally created mostly to store intermediate data.

In our experience, an efficient directory structure for random conical reconstruction is the following setup:

<main directory for the project> with the subdirectories.

<directory process> for storage of scripts (procedures, batch files), document files, results, and miscellaneous unique files.

<directories pair %%%%>, one directory for the scanned micrographs of each tilt pair. "%%%%" should be replaced by the last four digits of the micrograph number. Also the document files with particle coordinates and the scripts for particle windowing from each pair are stored here.

<directory not> for the storage of the complete series of all original windowed single-particle images from the 0° micrograph.

<directory tilt> for the storage of the complete series of all original windowed single-particle images from the tilt micrograph.

<directory ucent> for storage of the centered 0° images.

Directories <alia> <alib> <alic>,..., <ali(n)> for the storage of the 0° images after each rotational/translational alignment.

Directories <classi1> <classi2> <classi3>,..., <classi(n)> for each run of pattern recognition/classification, including support files, script files, and resulting images.

<directory tcent> for storage of the centered tilt images.

Directories <talia> <talib> <talic>,..., <tali(n)> for the storage of the tilt images after each alignment.

<directory threed> for storage of the three-dimensional volumes.

The basic idea of this structure is the following: Directory <process> is the directory where most of your processing is done, where your script/procedure, command files, and the results files are located. There might be some exceptions to this because the scripts you use may not have the flexibility built in to read and deposit data in different directories. In this case, you might have to execute those in the directories where the data are located. In our application using SPIDER (5.0) in combination with the scripts that we have developed over the years, this is true for the windowing of the images from the original micrographs (reason: scripts), and correspondence analysis (reason: program structure). However, distributing scripts and command files over many directories should be kept to a minimum. Having all the scripts and command files in one directory will help to keep track of the order in which the processing was executed. Avoid using command files twice with different parameters; rather, create a separate copy for each run.

2.3. Image Digitization

1. Scan the negatives. Scan the negatives always in the same direction and orientation; for example, emulsion down, scanning starting at the side of the negative opposite to the number going from left to right. If you change any of these parameters, you might change the handedness of the final structure (*see* **Note 3**). The pixel size used should be smaller than one third of the final resolution expected. A factor of one fourth is better. For example, if the magnification in the microscope is ×60,000, a scanning pixel size of 21 μm will result in a pixel size equivalent to 3.5 Å, which would allow for a final resolution of about 11 Å (*see* **Note 4**).
2. Convert the images to the format of the image processing system that you are using (*see* **Note 5**).
3. Create smaller copies of the images by binning, typically a factor of 3. The reduction is needed for the selection of particles.

2.4. Particle Extraction

2.4.1. Particle Selection

1. Display the two images of a tilt-pair side by side, first the image of the untitled specimen and then the tilt image. (In the example here, we use WEB; if other systems are used, you have to find the equivalent operations.) Move the two images such that you see the same area in both. Select four to five particles (absolute minimum number of particles is three), clicking on the center of the particle first in the left image and then in the right image. Make sure that the particles are well separated and are not located along one line. These first particle coordinates will be used to calculate the geometrical relation between the two micrographs (*see* **Note 6**).
2. After the first five particles are selected, go to the menu of the particle picking program and let it determine the tilt angle first. This angle should be within 1° or 2° of the goniometer reading; if the value is systematically off for a series of tilt-pairs, the reason might be that the goniometer reading is not well calibrated. Next, determine the direction of the tilt axis. Enter the start value for the direction of the tilt axis within 40° accuracy. Make sure the tilt axis direction is approximately the same for all images recorded with the same experimental geometry. There is a

180° ambiguity in the fitting of the tilt axis direction. The wrong direction can lead to a mirrored structure (*see* **Note 7**).

3. After determining the geometry, calculate the positions of the particles in the tilt image from the location of the particles in the 0° negative. An overlay of the calculated positions in the tilt image should be within a few pixels of the selected position. If the deviations are large, redo the particle picking (*see* **Note 8**).
4. Once the geometry is determined, go back to selecting particles. Once you select a particle in the 0° image, the program is now capable of predicting the approximate location of the tilted counterpart. Check that the cursor points to the center of the tilt particle; if necessary, correct the location and continue selecting first the 0° then the tilt particles. Any time during this process you can go back to the menu and redetermine the geometry (*see* **Note 9**).
5. Select the particles from all of the tilt-pairs available. For each tilt-pair, the result of the particle picking will be one document file containing the center coordinates of the particles in the 0° image, one document file containing the center coordinates of the particles in the tilt image, and one document file containing data common to the tilt-pair, like the tilt angle and the direction of the tilt axis of the 0° image and the direction of the tilt axis in the tilt image.

2.4.2. Windowing of the 0° Images, Contrast Normalization, Tilt Axis Rotation

All three steps are done combined into one processing step.

Input data needed

1. The document files containing the 0° particle coordinates.
2. The document file containing the directions of the tilt axis in the tilt and the 0° micrograph.
3. The micrograph.

Input parameters required:

1. The micrograph number.
2. The dimensions for the images of each particle, which should be approximately twice the particle diameter.
3. The number of the first image that is extracted from the micrograph.
4. A radius for a circular mask, which should be sufficiently large to include the apparently largest particle completely. Allow additional space for particles not perfectly centered.
5. The name of the images, including the directory where they are to be stored.
6. The name of an output document file that will contain a copy of the image coordinates in the micrograph and the micrograph number for each particle.

The extraction proceeds as follows:

1. Read micrograph.
2. Read direction of tilt axis.

3. Loop over the coordinates in the document file.
4. Read coordinates *x,y*.
5. Extract image at point *x,y* from the micrograph.
6. Rotate image so that the tilt axis is parallel to the *y*-axis.
7. Mask image, using a circular mask.
8. Determine image average $\bar{I}$ outside of this mask.
9. Normalize the (nonmasked) image using the equation $I_c = (I-\bar{I})/\bar{I}$, where I_c is the corrected image value, I is the original value of each image pixel, and $\bar{I}$ is the average of the image outside the circular mask.
10. End of loop (*see* **Notes 10** and **11**).

2.5. Image Alignment

2.5.1. Image Centration

Most image alignment algorithms require a good estimate for the rotation center of each particle. One of the most useful centration algorithms creates a rotationally averaged, centered template and uses this as a reference for translational alignment.

2.5.1.1. Create a Centered, Rotationally Averaged Reference

1. Add all of the 0° particle images, without any alignment (*see* **Note 12**).
2. Create an image of a disk with the approximate diameter of the averaged particles; if uncertain, make the disk slightly smaller.
3. Strongly low-pass filter this image to about half the disk diameter.
4. Cross-correlate this disk with the average image and determine the maximum of the cross-correlation function.
5. Shift the average image by the negative value of the location of the maximum relative to the correlation function's origin (*see* **Note 13**).
6. Rotationally average the centered particle average.
7. Apply a high-pass filter and a low-pass filter to the rotational average. (High pass is equivalent to one particle diameter; low-pass to 3 nm). (*See* **Note 14**.)

The result is the reference that is used for the centration of the image series.

2.5.1.2. Centration

1. Loop over all particles
2. Cross-correlate the particle with the reference.
3. Determine the location of the maximum of the cross-correlation function.
4. Shift the image by the coordinates of the cross-correlation peak. (Whether this shift has to be negative or positive depends on the program and how it was used. In SPIDER, operation CC, if the image is the first file and the reference is the second file, the shift must be negative.)
5. End of loop.

First, try the centration with about 50 particles; inspect the centered particles. If the centration is apparently unsuccessful, adjust the high-pass and low-pass filters and try again. Check how the cross-correlation function looks. If it looks too noisy, then a stronger low-pass filter will help. If there is a peak visible in the center, yet the function is lighter toward its corners, then a stronger high-pass filter of the reference may help (*see* **Note 15**). If your particles are elongated or L-shaped, this centration procedure might not perform quite satisfactory. As a centration reference, try using a low-pass filtered disk with the approximate size of the particle instead of the rotational average.

2.5.2. Rotational and Translational Alignment

The goal of the rotational and translational alignment is to position all particle images in the center of each image in identical orientations. In most image processing systems, this alignment is carried out by a sequence of rotational followed by translational alignments ***(46–48)***. An alternative algorithm for simultaneous translational/rotational alignment, currently only available as an extension to SPIDER V. 5.0, will also be explained ***(10,49)***. These algorithms will become available as stand-alone programs in the future.

At the start of the alignment, a reference is created. (For reference-free alignment, see **Subheading 2.5.2.2.**)

2.5.2.1. Selected Reference-Based Alignment

1. From the set of centered images, visually select a typical particle as first reference, or use the average created by reference-free alignment (*see* **Subheading 2.5.2.2.**).
2. Mask the reference image using a circular mask that contains the complete particle. This radius can now be smaller than the radius used when windowing the particles.
3. Apply a low-pass filter and a high-pass filter, the high-pass filter corresponding to the particle diameter and the low-pass filter corresponding to 3 nm.
4. Loop over all of the images.
5. Calculate the rotational cross-correlation.
6. Determine the maximum of the rotational cross-correlation function.
7. Rotate the image by the negative angle found in the rotational correlation. (The direction of the rotation again depends on the specific program used and the order of the input images.)
8. Calculate the translational cross-correlation between the resulting image and the reference.
9. Determine the location of cross-correlation maximum.
10. Shift the image by the negative of the location of the cross-correlation peak. (The direction of the shift depends on the program used and the order of the input images.)
11. End of loop.

For rotational alignment, the particle image is transformed into some form of polar coordinates, resulting in an image with coordinates r and φ. A one-dimensional cross-correlation function along the φ-coordinate is calculated and the location of the correlation maximum determines the relative rotation between the reference and each particle. In some systems, a subset of radial coordinates can be selected for this alignment. It is worthwhile to carefully select the best radii. The periphery of the particle, for example, has the most sensitive rotational information, the center has almost none, and the image outside the particle only adds noise to the correlation. This alignment should also be run first only over a few particles (~50) and it should be ascertained that the alignment is working successfully. In most cases, this can easily be done by visual inspection of the aligned images. The parameters that might need fine-tuning are the values of the radii used for rotational alignment and the low- and high-pass filters used for the reference. If the cross-correlation functions are too noisy, a stronger low-pass filter will help; if the translational cross-correlation function shows a clear peak near the center, yet very bright areas near the edges, a stronger high-pass filter might be required.

2.5.2.2. Reference-Free Alignment

The first reference in the above example was selected visually. A more objective reference can be found using the following procedure ***(50)***.

1. Loop over the image series.
2. Select the first, third, fifth, and so on. (i.e., every second) image in the series as reference. Mask, low-pass filter and high-pass filter them (same criteria as for the reference in the reference-based alignment).
3. Align the next image in the series to this reference using the rotational/translational alignment (*see* **Subheading 2.5.2.1., steps 5–10**).
4. Average the aligned image and the image used as reference.
5. End of loop.

The result is a series of averages from two images each. Apply the same procedure to the new image series.

6. Repeat until only one image is left. This is the image average obtained by reference-free alignment.

2.5.2.3. Simultaneous Rotational/Translational Alignment

1. Select a reference, apply mask, low-pass filter, and high-pass filter as in **step 2** of **Subheading 2.5.2.2**.
2. Calculate the two-dimensional Radon transform of the reference (coordinates p and φ).
3. Calculate the Fourier transform of the Radon transform along the radial coordinate p.
4. Loop over all of the images.
5. Calculate the two-dimensional Radon transform of every image.
6. Calculate the Fourier transform along p of all the Radon transforms of the images.
7. End of loop.

8. Calculate the three-dimensional (x,y,φ) cross-correlation functions between the reference transform and the transform of every image n and determine the locations of the maxima of these functions $(x_m,y_m,\varphi_m)_n$. All of this is done within one program. The output is a list of angles and shift vectors.
9. Loop over all original images.
10. Shift each image n by $+x_{mn}$, $+y_{mn}$.
11. Rotate the resulting image by $+\varphi_{mn}$.
12. End of loop.

The simultaneous rotational/translational alignment can also be used in the reference-free alignment. In many cases, even the precentering step can be skipped. If this is done, however, the average of the aligned images might not be in the image center, and if this is required, the average image needs to be centered separately. In general, the simultaneous translational/rotational alignment gives better results than an iteration of translational followed by rotational alignments.

2.6. Pattern Recognition and Image Classification

In most cases, the particles seen in a micrograph do not have the same apparent shape. To obtain a high-resolution average, the images must be classified into groups of images that show the same motif.

There are two distinct approaches to classification and which is best depends on the specimen and the intended result. The two major options that exist are a neural network approach using self-organizing maps for classification ***(51)*** or correspondence analysis ***(52,53)*** followed by a classification algorithm. The major difference between the two approaches is that self-organizing maps allow for an analysis of differences between the complete images, whereas correspondence analysis followed by classification allows focusing on specific changing features. For most projects, both types of technique should be applied and then it should be decided which of the approaches provides the best separation of the essential variations in the specimen.

2.6.1. Image Classification With Self-Organizing Maps (Available in XMIPP)

If you used a different image processing system for the rest of the processing, you have to make changes to comply with the syntax and file formats of XMIPP. If you used SPIDER previously, then the images do not need to be converted. The only file that needs to be created is the selection file that XMIPP requires. In the first column, this file contains the list of file names and, separated by a space, the number 1 or 0 in the next column. All images having a 1 in the second column will be processed. A detailed description of the complete procedure can be found in the XMIPP manual, including an example script that only requires the change of a few parameters:

1. The circular mask being used, which depends on the size of the particle.
2. The number of nodes displayed in the self-organizing map. This number should be large enough so that the final map contains nodes with no members. The best value requires test runs.
3. The radius of nodes that will be influenced by one new image in the calculation of the neural network. Typically, five nodes for the first iterations and later reduced to three nodes.
4. The number of iterations.
5. The image size used for the representation of the self-organizing map.

As for correspondence analysis, it can be advantageous to low-pass filter the images before using them in the neural network algorithm. This reduces the influence of high-frequency noise.

From the neural network map, node images can be directly selected and used as references for multireference alignment. These images can be found in the same directory in which the program was run. The name of the images depends on the input names; the number is the counter shown in the square in the neural network map. In newer versions of the program, the node images might have a step between the inside and outside of the mask used in the algorithm. In this case, remask the node image and specify the image average as the background before using them as references for a multireference alignment (*see* **Subheading 2.7.**).

2.6.2. Correspondence Analysis (Available in SPIDER, SUPRIM, IMAGIC, SPARK, EM)

The example follows the implementation in SPIDER and SUPRIM.

1. Low-pass filter all of the aligned images, typically to 3 nm or lower. If needed, also apply a high-pass filter, typically corresponding to the particle diameter.
2. Create a sharp mask that closely matches your particle, yet includes all possible variations. If in doubt, a circular mask of sufficiently large diameter will work.
3. Create the input matrix for correspondence analysis from your file series.
4. Run Correspondence analysis typically with 8–12 factors. The result is a coordinate file, in which each image is represented by the specified number of factors. Identical images have identical coordinates; the coordinates of very different images point to very different locations. For all of the following explanations, we assume that eight factors have been specified.
5. Analysis of the factor space. The eight-dimensional space can be analyzed in each possible combination of two dimensions. These two-dimensional representations are called maps, usually labeled by the factors they represent (e.g., "map factor 2 vs 5").

The maps can be represented as printable maps on which each image is identified with a number. Image points far apart indicate images with large differences in the features described by the two factors represented. Points close to each other represent very similar images relative to the feature represented by the two factors.

Each factor can be represented as an image, showing the feature that is changing in the direction of this factor. Each image can be reconstituted from the image that is in the origin of the factor space plus the factor images added with a weight corresponding to the coordinates of the image. If the reconstitution is done without the image representing the origin, then the reconstituted image shows the difference between the image at the origin and the image at a specific coordinate. These images are called differential reconstituted images. The features that are described by each factor can be shown as an image by reconstituting an artificial image with all factor coordinates except one set to zero. For example, the images at point (0,1,0,0,0,0,0,0) and (0,-1,0,0,0,0,0,0) show the feature described by factor 2 in its two extremes, missing on one side of factor 2 and present on the opposite side. Correspondence analysis allows the calculation of these images.

In SPIDER, the command for reconstituting images requires as input the factor(s) that should be used for reconstitution and the number of the image that should be reconstituted. The program looks up the coordinates of the specified image and uses these coordinates to calculate either the reconstituted or the differential reconstituted image. If the image numbers provided for the calculation of the reconstituted images is not within the range of the numbers present in the input file series (i.e., no coordinates can be looked up for this image), the program asks for the input of the coordinates. Entering 1 or –1 for the specified factors results in the creation of the extreme images, representing the feature change described by the specified factor(s).

6. Visualization of factor space. A visual representation of the two-dimensional maps can also be created from the file containing the image coordinates in factor space. For each visual map, one factor vs another, an empty large image is created and divided into typically 10×10 fields. One axis is assigned to the first factor represented; the other axis is assigned to the second. The coordinates of each image can be extracted from the coordinate file, and for each field, an average of all images whose coordinates fall into the same field can be calculated. This average is then inserted in the position of this field into the large originally empty image. The result is an image of the two-dimensional factor map that shows how the images change from one end of the map to the other. In SPIDER, there exists a single command for this operation ("ca vis").

   In addition to the visual representation of the factor maps, differential factor images can be reconstituted. As a first step in an eight-factor analysis, seven maps should be created: map factor 1 vs 2, 2 vs 3, 3 vs 4, 4 vs 5, 5 vs 6, 6 vs 7, and 7 vs 8. Both the visual maps and a gallery of the reconstituted factor images should be printed or displayed side by side on the screen. An examination of all these images can be extremely helpful for a better understanding of the variations in the specimen. For example, in negatively stained specimens, one of the lower factors often shows, on one side, the particle very bright, and on the other side, rather dark, but with no significant differences in other features. We often interpret this factor as a variability of stain thickness, which is not important for understanding the particles themselves. Other factors might describe a distinctly moving feature.

7. Classification. Correspondence analysis provides a set of points forming "clouds" in the eight-dimensional factor space. These points can be sorted into classes. It should be noted, however, that in a dataset of electron microscopical images, distinct classes can rarely be observed. In most cases, changes in the images are continuous (e.g., a flexible arm, a continuous rotation of the particle, etc.). Thus, the purpose of classification is, in most cases, to sort the images into groups of similar images, and the transitions can be fluent. There are many different classification algorithms. One of the best algorithms is the method of classification with moving centers developed by Diday ***(54)***, followed by a hierarchical ascendant classification. This algorithm is applied to the results of correspondence analysis.

The implementation of this algorithm in SPIDER allows the following parameters to be specified:

1. Factor numbers: 1, 3, 5, 7, and 8. (These should be those factors that, according to the inspection of the visual maps, are the most meaningful. For example, a stain-depth-describing factor should be omitted).
2. Number of starting classes (seeds) (typically three to five; the final classification can result in many more classes).
3. Number of iterations: between 5 and 100.
4. Number of repeats: typically 5.
5. Random or specific choice of seeds: random. (Seeds for starting the class formation can either be specified as specific coordinates, specific images, or created by a random generator.)

The algorithm in this example uses three to five points that will be randomly picked within the eight-dimensional space. The distances of each image to all seed coordinates are calculated and the image is assigned to the closest seed. In the next step, the centers of gravity of each group of images are calculated and used as new seeds. The distances of each image to the new seeds are determined and the images are reassigned. This process is iterated between 5 and 100 times, depending on the iteration parameter specified, each time starting with new seeds. After the iteration, final classes are formed and the classification reflects which particle images were combined most often in the same class and how often they were reclassified. The final number of classes can be more than 100. The program in SPIDER allows for a maximum of 99 classes, and all of the particles that do not fit into the 99 classes are assigned to class 100. The parameters have to be chosen such that the total number of classes is less than 100. More iterations result in fewer classes.

After this classification, the classes are arranged into a tree using a hierarchical ascendant classification algorithm. The tree should be interpreted as in any other hierarchical classification and similar classes can be merged by specifying a threshold for the class differences (i.e., a level for cutting the classification tree). How many classes are appropriate strongly depends on the specimen.

2.7. Multireference Alignment

For the following, we assume that classes have been obtained and the particles in each class are added to form an average image.

In the early steps of alignment, only one reference is available and all images are aligned such that they match best the reference. Having class averages allows a classification with more references and the combination of alignment and classification.

1. Recalculate the class averages, and for each average, use the same number of particle images. Calculating the averages from the same number of particles reduces the effect that the alignment/class assignment depends on the signal-to-noise ratio.
2. Align the references toward each other, choosing one of the class averages as the main reference. In this way, the new classes will not be based on different in-plane orientations.
3. Mask, low-pass, and high-pass filter all references, using the same parameters as earlier.
4. Align all images relative to all references and assign them to the reference with the highest cross-correlation coefficient. Use the alignment parameters (x,y-shift, rotation) that were found relative to this reference and shift and rotate each image.
5. Repeat either the self-organizing map or correspondence analysis combined with classification. The number of classes can be different from before.
6. Repeat the multireference alignment.
7. Repeat the complete process (classification/multireference alignment) a minimum of three times or until the class memberships of the particles are stable.

2.8. Three-Dimensional Reconstruction

2.8.1. Centration of the Tilt Images

All of the images of the tilted specimen can be centered without a separation of classes. There are two basic schemes for centering tilt images. Scheme 1 is most applicable for thin specimens, like standard negatively stained samples; scheme 2 is appropriate for thicker specimens, like frozen hydrated or deep-stain specimens. For thin specimens, the object can, in approximation, be considered as being a two-dimensional sheet, and stretching of the tilt image creates the same two-dimensional pattern as can be seen in a projection of the specimen without tilt *(55)*. As the specimen becomes thicker, the two patterns match less and less.

2.8.1.1. Scheme 1 (Thin Specimens)

1. Create a loop over all the images in the series.
2. Take one aligned image from the untilted series; read the tilt angle and the in-plane rotation.

3. Using the in-plane rotation angle, rotate the image back to its original orientation.
4. Take the corresponding tilt image.
5. Using the tilt angle, interpolate the image up in the x-direction to a dimension which is $n_{x0}/\cos(\theta)$, where n_{x0} is the original x-dimension and θ is the tilt angle.
6. Cut the image back to its original size.
7. Cross-correlate the last version of the 0° image and the last version of the tilt image.
8. Low-pass and high-pass filter the cross-correlation function.
9. Determine the location of the maximum of the cross-correlation function.
10. Multiply the x-direction of the peak coordinate by $\cos(\theta)$.
11. Apply the shift to the original tilt image.
12. Continue with the next image in the series.
13. End of loop

2.8.1.2. Scheme 2 (Thick Specimens)

1. Read the reference that was used for the centration of the 0° images, either an image of a low-pass-filtered disk or the rotationally averaged image average.
2. Interpolate down in the x-direction by $\cos(\theta)$.
3. Pad back to its original size; this is your reference for centration.
4. Loop over all tilt images.
5. Cross-correlate reference and tilt image.
6. Determine the location of the cross-correlation maximum.
7. Shift the tilt image using the coordinates of the cross-correlation maximum.
8. Continue with the next image.
9. End of loop

2.8.2. Assignment of Angles

1. Loop over all tilt images.
2. Read the angles that belong to this image from the document file created during the 0° alignment.
3. Enter the angles into the header of the image.
4. End of loop over images (*see* **Note 16**).

2.8.3. Class Separation

From the document file created by the last classification of the 0° image, create a separate document file for each class. For standard application in SPIDER, each document file contains as a line identifier (key) the image number followed in the second column by a counter of how many values will follow, and in the third column the value 1 if the image belongs to the class or the value 0 if it does not.

The next step depends on the reconstruction algorithm used. There exist many reconstruction algorithms, each with its own advantages and disadvantages. Here, we give two examples: reconstruction by weighted back-projection and reconstruction by a two-step Radon inversion algorithm. These algorithms are linear algorithms. They do not require any *a priori* information, yet they also do not allow including *a priori* information during the reconstruction process. *A priori* information, however, can be applied in a separate step. The advantage of first using a linear algorithm without any *a priori* information is that the results represent solely the information contained in the data. *A priori* information applied in a separate step should, if correct, only improve the resolution and not create any features that were not present in a reconstruction from measured data only.

2.8.4. Three-Dimensional Reconstruction Per Class

2.8.4.1. Weighted Back-Projection ***(17,42,56,57)***

1. Apply the weighting function to all the projections in one class (this is usually a single command).
2. Sum the weighted projections into a volume.

2.8.4.2. Two-Step Radon Inversion ***(8)***

1. Calculate the two-dimensional Radon transform of each projection.
2. Sum all two-dimensional transforms into a three-dimensional transform.
3. Invert the three-dimensional transform.

2.8.5. Projection Onto Convex Sets

Projection onto convex sets (POCS) allows the imposition of *a priori* data onto the reconstruction ***(58–60)***. The *a priori* information can be used to fill in data into the missing cone and reduce the artifacts originating from it. It is imperative that the *a priori* information be correct; otherwise artifacts could be created. One of the safest *a priori* information in three-dimensional electron microscopy is the limit in space of the three-dimensional object.

2.8.5.1. POCS by Applying a Mask in Real Space (Similar to Solvent Flattening)

1. Create a mask of the particle. Make sure the mask is sufficiently large. The *a priori* information imposed is the following: The molecular structure is confined to the inside of the mask, whereas outside the mask, no structure is present. If this mask is applied to negatively stained specimens, make sure to include the stain halo into the mask. Empty space is present only outside the stain shell. Create the mask following the same procedure used to create masks in two dimensions, by a sequence of low-pass filtrations and thresholding operations on the volume.
2. Calculate a three-dimensional Fourier transform of the reconstruction (without a low-pass filter).
3. Loop over typically 10 iterations.
4. Multiply the volume with the mask.

5. Fourier transform the masked volume.
6. Replace the data outside the missing cone with the data from the Fourier transform of the original volume.
7. Inverse Fourier transform the volume to obtain the new volume in real space.
8. End of loop.

2.8.5.2. POCS as Part of the Radon Inversion Algorithm *(11)*

Radon transforms have to comply with certain mathematical rules. Within the suite of programs for three-dimensional Radon inversion, a filter can be applied that enforces that the experimental three-dimensional Radon transform complies with these rules. The experimental transform normally does not comply because of contradictions within the data arising either from noise, the missing cone, or both. The mathematical rules can be imposed in one single step using, as a parameter, a generous estimate of the largest object dimension. This filter is available as a single command in SPARK and in a modified version of SPIDER (5.0).

2.8.6. Refinement of Projection Alignment

2.8.6.1. Two Types of Methods Exist for Refinement of the Alignment of Projections

Mathematically both are equivalent. One is the realignment by a combination of reprojection of the volume followed by two-dimensional alignments (called reprojection methods). The other is the refinement of the projection directions directly by correlation of the projection transform in the three-dimensional Radon or Fourier space. Here, we describe the refinement of projection in Radon/Fourier space ***(8,61)***.

1. Low-pass filter the volume to the measured resolution.
2. Calculate the three-dimensional Radon transform of the volume.
3. Calculate the one-dimensional Fourier transform of the three-dimensional Radon transform; apply any further high- and low-pass filters needed to optimize the reference.
4. Loop over all projections.
5. Calculate the two-dimensional Radon transform of the projection.
6. Calculate the one-dimensional Fourier transform of the two-dimensional Radon transform.
7. Correlate the two-dimensional Fourier-Radon transform of the projection with the three-dimensional Fourier-Radon transform of the volume. Input parameters for this correlation are as follows:
    a. Specify if this is a total search (i.e., over the complete angular range) or a refinement search over a limited angular range around the current position of the projection. Here, answer "subsearch" for refinement.

b. Specify the angular range for each Euler angle. The suggested range for a first refinement is ± 30° in 5° increments for all three angles.
c. Specify the range of translational refinement. In the first round, a range of five pixels is appropriate.

8. End of loop over projections.

The result of this process is a document file containing the values of the cross-correlation maximum of each projection and the corresponding angle and translational offset.

2.8.6.2. Correct the Position and Orientation of the Projections

1. Loop over all projections.
2. Shift all the original real space projections by the translational offset found in the refinement.
3. Update the header of each projection with the new Euler angles.
4. Recalculate the two-dimensional Radon transforms.
5. Add the two-dimensional Radon transforms into an empty three-dimensional Radon transform.
6. Invert the three-dimensional Radon transform to obtain the new volume.

For a first three-dimensional reconstruction from a single random conical series, one round of refinement should suffice. Additional iterations are most appropriate after several volumes have been merged.

2.8.6.3. Iteration of the Refinement Process

The refinement process should be iterated. As the iterations proceed, the angular search range and the angular increment should be reduced. The translational search can be limited to fewer pixels, because the translation should be well defined by now. Always leave one pixel translational alignment. As the alignments proceed, the low-pass filter of the reference volume should be relaxed. The best filter is to a resolution about 15% lower than the measured resolution of the volume (*see* **Note 17**).

2.8.6.4. A Variant to the Projection Alignment With Signal-to-Noise Ratio Correction

This procedure requires a well-filled three-dimensional Radon transform.

1. Calculate the three-dimensional Radon transform from a set of aligned images; do not apply either a nontomographic noise reduction filter or a Radon inversion.
2. Calculate the one-directional Fourier transform of this three-dimensional radon transform.
3. Radon transform and Fourier transform the projections.
4. If it was used in the calculation of the three-dimensional transform, subtract the Fourier/Radon transform of the current projection from the three-dimensional transform.

5. Cross-correlate projection transform to the reference transform. In this procedure, specify the signal-to-noise ratio of a single projection. The alignment program uses this value to calculate the signal-to-noise ratio of every line in the three-dimensional reference and corrects the cross-correlation value accordingly (*see* **Note 18.**)

2.9. Resolution Measurement

The most common technique for the determination of the resolution of a three-dimensional reconstruction is through the calculation of the Fourier shell correlation ***(62,63)***. A prerequisite is to calculate two independent volumes of the same structure to carry out a comparison.

2.9.1. Calculation of the Fourier Shell Correlation

1. Split the projection set into two half-sets, assigning the first projection of the series to set 1, the second to set 2, the third again to set 1, the fourth in the series to set 2, and so forth. (Because the series of projections might not be numbered continuously, a split between even- and odd-numbered projection files might not split the dataset into two equal halves.)
2. Calculate two volumes.
3. Calculate a three-dimensional mask.
4. Apply a strong low-pass filter to the mask; use a cutoff mask of four to five Fourier units. The mask must be very smooth to minimize its influence on the resolution measurement. The low-pass-filter radius must be substantially lower than the expected resolution value. It is acceptable if the smoothed mask also reaches up to the edge of the image or shows ripples originating from the low-pass filter.
5. Multiply each volume with the mask.
6. Fourier transform both volumes.
7. Calculate the Fourier shell correlation function.

2.9.2. Determination of the Resolution

The resolution value determined from the Fourier shell correlation curve depends on the criterion used. Currently, there are a large number of different criteria used simultaneously and any resolution value should include the criterion that was used in its determination.

The criteria are listed below. Criteria 1–3 are based on a comparison with a predicted noise correlation function. In this case, a theoretical function is calculated that shows the Fourier ring correlation that would be obtained from volumes containing pure noise. This curve starts at 1 and falls off with the square of the radius. It is $1/\sqrt{n}$, where n is the number of points compared within a shell (*see* **Note 19**).

Criterion 1: Intersection of volume shell correlation with twice the noise correlation function (extremely optimistic criterion, essentially shows the resolution that would be possible to obtain if sufficient data were available to calculate a noise-free reconstruction).

Criterion 2: Intersection of the volume shell correlation with three times the noise correlation function (still a quite optimistic criterion).

Criterion 3: Intersection with five times the noise correlation function This criterion has been shown in specific cases to be equivalent to the 45° phase residual criterion ***(64,65)***.

Criterion 4: Intersection with the value 0.5. This criterion is conservative for asymmetrical particles and has been shown to correspond to a signal-to-noise ratio of 1 in each of the two volumes. Thus, the signal-to-noise ratio of the complete volume should be 1.4 ($\sqrt{2}$).

If symmetry was enforced, then the criteria based on a comparison with the noise correlation functions should be used (criteria 1–3), where the noise correlation is multiplied additionally with the square root of the enforced symmetry.

2.9.3. Interpretation of the Fourier Shell Correlation Curve

The Fourier shell correlation curve not only provides the information for a specific resolution value but also contains information about the quality of the dataset. If the curve is very steep (i.e., the difference in resolution when the different criteria are applied is minimal), then the information in the data is exhausted. Higher resolution can then only be achieved from better data. If the curve is very flat, and thus the difference between the different criteria is large, then increasing the size of the dataset should lead to a higher resolution—in the best case, to the resolution predicted by twice the noise correlation criterion.

2.10. Merging of Volumes With Different Orientations

After classification of the 0° images, volumes have been calculated separately for each class. If you apply POCS, the following steps will be easier to do. First, inspect the volumes and decide if they show the same molecule viewed from different angles. Mainly check that all subunits in a multisubunit complex are present. Please note that some differences might be caused by the missing cone in a reconstruction from a conical dataset.

Before attempting to merge volumes, do the following. First, low-pass filter each volume to its resolution; second, compare the volumes and decide which ones show the same structure. Merge only those volumes that show the same structure.

The merging process proceeds in several steps ***(44)***. The volumes are first aligned relative to each other. Make sure that the angles are realistic. Volumes with small angular differences are merged first; volumes with large angular differences are merged later. If the volumes originate from a random conical tilt series with a tilt angle of, for example, 45°, then there is not sufficient overlap in the data to reliably align a volume that is tilted by a large angle relative to the reference. Applying the merging procedure as described, the missing cone will be filled in steps and will not compromise the alignment.

2.10.1. Alignment of Volumes

1. Select one volume as reference and apply low- and high-pass filters.
2. Calculate the three-dimensional Radon transform of the reference volume.
3. Calculate the one-directional Fourier transform of the three-dimensional Radon transform (Fourier/Radon transform).
4. Calculate the 0° projections of all other volumes that should be merged.
5. Calculate the two-dimensional Radon transforms of the 0° projections.
6. Calculate their one-directional Fourier transforms.
7. Align each two-dimensional Fourier-Radon transform of the 0° projections to the three-dimensional Fourier-Radon transform of the reference volume using a full-range search and a 12°–15° increment.
8. Iteratively refine the alignment using a search range of approx ±1.5 times the last increment and, accordingly, smaller search increments. A 3° increment is sufficient for the last refinement.

2.10.2. Merging

Start with the volumes whose 0° projection shows the smaller out-of-plane orientations (angle θ). Ignore all volumes whose orientation deviates by more than the tilt angle of the tilt-pair used for the random conical reconstruction. If all volumes show a larger angle than this angle, chose a new reference and repeat the alignment process. For example, if the tilt angle for data collection was 50°, merge only volumes tilted up to about 30°, to be safe.

1. Choose the projection set of the first volume that is to be merged.
2. Update the projection angles to the new orientation. Your system should have a command for doing this. In SPIDER, the command is "vo ceul" with option external. Enter the Euler angles of the projection first and the rotation angles of the volume second.
3. Calculate a reconstruction from the updated projections by themselves.
4. Check visually that the new volume now has the same orientation as the reference. If the new volume is blurred and it was not before, the updating of the angles went wrong. If it is rotated in the wrong orientation, then either the alignment did not work or the volume rotation angles were entered either in the wrong order or with the wrong sign, or both.
5. Chose the projection set of the next volume, update angles, reconstruct, and compare.
6. Do **steps 1–4** for all the volumes that are to be merged.
7. Combine the projection sets of all volumes that show the correct orientation into one series.
8. Calculate a reconstruction from this combined series.
9. Run a translational/rotational refinement of this projection series; the same as for a single set (*see* **Subheading 2.8.6.**).

10. Use the combined volume as a reference and repeat the alignment of the 0° projection from those volumes that were not included in the combined structure. If the angular range is filled now and no significant missing cone is left, then all datasets, even those with larger angles, can be merged now.
11. Always follow the merging with angular refinements.

3. Three-Dimensional Projection Alignment

If it was possible to combine all datasets, because all of them showed the same three-dimensional structure, then the reconstruction can be improved by adding data from single micrographs. In many cases, 0° micrographs might be appropriate if the particle shows a good distribution of orientations; otherwise, tilt images should be collected to obtain a sufficiently complete angular coverage.

If more than one conformation of the particle was present, then the safest continuation of the reconstruction process is adding more data from random conical tilt-pairs. Techniques for separating different conformations or different particles by comparison to three-dimensional volumes are only in their very first stages of development and are not reliable yet. (*See* **Note 20**.)

The projection alignment *(8)* described here requires a three-dimensional reference volume. This volume, however, should always be available, because reconstructions and classifications as described earlier are a prerequisite for establishing the structural homogeneity of the sample.

1. First, select and extract the series of new images from micrographs.
2. Calculate the three-dimensional Radon transform of the reference volume.
3. Calculate the one-directional Fourier transform of the reference Radon transform (three-dimensional Fourier/Radon transform).
4. Loop over all images.
5. Calculate the two-dimensional Radon transform of the image.
6. Calculate the one-directional Fourier transform of the image (two-dimensional Fourier/Radon transform).
7. Correlate the two-dimensional Fourier Radon transform of the image with the three-dimensional Fourier-Radon transform of the reference. Use a complete search in all angles, starting with an angular increment of 15°–20° and translation range of five pixels (if you picked the particles rather precisely, otherwise increase to a range of seven pixels). Check in the output data file that the cross-correlation maximum is not at the end of any search range; otherwise, increase the range.
8. Translate image.
9. Enter the new angles into the image header.
10. End of the loop over images.
11. Calculate a three-dimensional volume from the aligned images.
12. Check if the volume is correct.
13. Refine projection directions as described previously.

4. Achievement of High Resolution

The resolution of the three-dimensional reconstruction depends on the size of the dataset and on the quality of the images. With careful alignment of the microscope and under the proper imaging conditions (*see* Chap. 19), it is possible to obtain large datasets of frozen hydrated particles reaching to resolutions of 0.7–0.8 nm even when using a microscope equipped only with a LaB_6 filament. However, high resolution requires the correction of the transfer function.

Images of single particles are recorded at a defocus that allows their visual recognition in the micrograph. For 0° images, typical defoci are in the range 1.5–3.5 μm. The center of a tilt image is typically at about 1.5 μm, so that the complete micrograph is in underfocus. The values given are for an accelerating voltage of 100 kV. Visibility of single particles at higher voltages requires a larger defocus.

4.1. Correction of the Transfer Function

Images collected at 100 kV with a defocus of 1.5 μm show the first 0 of the transfer function at approximately 3 nm. To obtain high-resolution reconstructions, the phase changes of the transfer function must be corrected ***(66,67)***. Some image processing systems contain closed programs for the determination of the parameters of the transfer function; others contain elements that can be combined in a script to perform the same task. A prerequisite for the fitting procedure is a program that can calculate the power spectrum for any combination of defocus, astigmatism, and strong scattering (amplitude) contrast ***(68)***, with and without envelope function. For correction of the transfer function in tilt images, additionally a program is required that can calculate the parameters of a plane in space in any direction, usually done by a least squares fit. In addition, all other typical image processing tools (Fourier transform, correlations, mask, filter, value comparisons, etc.) must be available. Fitting and correction follows the methods described in **ref. *44***.

4.1.1. Fitting of the Transfer Function to 0° Images

4.1.1.1. Calculate the Averaged Periodogram of the Complete Micrograph

1. Select a window size (typically 256 × 256 pixels).
2. Window frames from the micrograph starting in the upper left corner and moving by half the window dimension. Continue the next row, moving down by half a window dimension. From this process, exclude areas where the micrograph is obviously "bad", like areas with large nontransparent contamination. Calculate the power spectrum of each window and average all of the power spectra. This is the averaged periodogram.
3. High-pass filter the periodogram with a filter radius corresponding to approximately 100 pixels. (This is done instead of a background subtraction). The parameter might have to be adapted for the best convergence of the procedure.

4. Mask out the center of the periodogram using a mask whose radius is approximately the distance of the first maximum of the transfer function from the origin.

4.1.1.2. Fit the Values of Defocus and Astigmatism

To begin, assume an amplitude contrast parameter of 0.1 for frozen hydrated images and 0.2 for stained images. Assume a value of 0 for the astigmatism.

Iterate over the defocus value. Select a value that is within 1 µm of the estimated defocus. (The defocus value that you were aiming for when you collected the images is a good starting value.) Perform a search in 10-nm steps, over a range that is such that the first minimum of the theoretical transfer function cannot fall into the second minimum of the experimental function and vice versa. Experimentally determine the parameters for the theoretical envelope function that matches the falloff of the transfer function in the periodogram. If unsure, use an envelope that shows less falloff than experimentally observed. Chose this parameter such that the falloff is small enough to fit to all micrographs that are being corrected.

Defocus:

1. Start a loop over defocus values.
2. Calculate the theoretical transfer function. Apply the same high-pass filter that had been applied to the periodogram.
3. Cross-correlate the periodogram and the theoretical transfer function. Use a normalized cross-correlation program (perfect fit =1, less than perfect fit <1).
4. In a table, store the iteration step, cross-correlation coefficient, the defocus value, the astigmatism value, astigmatism direction, the amplitude contrast parameter, and the micrograph number.
5. Check if the cross-correlation value is higher than any previous value. If yes, store all of the parameters, including the iteration step.
6. Continue with the next defocus value.
7. End of loop.
8. Write the parameters that resulted in the highest cross-correlation at the end of the table (document file).
9. Calculate the theoretical transfer function that belongs to the highest cross-correlation.

Astigmatism:

10. Loop over the astigmatism values; use defocus values determined previously.
11. Starting with 0 nm, go in steps of 1–10 nm up to a value slightly larger than expected.
12. Loop over the direction of the astigmatism. Steps of 10° from 0° to 170° are usually sufficient.
13. Calculate the theoretical transfer function. Apply the same high-pass filter as had been applied to the periodogram.

14. Cross-correlate the periodogram and the theoretical transfer function. Use a normalized cross-correlation program.
15. In a table, store the iteration step, cross-correlation coefficient, the defocus value, the astigmatism value, the amplitude contrast parameter, and the micrograph number.
16. Check if the cross-correlation value is higher than any previous value. If yes, store all of the parameters, including the iteration step.
17. Continue with the next direction of the astigmatism.
18. Continue with the next value of the astigmatism.
19. End of the loop over astigmatism.
20. Calculate the theoretical transfer function with the parameters that yielded the highest cross-correlation.

Refinement:

21. With these values, start a second refinement of the defocus followed by astigmatism correction. A repeat of three times is usually sufficient.

To check the performance of the fit, display the periodogram with your image display program; to its side and within the same window, show the theoretical transfer function. Display the theoretical transfer function underneath the periodogram also. If you use WEB, from the edit menu select the option bar with a length of approximately twice the periodogram dimension. Position the bar in the minima of the periodogram. This bar then also should cross the minima of the theoretical transfer function. Do the same procedure with vertical bars. If the minima in all of them coincide, the defocus and astigmatism values are correct. If the minima fit well only at low radii and there is an increasing discrepancy at higher resolutions, then the value for the amplitude contrast is incorrect. Run the same procedure again with an outer iteration over amplitude contrast values, or if you have a good guess, run the procedure a few times with different specified values. The amplitude contrast values should only need to be fitted once for the same type of specimen and can be used for all other images from the same specimen without fitting.

4.1.2. Correction of the Transfer Function in 0° Images

1. Calculate the theoretical transfer function with the fitted parameters. This function will have values between +1 and –1. Do not apply an envelope function. Modify the transfer function such that it has a value of constant –1 at radii below the first maximum and does not start at 0. Some programs have this option built in. Otherwise, you will have to mask the center of the Fourier transform and replace the circle below the first maximum with –1. This works only if the astigmatism is not too large.
2. Loop over the images of each micrograph and multiply the Fourier transform of each image with the transfer function.

This procedure is equivalent to flipping the signs in each band with a smooth function.

4.1.3. Fitting of the Transfer Function to Tilt Images

The fitting of the transfer function to tilt images follows a procedure similar to the fitting of the function to 0° images.

1. Split the micrograph into a number of quadrants, typically 3 × 4.
2. For each quadrant, determine the transfer function parameters in the same way as was done for the 0° micrographs.
3. For each quadrant, collect the values of astigmatism, astigmatism direction, defocus, and amplitude contrast together with the center coordinate of the quadrant.
4. Average the astigmatism values for all quadrants. If some values (most of the time those of the corner quadrants) deviate strongly, leave them out of the average.
5. Using the x,y-coordinates of the quadrants and the defocus as the z-coordinate, fit a plane to this set of values. Save the parameters of the equation of this plane.

4.2. Correction of the Transfer Function in the Tilt Images

1. For each single image, look up the micrograph number and the coordinates from where the particle was picked.
2. Use the particle coordinates and the equation of the plane determined in the fitting procedure to calculate the defocus at the location of this particle.
3. Using this defocus value and the other transfer function parameters, calculate the theoretical transfer function at this position in the image, again without envelope and, as above, with a value of constant –1 at radii lower that the radius of the first maximum.
4. Multiply the Fourier transform of the particle image with this modified theoretical transfer function.

The corrections can be applied either to the aligned images or to the images at any stage in the processing. It is recommended to apply the corrections either to the first windowed and contrast normalized images or to the images after the alignment. If the correction is applied at an intermediate step, there is a greater chance that the correction might get lost when some of the processing steps are repeated.

4.3. Achieving High Resolution

1. Using the corrected images, recalculate a three-dimensional reconstruction.
2. Carry out an angular and translational refinement as described in **Subheading 2.8.6.**
3. High-pass filter the last reconstruction using a value of approximately one-third of the particle diameter.
4. Low-pass filter the volume to a resolution about 15% lower than the last measured resolution.
5. Repeat the angular refinement, starting with an angular range of approx 15° and increments of about 3° and reduce the range stepwise down to 5° and an increment of 1°. Iterate until stable.

6. Check the cross-correlation coefficients. Remove those particles from the reconstruction that result in the lowest 5% of the cross-correlation. By removing the worst particles, you will observe an improvement in resolution.
7. Run one more iteration of the refinement.

5. Notes

1. Typically, one can collect about 200 particles from each tilt-pair. For a medium-resolution structure (a particle with a 30-nm diameter at a resolution of slightly better than 3 nm), 5000 particles are usually sufficient. For heterogeneous samples, the number of particles should be such that all major conformations are represented by approx 1000 images. The same number applies for a first three-dimensional inspection of a homogeneous sample. Thus, between 5 and 25 tilt-pairs suffice for a medium-resolution reconstruction.
2. The tilt image shows a larger area than the 0° image. If the match is only partial, decide if you want to process only parts of the micrographs or if you want to leave out the micrograph.
3. Check at least once your complete experimental setup, the direction of the tilt axis, the direction of the tilt angle that you call positive, and the scanning geometry. At least once, go through all of the steps and make sure that you obtain the correct handedness in the final result. This can be done in a Gedankenexperiment, but if you are not quite sure, take a well-known molecule and go through the complete reconstruction process to make sure that you obtain the correct handedness. Once you have done this, do not change any of the geometric parameters without rechecking the handedness.
4. In theory, a pixel size of 3.5 Å is sufficient for a reconstruction with a resolution of about 7 Å; in practice, because of possible interpolation errors, at best a resolution of 11–14 Å can be expected. Make sure that the data you obtain from the scanner are optical densities (maybe with a scaling factor) and that no other correction curve has been applied. Only the optical density of the film is proportional to the number of electrons recorded. If your scanner provides transparencies, convert them to optical densities. Make sure that in addition to the proper resolution, your scanner has a sufficiently large dynamic range. Quantization of densities is an additional source of noise and should be kept to a minimum. A 14- to 16-bit transparency range can be expected from most modern scanners and should suffice.
5. We have used SPIDER (version 5.0) and WEB (the non-Java version), but many other systems are available and should be able to carry out most of the calculations. The most important differences can be found in the interactive programs for particle picking, in the reconstruction algorithms — the last step of processing — and in the pattern recognition algorithms used for classification. You might consider using more than one system. We regularly use both SPIDER and XMIPP.
6. If you cannot identify five particles uniquely in both images, you can change the areas that you are comparing. In extreme cases, it is helpful to identify the first particles in photographic prints of the negatives.
7. The direction of this axis can be slightly different in the 0° image and the tilted image if the negatives were not perfectly aligned in the scanner. Thus, two angles

are determined. Because the micrographs are projections and the coordinates that are collected during particle picking lie in a plane, the direction of the tilt axis is ambiguous by 180°. Once the first direction has been determined, make sure that it is the same in all tilt-pairs, by entering an appropriate start value. Otherwise, you might obtain a mixture, and from one data portion, you would get the enantiomorphous structure to the one you will get from the other portion. They cannot be combined without correcting all of the angles.

8. The particle picking program displays information at the bottom of the display screen. During the fitting of the tilt angle, a list of points is presented that might be mismatched. For the determination, of the tilt angle, the program forms all possible triangles with all possible combinations of points in the 0° image and in the tilt image. The area of each triangle should be smaller in the tilt image than in the 0° image by the cosine of the tilt angle. If the triangle area in the tilt image is larger than in the 0° image, the program issues a warning that one of the points might be wrong. A mismatch most often occurs when the three points used for forming a triangle are positioned along a line. In this case, the area in the 0° image is close to zero and a minor inaccuracy in picking the particle position in the tilt image creates a conflict. The parameter "minimal area for tilt angle determination" protects against too many conflicts. Triangles that have a smaller area than the minimal area specified will not be used in the determination of the tilt angle. If warnings appear at the bottom of the screen, check the mismatched points. If these points seem correct, increase the value for the minimum area.
9. As more coordinates are available, the angles will become more accurate. With about 50 particles, the values are within less than 0.5° of the true values. For negatively stained specimens, the tilt angle is usually within 1° of the goniometer reading. Images from frozen hydrated specimens might have larger deviations. The algorithm for the determination of the tilt angle assumes that the 0° image is truly at 0°. If the supporting carbon is warped, this condition is not fulfilled. However, the tilt angle can be refined at a later stage in the reconstruction. For frozen hydrated specimens, deviations up to 15° from the goniometer reading are acceptable. If the difference is larger, the tilt-pair should be discarded from the random conical dataset.
10. The basic rotation of the images such that the tilt axis direction is parallel to one axis of the images eliminates the need to keep track of this direction for every image. It simplifies the alignment and makes it easier to add further data to a dataset.
11. There are a number of different image normalization procedures, all of them being approximations to the true purpose of normalization. The simple formula in **step 9** is based on the assumption that there is empty space around each particle and that the only effect that needs to be corrected for is the exposure of the micrograph. There exist normalizations based on the image variance; however, if analyzed closer, these are dependent on the resolution and on the imaging conditions, having the tendency to assign lower weights to higher-resolution images and vice versa. If the micrograph is very evenly illuminated, then instead of separate values

of the image average for each particle, one single average value for the whole micrograph can be used. For all methods that are based on Radon transforms, it is essential that the background surrounding a particle has an average of zero.

12. Because the particles were selected approximately centered, but not aligned, this creates a blurred average image. This image will be approximately centered; however, it might have an offset resulting from a systematic offset during the selection procedure.
13. The direction of the shift varies and depends on the correlation program used. In SPIDER (5.0), the direction of the shift is negative if the image is entered first and the reference image is entered as the second image in the cross-correlation program.
14. The low-pass filter is intended to reduce the effect of noise and, as a rule of thumb, should be to about 3 nm resolution. The high-pass filter should approximately correspond to the diameter of the particle.
15. The cross-correlation is carried out through a multiplication of the Fourier transforms of the image and the reference. Thus, if the reference is low-pass filtered, so will the cross-correlation function, because all Fourier components that are 0 in the reference will also be 0 in the cross-correlation function.
16. For certain reconstruction algorithms, the angle might have to be stored in document files separate from the image series.
17. Do not apply many consecutive shifts to one image. Every image shift uses an interpolation that might reduce the resolution of the image. Instead, add up all of the shifts and apply then to the starting image. Rotations of the tilt images that are used for the reconstruction are only applied as a parameter in the image header for use by the reconstruction algorithm and, therefore, multiple rotational alignments do not deteriorate the image.
18. If the reference volume is calculated from a real space volume or if the reference Radon transform has been filtered, a prediction of the signal-to-noise ratio of every single line in the transform can no longer be calculated. In this case, you can avoid the problem of the varying signal-to-noise ratio by calculating the volume from a subset of projections that are evenly distributed in the angular space. Which of the two approaches performs better might need to be tried for different situations.
19. The comparison with the noise correlation curve assumes that all points in the compared Fourier transforms are statistically independent. This, in general, is not the case. The Fourier transform is convoluted with the shape function of the molecule, thus creating dependencies. A simple example might illustrate the problem. When the shell thickness is increased from one to two pixels, then the corresponding noise correlation curve is lowered by a factor of 2 because now, at the same Fourier radius, four times as many points are being compared. The Fourier shell correlation function of the volumes, however, does not change significantly.
20. There are a few techniques that can be considered for separating different particles. If the volumes obtained for different classes of particles show a good angular coverage when reconstructed from 0° projections only, then it might be

possible to just collect additional 0° images, coclassify them with the original dataset, and separate them into the same classes. This could best be achieved by using correspondence analysis and coprojecting the new images as inactive images within the old classification. However, coprojection, even though it is one of the great advantages of correspondence analysis vs other techniques, is not implemented in all systems. The technique should be available in SUPRIM and, possibly, in SPARK.

References

1. Frank, J. (1996) *Three-Dimensional Electron Microscopy of Macromolecular Assemblies*, Academic, San Diego, CA.
2. Hoppe, W., Langer, R., Knech, G., and Poppe, C. (1968) Proteinkristallstrukturanalyse mit Elektronenstrahlen. *Naturwissenschaften* **55,** 333.
3. DeRosier, D. J. and Klug, A. (1968) Reconstruction of three dimensional structures from electron micrographs. *Nature* **217,** 130.
4. Crowther, R. A., DeRosier, D. J., and Klug, A. (1970) The reconstruction of a three-dimensional structure from projections and its application to electron microscopy. *Proc. R. Soc. London Ser.* **A 317,** 319–340.
5. Radon, J. (1917) Über die Bestimmung von Funktionen durch ihre Integralwerte längs gewisser Mannigfaltigkeiten. *Ber. Verh. König. Sächs. Ges. Wiss. Leipzig, Math. Phys. K1.* **69,** 262–267.
6. Deans, S. R. (1983) *The Radon Transform and Some of Its Applications*. Wiley, New York.
7. van Heel, M. (1987) Angular reconstitution: a posteriori assignment of projection directions for 3D reconstruction. *Ultramicroscopy* **21,** 111–124.
8. Radermacher, M. (1994) Three-dimensional reconstruction from random projections: orientational alignment via Radon transforms. *Ultramicroscopy* **53,** 21–36.
9. Serysheva, I., Orlova, E., Chiu, W., Sherman, M., Hamilton, S., and van Heel, M. (1995) Electron cryomicroscopy and angular reconstitution used to visualize the skeletal muscle calcium release channel. *Nature Struct. Biol.* **2,** 18–24.
10. Radermacher, M. (1997) Radon transform techniques for alignment and 3D reconstruction from random projections. *Scan. Microsc.* **11,** 171–177.
11. Lanzavecchia, S., Bellon, P. L., and Radermacher, M. (1999) Fast and accurate three-dimensional reconstruction from projections with random orientations via Radon transforms. *J. Struct. Biol.* **128,** 152–164.
12. Hoppe, W., Schramm, H. J., Sturm, M., Hunsmann, N., and Gaßmann, J. (1976) Three-dimensional electron microscopy of individual biological objects. I. Methods. *Z. Naturforsch.* **31a**, 645–655.
13. Hoppe, W., Schramm, H. J., Sturm, M., Hunsmann, N., and Gaßmann, J. (1976) Three-dimensional electron microscopy of individual biological objects. II. Test calculations. *Z. Naturforsch.* **31a**, 1370.
14. Hoppe, W., Schramm, H. J., Sturm, M., Hunsmann, N., and Gaßmann, J. (1976) Three-dimensional electron microscopy of individual biological objects. III. Experimental results on yeast fatty acid synthetase. *Z. Naturforsch.* **31a**.
15. Frank, J. (ed.) (1992) *Electron Tomography,* Plenum, New York.

16. Radermacher, M. (1988) Three-dimensional reconstruction of single particles from random and nonrandom tilt series. *J. Electron. Microsc. Tech.* **9,** 359–394.
17. Radermacher, M., Wagenknecht, T., Verschoor, A., and Frank, J. (1986) A new 3-dimensional reconstruction scheme applied to the 50s ribosomal subunit of *E. coli. J. Microsc.* **141,** Rp1–Rp2.
18. Radermacher, M., Wagenknecht, T., Verschoor, A., and Frank, J. (1987) Three-dimensional reconstruction from a single-exposure, random conical tilt series applied to the 50S ribosomal subunit of Escherichia coli. *J. Microsc.* **146,** 113–136.
19. Goncharov, A. B. (1987) Methods of integral geometry and determination of mutual orientations of arbitrarily arranged in the plane identical particles from their projections onto the line. *Dokl. Acad. Sci. USSR* **293,** 355–358 (in Russian).
20. Goncharov, A. B. and Gelfand, M. S. (1988) Determination of mutual orientation of identical particles from their projections by the moments method. *Ultramicroscopy* **25,** 317.
21. New, P. F., Scott, W. R., Schnur, J. A., Davis, K. R., and Taveras, J. M. (1974) Computerized axial tomography with the EMI scanner. *Radiology* **110,** 109–123.
22. Edholm, P. (1960). The tomogram: its formation and content. *Acta Radiol.* **193(Suppl.),** 4–109.
23. Cormack, A. M. (1964) Representation of a function by its line integrals, with some radiological applications II. *J. Appl. Phys.* **35,** 2908–2913.
24. Cormack, A. M. (1963) Representation of a function by its line integrals, with some radiological applications. *J. Appl. Phys.* **34,** 2722–2727.
25. Gilbert, P. (1972) Iterative methods for the three-dimensional reconstruction of an object from projections. *J. Theor. Biol.* **36,** 105–107.
26. Lewitt, R. M. and Bates, R. H. T. (1978) Image reconstruction from projections: I. Theoretical considerations. *Optik* **50,** 19–33.
27. Lewitt, R. M., Bates, R. H. T., and Peters, T. M. (1978) Image reconstruction from projections: II: Modified back-projection methods. *Optik* **50,** 85–109.
28. Lewitt, R. M. and Bates, R. H. T. (1978) Image reconstruction from projections: III: Projection completion methods (theory). *Optik* **50,** 189–204.
29. Lewitt, R. M. and Bates, R. H. T. (1978) Image reconstruction from projections: IV: Projection completion methods (computational examples). *Optik* **50,** 269–278.
30. Bates, R. H. T. and Heffernan, P. B. (1980) Image reconstruction from projections V: Blurring due to object movement. *Optik* **56,** 101–112.
31. Hefferman, P. B. and Bates, R. H. T. (1982) Image reconstruction from projections. VI: Comparison of interpolation methods. *Optik* **60,** 129–142.
32. Garden, K. L. and Bates, R. H. T. (1984) Image reconstruction from projections VII: Interactive reconstruction of piecewise constant images from few projections. *Optik* **68,** 161–172.
33. Cormack, A. M. and Koehler, A. M. (1976) Quantitative proton tomography: preliminary experiments. *Phys. Med. Biol.* **21,** 560–569.
34. Cormack, A. M., Koehler, A. M., Brooks, R. A., and Dichiro, G. (1977) Proton tomography. *J. Computer-Assist. Tomogr.* **1,** 265–265.

35. Cormack, A. M. and Laureate, N. (1980) Physics of computerized-tomography. *Med. Phys.* **7,** 445–445.
36. Cormack, A. M. (1980) Early Two-dimensional reconstruction (CT scanning) and recent topics stemming from It—Nobel Lecture, December 8, 1979. *J. Computer Assist. Tomogr.* **4,** 658–664.
37. Cormack, A. M. (1992) 75 Years of Radon-transform. *J. Computer. Assist. Tomogr.* **16,** 673–673.
38. Budinger, T. F., Gullberg, G. T., and Huesman, R. H. (1979) Emission computed tomography, in *Image Reconstruction from Projections* (Herman G. T., ed.),Topics in Applied Physics Vol. 32, Springer-Verlag, Berlin, p. 147–246.
39. Colsher, J. G. (1980) Fully Three-dimensional positron emission tomography. *Phys. Med. Biol.* **25,** 103–115.
40. Colsher, J. G. (1977) Iterative three-diminsional image reconstruction from tomographic projections. *Comput. Graphics Image Process.* **6,** 513–537.
41. Penczek, P., Marko, M., Buttle, K., and Frank, J. (1995) Double-tilt electron tomography. *Ultramicroscopy* **60,** 393–410.
42. Radermacher, M. and Hoppe, W. (1978) 3-D reconstruction from conically tilted projections. Proceedings of the 9th International Congress on Electron Microscopy, Vol. 1, pp. 218–219.
43. Radermacher, M. and Hoppe, W. (1980) Properties of 3-D reconstruction from projections by conical tilting compared to single axis tilting compared to single axis tilting. Proceedings of the 7th European Congress on Electron Microscopy, Vol. 1, pp. 132–133.
44. Radermacher, M., Ruiz, T., Wieczorek, H., and Grüber, G. (2001) The structure of the V1-ATPase determined by three-dimensional electron microscopy of single particles. *J. Struct. Biol.* **135,** 26–37.
45. Frank, J., Radermacher, M., Penzcek, P., et al. (1996) SPIDER and WEB: Processing and visualization of images in 3D electron microscopy and related fields. *J. Struct. Biol.* **116,** 190–199.
46. Langer, R., Frank, J., Feltynowsky, A., and Hoppe, W. (1970) Anwendung des Bilddifferenzverfahrens auf die Untersuchung von Strukturänderungen dünner Kohlefolien bei Elektronenbestrahlung. *Ber. Bunsenges. Phys. Chem.* **74,** 1120–1126.
47. Frank, J., Goldfarb, W., Eisenberg, D., and Baker, T. S. (1978) Reconstruction of glutamine synthetase using computer averaging. *Ultramicroscopy* **3,** 283–290.
48. Steinkilberg, M. and Schramm, H. J. (1980) Eine verbesserte Drehkorrelationsmethode für die Strukturbestimmung biologischer Makromoleküle durch Mittelung elektronenmikroskopischer Bilder. Hoppe-Seylers Zeitschrift für *Physiol. Chem.* **361,** 1363–1369.
49. Radermacher, M. (2001) Appendix: simultaneous rotational and translational alignment. *J. Struct. Biol.* **135,** 35–37.
50. Marco, S., Chagoyen, M., de la Fraga, L. G., Carazo, J. M., and Carrascosa, J. L. (1996) A variant to the random approximation of the reference-free alignment algorithm. *Ultramicroscopy* **66,** 5–10.

51. Marabini, R. and Carazo, J. M. (1994) Pattern-recognition and classification of images of biological macromolecules using artificial neural networks. *Biophys. J.* **66,** 1804–1814.
52. van Heel, M. and Frank, J. (1981) Use of multivariate statistics in analysing the images of biological macromolecules. *Ultramicroscopy* **6,** 187–194.
53. Frank, J. and van Heel, M. (1982) Correspondence analysis of aligned images of biological particles. *J. Mol. Biol.* **161,** 134–137.
54. Diday, E. (1971) La methode de nuees dynamiques. *Rev. Statist. Appli.* **19,** 19–34.
55. Guckenberger, R. (1982) Determination of a common origin in the micrographs of tilt series in three-dimensional electron microscopy. *Ultramicroscopy* **9,** 167–174.
56. Gilbert, P. F. (1972) The reconstruction of a three-dimensional structure from projections and its application to electron microscopy. II. Direct methods. *Proc. R. Soc. London B: Biol. Sci.* **182,** 89–102.
57. Radermacher, M. (1992) Weighted backprojection methods, in Electron tomography (Frank, J., ed.), Plenum, New York, pp. 91–116.
58. Levi, A. and Stark, H. (1983) Signal restoration from phase by projections onto convex sets. *J. Opt. Soc. Am.* **73,** 810–822.
59. Levi, A. and Stark, H. (1984) Image restoration by the method of generalized projections with application to restoration from magnitude. *J. Opt. Soc. Am.* **A1,** 932–943.
60. Carazo, J. M. and Carrascosa, J. L. (1987) Restoration of direct Fourier three-dimensional reconstructions of crystalline specimens by the method of convex projections. *J. Microsc.* **145,** 159–177.
61. Ruiz, T., Kopperschläger, G., and Radermacher, M. (2001) The first three-dimensional structure of phosphofructokinase from *Saccharomyces cerevisiae* determined by electron microscopy of single particles. *J. Struct. Biol.* **136,** 167–180.
62. Saxton, W. O. and Baumeister, W. (1982) The correlation averaging of a regularly arranged bacterial cell envelope protein. *J. Micros.* **127,** 127–138.
63. van Heel, M., Keegstra, W., Schutter, W., and van Bruggen, E. J. F. (1982) Arthropod hemocyanin structures studied by image analysis, in *Life Chemistry Reports, Suppl. 1, The Structure and Function of Invertebrate Respiratory Proteins*. Embo Workshop, Leeds (Wood, E. J., ed.), pp. 69–73.
64. Crowther, R. A. (1971) Procedures for three-dimensional reconstruction of spherical viruses by Fourier synthesis from electron micrographs. *Philos. Trans. R. Soc. London* **261,** 221–230.
65. Frank, J., Verschoor, A., and Boublik, M. (1981) Computer averaging of electron micrographs of 40S ribosomal subunits. *Science* **214,** 1353–1355.
66. Thon, F. (1966) Zur Defukussierungsabhängigkeit des Phasenkontrastes bei der elektronenmikroskopischen Abbildung. *Z. Naturforsch.* **21a,** 476–478.
67. Frank, J., Bussler, P., Langer, R., and Hoppe, W. (1970) Einige Erfahrungen mit der rechnerischen Analyse und Synthese von Elektronenmikroskopischen Bildern hoher Auflösung. *Bericht. Bunsen-Gesellsch. Phys. Chem.* **74,** 1105–11115.
68. Typke, D. and Radermacher, M. (1982) Determination of the phase of complex atomic scattering amplitudes from light optical diffractograms of electron microscope images. *Ultramicroscopy* **9,** 131–138.

21

A New Microbiopsy System Enables Rapid Preparation of Tissue for High-Pressure Freezing

Dimitri Vanhecke, Peter Eggli, Werner Graber, and Daniel Studer

Summary

A microbiopsy system was developed to overcome long sampling times for tissues before they are cryo-fixed by high-pressure freezing. A commercially available biopsy gun was adapted to the needs of small-organ excisions, and biopsy needles were modified to allow small samples (0.6 mm × 1.2 mm × 0.3 mm) to be taken. Specimen platelets with a central slot of the same dimensions as the biopsy are used. A self-made transfer device (in the meantime optimized by Leica-Microsystems [Vienna, Austria]) coordinates the transfer of the excised sample from the biopsy needle into the platelet slot and the subsequent loading in a specimen holder, which is then introduced into a high-pressure freezer (Leica EM PACT; Leica Microsystems, Vienna, Austria). Thirty seconds preparation time is needed from excision until high-pressure freezing. Brain, liver, kidney and muscle excisions of anesthetised rats are shown to be well frozen.

Key Words: Cryo-fixation; electron microscopy; fast sampling; freeze substitution; high-pressure freezing; microbiopsy system; rat tissue

1. Introduction

Chemical fixatives (e.g., aldehydes and osmium tetroxide) have negative effects on the native structure and antigenicity of biological specimens. They withdraw osmotic gradients during fixation and, therefore, affect ultrastructure *(1,2)*. The search for alternative fixation methods started soon after the commercialization of electron microscopy *(3,4)*. Cryo-fixation, or physical fixation, turned out to be the best alternative, and numerous cryofixation techniques were developed based on vitrification of the specimen (reviewed in **ref. 5**). Using ambient or high-pressure cryotechniques, the fixation occurs within milliseconds (usually 0.1–100 ms, depending on the size of the sample), which is much faster than chemical fixation. The major drawback of ambient cryo-fixation methods is the small fixation depth; under optimal physical conditions at ambient pressure,

From: *Methods in Molecular Biology, vol. 319: Cell Imaging Techniques: Methods and Protocols*
Edited by: D. J. Taatjes and B. T. Mossman © Humana Press Inc., Totowa, NJ

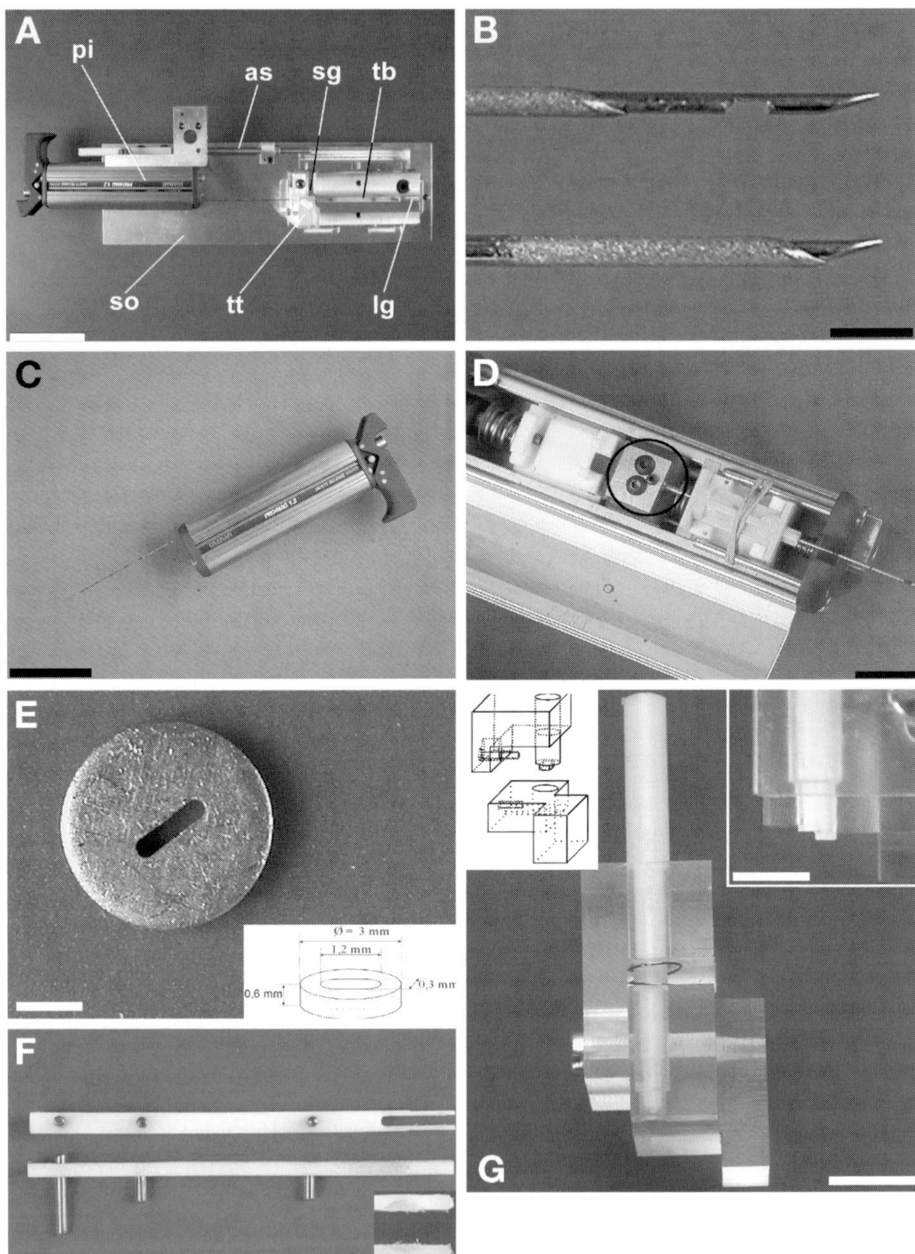

Fig. 1. Overview of the microbiopsy tools involved. **(A)** Overview of the microbiopsy transfer station. One can see the raised part, called the "tissue transfer bench" (tb), with a long axis groove (lg), and the transfer tool (tt) mounted in the short axis groove (sg). The remainder of the microbiopsy transfer station is called the "socket" (so) and provides

the fixation depth rarely exceeds 20 μm *(5–8)*. High-pressure freezing at about 2000 bar, however, allows a 10-fold increase of vitrification depths *(7,8)*.

The fast fixation performed by cryo-fixation is only profitable when post-mortem autolytic processes, which start immediately after cell death, are kept to a minimum. Therefore, the time between tissue excision and cryo-fixation is crucial and should be as short as possible *(9)*. For this reason, a microbiopsy system for the Leica EM PACT high-pressure freezing machine *(10)* was developed. The entire procedure from excision until freezing lasts about 30 s. The microbiopsy system presented herein is available from Leica Microsystems (Vienna, Austria).

2. Materials

1. Microbiopsy system (*see* **Fig. 1A**).
2. Biopsy needles (Manan, Northbrook, IL) (*see* **Fig. 1B**).
3. Biopsy gun (Pro-mag 1.2 Manan biopsy gun, Manan, Northbrook, IL) (*see* **Fig. 1C,D**).
4. Biopsy platelets (*see* **Fig. 1E**).
5. Platelet transporter (*see* **Fig. 1F**).
6. Tissue transfer tool (*see* **Fig. 1G**).
7. Anesthetised rat.
8. Leica EM PACT high-pressure freezing machine (Leica Microsystems, Vienna, Austria).
9. Styrofoam box (10 cm × 5 cm × 5 cm); sb in **Fig. 2A**.
10. Transport container (1-L cryogenic Dewar); cd in **Fig. 2A**.
11. Torque wrench (a torque force of 35 cNm is required) (*see* **Fig. 2A**).
12. Pod of flat specimen holder system of Leica EM PACT high-pressure freezing machine (Leica Microsystems, Vienna, Austria); pd in **Fig. 2A–D**.

Fig. 1. *(Continued)* a location for the biopsy pistol (pi), which is aligned for biopsy transfer with a screw (as). (Bar=5 cm.) **(B)** The relative positions of the inner and the outer needles before tissue excision (top picture) and after outer-needle release (bottom picture). (Bar=2 mm.) **(C)** The Pro-mag 1.2 Manan biopsy gun. (Bar=5 cm.) **(D)** Detail of the inner mechanism of the modified Pro-mag 1.2 biopsy gun. In the circle, the modification is shown: the inner needle is mounted in a fixed position. (Bar=2 cm.) **(E)** Depiction of the biopsy platelet, with indications of its dimensions (schematic). (Bar=1 mm.) **(F)** The platelet transporter depicted as seen when mounted on the tissue transfer bench (top) and in side view (bottom), revealing the small rod at the rear end used for the perfect positioning of the transporter. A detail of the platelet clamp is shown in the inset. (Bar=1 cm; inset bar=0.25 cm.) **(G)** The tissue transfer tool shows the ridge at the rod ending (right inset). This ridge is needed to transfer the tissue. An assembly scheme of the tissue transfer tool is depicted (left inset), providing an easier three dimensional view. (Bar=1 cm.)

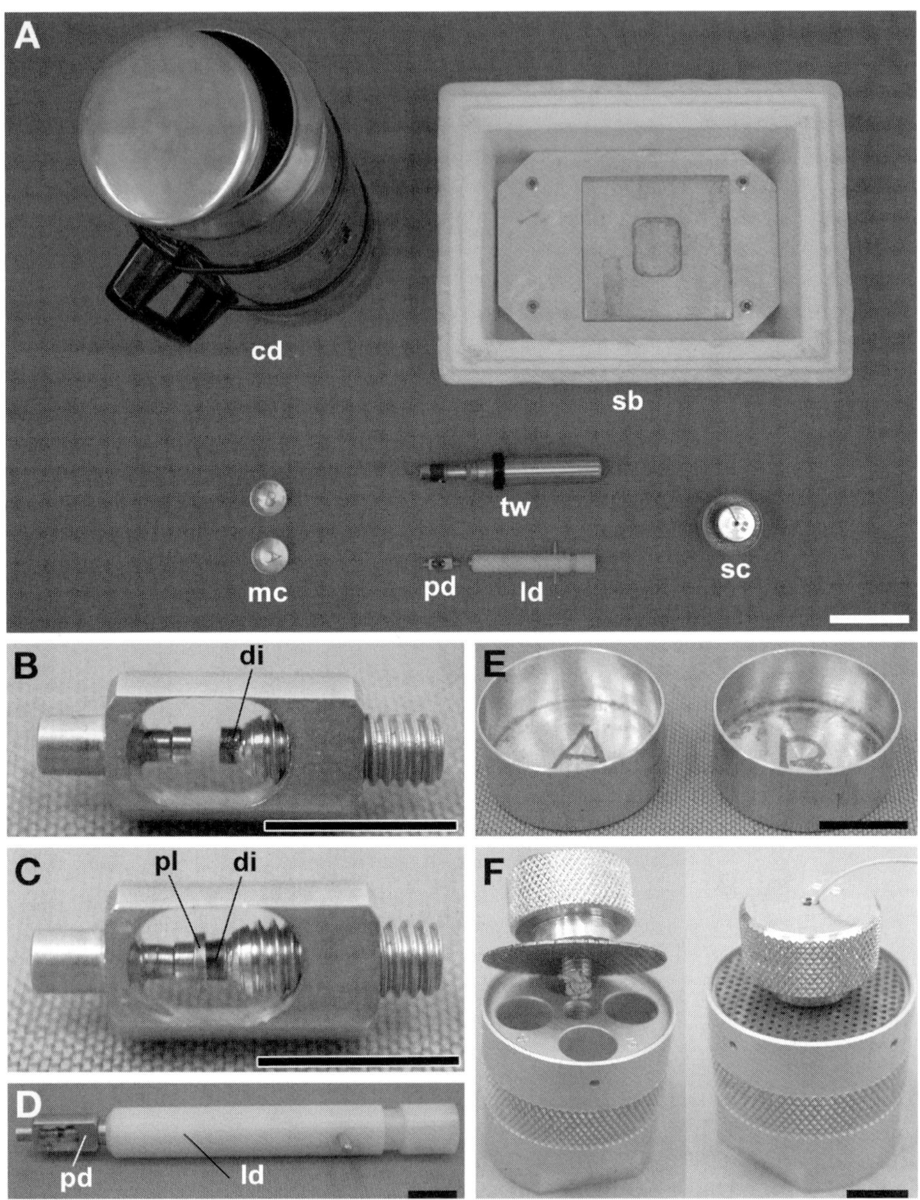

Fig. 2. Overview of the cryo-tools involved. **(A)** High-pressure frozen samples are transferred from the Leica EM PACT high-pressure freezing machine in small metal containers (mc) to a styrofoam box (sb), both filled with liquid nitrogen. Submersed in the liquid nitrogen, the platelets are transferred from the metal containers to a homemade storage container (sc). The storage containers can be held temporarily in a cryogenic Dewar (cd) for a few hours before they are transferred to liquid nitrogen storage or,

13. Numbered small metal containers (diameter: 2 cm; height: 0.5 cm); mc in **Fig. 2E**.
14. Homemade storage containers (ϕ = 2.5 cm, *h* = 5 cm) for holding samples in liquid nitrogen; sc in **Fig. 2A,F**.
15. Approximately 20 L of liquid nitrogen (LN_2).
16. Forceps.
17. Stereomicroscope.
18. Rat anestheticum mixture: 1 part Prequillan (10 mg/mL acepromazinum [Fatro S.p.A., Ozzano Emilia, Bologna, Italy]), 4 parts 0.9% NaCl, 5 parts Xylapan (20 mg/mL xylazinum), and 10 parts Narketan (100 mg/mL Ketaminum; both from Chassot AG, Belp, Switzerland).
19. Fluothane, 0.5 mL/L volume (Astrazeneca, Wedel, Germany)

3. Methods

3.1. Anesthesia of the Rats (see *Note 1*)

The rats were relaxed by enclosing the cage with a plastic bag containing two paper tissues soaked with Fluothane for 5 min. Deep anesthesia was achieved by injection of 120–140 μL of anesthetic mixture per 100 g body weight.

3.2. Preparation of the Microbiopsy Transfer Station

The platelet transporter (*see* **Fig. 1F**) controls the perfect positioning of the platelet (*see* **Note 2**) in the transfer bench. The transporter can be positioned at three different locations on the transfer bench (**Fig. 3A**, positions A, B, C) and holds the biopsy platelet in a clamp.

1. Install the platelet transporter in the groove of the long axis (lg in **Fig. 1A**) of the tissue transfer bench.
2. Place the platelet transporter (*see* **Fig. 1F**) in position A (most retracted position) (*see* **Fig. 3A**).
3. Load a biopsy platelet in the clamp of the platelet transporter (*see* **Fig. 3**).
4. The slot of the biopsy platelet must be parallel with the direction of movement of the platelet transporter.

Fig. 2. *(Continued)* alternatively, samples can continue immediately with subsequent treatments. The pod (pd) of the flat specimen holder system of the Leica EM PACT high-pressure freezing machine with a loading device (ld) and a torque wrench (tw) are depicted. **(B)** The pod of the flat specimen holder system of the Leica EM PACT is shown. **(C)** The diamond (di) seals the biopsy platelet (pl). **(D)** Pod (pd) fixed to the loading device (ld) is shown. **(E)** Small metal containers are used to transfer platelets from the liquid nitrogen bath of the Leica EM PACT to a liquid-nitrogen-filled styrofoam box. **(F)** Open and closed homemade storage containers. (Bar in A = 5 cm; bars in B–F = 1 cm.)

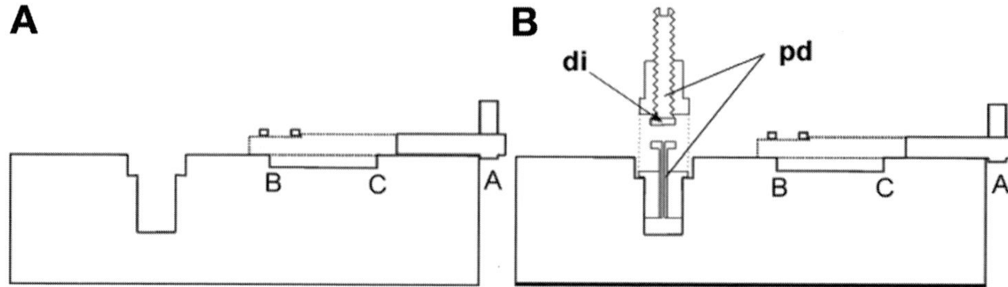

Fig. 3. Schematic representations of the tissue transfer bench before transfer. **(A)** A side-view scheme of the tissue transfer bench, with the platelet transporter in the fully retracted position (position A) and the biopsy platelet in the clamp. **(B)** A side-view scheme of the Leica EM PACT pod (pd), with indication of the diamond (di).

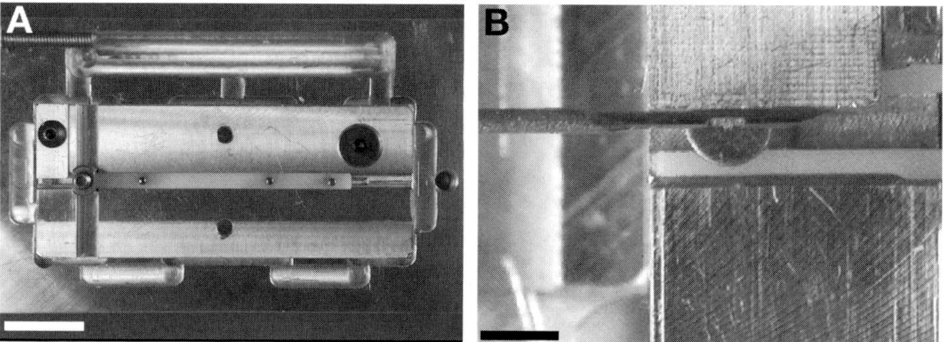

Fig. 4. Schematic representations of the tissue transfer bench before transfer. **(A)** The tissue transfer bench with the platelet transporter in position B is shown. (Bar = 2 cm.) **(B)** A detail of the tissue transfer bench, revealing the platelet transporter clamp with platelet. The perfect match between the needle notch (still empty) and the empty platelet slot is obvious (*see also* **Fig. 5B**). (Bar = 0.25 cm.)

5. Install the pod of flat specimen holder system (*see* **Note 3** and **Fig. 3B**) in the central opening (where the two grooves cross).
6. Open the screw of the pod (*see* **Fig. 3B**).
7. Move the platelet transporter, with the platelet in its clamp, in the direction of the pod, to position B. The biopsy platelet and the two arm rests of the clamp go through the pod (*see* **Fig. 4A**).
8. Insert the biopsy gun (*see* **Note 4**), without excised tissue, in the socket to check if the needle notch is perfectly aligned above the slot of the biopsy platelet (*see* **Fig. 4B**). If this is not the case, adjust the position of the gun using the long screw (as in **Fig. 1A**) by the side of the socket (also in **Fig. 1A**). This needs to be done only once per needle.

With the biopsy platelet installed properly, the gun accurately aligned in the socket, and the platelet transporter in position B, the notch of the biopsy needle appears to be exactly above the biopsy platelet slot (*see* **Fig. 4B**). Excised tissue can now be transferred from the needle notch (*see* **Note 5**) to the biopsy platelet.

3.3. Excision of the Biopsy

It is critical for tissue preservation that from the moment the sample is excised, all steps should be executed as quickly as possible.

1. Retract the outer tube of the gun by pulling on the lever at the back of the gun.
2. Position the needle notch at the place of interest in the tissue by inserting the needle in the organ of the anesthetised rat. The tissue fills the needle notch.
3. Release the cutting tube by pushing the button at the rear of the gun.
4. The tissue in the notch is excised.
5. Quickly remove the biopsy needle from the tissue and retract the outer needle once more by pulling the lever. The needle notch, with its content, is now exposed.
6. Quickly dip the needle tip in 1-hexadecene. This will avoid drying of the sample.
7. Place the biopsy gun in the socket of the microbiopsy transfer station.

3.4. Transfer of Tissue

The transfer of the tissue from the needle notch to the platelet slot is carried out with the help of the tissue transfer tool (*see* **Note 6**). The use of a stereoscopic microscope during this step makes the transfer very easy.

1. Install the tissue transfer tool in the short axis groove (sg in **Fig. 1A**) of the tissue transfer bench. This tool is properly installed when the small ridge fixes the gun needle against the bench (*see* **Fig. 5A**). The tissue transfer tool is then positioned precisely above the tissue in the needle notch (*see* **Fig. 5A,B**).
2. Press down the rod of the tissue transfer tool. The piece of tissue is transferred through the biopsy needle notch into the slot of the biopsy platelet (*see* **Fig. 5C**).
3. Remove the tissue transfer tool by sliding it into the groove. To and fro movements should be avoided because there is a risk of withdrawing the tissue piece from the platelet slot.
4. The biopsy platelet is now filled with tissue (*see* **Fig. 6A**).
5. Place the platelet transporter in position C. The platelet is located exactly under the diamond of the pod (*see* **Figs. 5D** and **6B**).
6. Close the pod screw with a torque wrench (*see* **Figs. 2A** and **7A**). Apply a force of 35 cNm.
7. Remove the platelet transporter; the clamp will detach easily.
8. Finally, fix the pod to the loading device (ld in **Fig. 2A,D**) of the Leica EM PACT high-pressure freezing machine (*see* **Fig. 7B**).

From now on, all further steps involve methods explained elsewhere. Only brief descriptions will be given.

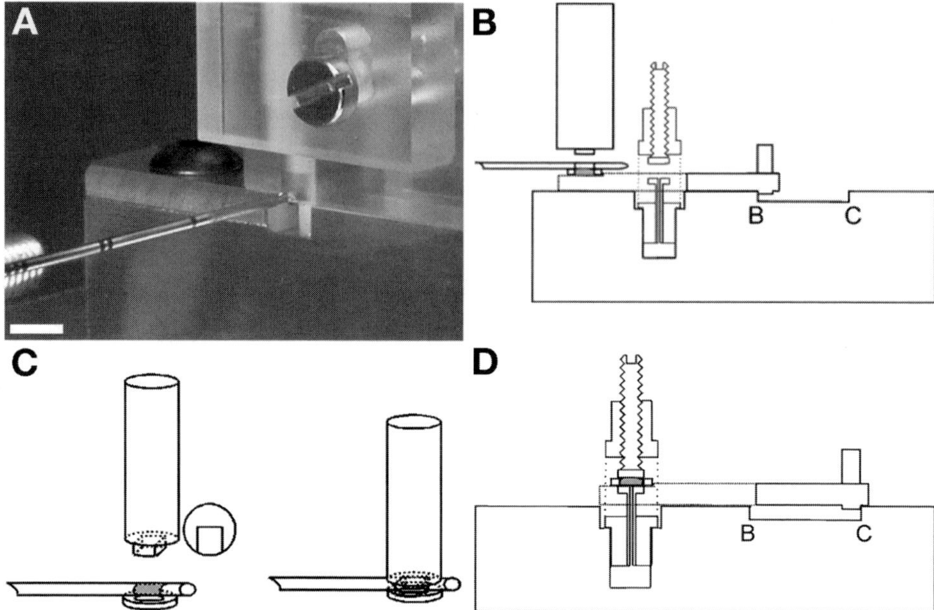

Fig. 5. **(A)** A detail of the tissue transfer bench, just prior to the tissue transfer. The tissue transfer tool is correctly positioned on the tissue transfer bench. The needle is pushed against the ridge. (Bar = 0.5 cm.) **(B)** A side-view scheme of the tissue transfer bench after the sample was transferred from the biopsy needle notch into the slot of the biopsy platelet. For simplicity, only the rod is depicted. **(C)** A schematic representation of the function of the tissue transfer tool. **(D)** A side-view schematic of the tissue transfer bench with the platelet transporter in position C. The biopsy platelet is now correctly aligned and fixed into the pod of the Leica EM PACT.

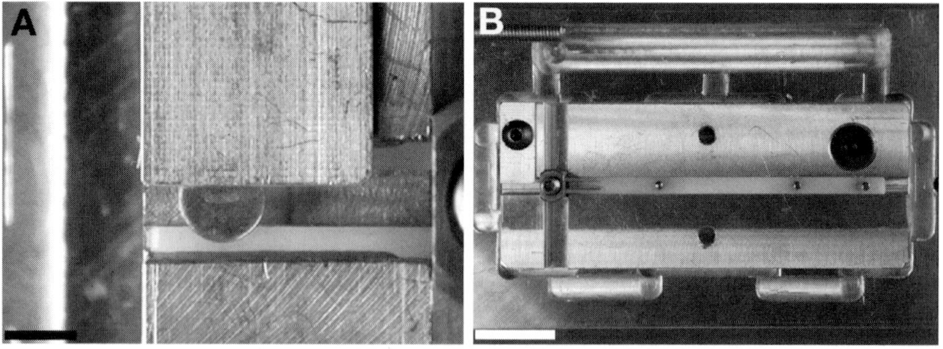

Fig. 6. **(A)** Muscle tissue in the platelet slot after the transfer. (Bar = 0.25 cm.) **(B)** The transporter in position C, ready to be fixed into the Leica EM PACT flat holder pod. (Bar = 2 cm.)

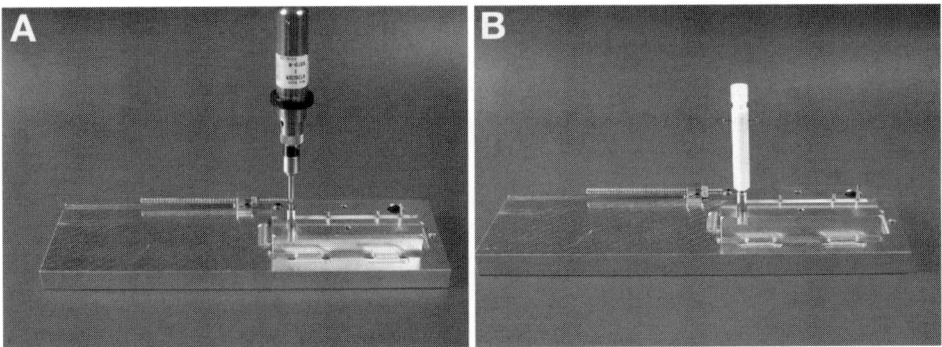

Fig. 7. **(A)** Closing the pod with the torque screw (torque: 35 cNm). **(B)** Fixing the pod into the loading device (L) of the Leica EM PACT.

3.5. Cryo-Fixation in the Leica EM PACT High-Pressure Freezing Machine

Once the tissue is cryo-fixed by high-pressure freezing, handle the tissue with care when using forceps (*see* **Note 7**). The manipulations are described in detail in **ref. *10***.

1. Rotate the pod in the loading device until the flat surfaces of the pod are perpendicular to the metal rods of the loading device.
2. Position the loading device in the Leica EM PACT loading arm.
3. Close the arm manually.
4. Push the "LOCK" button on the touch screen of the "preparation" menu.
5. Push the "START" button to freeze.
6. The sample is high-pressure-frozen and stored in the LN_2 bath.
7. Retract the platelets from the pod by loosening of the pod screw and collect them in the small containers (diameter: 1 cm; height: 0.5 cm).

The small containers (*see* **Fig. 2A,E**) are used to transport the samples from the LN_2 bath of the Leica EM PACT freezing machine to a styrofoam box (10 cm × 5 cm × 5 cm; *see* **Fig. 2A**), which is filled with LN_2. In the styrofoam box, the samples are transferred from the small containers to a closable storage container (*see* **Fig. 2A,F**). A small cryogenic Dewar (*see* **Fig. 2A**) is used to transport the containers to a large LN_2 storage system.

3.6. Follow-Up Methods

3.6.1. Freeze-Substitution

Freeze-substitution was carried out in an AFS freeze-substitution system (Leica Microsystems, Vienna, Austria) using Eppendorf tubes filled with the substitution medium. To avoid poor substitution because of 1-hexadecene (acetone does not dissolve 1-hexadecene at low temperatures), samples surrounded with it were cleaned by softly scratching the samples with forceps in liquid nitrogen.

A substitution medium of 2% osmium tetroxide in acetone (calcium chloride dried) was applied ***(10)*** for 26 h at 183 K, 8 h at 213 K, and 8 h at 243 K. Temperature rises of 2°C/h were used, resulting in two temperature slopes of 15 h. Afterward, the biopsy platelets were removed and the samples put on ice (273 K) in dry acetone for 1 h. The removal of biopsy platelets was carried out by holding the platelet with forceps and pushing the sample out of the slot with a piston of appropriate size.

3.6.2. Embedding

Samples were embedded in Epon 812 stepwise (12 h in 30% Epon in acetone at 277 K, 12 h in 70% Epon in acetone at 277 K, and 12 h in 100% Epon at room temperature). Embedding was completed by polymerization for 3 d at 333 K using fresh resin.

3.6.3. Electron Microscope Preparation

The Epon blocks were sectioned with diamond knives (Diatome, Biel, Switzerland) using an ultramicrotome (UCT; Leica Microsystems, Vienna, Austria). Ultrathin (50 nm) sections were collected on 150-mesh grids and contrasted with uranyl acetate and lead citrate.

3.7. Results

Brain, muscle, liver, and kidney biopsies were taken from anesthetised rats. Untrained, but well-informed, persons were able to perform all steps (from excising until the tissue is cryo-fixed by high-pressure freezing) in a time period of about 30 s.

The samples were well frozen, and structural preservation was excellent, as shown in **Figs. 8–13**.

3.8. Discussion

A newly developed microbiopsy system allows very quick excision of soft animal tissue for high-pressure freezing. Moreover, it allows excision of small organs (e.g., rat kidney). It was pointed out ***(9)*** for mouse brain tissue that sampling times of 30 s result in a different morphological preservation compared to longer sampling times, reflecting artifact formation during longer preparations. The system presented allows cryo-fixation of animal tissues within 30-s time periods. The 30 s from excision to cryo-immobilization are still a long time period for certain samples (e.g., brain). However, to date, it is the best approach for preparing bulk tissue samples for cryo-fixation by high-pressure freezing. The good quality of the micrographs is proof that the system works.

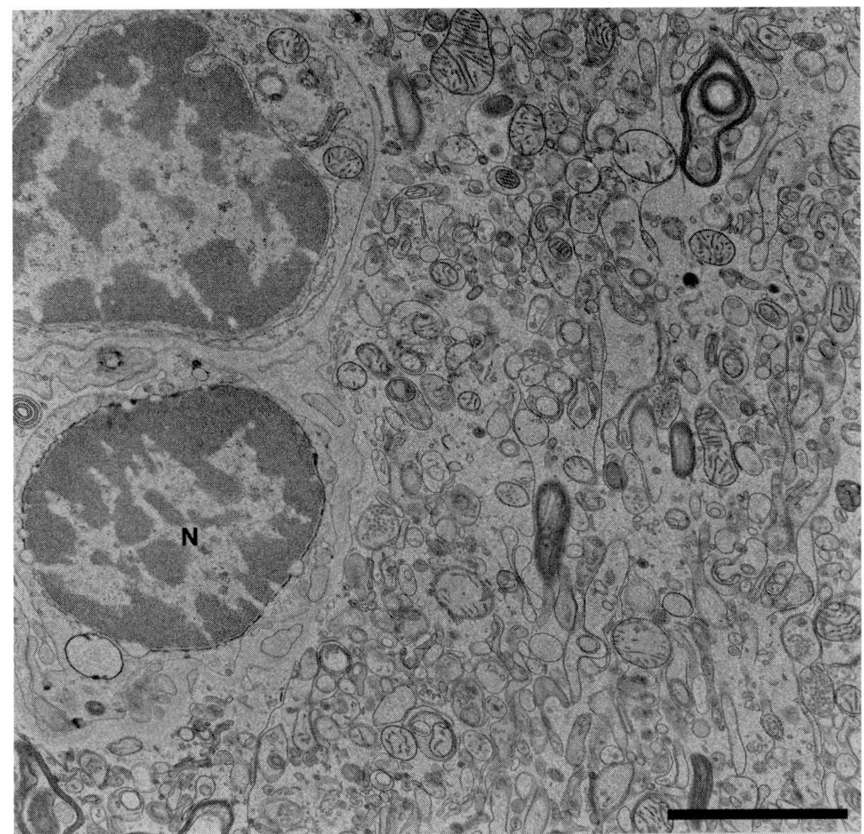

Fig. 8. Electron micrograph overview of a brain biopsy. Two cells with nuclei (N) are shown. (Bar = 2 μm.)

4. Notes

1. Although anesthesia can cause alterations on a cellular level, we favor the use of these drugs over the excision from both dead and conscious animals. Because autolytic reactions commence immediately after death, the excised tissue will not represent the native state. On the other hand, taking biopsies from conscious animals demands an expertise that is not available in every lab. Moreover, from the viewpoint of animal ethics, it is better to anesthetize the animals before taking any biopsies.
2. Biopsy platelets (*see* **Fig. 1E**) are the link between the microbiopsy system and the Leica EM PACT high-pressure freezing machine (Leica Microsystems, Vienna, Austria).The excised tissue in the needle notch fits exactly in the slot of the biopsy platelet, and the platelet is easy to mount in the flat specimen holder system of the Leica EM PACT.

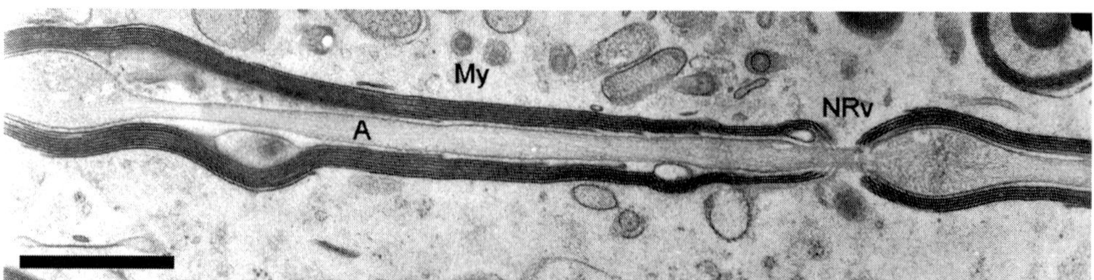

Fig. 9. Longitudinal section through an axon in rat brain, revealing the darker-stained myelin sheath (My) wrapped around the axon (A). A node of Ranvier (NRv) can be seen at the interruption in the myelin sheath. (Bar = 500 nm.)

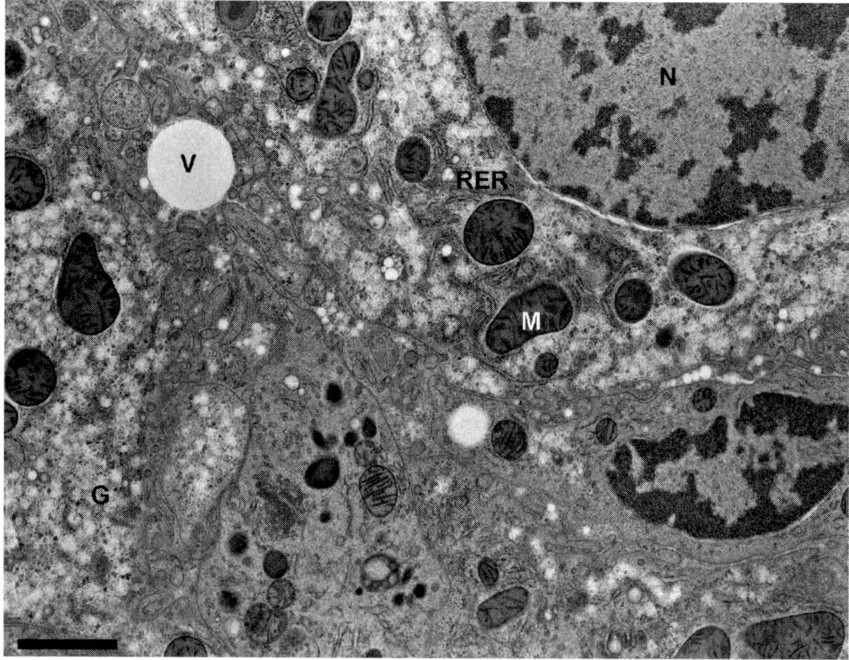

Fig. 10. Cross-section through rat liver. Mitochondria (M) are dark stained and glycogen (G) and vesicles (V) are apparent. Rough endoplasmic reticulum (RER) is present around the nucleus (N). (Bar = 2 µm.)

3. The pod consists of an alloy body with a central opening and an apical screw. On the tip of the screw, a diamond is mounted on a ball joint. The biopsy platelet has to be placed between a small table and the diamond screw. High pressure will be applied to the biopsy platelet slot via a small outlet in the center of the table. By closing the screw properly with a torque wrench (35 cNm applied), the biopsy

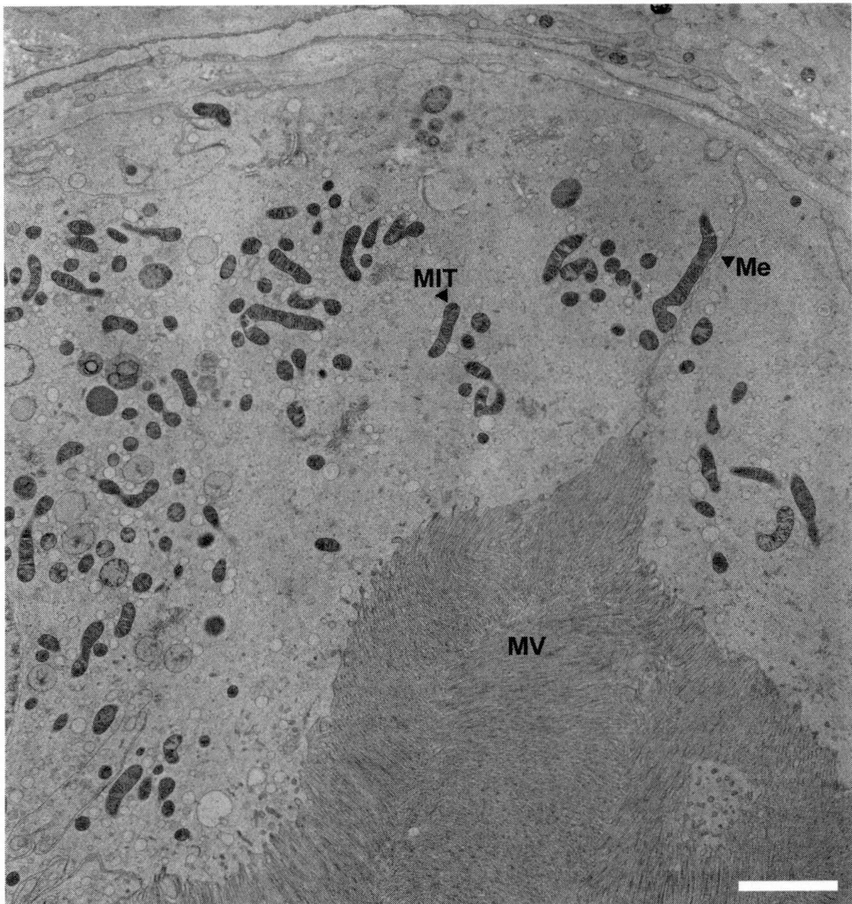

Fig. 11. Rat kidney overview: part of a cross-section through a tubule. In the parietal cytoplasm, darkly stained mitochondria (MIT) are present. A cell membrane (Me) borders two neighboring cells. The central area of the tubule is filled with cellular protrusions, microvilli (MV). (Bar = 4 μm.)

platelet slot is tightly sealed with the diamond and can withstand the more than 2000 bar pressure during the high-pressure freezing process. It is important that the biopsy platelet is accurately centered in the pod. If not, pressure losses will occur during freezing and high pressure freezing will fail.

4. The biopsy gun is a modified version of a Pro-mag 1.2 Manan biopsy gun (Manan, Northbrook, IL). When triggering and releasing the original Pro-mag 1.2 Manan biopsy gun (*see* **Fig. 1C,D),** the inner needle is inserted into the tissue, immediately followed by the outer needle. Because of the extrusion of the inner needle, difficulties arise when one wants to obtain a biopsy from a particular, well-defined area or when small organs (e.g., mouse kidney) are subjected to excision. The modification

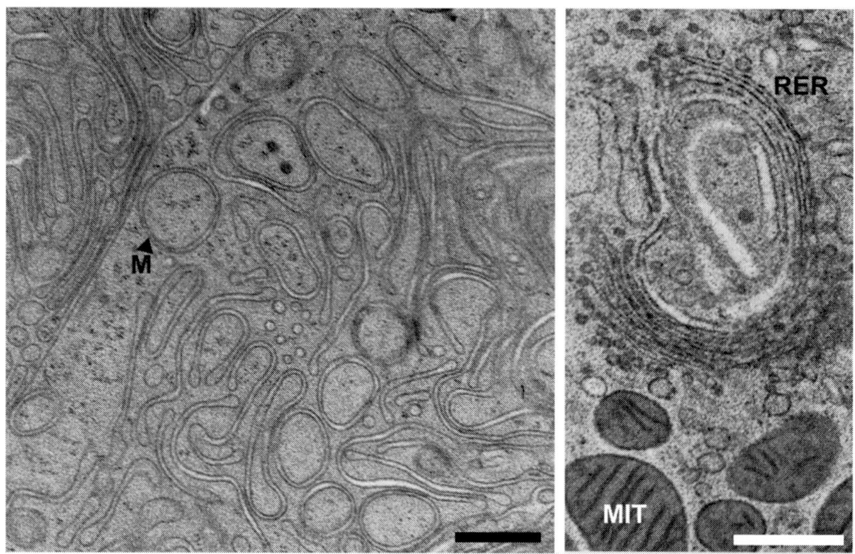

Fig. 12. Rat kidney ultrastructure. In the left panel, a section through the filter apparatus with smoothly outlined membranes (M); in the right panel, rough endoplasmic reticulum (RER) with surrounding mitochondria (MIT). (Bars = 500 nm.)

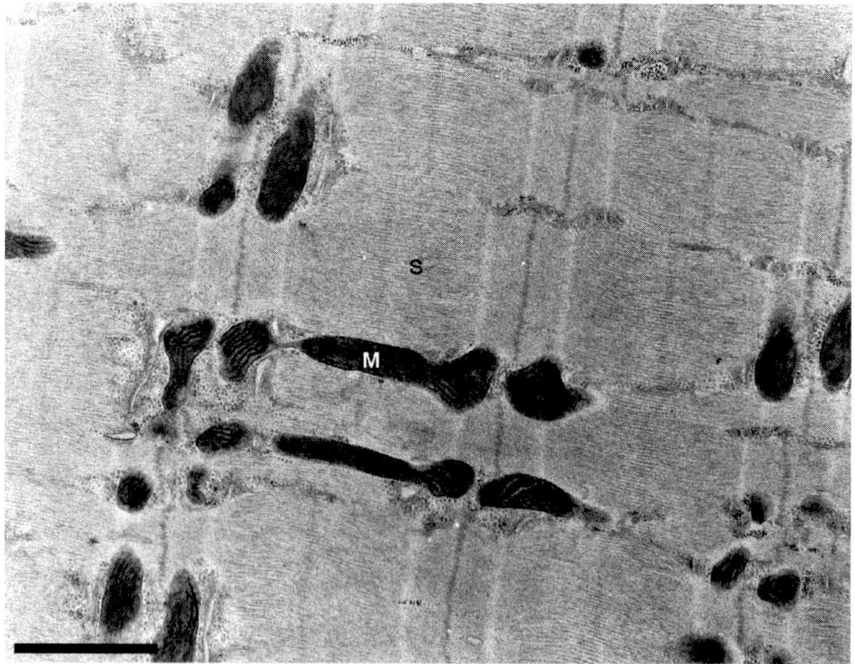

Fig. 13. Rat muscle ultrastructure shown in longitudinal section. The striped appearance of the sarcomeres (S) can be observed. The mitochondria (M) are well preserved and darkly stained. (Bar = 1 μm.)

we applied consists of fixing the inner needle to the gun (*see* **Fig. 1D**). The needle is inserted in the tissue; the notch positioned at the place of interest is filled with tissue. When releasing the cutting mechanism, only the outer needle moves forward and dissects the piece of tissue filling the notch.

5. The modified biopsy needles (*see* **Fig. 1B**) are based on Manan 20Ga.TWx8cm (Manan, Northbrook, IL) needles. The modified needles consist of an inner needle (diameter of 0.6 mm) with a notch and are surrounded by a tube with a sharp edge. The notch is 1.2 mm in length and half of the needle diameter deep.
6. The commercially available tissue transfer tool as developed by Leica differs from the device described here, but both tools are based on the same principle. The body of the tool ensures the position of a rod or ledge exactly above the needle notch. With a vertical movement, the tissue is pushed out from the needle directly into the platelet slot.
7. Forceps should always be precooled when handling cryo-fixed samples. When frozen samples are handled with warm (room temperature) tweezers, recrystallization of ice could occur, resulting in severe deterioration of the sample. Holding the tips of the forceps in liquid nitrogen until boiling stops (approx 20 s) is sufficient.

References

1. Van Harreveld, A., Crowell, J., and Malhotra, S. K. (1965) A study of extracellular space central nervous tissue by freeze substitution. *J. Cell Biol.* **25,** 117–137.
2. Studer, D., Hennecke, H., and Müller, M. (1992) High-pressure freezing of soybean nodules leads to an improved preservation of ultrastructure. *Planta* **188,** 155–163.
3. Fernández-Morán, H. and Dahl, A. O. (1952) Electron microscopy of ultrathin frozen sections of pollen grains. *Science* **116,** 465–467.
4. Riehle, U. (1968) Schnellgefrieren organischer Präparate für die Elektronenmikroskopie. *Chem. Ing. Tech.* **40(5),** 213–218.
5. Steinbrecht, R. A. and Zierold, K. (1987) *Cryotechniques in Biological Electron Microscopy,* Springer-Verlag, Berlin.
6. Escaig, J. (1982) New instruments which facilitate rapid freezing at 83 K and 6 K. *J. Microsc.* **126(3),** 239–249.
7. Sartori, N., Richter, K., and Dubochet, J. (1993) Vitrification depth can be increased more than 10 fold by high pressure freezing. *J. Microsc.* **172(1),** 55–61.
8. Studer, D., Michel, M., Wohlwend, E. B., and Buschmann, M. D. (1995) Vitrification of articular cartilage by high-pressure freezing. *J. Miscrosc.* **179(3),** 321–332.
9. Van Harreveld, A. and Crowell, J. (1964) Electron microscopy after rapid freezing on a metal surface and substitution fixation. *Anat. Rec.* **149,** 381–385.
10. Studer, D., Graber, W., Al-Amoudi, A., and Eggli, P. (2001) A new approach for cryofixation by high-pressure freezing. *J. Microsc.* **203(3),** 285–294.

Index

CARDIFF UNIVERSITY
PRIFYSGOL CAERDYDD